THE SPACE SHUTTLE

Roles, Missions and Accomplishments

WILEY-PRAXIS SERIES IN SPACE SCIENCE AND TECHNOLOGY

Series Editor: John Mason, B.Sc., Ph.D.

This series reflects the significant advances being made in space science and technology, including developments in astronautics and space life sciences. It provides a forum for the publication of new ideas and results of current research in areas such as spacecraft materials, propulsion systems, space automation and robotics, spacecraft communications, mission planning and management, and satellite data processing and archiving.

Aspects of space policy and space industrialization, including the commercial, legal and political ramifications of such activities, and the physiological, sociological and psychological problems of living and working in space, and spaceflight risk management are also addressed.

These books are written for professional space scientists, technologists, physicists and materials scientists, aeronautical and astronautical engineers, and life scientists, together with managers, policy makers and those involved in the space business. They are also of value to postgraduate and undergraduate students of space science and technology, and those on space-related courses (including psychology, physiology, medicine and sociology) and areas of the social and behavioural sciences.

For further details of the books listed below and ordering information, why not visit the Praxis Web Site at http://www.praxis-publishing.co.uk

METALLURGICAL ASSESSMENT OF SPACECRAFT PARTS, MATERIALS AND PROCESSES
Barrie D. Dunn, Head of Metallic Materials and Processes Section, ESA-ESTEC, Noordwijk, The Netherlands

SATELLITE CONTROL: A Comprehensive Approach
John T. Garner, Aerospace Consultant, formerly Principal Ground Support Engineer, Communications Satellite Programmes, ESA-ESTEC, Noordwijk, The Netherlands

SOLAR POWER SATELLITES: A Space Energy System for Earth
Peter E. Glaser, Vice President (retired), Arthur D. Little, Inc., USA, Frank P. Davidson, Coordinator, Macro-Engineering Research Group, Massachusetts Institute of Technology, USA, Katinka I. Csigi, Principal Consultant, ERIC International, USA

THE MIR SPACE STATION: A Precursor to Space Colonization
David M. Harland, formerly Visiting Professor, University of Strathclyde, UK

THE SPACE SHUTTLE: Roles, Missions and Accomplishments
David M. Harland, formerly Visiting Professor, University of Strathclyde, UK

LIVING AND WORKING IN SPACE: Human Behavior, Culture and Organization, Second edition
Philip Robert Harris, Executive Editor, *Space Governance* Journal; Vice President, United Societies in Space, Inc., USA

THE CHINESE SPACE PROGRAMME: From Conception to Future Capabilities
Brian Harvey, M.A., H.D.E., F.B.I.S.

THE NEW RUSSIAN SPACE PROGRAMME: From Competition to Collaboration
Brian Harvey, M.A., H.D.E., F.B.I.S.

Forthcoming titles in the series are listed at the back of the book.

THE SPACE SHUTTLE

Roles, Missions and Accomplishments

David M. Harland
formerly Visiting Professor, University of Strathclyde, UK

JOHN WILEY & SONS
Chichester • New York • Weinheim • Brisbane • Singapore • Toronto

PRAXIS

Published in association with
PRAXIS PUBLISHING
Chichester

The White House,
Eastergate, Chichester,
West Sussex, PO20 6UR, England

Published in 1998 by
John Wiley & Sons Ltd
in association with Praxis Publishing Ltd

Wiley Editorial Offices

John Wiley & Sons Ltd, Baffins Lane,
Chichester, West Sussex, PO19 1UD, England

John Wiley & Sons, Inc., 605 Third Avenue,
New York, NY 10158-0012, USA

Wiley-VCH Verlag GmbH, Pappelallee 3,
D-69469 Weinheim, Germany

Jacaranda Wiley Ltd, G.P.O. 33 Park Road, Milton,
Queensland 4001, Australia

John Wiley & Sons (Asia) Pte Ltd, 2 Clementi Loop #02-01,
Jin Xing Distripark, Singapore 12981

John Wiley & Sons (Canada) Ltd, 22 Worcester Road,
Rexdale, Ontario, M9W 1L1, Canada

Library of Congress Cataloguing-in-Publication Data
Harland, David M.
The space shuttle : roles, missions and accomplishments David M. Harland
p. cm. — (Wiley–Praxis series in space science and technology)
Includes bibliographical references (p.) and index.
ISBN 0-471-98138-9 (cloth : alk. paper)
1. Space Shuttle program (U.S.)—History. 2. Space shuttles—History.
I. Title II. Series.
TL789.8.U62S634 1998
387.8'0973—dc21 97-43764
CIP

A catalogue record for this book is available from the British Library

ISBN 0-471-98138-9

Printed and Bound in Great Britain by MPG Books Ltd, Bodmin

I have decided today that the United States should proceed at once with the development of an entirely new type of space transportation system designed to help transform the space frontier of the 1970s into familiar territory, easily accessible for human endeavour in the 1980s and '90s. This system will center on a space vehicle that can shuttle repeatedly from Earth to orbit and back. It will revolutionise transportation into near space, by routinising it.

Richard Nixon, President of the United States of America, 5 January 1972

From a pilot's standpoint, you could not ask for a more superb flying machine

Bob Crippen, Pilot, STS-1

The Space Shuttle did more than prove our technological abilities, it raised our expectations once more; it started us dreaming again...

Ronald Reagan, President of the United States of America,
in his Address to a Joint Session of Congress, 28 April 1981

Politicians are a strange bunch of critters

John Young, Commander, STS-1, upon hearing that Ronald Reagan, having first lauded the shuttle as a "brave adventure", promptly cut NASA's budget

Space, the final frontier...

James T. Kirk, Captain, Starship Enterprise

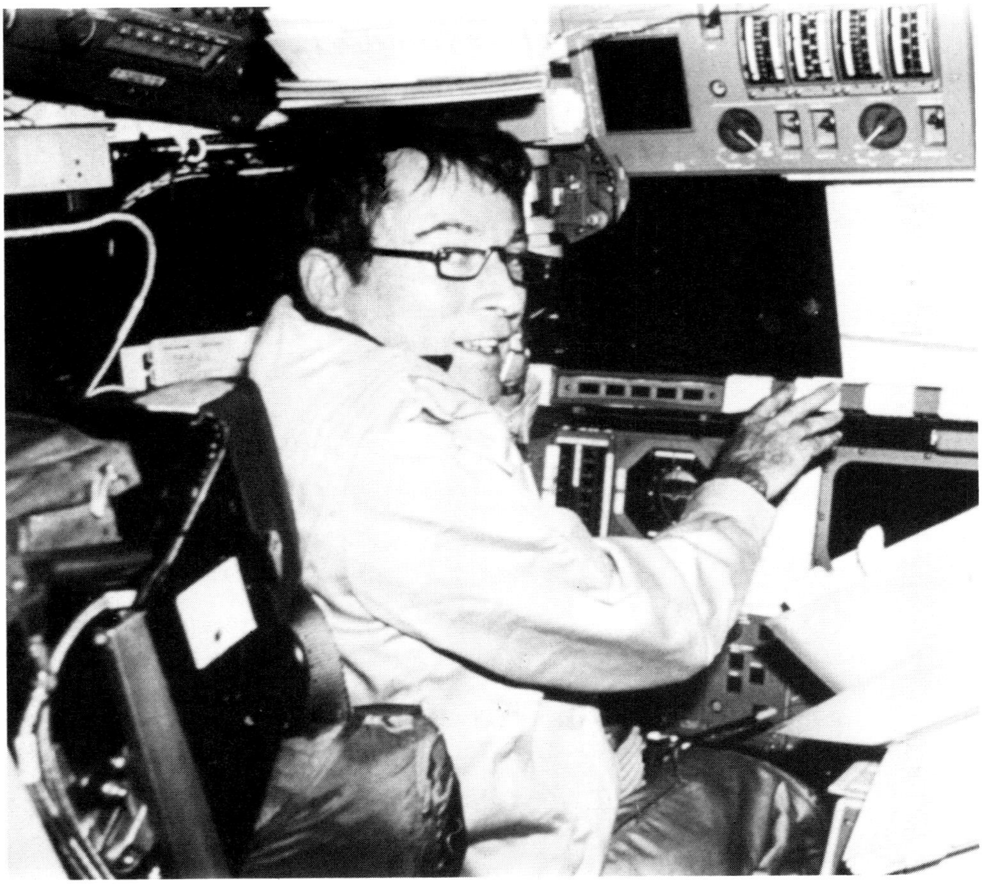

John W. Young – The Astronauts' Astronaut

Table of contents

Foreword

When NASA began work on the Space Shuttle in the late 1960s, few people recognized how important a part of American life it would become over the next 30 years. *The Space Shuttle: Roles, Missions and Accomplishments*, by David M. Harland, goes a long way towards capturing the essence of the Shuttle's place in that history. It is an important contribution to the historical literature of the Space Shuttle and its uses.

In many respects the idea of a reusable Space Shuttle dates at least to the theoretical rocketplane studies of the 1930s by Austrian aerospace designer, Eugen Sänger. By the 1950s it had become an integral part of Wernher von Braun's master plan for space exploration: an orderly set of stages aimed at creating a permanent space station serviced from the Earth by a reusable winged vehicle or shuttle, leading to a colony on the Moon, and finally a human expedition to Mars. This model gained increased legitimacy in the 1968 feature film, *2001: A Space Odyssey*, in which the stunningly picturesque wheeled space station was reached from Earth by a reusable winged space shuttle.

These were the ideals that motivated NASA engineers in the 1960s, as they pursued the dream of a permanent presence in space, made sustainable by a reusable winged vehicle providing routine access to space at an affordable price. Some NASA officials compared the methods of launching into orbit used on Project Apollo with operating a railroad and then throwing away the locomotive after every trip. A reusable Space Shuttle, they argued, would make the trip much more cost effective. Studies which NASA conducted in the mid-1960s found that reusable space technology was within reasonable grasp, more evolutionary than revolutionary, and that a hefty investment of research and development funds could yield a substantial reduction in operations costs. Flying 30 or more times a year, such a system would be an economical alternative to the use of large 'throw-away' launchers like the Saturn 5. All of the spacefaring nations have eventually accepted that paradigm as the *raison d'être* of their human spaceflight efforts in the latter part of the twentieth century.

The goal of efficient operations in a heavy-lift launcher – especially with the decision for budgetary reasons to terminate the Saturn 5 booster production line in mid-1968 after the completion of 15 launch vehicles – prompted NASA's commitment to the Space Shuttle as a continuation vehicle for human spaceflight. Once it was under way, NASA leaders believed, they could also move forward with a space station, which the Space Shuttle could both place in orbit and support logistically. In addition, and this was in part serendipity from the NASA perspective, because of the Space Shuttle's size and versatility a portion of its payload bay could be used to haul scientific and applications satellites of all types into orbit for all users. The Space Shuttle was to be, essentially, the achievement of 'one-size-fits-all', in this instance the vehicle providing all orbital services required by users. This type of standardization has long been an important part of American mass production; the Model-T automobile and the F-111 fighter-bomber being examples of how it was supposed to work.

Although the development program was risky, between 1972, when President Nixon approved the effort, and 1981, when the first orbital flight took place, a talented group of scientists and engineers worked to create the world's first reusable space vehicle. Since that first flight the various orbiters - Atlantis, Columbia, Discovery, Endeavour, and Challenger (lost in 1986 during the only Space Shuttle accident ever to take place) - have made nearly 100 flights into space. Throughout, the vehicle has been a workhorse of space exploration for projects both international and domestic.

The Space Shuttle has launched numerous scientific satellites, including the Magellan spacecraft to Venus, the Galileo probe to Jupiter, and the international Ulysses spacecraft to study the Sun. Each also undertook scientific and technological experiments ranging from the release of experiments into space, through the continued flights of the European Space Agency's 'Spacelab', to a dramatic three-person EVA in 1992 to retrieve an Intelsat communications satellite: this had been stranded in low orbit by its expendable launcher and the astronauts fitted a new rocket motor to enable it to complete the journey to its operating station in geostationary orbit. The Shuttle also deployed the Gamma-Ray Observatory, the Hubble Space Telescope, and the Upper Atmosphere Research Satellite. It has also demonstrated its usefulness in two complicated servicing missions of the Hubble Space Telescope, in 1993 and 1997.

Between April 1981 and the end of 1997, the Space Shuttle carried approximately 1,000 tonnes of cargo and more than 750 major payloads into orbit, including more than 300 for NASA, more than 140 for the Department of Defense, and more than 100 for commercial interests. Through 1997, astronaut crews have also conducted more than 50 extravehicular activities (EVAs) and Shuttle crews are actively preparing for the EVAs necessary to build the International Space Station in orbit beginning in late 1998. Through all of these activities, a good deal of realism about what the Space Shuttle can and cannot do has now emerged.

At the end of the twentieth century, the Space Shuttle enjoys the same plaudits and suffers from the same criticisms that have been made clear since not long after the program began in 1981. It remains the only vehicle in the world with the dual capability to deliver and return large payloads to and from orbit. The design, now more than two decades old, is still state of the art in many areas, including computerized flight control, airframe design, electrical power systems, thermal protection system, and main engines. It is also the most reliable launch system now in service anywhere in the world, with a success-to-failure ratio of better than 98 per cent.

David M. Harland's important new study of the Space Shuttle fills in many of the details of its myriad uses over the nearly two decades of its operational life. It provides an overview of the variety of missions and the unique capabilities of this remarkable machine. As such, it is one of the critical building blocks in the furtherance of historical knowledge about the Space Age and the place of NASA and the Space Shuttle within that history.

Roger D. Launius, Ph.D., Chief Historian
NASA, Washington DC, March 31, 1998

Author's preface

John Young (see Frontispiece) was a real survivor. He had become an astronaut in 1962, and had first flown in space in 1965, along with Gus Grissom on the first Gemini mission, and so became one of the first men to ride the Titan missile. Grissom had gone on to be assigned to command the first Apollo mission, but he and his crew had been killed when their capsule caught fire during a countdown demonstration. Young had been on the back-up crew. By that time, he had commanded his own mission, Gemini 10, and on that occasion had been accompanied by Michael Collins. A few months later, Young tended to Apollo 10's command module while his colleagues rehearsed the initial stage of the descent to the lunar surface, and then the ascent and rendezvous phase. This paved the way for Apollo 11. On this mission, Collins looked after the command module while Neil Armstrong and Buzz Aldrin flew down to the lunar surface. After serving as back-up commander for Apollo 13, which aborted on the way to the Moon, Young, as commander of Apollo 16, spent three days exploring the mountainous region known as Descartes.

Walking on the Moon was effectively the end of an astronaut's career; he was unlikely to be sent back, and, in any case, there were too few flights available to permit this luxury. As a result, most moonwalkers retired soon after returning; but not Young, who stayed on and became Chief of the Astronaut Office. By 1981 Young was 51 years of age, and he was NASA's most experienced pilot. Nevertheless, his vast experience could not have fully prepared him for what he was about to do.

Bob Crippen had been an astronaut almost as long as Young, but this momentous day – 12 April, the twentieth anniversary of Yuri Gagarin's pioneering orbit of the Earth – was to see him make his first flight.

Because no part of the shuttle's complex configuration had ever flown before, its first test flight would be, potentially, the most dangerous space mission ever attempted. Failure could strike any component at any time. This was *flight test* in the classic sense of the pilot who straps into a new aircraft and takes it up *to see if it will fly*. One million people had congregated at the Kennedy Space Center to watch. Millions more watched on television. The tension mounted as the clock ran down to its fateful moment with destiny; all the more so because *no-one* – not even Young and Crippen – really, truly, knew what to expect.

In the event, as Columbia lifted off, Crippen's heart rate soared and he yelled, "What a ride!" In contrast, Young's heart maintained a steady seventy beats per minute. Two days later, as Columbia began its final approach to the vast dry lake at Edwards Air Force Base, an overjoyed Crippen exclaimed, "What a way to come to California!" During the roll-out,

Young, ever the taciturn test pilot, casually enquired of the control tower, "Do you want me to put it in the hangar?" After a *textbook* flight, NASA pronounced its new spacecraft to be "a magnificent flying machine". Amazingly, it planned to make such hair-raising missions routine!

This book outlines the operations of the shuttle since its historic first test flight. This is done in a *thematic* way, rather than chronologically, because NASA has used the shuttle to pursue several programmes in parallel, and it is only by separating these out that they can be truly appreciated. However, a chronological mission log has been incorporated for ease of reference. Also, a comprehensive glossary of terms has been included so that the main text is not littered with unnecessary technical detail. Although the shuttle was conceived of as a 'space truck' to ferry satellites into low orbit, and its contribution could be said to have ended with the act of deployment, the shuttle was not an end in itself, so this book will follow up and relate what each satellite went on to do after its release. In addition to these high-profile satellites, a wide variety of secondary payloads were carried, and these will also be discussed. The thematic approach was chosen because it readily facilitates a broad, yet also focused, coverage. Overall, the objective of this book is to put shuttle operations into context – to ask difficult questions designed to assess the programme critically, yet fairly. Amongst the many questions posed are: Was the shuttle really necessary? How has the programme evolved with time and experience? Was the Challenger accident avoidable? What was the significance of the microgravity work? What was achieved? What happened to the much-hyped commercialisation? What role has the Russian Mir space station played in NASA's programme? Is the International Space Station necessary? Will it complement the shuttle? Does the shuttle have a long-term future?

Kelvinbridge, Glasgow
March 1998 David M. Harland

Acknowledgements

I would like to thank Roeluf Schuiling and Craig Covault for their tremendous efforts in print; Brian Harvey, Joseph Loftus of the Johnson Space Center, William McLaughlin of the Jet Propulsion Laboratory, and Roland Dore of the International Space University for their comments in the formative stage; Neville Kidger, Flo McGuire, Alex Williams, Alasdair Downes and Sarah Dougan for their assistance in sourcing material and in preparing the manuscript; NASA for all of the photographs and most of the line diagrams; Roger Launius, NASA's Chief Historian, for kindly providing the Foreword; and, of course, John Mason and Clive Horwood of Praxis Publishing for their continuous support.

List of illustrations and tables

Frontispiece: John Young on duty in the commander's seat of Columbia

PART 1: OPERATIONS

Chapter 1

Chapter 2

Chapter 4

Chapter 11

Chapter 12

Chapter 13

Chapter 14

Chapter 15

Chapter 16

PART 4: OUTPOST

Chapter 17

Chapter 18

PART 5: CONCLUSIONS

Chapter 19

Tables

Part 1: Operations

1

Origins

In March 1952, *Collier's*, the mass-market magazine published in New York, included an article entitled *Across The Last Frontier* by Wernher von Braun. Although this German rocket pioneer had built the V-2 rocket during the Second World War to serve as a long-range ballistic weapon, he considered his rocket to be the key which would open the door to orbital flight, and he advanced his vision of the human colonisation of space in this and half a dozen other articles which *Collier's* published over the next two years. However, it was Chesley Bonestell's magnificent paintings of winged space planes, massive rotating space stations and ungainly interplanetary transport vessels illustrating von Braun's prose that captured the imagination of a generation overwhelmed by the rapid pace of technological advancement in jet engines, rockets and atomic bombs resulting from the war.

The V-2 could climb to the fringe of space on a ballistic trajectory, but it would always fall back to Earth; it did not have nearly enough energy to attain a speed of 8 km/s, to achieve orbital velocity. The Bell X-1 rocket-powered aircraft had broken the 'sound barrier' during a high-altitude soaring arc, but it was confined to the atmosphere. A chemical-powered rocket able to attain orbit *might* be impracticably enormous. The atom was seen as the solution to all problems in need of a vast supply of energy, but the prospect of an atomic rocket must have appeared rather remote. In fact, the idea of 'spaceships' was widely considered to be so far beyond what was practicable that it belonged to the realm of science fiction, if not sheer fantasy. The launch of Sputnik on 4 October 1957 transformed the situation and, almost overnight, von Braun's articles seemed to be prophetic rather than fantastic.

Although by the end of the 1950s the high-performance X-15 rocket plane, continuing the hybrid aircraft evolution pioneered by the X-1, was generally regarded as the best route to orbital flight, the US Air Force planned the X-20 as a hybrid combining a spaceplane with a vertically-launched rocket. Dubbed the DynaSoar, it was to soar hypersonically on the fringe of the atmosphere, circling the globe to undertake reconnaissance and to deliver a nuclear weapon from far beyond the reach of air defences. Their modes of launch were different, but the X-15 and the X-20 were similar in that they were both aircraft which were to make a runway landing, be refurbished and be relaunched. It would take time to develop the technology and to scale it up to match von Braun's vision, but the door was now open, and the logic was impeccable: the spaceplane would build the station which would serve as a base for mounting expeditions to destinations beyond.

But in the rapid transition from fantasy to reality, there was no time to pursue the most logical evolutionary route. In reacting to the shock of Sputnik, the newly-established US National Aeronautics and Space Administration (NASA) set out to send a human being into orbit ahead of the Soviets. It was a race! There was no time to upgrade the X-15 sufficiently to attain orbit and survive re-entry, and the X-20 was still on the drawing board, so the only option was to develop a far cruder ballistic-return capsule. NASA saw its Mercury man-in-space-soonest project as a 'crash' response to a national crisis, and the inelegant capsule as an expedient that would serve until a spaceplane became operational. The way to the planets was still by way of von Braun's station and, left to itself after Mercury, NASA could well have followed this strategy.

Fate intervened once again, however. With Yuri Gagarin's single orbit of the Earth, on 12 April 1961, NASA lost the race! Within six weeks, President John F. Kennedy decided that it was time to "take longer strides", and he set a new goal of staggering audacity – nothing less than a lunar landing. As this was to be another 'crash' programme, to be achieved within the decade, the spaceplane and way-station scenario was impracticable. Yet the *direct* route to the Moon would still require a rocket with unprecedented lifting power. Atomic power was not yet an option, so it would have to be a chemical rocket. So it was that von Braun turned his attention to the development of the Saturn 5. When he planted 'Old Glory' on the lunar surface on 21 July 1969, Neil Armstrong met Kennedy's challenge.

By that time, to the public, spaceflight meant a ballistic capsule on a throw-away rocket, and von Braun's winged spaceplane and the wheel-in-space now appeared to have been irrelevant distractions. Having perfected its technology, NASA eagerly looked forward to exploiting it with a multifaceted Apollo Applications programme.

COST-BENEFIT

Yet even as NASA basked in the glory of Apollo, the new Nixon administration, eager to cut federal spending, ordered production of the Saturn 5 to cease. Nixon had no desire to kill off NASA; quite the reverse, in fact. He simply insisted that NASA devise a more cost-effective technology, so that America could consolidate its leadership in space.

In energy terms, the most costly part of a trip into space is attaining low orbit, and most of the Saturn 5's prodigious energy went into climbing out of the Earth's 'gravity well'; all subsequent manoeuvres – even setting off for the Moon – were relatively inexpensive. Cutting the cost of spaceflight therefore meant reducing the cost of achieving orbit. In developing its response to Nixon's demand, NASA proposed a 'modular' approach using a reusable spaceplane whose initial mission would be assembling a 12-person station in low orbit. In the fullness of time, this would serve as a base from which to mount an expedition to Mars. NASA saw this station and its transportation infrastructure as an integral system, but when it costed the plan it became clear that it would have to choose one or the other, because it could not have both.

A space station built using existing expendable rockets would be the simpler task, and it would form a logical follow-on to the Skylab orbital workshop, scheduled for 1973, which was the only part of the Apollo Applications programme to survive Nixon's budgetary axe. It would take longer to develop the innovative spaceplane, but, because it would reduce the cost of access to orbit, developing the transportation infrastructure would ensure the programme's long-term future. For the 'Space Shuttle' to achieve a low operating cost, it would require both almost total reusability and a *very* high flight rate. In fact, the cost-benefit analysis that had shown that the shuttle could recoup its development had envisaged flying on a *weekly* basis, and maintaining this would impose an unprecedented operational strain on NASA, which was accustomed to mounting expeditionary missions. In effect, to run the shuttle, it would have to transform itself into an airline. Would there be sufficient work to sustain such a flight rate? The only option was to phase out all existing rockets and assign all payloads to the shuttle. This *shuttle-only* policy was formally recognised by the designation of the shuttle as the National Space Transportation System (NSTS).

In fact, the statistics were broadly supportive of the case for a weekly flight rate for the NSTS. Table 1.1 shows the Royal Aircraft Establishment's *Table of Earth Satellites* record of rocket launches during the shuttle's formative years.

Table 1.1. Annual satellite launch rates

	1969	1970	1971	1972	1973	Total
US	38	29	32	27	23	149
USSR	68	79	81	70	83	381
Total	106	108	113	97	106	530

Launches attributed to the US included all services provided for Intelsat and for foreign institutions. In 1972, the Federal Communications Commission opened up the market for satellite broadcasting, so it was expected that a proliferation of commercial satellites would not only inflate the launch rate but also create a market in which the shuttle could generate revenue.

WAYS AND MEANS

For Mercury, NASA had selected a ballistic capsule because, for the limited mass that a contemporary rocket could put in orbit, it was a more effective solution than a spaceplane. As aerodynamic vehicles, the streamlined X-15 and X-20 had higher lift-to-drag ratios. The development of an orbital equivalent would have required a heavyweight rocket, but this was not a serious issue because larger rockets were under development. A far more serious issue, however, for which there was no evident solution, was the insulation of the skin of the craft from the thermal stress of hypersonic re-entry. A conical capsule that penetrated the atmosphere blunt-end forward was a far more manageable task, because the area in need of thermal protection could be minimised and, because such a capsule would fly only once, an ablative shield could be used. By flaking off into the slipstream, this carried away the heat and prevented it from penetrating the structure. The conic angle was defined to ensure that the superheated plasma that curled around from the base did not come into contact with the side of the capsule. On the other hand, the capsule had such a low lift-to-drag ratio that it could not fly, so its 'pilot' was obliged to make an ignominious parachute descent. It was by this logic that the world came to draw a clear distinction between aircraft and spacecraft. And yet, as the Apollo astronauts flew their spidery lunar landers into ever-more mountainous terrain, the spaceplane was announced as the way forward!

But why this about-turn? A spaceplane had been the *obvious* way forward, because it built on the ever-faster, ever-higher trend in post-war aircraft development, but the race to be the first to send an astronaut into orbit had obliged NASA to employ a ballistic capsule because the available rocket could lift it, a man could fit into it, and it could be shielded for re-entry. Gemini and Apollo, driven by another deadline, had used the same technology for the same reasons. In pursuing Apollo, however, a truly enormous rocket with a prodigious lifting power had been built, and this had opened up options for spacecraft development. Could it launch a spaceplane?

The Air Force's X-20 had been cancelled early on, long before it could enter flight test. The X-15 continued to explore hypersonic flight through to 1968, but no attempt was made to upgrade it for orbital flight, a step which would have required fitting it with a substantial booster and a yet-to-be-devised thermal protection system. NASA had studied the subsonic descent characteristics of a number of lifting-bodies, so-called because they achieved a high lift-to-drag ratio from the squat shape of the bodies of their wingless airframes. As with the X-15, they were released by a bomber. Because they could *fly*, they could manage their energy, and could line up to land on a runway. Although NASA had never launched a lifting-body on a sub-orbital trajectory to assess its hypersonic characteristics, and the lack of such data was a matter of serious concern, it was natural for NASA to use its lifting-body experience as a starting point for the design of the shuttle.

It so happens that it is easier for a *large* high lift-to-drag aircraft to accommodate the thermal stresses of hypersonic flight than it is for a smaller one. This is because the heating rate and the total heat load are functions of mass and size, and as the size and payload of a vehicle are increased the surface area increases faster than the mass. Clearly, although some form of reusable thermal protection still had to be devised for the shuttle, the thermal stress diminished as the airframe was scaled up. Furthermore, the overriding mass constraint that had inhibited spaceplane development was alleviated by the availability of powerful rocket engines, such as those which powered the Saturn 5.

A spaceplane was therefore now a possibility, but was it strictly necessary? Could not a ballistic system be scaled up too? As it turned out, although the ballistic capsule was ideally suited to sending small crews on short missions, it could not accommodate much cargo on the return trip. The later Apollos carried a battery of cameras but, because these instruments were in the service module, they had to be jettisoned at the end of the mission. Apollo was used to fly to and return from Skylab, but this was feasible only because the station carried all the apparatus that would be required for the research programme. Each crew could ferry up a certain amount of cargo – such as spare parts and perishables – but these items had to be small as well as lightweight. This was because its shape was defined by the plasma sheath which formed on re-entry. The ballistic-capsule was adequate for crew transport to and from orbit, to staff a space station, and for brief trips to the Moon, but it was not a platform for scientific instruments and it offered no potential as a cargo hauler. What NASA had in mind for the shuttle was a payload bay large enough to carry a satellite, a battery of instruments, a research laboratory or a module for a space station. It was to be a space *truck*. For this, it required a different type of vehicle. It was also clear that it needed a *stretched* lifting-body, not an X-20 style orbital bomber.

This logic dictated that the spacecraft developed at great expense to fly to the Moon would have to be written off as a technological *cul-de-sac*, and that an essentially new, yet ironically rather familiar, line of development would have to be pursued instead.

Despite the existence of powerful rocket motors and the data on lifting bodies, the basic configuration of the shuttle was essentially wide open. In fact, the design was sensitive to a wide range of factors, ranging from the lack of knowledge of the hypersonic characteristics of a lifting-body; the use, for the first time, of an aerodynamic control system; the fact that a non-ablative thermal insulation had yet to be developed; the fact that no high-performance rocket engine had ever been designed to be reusable; the likelihood that this engine would have to be throttleable across a wide range of thrust; and, not least, to the shuttle's operational requirements. On the other hand, by this time NASA and its primary contractors had a vast base of knowledge of spacecraft development, so the aerospace industry was exceptionally well-skilled for the daunting task of designing a reusable spaceplane. Nevertheless, NASA took *three years* to settle upon the final configuration.

All of the major aerospace players submitted proposals because with such an enormous contract on offer no company could afford to *elect* out and, indeed, companies were aware of the long-term prospects of being *selected* out. A wide variety of possible configurations were suggested, therefore, with varying degrees of reusability. About the only aspect upon which they all agreed was that a single-stage-to-orbit vehicle was impracticable. This meant that a composite vehicle would be needed, and this would complicate the aerodynamics during launch and require a way to safely separate the elements, both under normal circumstances and in an emergency.

The preferred option was a fully-reusable vertically-launched stack with a large fly-back booster and a small orbiter, both piloted, mated either belly-to-belly or piggyback, with the stack separating at high altitude and the booster returning to the launch site while its charge flew on to orbit. Unfortunately, the more the fly-back booster was studied, the more certain it became that the separation manoeuvre would have to be performed at a higher altitude and at a higher speed than originally thought. This transformed the booster into a hypersonic aircraft of unprecedented size, and the cost of developing it became comparable with that of a massive hypersonic airliner, a vehicle which the aerospace industry had already ruled out as impracticable. Deleting the fly-back booster meant withdrawing from total reusability, and this, in turn, meant increasing the overhead once the shuttle became operational, but there did not seem to be any alternative.

One major factor guiding the design was the need to meet the operating requirements of the Air Force. This was because NASA had had to secure the support of the Department of Defense in order to make a skeptical Congress accept the shuttle. The Air Force – which was responsible for launching classified satellites – had reluctantly agreed to abandon its proven expendable rockets, but it imposed a stiff penalty. In addition to requiring that the shuttle be capable of flying sufficiently frequently to deal with its planned manifest, it insisted that the shuttle be sufficiently large and sufficiently powerful to carry the most outsized satellite that it was every likely to produce. Specifically, it demanded that the shuttle be able to launch a satellite which was twice as heavy and three times as large as would fit on the Titan 3, then the most powerful rocket in its inventory. This meant that the orbiter had to be *much* larger than NASA had originally envisaged, with an 18-metre long payload bay and a capacity of 30 tonnes. Furthermore, because the Air Force insisted on being able to mount X-20 style single-orbit flights, it required the shuttle to be capable of manoeuvring extensively during re-entry, using hypersonic soaring to fly 2,000 km from the orbital ground track in order to counter the Earth's rotation and return to the launch site. This required the orbiter to have a large delta-shaped wing, to hit the atmosphere at a low angle-of-attack, and to fly a steep banking turn at extreme altitude. NASA had planned an orbiter with short, stubby, straight wings which would hit the atmosphere at a high angle-of-attack, take most of the heat on its belly, make only slight hypersonic manoeuvres, and then make a straightforward approach. The stubby wing would have given the orbiter excellent subsonic glide characteristics, but a delta wing is a high-speed configuration which has little to recommend it in subsonic flight. Consequently, an orbiter designed to suit the Air Force would not just need to endure far greater structural and thermal loads during re-entry, its poor subsonic gliding characteristics would mean that it would make an exceptionally steep descent, and it would touch down at a correspondingly higher speed.

It was the scaling-up of the orbiter to accommodate the Air Force's requirements that ruled out the fly-back booster. NASA considered mounting the orbiter on top of the first stage of a Saturn 5, and Boeing, its manufacturer, suggested making its rocket reusable, but this idea was soon ruled out. Meanwhile, the orbiter itself was a cause for concern. How would the cryogenic tanks be integrated into its structure? How would residual propellant be vented in orbit? How would the tanks be rendered inert? After much consideration, it was decided to use an *external* tank which would be jettisoned. Because the tank could not be discarded until the orbiter's engines shut down, and this could not happen until it had reached orbit, it was evident it would not be possible to reuse the tank, and the provision of a new tank for each mission would add to the operational overhead. Given the external tank, the decision to use thrust-augmentation for the launch phase was logical.

Augmentation meant strapping auxiliary motors around the core of a rocket, so that they would increase the launch-thrust of the host. For simplicity, the augmentation motors were usually solid rockets. The Air Force had developed an enormous solid motor to augment its Titan 2, thereby creating the Titan 3, but even this would not have the power to augment an orbiter enlarged to match the Air Force's 30-tonne payload requirement, so a more powerful motor was ordered. These motors (there would be two) would parachute into the ocean, and be recovered and reused. Their refurbishment might take a considerable time in terms of the number of flights, given a high flight-rate, and be costly, but they would be reusable. Some expressed concern that the use of solid rockets represented an unacceptable risk because they could not be shut down, but the Air Force had intended to launch its Manned Orbiting Laboratory on a Titan 3, with a pair of astronauts riding on top in a Gemini capsule, and this project had been cancelled only in 1969, so the use of solid rockets was of less concern to the Air Force.

In March 1972, Richard Nixon formally ordered NASA to develop the National Space Transportation System, and it revealed the configuration for the shuttle's stack. It would obviously take at least five years to develop such an innovative vehicle, but it was confidently expected that the shuttle would become operational in time to refurbish Skylab so that it could serve as an interim orbital base until the planned modular space station could be built. However, the full magnitude of the technology, systems and integration problems soon became evident. Almost every aspect of the vehicle's operation raised serious issues: the stresses that the stack would endure at launch; throttling down the engines to cope with the aerodynamic pressure associated with flying through the sound barrier; releasing the solid rockets; the greatest acceleration to which the orbiter should be subjected as it continued to climb; the hypersonic aerodynamics of re-entry manoeuvring and the use of the flight control systems; a radiative thermal protection system; and last, but by no means least, devising and facilitating abort options for the various phases of the ascent.

It is not the purpose of this book to dwell on the *development* of the shuttle; rather, its focus is the *operation* of the shuttle. It is sufficient to note that the problems encountered in making the lightweight, extremely high-pressure, main engine reliable (it had a tendency to misfire), and in making the brittle ceramic silica tiles stick to the flexible airframe (there were 30,000 individually shaped tiles; they had to be applied by hand, and a skilled worker could apply them at a rate of only two or three per week) made a mockery of the carefully plotted development cycle, with the result that flight testing was progressively slipped from 1978 to 1981.

Fig. 1.1. Inspection of a Space Shuttle Main Engine (SSME). Compared with the mighty F-1 engines of the Saturn 5, the SSME is tiny. However, it was designed to be throttled across a wide range of thrust and, of course, it is reusable.

THE STACK

The shuttle comprises the delta-winged orbiter, the external tankage, and two boosters, all bolted together in a configuration referred to as the *stack*.

With a dry mass of about 65 tonnes, a wingspan of 24 metres, and a 37-metre airframe, the orbiter is dimensionally comparable with a DC9 airliner. It does not *look* like an airliner, however. Its profile is dominated by the engines clustered beneath its vertical tail assembly. When the three Space Shuttle Main Engine (SSME) units are running at 100 per cent of their rated power, they each deliver 170,000 kg of thrust. Uniquely for such powerful rocket engines, they can be throttled down to 65 per cent, and up to 109 per cent. The force is imparted to a thrust frame in the aft bay which distributes it partly through the lower attachment points to the external components, and partly to the orbiter's mid-structure that holds the 18-metre long, 4.5-metre wide cylindrical payload bay. A trio of Auxiliary Power Units (APUs) in the aft bay provide power to gimbal the SSMEs during the ascent, and to drive the flight control surfaces during the descent.

During periods of high activity such as ascent and descent, when most of the orbiter's systems are operating and both primary and backup systems are active, excess heat is shed by a flash evaporator system. The payload bay doors have to be opened within an hour or so of reaching orbit to expose the radiators mounted on their inner surfaces, so that excess heat can be dissipated. If the doors fail to open, the flight has to be cut short. In space, fuel cells combine oxygen and hydrogen in such a way as to generate electricity and yield water as a by-product. The cryogenic reactants are stored in tanks beneath the payload bay floor.

A pair of Orbital Manoeuvring System (OMS) engines are mounted in pods on either side of the base of the tail. At 2,670-kg thrust, these engines can be fired either together or singly. They are gimballed to deliver their force through the orbiter's centre of mass, even when fired singly. The OMS is the primary system for making the deorbit burn. The orbiter has smaller Reaction Control System (RCS) engines for attitude control. There are 14 thrusters clustered in the nose and 12 in each of the OMS pods. Most produce 400 kg of thrust and are used for general translational and rotational manoeuvres, but six deliver only 11 kg of thrust and are used for fine control. All of the orbiter's engines burn monomethyl hydrazine in nitrogen tetroxide, a hypergolic mixture that eliminates the need for an igniter.

The orbiter is able to use an aluminium skin because the Thermal Protection System (TPS) will protect it from the frictional heat of re-entry. An innovative ceramic tile was developed to protect the orbiter. It sounds deceptively trivial to describe these tiles as silica coated with borosilicate, but their development represented a significant advance in state-of-the-art materials processing. An ablative imposes an order of magnitude greater weight penalty than the super-lightweight tiles so, in addition to the fact that it would not have been reusable, it would have been impracticable to have coated the whole of the orbiter's surface with an ablative. A tile radiates its surface heat so efficiently that only a few centimetres of material is required to insulate the orbiter. Several forms of Reusable Surface Insulation (RSI) were manufactured, differing mainly in the additives employed, to cater for the range of thermal stresses expected. Areas subjected to temperatures of

Fig. 1.2. Space shuttle orbiter Enterprise separates from the SCA 747 to conduct an approach and landing test. In this manoeuvre the 747 dives steeply away. Note the chase plane beyond and behind the 747's customised tail assembly.

650 °C were protected by white tiles, and those that had to withstand 1,200 °C were protected by black tiles. The nose cone and the leading edges of the wings need to endure the highest temperatures, peaking at 1,600 °C, so they were coated by a Reinforced Car-

Table 1.2. Approach and landing tests – captive flights

#	Date	Enterprise crew
1	18 Feb 1977	unoccupied
2	22 Feb 1977	unoccupied
3	25 Feb 1977	unoccupied
4	28 Feb 1977	unoccupied
5	2 Mar 1977	unoccupied
6	18 Jun 1977	Haise and Fullerton
7	28 Jun 1977	Engle and Truly
8	26 Jul 1977	Haise and Fullerton

Table 1.3. Approach and landing tests - free flights

#	Date	Enterprise crew	duration (min)	altitude (m)	landing (km/h)
1	12 Aug 1977	Haise and Fullerton	5.3	7,265	340
2	13 Sep 1977	Engle and Truly	5.5	7,878	360
3	23 Sep 1977	Haise and Fullerton	5.6	7,484	354
4	12 Oct 1977	Engle and Truly	2.6	6,788	368
5	26 Oct 1977	Haise and Fullerton	2.0	5,760	352

bon–Carbon (RCC) laminate of graphite cloth impregnated with resin. Applying the tiles presented a problem in itself. The orbiter's surface flexes in response to dynamic loads, but the rigid ceramic tiles were brittle. They had to be *glued* on. A strain-isolating pad was first bonded to the metal to absorb the flexure, then the tile was stuck to the pad. It took a while to get this bonding process right, but once the problems were resolved, the tiles were able to withstand both the orbiter's structural dynamics and the aerodynamic loads of the airflow.

The Solid Rocket Booster (SRB) is a 46-metre long, 3.7-metre wide cylinder. It has a conical nose cap containing a parachute, and a flared skirt at the rear, around the nozzle. The metal casing weighs 83 tonnes, but this rises to 586 tonnes when it is filled with the atomised aluminium powder that is burned with ammonium perchlorate oxidiser. The propellant does not actually fill the motor; it forms an annular column within the casing, the shape of which is designed to control the combustion rate so as to vary the thrust according to a predefined profile. Each of the shuttle's two SRBs delivers 1,300 tonnes of thrust. An auxiliary power system gimbals the nozzles to vector the thrust in order to steer the stack along a specific trajectory up through the atmosphere.

The 42-metre long External Tank (ET) is basically an 8.4-metre diameter cylinder, but it is actually two tanks linked by an intertank ring. Although its dry mass is just 40 tonnes, it carries 600 tonnes of oxygen and 100 tonnes of hydrogen for launch. The hydrogen tank forms the lower two thirds of its length, and the oxygen tank is at the top. This ratio derives from the fact that an oxygen atom is more massive than a hydrogen atom. The outer surface of the ET is covered with cork and a thick coat of foam insulation to prevent the build-up of 'hot spots' which might set up convection currents in its cryogenic contents (the oxygen is at −180 °C and the hydrogen is at −253 °C). The propellants are fed through pipes running down the outside of the hydrogen tank and then into the orbiter's aft bay, whereupon high-pressure, lightweight turbopumps feed the orbiter's engines. At its peak thrust, the orbiter draws propellant at the rate of 500 kg per second. The ET is the main load-bearing element of the stack; both the orbiter and the SRBs are bolted to it. The SRB is fastened to the ET's intertank ring by an attachment point near the top of its casing, and to the circumferential support ring at the ET's base by an attachment point at its lowest field-joint, just above the nozzle segment.

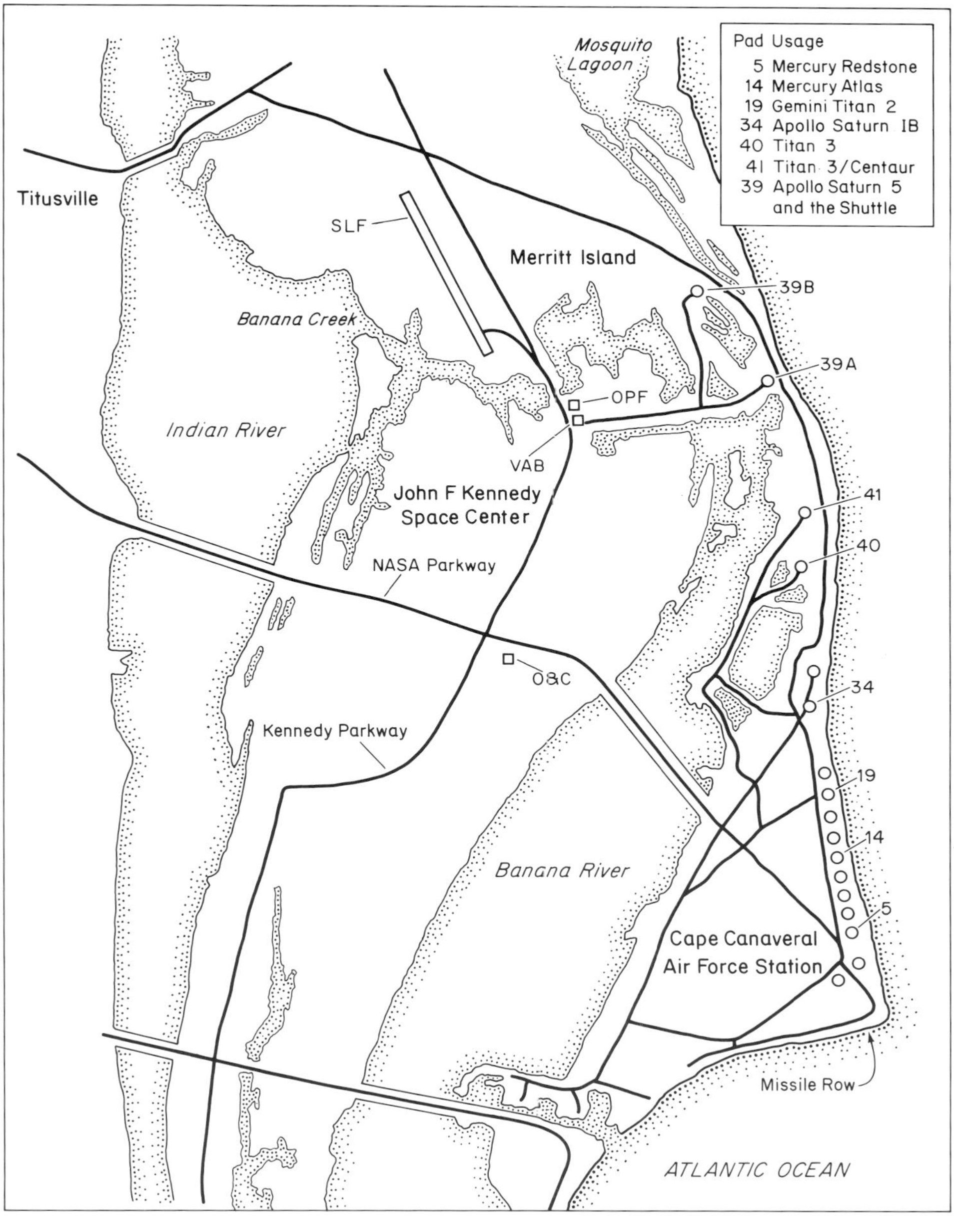

Fig. 1.3. A map of Cape Canaveral and Merritt Island. Although launches may take place across a wide range of azimuths out over the North Atlantic, safety considerations deny launches from Florida access to polar orbit.

GLIDE TRIALS

Although the orbiter had been designed for a gliding descent, its delta wing was shaped to permit dynamic soaring during the hypersonic phase of the re-entry; at subsonic speeds this wing had the aerodynamics of a *brick*. As a result, the orbiter made a much faster and much steeper final descent than a conventional aircraft (fully an order of magnitude steeper than an airliner, in fact), and it did not level out until a few seconds before landing.

To give the astronauts an opportunity to practise flying this unique glider, a Boeing 747 called the Shuttle Carrier Aircraft (SCA) was adapted to carry an orbiter on its back, and this took the orbiter Enterprise to high altitude and then released it so that the orbiter could rehearse the subsonic phase of the return from orbit. This programme, called the Approach and Landing Test (ALT) was conducted by Fred Haise, Gordon Fullerton, Joe Engle and Dick Truly. Only Haise had flown in space. This had been on Apollo 13, so his one and only chance to walk on the Moon had been frustrated. The Enterprise was delivered to Edwards Air Force Base (AFB) in California on 31 January 1977. The programme started with a series of captive flights in which the inert orbiter remained on the back of the 747

Fig. 1.4. This view of the John F Kennedy Space Center is dominated by the Vehicle Assembly Building (VAB). The Launch Control Center (LCC) is dwarfed in the foreground, the Orbiter Processing Facility (OPF) is to the left, and the Shuttle Landing Facility (SLF) is beyond that. The pads of Launch Complex 39 are three miles away, off-frame to the right. Note the Saturn 5 service structure (right) parked as a monument to a bygone era.

(b)

(a)

Fig. 1.5. Preparing the shuttle: (a) in the Orbiter Processing Facility (OPF), and (b) in the Vehicle Assembly Building (VAB).

(a)

(b)

Fig. 1.6. The roll-out: (a) egressing the Vehicle Assembly Building (VAB), (b) on the crawlerway, (c) approaching Pad 39, and (d) a rare view of two shuttles on the pad.

(c)

(d)

to determine the dynamics of the combination in all phases of the SCA's flight profile. Haise and Fullerton first rode in the orbiter on 18 June 1977. They made the first gliding descent to the dry lake on 12 August. On their final flight, they flew Enterprise down to a landing on the concrete runway.

The ALT programme was completed on schedule, but the date for the first *orbital* test flight repeatedly slipped due to on-going development problems with the SSME and the TPS.

LAUNCH FACILITIES

Rather than build new facilities – as it had in the past whenever it had introduced a new rocket – NASA opted to modify its Saturn 5 facilities to handle the shuttle. As it happened, whereas all its earlier launch facilities were located on the narrow coastal strip alongside the Canaveral Air Force Station, which marked the point of origin for missiles fired down the Eastern Test Range (ETR), the shuttle's facilities are across the Banana River on the Merritt Island Launch Area. The centrepiece is not the launch pads of Launch Complex 39, but the Vehicle Assembly Building (VAB); this 160-metre per side cube built to accommodate up to four Saturn 5s was re-equipped to prepare two shuttles.

Shuttle assembly starts by erecting the two SRBs, which is done on the Mobile Launch Platform (MLP). This had extra vents cut in it to pass the SRB efflux through to the flame trench. First the nozzle segment is bolted to the platform, then the three motor segments are added, and finally the nose is put on top. The segments are mated by an annular tang-and-clevis mechanism and a ring of 180 bolts. Once both SRBs are in place, the ET is mounted between them. All is now ready for the orbiter. This is prepared in the Orbiter Processing Facility (OPF), a new building close alongside the VAB. The hangar can 'turn around' two orbiters simultaneously. Payloads can be exchanged and an orbiter stripped down and then rebuilt, as necessary, to refurbish it for flight, but all processing must be done in a horizontal configuration. Once ready, the orbiter is towed to the VAB, where a crane lifts it, turns it to vertical, and then mates it with its ET. Payloads which need to be installed once the orbiter is vertical can now be installed. Payloads are stored in Operations and Checkout (O&C) Building several kilometres south of the VAB. This also has living facilities for astronauts preparing for a mission. They leave in the Astrobus to go to the pad about four hours before a launch is due.

The Crawler–Transporter (CT), affectionately known as 'the tortoise', is unchanged. It drives into the VAB, retrieves the loaded MLP, and then moves down The Crawlerway at a top speed of only 1 km/h. On its own, the crawler weighs 2,750 tonnes; it is twice as much fully loaded. Once over the Banana River causeway, it either continues east for 39A, or it makes a left turn and heads north to 39B. Each metre of its 6-km journey consumes a litre of fuel. At the pad, the CT offloads the MLP with an accuracy of just a few centimetres, and then withdraws.

Although the shuttle stack is only 56 metres tall, it is fully 24 metres from the tip of the orbiter's tail to the far side of the ET. It does not need the 120-metre tall tower required to access to the huge Saturn 5, so the lower third was discarded, and rather than ferry the rest back and forth on the MLP, it was permanently mounted on the pad as the Fixed Service Structure (FSS). The swing arm with the 'white room' at its end was retained for access to the orbiter's side hatch. A new unit, the Rotating Service Structure (RSS), was added to protect the stack from the elements. This incorporates a clean room to enable payloads to be installed after the shuttle has been checked. When ready, it swings 120 degrees away from the orbiter. A sound-suppressant system pumps water onto the pad to absorb the low-frequency acoustic shock which accompanies SRB ignition, to prevent the reflection from the concrete striking the orbiter, deflecting its flight control surfaces and damaging their actuators. Nevertheless, the sound level around the periphery of the launch complex rises almost instantaneously to 160 decibels.

The Launch Control Complex (LCC) alongside the VAB, with its independent firing rooms, was modified to deal with the shuttle's launch process. As soon as the shuttle clears the tower, it is 'handed over' to the Johnson Space Center in Houston.

A completely new facility for a spacecraft is the Shuttle Landing Facility (SLF), a 5-km long runway. The dry lake of Edwards is available for orbiters unable to return to Florida. In the event of an abort, an orbiter will try an emergency descent to any suitably-equipped commercial airport.

2

Shuttle operations

In March 1978, John Young was named as the commander of STS-1. A veteran of two Gemini and two Apollo missions, Young was one of the dozen men to have walked on the Moon. Although Bob Crippen was a rookie in terms of flight experience, his expertise was the orbiter's computer system, so he was assigned to fly with Young. In shuttle parlance, Young was the mission commander (CDR), and he flew the left seat, and Crippen, who sat in the right seat, was the shuttle pilot (PLT). Neither actively participated in the ALT phase of testing; they concentrated on the preparations to launch Columbia, the first space-worthy orbiter of the fleet.

TEST FLIGHTS

Although the Saturn 5 was designed to launch into space, as it stood on the pad it had the majestic elegance of a monument. The shuttle stack, in contrast, resembled a bird that was eager to take to the air. On 10 April 1981, the countdown incorporated a number of 'planned holds' to allow time to catch up on minor problems. The last such hold was at T-9 (T minus 9) minutes. When the count resumed, the pace picked up.

At the T-31 seconds point, when control passed to the orbiter, a timing fault in the General Purpose Computer (GPC) caused the count to be halted and, soon thereafter, led to the launch having to be scrubbed. The shuttle was a fly-by-wire vehicle, so its computer system was so important that it had been made multiply-redundant. One processor could fly the orbiter, but there were four identical processors operating in parallel so that there would still be a level of redundancy even if two dropped off-line. They were all to run the same programme at critical phases of the mission, compare their output and 'vote down' any unit that yielded a spurious result. Furthermore, to guard against a generic fault in the four primary processors, a fifth processor, built by a different manufacturer and programmed independently, served in an overwatch capacity. However, a timing error had prevented the various machines interfacing properly, and the launch had had to be abandoned. This was a straightforward problem to remedy, however, so the count was recycled, and the launch set for 12 April which, as it happened, was the twentieth anniversary of Yuri Gagarin's pioneering first orbital mission.

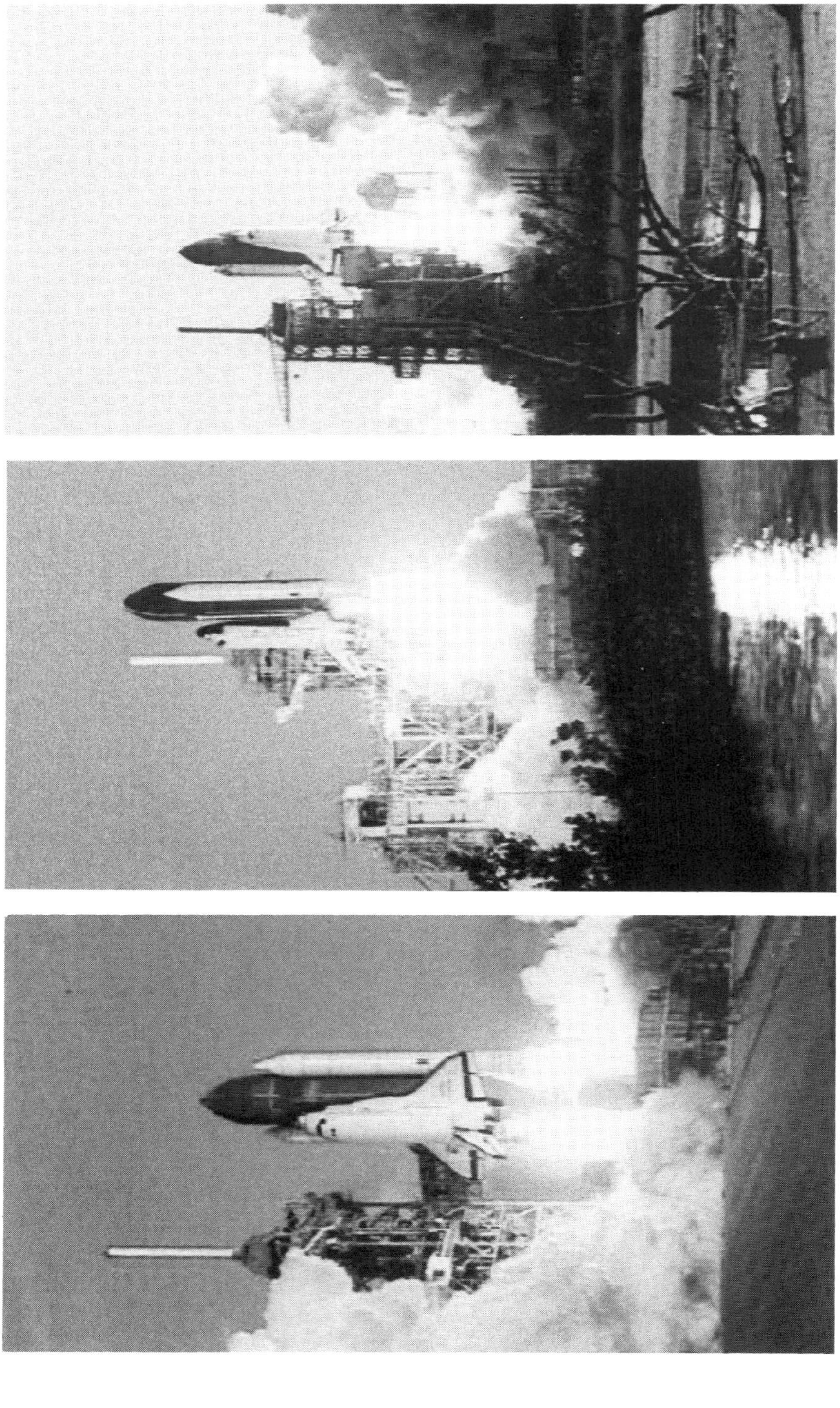

Fig. 2.1. Launch! In contrast to the slow and stately departure of the Saturn 5 rocket which sent Apollo to the Moon from the same Launch Complex, the shuttle stack rockets off the pad, and within a matter of seconds has cleared the tower.

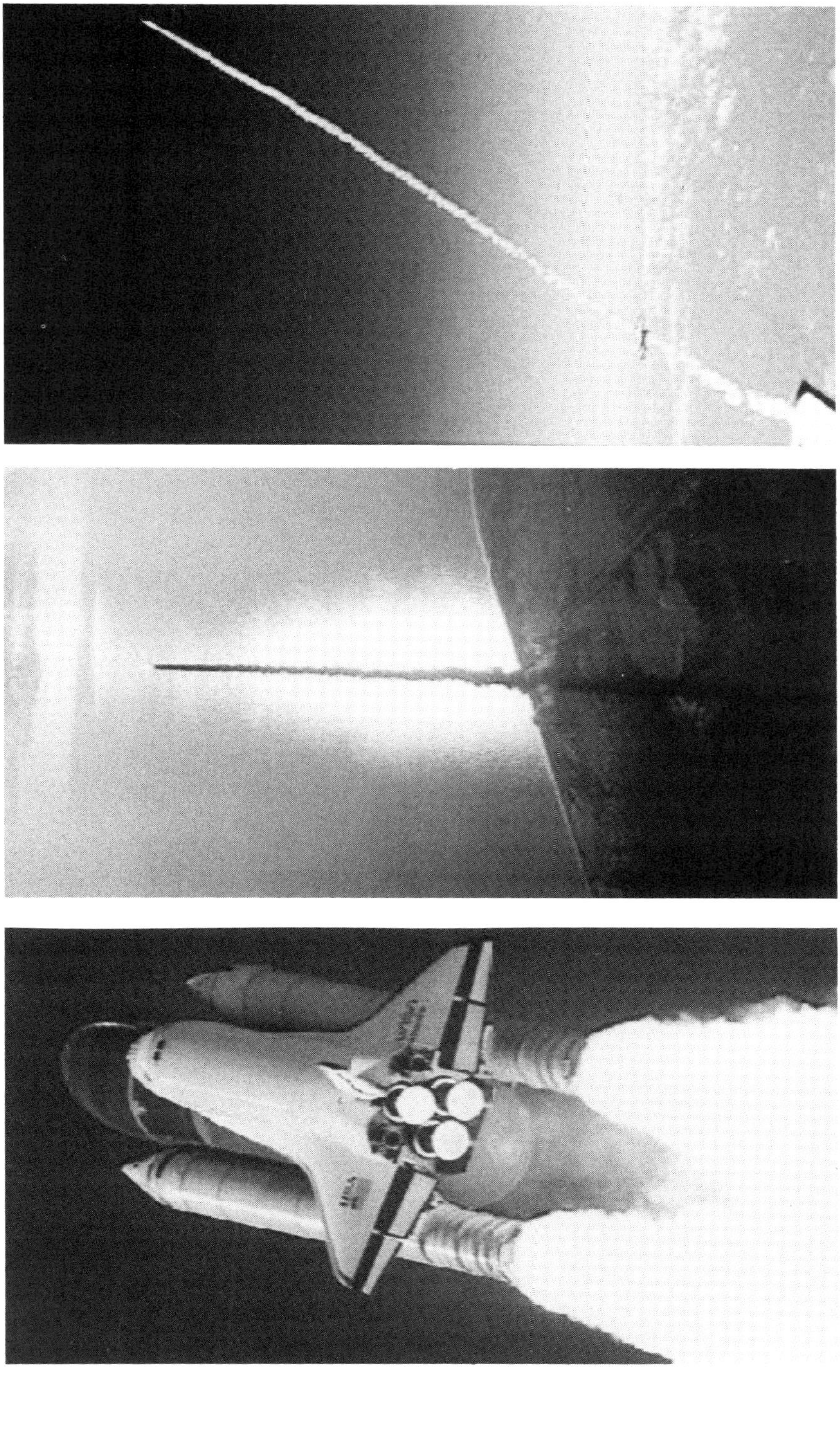

Fig. 2.2. Up and away. As soon as the shuttle stack has cleared the tower, it rolls onto the desired azimuth and then pitches onto its back for the long climb to orbit. Only NASA weather-monitoring aircraft are allowed near the shuttle's trajectory.

As the clock ran down through the final few seconds in the countdown, nobody really knew what to expect. There was a real chance that the vehicle would be destroyed in some catastrophic failure. Only after prolonged testing had the SSMEs finally proved sufficiently reliable for operational use, and they had been fired for 20 seconds for the Flight Readiness Firing (FRF) on 20 February 1981. However, the SRBs – the most powerful solid rocket motors ever developed – had been tested only when mounted on a horizontal frame, never upright, so the first time that their casings would be subjected to the stress of flight would be on the day that they were required to lift STS-1. This mode of engineering development, called *all-up* testing, made the shuttle the first US rocket to carry a crew on its first test flight. It was a risky strategy.

At the T–31 seconds point, the vehicle's internal sequencer took command and, starting at T–6.6 seconds, at 120-millisecond intervals, it ignited the three SSMEs, ran them up to full power, and then monitored their performance for any indicator heading out of toler-

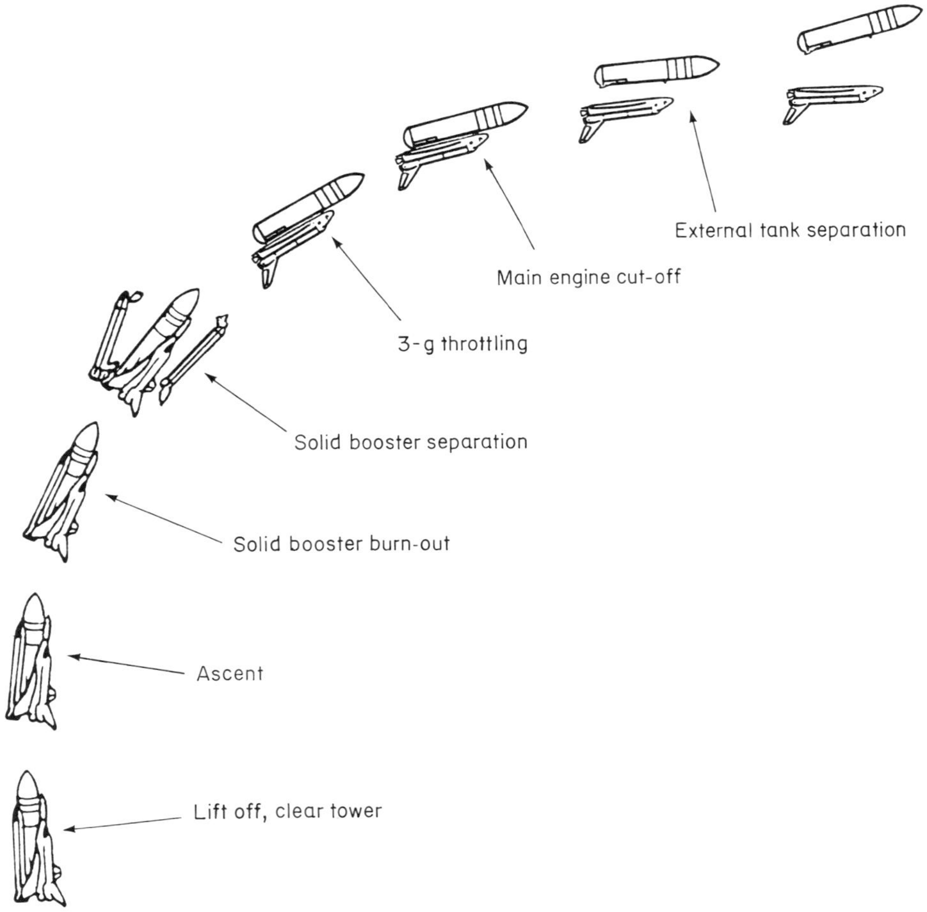

Fig. 2.3. A schematic of the climb to orbit, listing the major milestones.

ance. If the ultra-high-performance hydrogen-burning engines survived the first few seconds, they would probably run smoothly for the ascent to orbit. Although the orbiter's engines had ignited, it did not represent a commitment to launch; the stack remained bolted to the pad. Young and Crippen felt the orbiter vibrate as the SSMEs started, and they felt the pitching motion as the thrust pushed the stack forward on its pedestal. This oscillation, called 'the twang', had an amplitude of about two metres, and a period of several seconds, so the stack had yet to revert to vertical when the count reached T=0 and the command was sent to start the SRBs. This fired an igniter at the top of each motor, blasting a jet of extremely hot flame down the length of the motor's core to initiate solid propellant combustion. The signal also detonated the pyrotechnics in the massive bolts that clamped the SRBs to the MLP, to release the stack, because, one way or another, the shuttle was going to leave the pad. The response was immediate. In an instant, the SRBs more than tripled the thrust. In contrast to the slow and stately departure of the Saturn 5, the shuttle rocketed off the pad and, within a matter of seconds, it had not only cleared the tower but also completed its roll manoeuvre and was heading out over the Atlantic. Nothing had prepared the astronauts for the 'feel' of the ascent. Young was a man of few words, but Crippen, riding a rocket for the first time, exuberantly exclaimed: "What a ride!"

The stack weighed 2,000 tonnes. This was 1,000 tonnes lighter than a Saturn 5. But the shuttle's 3,100 tonnes of thrust was only marginally less than that of the Saturn 5, so the shuttle's acceleration was much more pronounced. In essence, the SRBs lifted the stack and the SSMEs balanced the offset centre-of-mass of its asymmetric configuration. As soon as it cleared the tower, the shuttle rolled onto the desired azimuth, and as it climbed it gradually pitched over so that it ended up flying on its back. As it accelerated towards the speed of sound, and the aerodynamic stresses on the stack reached their peak, the SSMEs throttled back to 65 per cent and the SRBs, by virtue of the configuration of their propellant, reduced their thrust. This phase lasted for about 20 seconds. Thereafter, the thrust rapidly piled back on. The SRBs exhausted their propellant after only two minutes, and they were jettisoned at an altitude of 45 km. After following a ballistic arc, they made a parachute descent and splashed down in the ocean. They were to be recovered, refurbished, and reused. The SRB-phase of the ascent had been essentially a vertical climb out of the atmosphere. The orbiter, running on its SSMEs, continued this climb, but it also pitched further back and started to increase its horizontal speed. As it consumed the ET's propellant, the shuttle rapidly built up acceleration, and when it reached 3 g its engines were progressively throttled in order not to put excessive structural loads on the vehicle. Eight and a half minutes after launch, the SSMEs were finally shut down. After Main Engine Cut Off (MECO), pyrotechnics jettisoned the spent ET, which re-entered the atmosphere over the Pacific and was destroyed. After briefly firing its OMS engines to avoid the same fate, Columbia was in orbit.

Prior to the launch, Young is said to have opined that there was only a 50/50 chance of launching into orbit and returning to Earth without a mishap. ALT had proven the orbiter's aerodynamics in the subsonic phase of the descent profile, but the only way to assess its hypersonic flight characteristics was by computer extrapolation of wind tunnel data, so, for the flight controllers, the plasma-induced radio blackout that would accompany Columbia's re-entry promised to be the most worrisome part of the flight. Their anxiety

Fig. 2.4. Retrieving a Solid Rocket Booster (SRB) from the Atlantic Ocean. The spent casings are towed to the Kennedy Space Center, disassembled, refurbished and, eventually, reused.

was heightened when the payload bay doors were swung open to expose the radiator each carried on its inner surface, and it was found that some of the tiles which were to protect the orbiter from the thermal stress of re-entry had come loose from the pods housing the orbital manoeuvring system (OMS) engines to either side of the vertical tail. The most crucial area was the belly of the vehicle. If a large number of tiles had been ripped off the belly, the vehicle might not survive the re-entry. It turned out that the acoustic shock accompanying the engine ignition had shaken loose some of the stuck-on tiles. This shock had been predicted, and the pad had been flooded with water in order to suppress the reflection of the sound, but there had evidently been a miscalculation. It was impossible to tell whether any tiles had been ripped off the belly of the orbiter. Even a few missing tiles could spell danger, because they might peel off like a zipper and expose a wide area of the thin aluminium skin beneath. It has been claimed that Columbia turned its belly to face an imaging reconnaissance satellite so that the risk could be assessed, but, if this was indeed done, this historic image has not yet been released.

With Columbia in its in-orbit configuration, Young and Crippen gave a televised tour of their split-level cabin, showed off the cavernous payload bay, and tracked the Earth drifting by; and Crippen, in space at last, enjoyed weightless antics. What struck the viewers most though, was that as the commander of the world's first true spaceship monitored his flight deck controls, he wore glasses.

After two days in orbit, Columbia was set tail-forward for the deorbit manoeuvre which was performed by the OMS engines, above the Indian Ocean. This caused its trajectory to dip into the atmosphere on its next low pass, half a world away. Columbia turned forward and pitched its nose high so that the intense heat of re-entry would be borne by the nose,

Fig. 2.5. Touchdown. Although the orbiter's steep delta-shaped wing is well-suited to hypersonic flight during re-entry, it offers little scope for manoeuvring during the subsonic approach. After re-entry, a series of tight turns are made in the vicinity of the landing site to line up with the runway with precisely the right amount of energy for a safe landing. After a far steeper descent than a conventional aircraft, the orbiter performs a brief flare manoeuvre to cut its rate of descent and to deploy its landing gear. Having no engines, there is little margin for error because the orbiter cannot climb to fly around for a second approach.

the leading edge of the wing, and the belly. The re-entry profile was straightforward; no attempt was made to assess crossrange performance, which would be left to later flights.

As the orbiter dug deep into the atmosphere at Mach 25, Young and Crippen observed the sky beyond the flightdeck's wraparound window turn from the intense black of space to fluorescent pink as the tenuous air was pushed aside and turned into an ionised shock-wave which blacked out radio communication for 11 minutes.

Throughout the descent, the 3-axis accelerometers of the Aerodynamic Coefficient Identification Package (ACIP) recorded the orbiter's hypersonic and supersonic flight characteristics in order to verify the predictions of the computer models.

The only aspect of the descent to be frustrated was the attempt by a high-flying C-141 transport aircraft to use a thermal imaging camera to track the orbiter as it began re-entry; this InfraRed Imagery of the Shuttle (IRIS) effort gave no usable data.

Upon becoming subsonic, Columbia flew the ALT-proven gliding approach, and made a textbook landing on the dry lake at Edwards AFB. Crippen was as enthusiastic as ever: "What a way to come to California!"

After Columbia had come to a halt, the fleet of recovery vehicles drew up and set about *safing* the vehicle. Only after the engine bay had been vented of propellant fumes were the crew allowed to disembark. As he bounded down the stairway, Young, the laconic veteran

Fig. 2.6. The Edwards dry lake. During the Approach and Landing Tests (ALT) and the first few operational missions, the shuttle landed in California, where the vast expanse of the dry lake at Edwards Air Force Base offered runways on a wide variety of azimuths and was free of obstacles.

moonwalker, grinned broadly and pumped his fist in the air. "From a pilot's standpoint", Crippen concluded, "you could not ask for a more superb flying machine."

It had been hoped to launch the second flight in September, but the date slipped, and on 9 October a spillage while pumping propellant into the orbiter's forward Reaction Control System (RCS) meant that the adjacent tiles had to be replaced. The 4 November countdown was scrubbed by a faulty fuel cell. Auxiliary Power Unit (APU) problems led to the attempt on 11 November being abandoned, but things ran smoothly the next day, and the age of the reusable spaceship became a reality. This time Columbia was flown by Joe Engle and Dick Truly. When they opened the bay doors they were delighted to see that there was no sign of damage to the tiles; the upgraded sound-suppression system had reduced the shockwave to acceptable levels.

On this occasion, the bay carried much more than an instrumentation package: NASA's Office of Space and Terrestrial Applications (OSTA) had supplied a pallet of instruments to study the Earth. This package included a terrain-mapping radar and a carbon monoxide monitor. The most impressive item on the manifest, however, was the Remote Manipulator System (RMS). This 16-metre long, triple-jointed Canadian-built 'arm' had both automatic and manual modes of operation. Many of the tasks to be assigned to the shuttle

Fig. 2.7. The recovery vehicles attend to the orbiter. Before the crew can egress, the recovery team vents the aft bay to clear any propellant gases that may have accumulated there from the Orbital Manoeuvring System (OMS). Once this has been done, and the crew has left, the orbiter is towed away.

would rely upon this manipulator, which was remotely-controlled from a station on the aft flight deck. Although it suffered a few teething problems, it was successfully put through its paces, and the faults were soon rectified. It had been difficult to test on Earth because, set up for use in space, it could not actually support its own weight. In return for supplying the arm, Canada was to be allowed to fly its own astronauts on the shuttle.

An ambitious five-day flight had been intended, but this was cut to two days when a fuel cell had to be shut down. This technology had been developed for Gemini, then used for Apollo. It combined cryogenic hydrogen and oxygen to produce electricity, with water as a by-product. The water was periodically vented. If illuminated by the Sun, such a water dump was a spectacular sight. The orbiter had three fuel cells, but they were its only source of power, and the rule was that if one failed the flight had to be curtailed. In this case, one of the fuel cells had become flooded with water. The early return prompted the cancellation of the plan to test the new spacesuit by having Engle depressurise the airlock. Although the return to Earth was routine, the IRIS aircraft once again failed to track the orbiter's re-entry.

It had once been intended to make six test flights before making the shuttle operational, but even before Columbia had made its first flight this test phase had been cut back to four,

Fig. 2.8. Mission control, Houston. This view is taken looking over the Flight Director's station as a mission draws to a close (note the landing orbiter on the viewscreen).

and upon STS-2's return NASA announced that the remaining tests were to take place in March and June 1982. When STS-3 launched on the appointed date, it was taken to be an omen that the programme would indeed be able to stick to a formal schedule. Fred Haise had retired, so Jack Lousma and Gordon Fullerton flew Columbia for its third outing. They used the arm to lift the 160-kg drum-shaped Plasma Diagnostics Package (PDP) off its bay mount and held it out to measure the physical properties of the ionosphere in the immediate vicinity of the orbiter. A fault in the arm preempted the plan to sweep the far larger Induced Environment Contamination Monitor (IECM) around the bay, to assess the extent to which outgassing 'polluted' the environment in which instruments carried in the bay would be called upon to operate, so this was limited to *in situ* measurements. It was while studying the extent to which the orbiter interacted with the ionosphere that the effect promptly dubbed 'shuttle glow' was first observed. To assess the vehicle's thermal characteristics, Columbia was held in various orientations with respect to the Sun in order to verify that the thermal protection tiles obviated the need to roll the orbiter in 'barbecue mode', as had been necessary with earlier spacecraft, to even out thermal stresses. This meant that the orbiter could safely spend extended periods in arbitrary orientations, serving as a science platform.

The excitement on STS-3 came at the end of the mission. Half-way through the planned seven-day flight, it became clear that the weather at Edwards would have degraded to such an extent that it would not be practicable to land there. The Kennedy Space Center's Shuttle Landing Facility (SLF) was not a viable option, because this was a 100-metre wide strip of concrete, and at this stage in the test programme the vast expanse of a dry lake was deemed necessary for safety. The only other immediately available desert site was at White Sands in New Mexico. A landing at White Sands would pose no problem for the orbiter, but there was no recovery equipment there to tend to it, so, during the final days of the mission, two special trains transported the recovery crew's vehicles from Edwards to the Northrup Field at White Sands. It was a hectic time on the ground, but in space life aboard Columbia had become routine, and studies of plant and insect adaptation to microgravity were underway. Unfortunately, an hour before the scheduled deorbit time, a sandstorm swept across White Sands, and John Young, flying a chase plane overhead, ordered a 24-hour postponement. Columbia came down without incident the next day. It would not to be the last flight to be extended to wait out the weather. This time, the IRIS aircraft was able to pick up the orbiter as it began its re-entry, and its imagery confirmed the predicted distribution of thermal stress across the vehicle's belly.

Ken Mattingly and Hank Hartsfield took Columbia up for the final test flight exactly on schedule. The event was marred only by the fact that faulty parachutes prevented the SRBs from being recovered. With missions seemingly routine, the Press focused its attention on the 'classified' payload carried in the payload bay for the Department of Defense. Although it was not revealed until much later, the cover on the aperture of the CIRRIS missile-tracking telescope failed to release. However, this time the arm was able to assess the extent to which outgassing polluted the bay using the IECM, and Mattingly was able to make the airlock test to clear the way for the spacewalk assigned to the next flight. During the re-entry, a series of Programmed Test Inputs (PTI) began to 'stretch the envelope' by assessing the hypersonic characteristics of the orbiter; the nose was pitched down, reducing the angle of attack, so as to assess the increased thermal stress. For the first time, Columbia landed on the concrete at Edwards. On this Independence Day, half a million people had driven out to the high desert to witness the final landing of the OFT programme. Amongst them was President Ronald Reagan. He declared the National Space Transportation System to be *operational*.

This decision was criticised by some as being premature. It was pointed out that a new aircraft typically had hundreds of test flights prior to entering service. Others compared the shuttle with the X-15 rocket plane with which NASA had explored hypersonic flight; it made almost 200 flights without being made operational. But the X-15 had no operational role; it was purely a research aircraft. Although the shuttle was a research aircraft, in that it was a step beyond the state of the art, it was specifically meant to provide a transportation service to and from low orbit, for the nation; it could not possibly make hundreds, or even dozens, of test flights. A fairer comparison would be the Saturn 5 rocket, which sent astronauts out to orbit the Moon on only its third flight. The decision to declare the shuttle operational was political. This cleared the way for commercial operations. The only real effect on operations was the increase in the crew complement, and the resultant removal of the ejector seats that had offered the two-man test crews a means of escape during the early phase of the ascent. From now on, in the event of trouble during

Table 2.1. Orbital test flights

Flight	Orbiter	Launch	Return	Mission
STS-1	Columbia	12 Apr 1981	14 Apr	instrumentation
STS-2	Columbia	12 Nov 1981	14 Nov	instrumentation; mapping radar
STS-3	Columbia	22 Mar 1982	30 Mar	instrumentation; science experiments
STS-4	Columbia	27 Jun 1982	4 Jul	instrumentation; missile tracking

the ascent, the astronauts would be committed to accompany their vehicle to its fate. On the other hand, NASA had exhaustively explored abort options to give the orbiter the best possible chance of survival, so as Reagan's words echoed across the desert, NASA was looking forward to increasing the pace of operations, to fulfil its promise of routine and cheap access to low orbit.

THE DREAMTIME

With the last two test flights having launched on schedule, NASA had high hopes that it would be able to build up the flight rate and expand its operations by pursuing independent programmes. The flight schedule – or the *manifest* as NASA preferred to call it, in order to convey the impression of a commercial transport operation – interleaved these many strands of activity.

Table 2.2. Shuttle manifest, *c*. mid-1982

Flight	Date	Orbiter	Mission
STS-5	Nov 1982	Columbia	SBS, Anik
STS-6	Jan 1983	Challenger	TDRS-A
STS-7	Apr 1983	Challenger	SPAS-1; Palapa; Anik; OSTA-2
STS-8	Jul 1983	Challenger	TDRS-B; Insat
STS-9	Sep 1983	Columbia	Spacelab 1
STS-10	Nov 1983	Challenger	classified; cancelled
STS-11	Dec 1983	Columbia	LDEF; SolarMax repair
STS-12	Jan 1984	Discovery	TDRS-C; Palapa-B2
STS-13	Mar 1984	Challenger	classified
STS-14	Apr 1984	Columbia	OAST; Anik; RCA
STS-15	May 1984	Discovery	TDRS-D; Westar
STS-16	Jun 1984	Columbia	SBS; Leasat
STS-17	Jul 1984	Challenger	Spacenet; Westar
STS-18	Aug 1984	Discovery	Arabsat; Telstar; Leasat
STS-19	Sep 1984	Columbia	Spacelab 3
STS-20	Oct 1984	Challenger	OSTA-4; RCA; Intelsat
STS-21	Nov 1984	Columbia	Spacelab 2
STS-22	Nov 1984	Discovery	classified
STS-23	Jan 1985	Challenger	HST; LDEF retrieval
STS-24	Jan 1985	Columbia	OSS; Arabsat; Intelsat
STS-25	Feb 1985	Discovery	classified
STS-26	Apr 1985	Atlantis	Spacelab D1
STS-27	Apr 1985	Columbia	classified
STS-28	May 1985	Challenger	Telstar; RCA; GStar; Leasat
STS-29	Jun 1985	Atlantis	OAST; Satcom; Spacenet
STS-30	Jul 1985	Columbia	classified

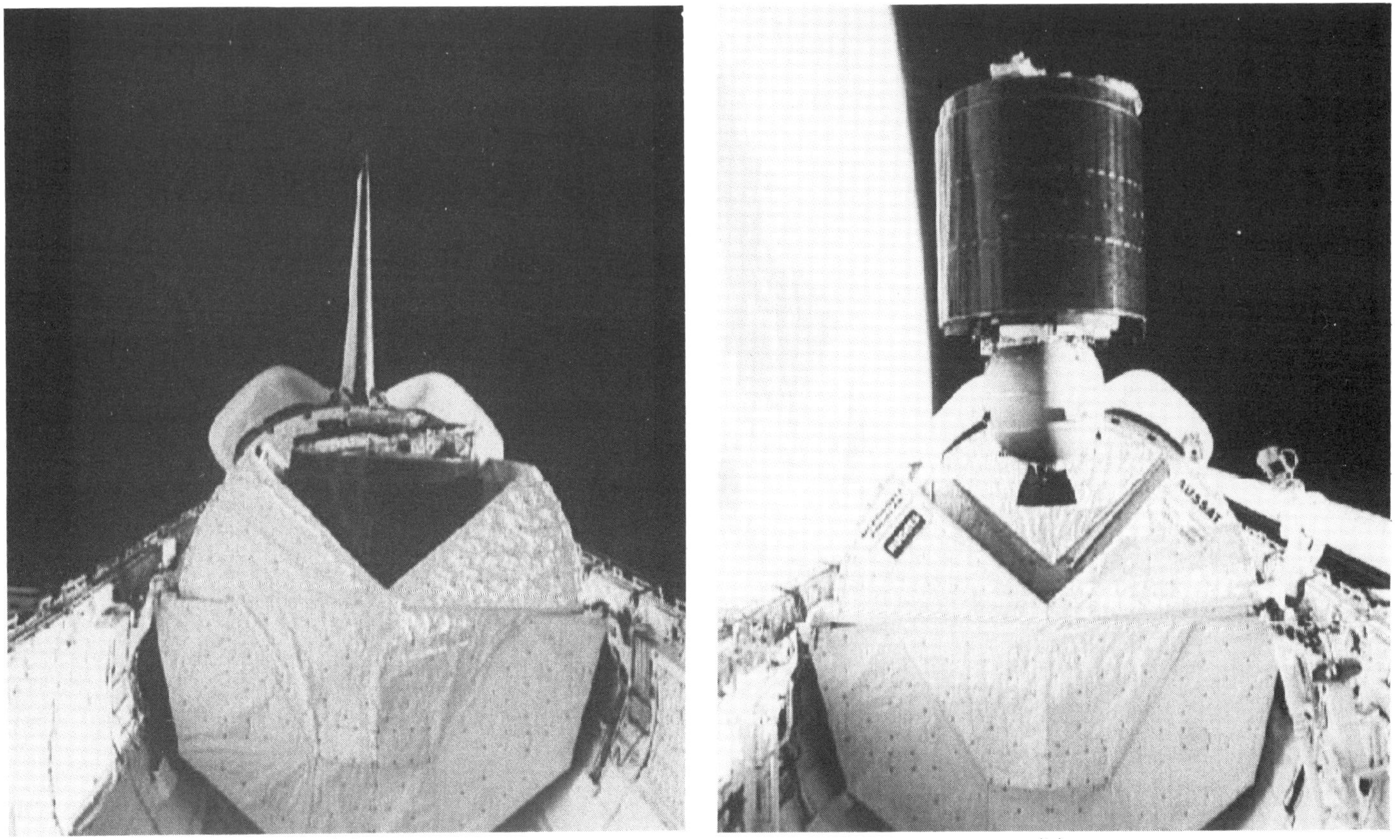

(a) (b)

Fig. 2.9. The deployment of an HS-376: (a) opening the sunshade prior to spin-up and (b) spring ejection. The HS-376 is the most successful commercial communications satellite on the open market. It is designed to spin for stability, so its solar cells are wrapped around its body. Spinning also distributes the thermal stress. The clamshell sunshade protects the satellite whilst it is immobile on its payload bay mount. An electric motor on the mount spins the satellite just prior to its release. The large spheroidal package at the base of the satellite is the Payload Assist Module (PAM), the kick-motor that injects the satellite into Geostationary Transfer Orbit (GTO); once this has been achieved the motor is jettisoned. The satellite has a small rocket engine of its own for the circularisation manoeuvre at the top of the elliptical transfer orbit to enter the operational '24-hour' Geostationary Orbit.

It had been intended to send the first Tracking & Data Relay System (TDRS) satellite up to geostationary orbit on the first operational flight, but a commercial satellite deployment mission had been given priority for Columbia's fifth flight. As expected, it took off exactly on time, on 11 November 1982. In addition to Vance Brand and Bob Overmyer, the flight crew, Mission Specialists Joe Allen and Bill Lenoir rode in fold-away seats at the rear of the flight deck. The first spacewalk of the programme had to be cancelled because the new extravehicular suits were faulty, but the dispatch of a pair of HS-376 communications satellites, the mission's main objective, was achieved without incident. PTIs were made during re-entry. It had been planned to attempt a crosswind landing at Edwards, but the salt flat was waterlogged and a normal landing was made on the runway instead. The touchdown was so smooth that Overmyer was prompted to ask the pilot of a chase plane to verify that the gear was running along the concrete. After this remarkable mission, Columbia was sent back to Rockwell to be refitted. A new orbiter – Challenger – was ready for service.

STS-6 was set for 20 January 1983, but the FRF on 19 December revealed a hydrogen leak in one of the SSMEs. The follow-up firing on 25 January revealed leaks in all three engines, so the launch slipped to April. This double delay was a disappointing disruption of a programme which had begun to seem routine. However, once the engine problems had been overcome, Challenger was dispatched without incident. Although the shuttle was now operational, its characteristics were still being tested. The push for performance was driven by the Air Force's demand that the shuttle meet the promised mass-to-orbit capacity and the crossrange capability during re-entry. For the first time, following the period of maximum aerodynamic pressure – or Max-Q, as it is dubbed – as the stack passed through Mach 1, the SSMEs were throttled to 104 per cent of their rated thrust. This began the process of 'stretching the envelope' of main engine performance. The goal was to work up to a routine 109 per cent of rated thrust; the 100 per cent level was simply the benchmark in the specification, not the engine's maximum performance. It would take 109 per cent to lift the heavy military payloads into polar orbit. Critics have argued that this unnecessarily risked the crews' lives, but incremental development was standard practice in the aerospace industry, and for a vehicle as expensive to operate as the shuttle there was no viable alternative.

The first item of business upon reaching low orbit was to dispatch the first of NASA's TDRS satellites. Far larger than the commercial HS-376 drum-like spinners, the folded-up package was mounted on a two-stage Inertial Upper Stage (IUS) rocket, and it was carried

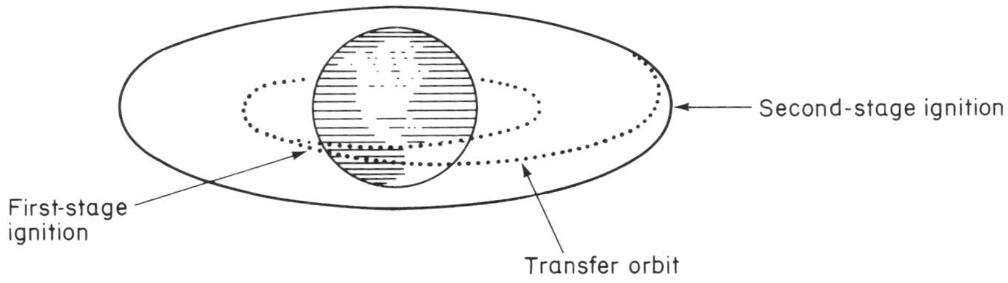

Fig. 2.10. A schematic of the Geostationary Transfer Orbit (GTO) manoeuvre sequence.

(a)

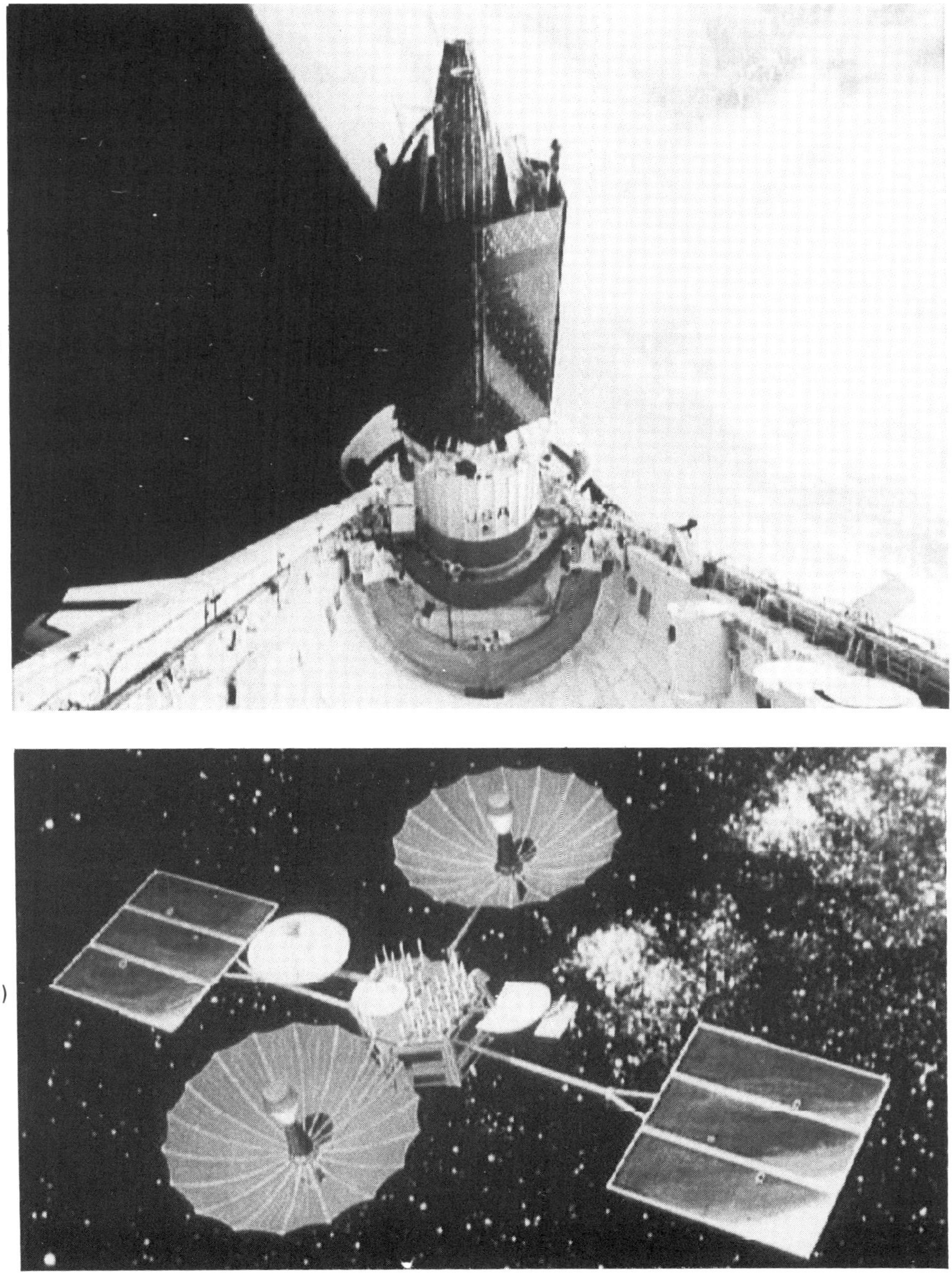

(b)

Fig. 2.11. A Tracking and Data Relay System (TDRS) satellite: (a) being ejected on an Inertial Upper Stage (IUS), and (b) an artist's impression of its operational configuration.

lengthwise in the bay. Although the deployment system had not been tested before, the annular tilt-table elevated the 17-tonne stack and spring-ejected it across the top of the orbiter's cabin. The IUS fired its big engine precisely on time to boost into geostationary transfer orbit. A few hours later, when the second-stage motor fired to circularise this high orbit, things went badly wrong. Although the satellite was later able to manoeuvre itself onto its assigned station, the failure of the IUS threw the programme into chaos. While this crisis played itself out in high orbit, Story Musgrave and Donald Peterson made NASA's first spacewalk since the heady days of Skylab.

The HS-376 commercial satellites had been boosted up to geostationary orbit by a small Payload-Assist Module (PAM), but this could only handle payloads up to 1.2 tonnes. The 2.2-tonne TDRS required the more powerful IUS. This rocket had been developed by the Air Force specifically to deliver its large shuttle-deployed satellites to geostationary orbit. It was meant to be the most reliable inter-orbit transfer stage ever made. Its flight-control systems were multiply-redundant, yet a straightforward mechanical failure had crippled it. It clearly could not be used until the fault had been rectified, but it did not mean, of course, that the shuttle was out of business. Grounding the IUS meant only that the payloads requiring it had to be pushed down the manifest. Although this opened up slots for commercial and scientific payloads, advancing missions complicated payload integration and led to dynamic scheduling which became ever more spontaneous.

Fig. 2.12. A startling view of the glow from the hot exhaust from an Orbital Manoeuvring System (OMS) engine firing in space.

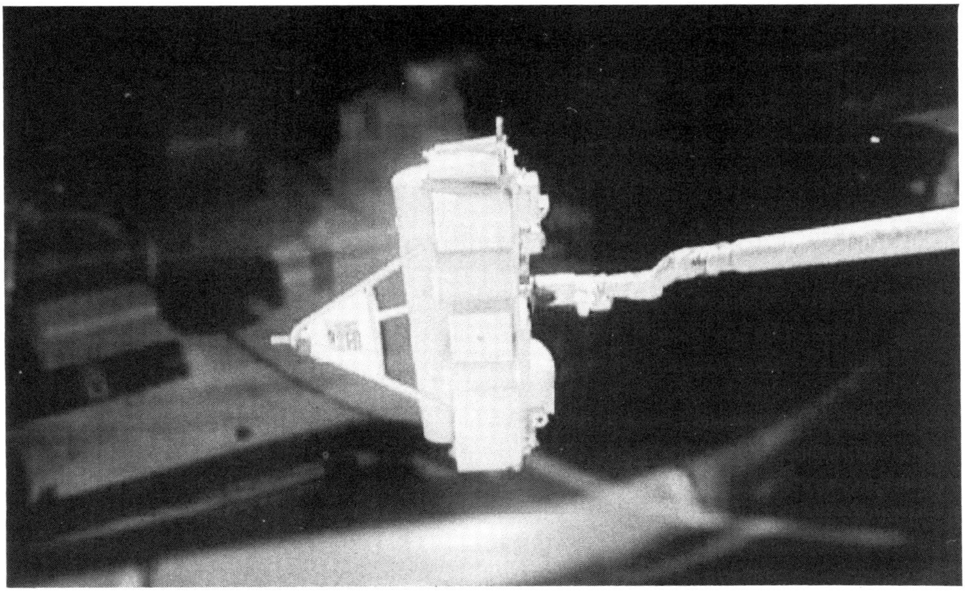

Fig. 2.13. The deployment of the SPAS free-flyer. The 15-metre long Remote Manipulator System (RMS) arm lifts the package off its mount in the payload bay, aligns it as required, and then releases it. Free-flyers are designed to be retrieved, returned to Earth, refurbished and reflown.

Fig. 2.14. The Challenger, as seen by the SPAS free-flyer. This was the first time that an orbiter was able to be viewed as a distinct vehicle flying in space.

It had been hoped to fly Challenger again in June and in August – the latter to deploy the second TDRS satellite. As soon as two relays were in service, Columbia was to make its return with the first Spacelab, and 'STS-10' in October was to carry the first satellite for the Air Force. The second TDRS and the classified satellite, which would also require an IUS, were immediately pulled from the manifest, and Spacelab was placed on hold, awaiting the outcome of the effort to rescue TDRS-A. The Spacelab science activities would be degraded without two relays, but it was a crucial international-crew mission and NASA was loath to cancel it.

So long as there was just one serviceable orbiter, the flight rate was difficult to increase because the rate was determined by the turnaround time, and although a management team had been formed specifically to streamline this process, there was only so much slack in the system. Nevertheless, allowing for the delays in its launch as STS-6, Challenger was launched on STS-7 more or less on schedule. It successfully deployed two commercial satellites, and, in deploying and retrieving the SPAS free-flyer, demonstrated the rendezvous which would be used to retrieve and repair the ailing SolarMax solar satellite. In commanding this demanding mission, Bob Crippen became the first astronaut to fly the

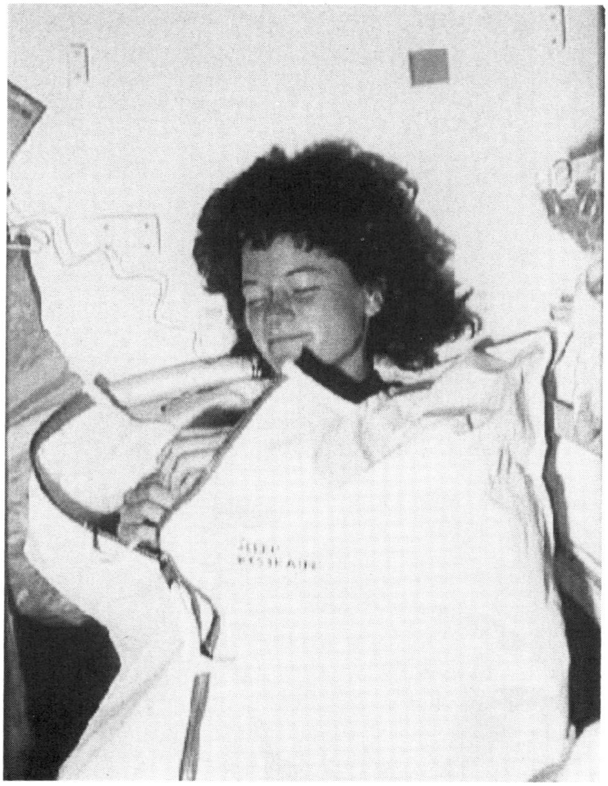

Fig. 2.15. America's first spacewoman: Sally Ride.

shuttle twice. He had hoped to make the first landing at the Kennedy Space Center, but out-of-tolerance weather forced him to divert to California.

Spacelab remained on hold while TDRS-A slowly manoeuvred towards its station. It was decided to fly Challenger on schedule, with the Payload Flight Test Article (PFTA) in place of TDRS-B. The primary item of business was to deploy an Indian satellite. Once this had been done, the RMS lifted the PFTA out of the bay. Its ballasted dumbbell frame was a representation of the size and mass of a large satellite, and it was swung about to assess the stresses on the arm's joints. The arm performed satisfactorily with this 3.3-tonne mass on its end, but it flexed alarmingly when the orbiter manoeuvred. As an engineering test, the PFTA was a valid objective, but it was improvisation. The orbit constraints for the Indian satellite had required the first night launch, which was spectacular, and this, in turn,

Fig. 2.16. As STS-8 blasted off, it became the programme's first night launch.

mandated the first landing in darkness, which was assigned to Edwards for an increased margin of safety. It was completed without incident, but after the SRBs had been retrieved a shock discovery was made: a wide segment of the 75-mm thick insulation which lined the aft nozzle of one of the boosters had been almost completely eroded by the exhaust plume. The burn-through, which was the first sign of significant damage on an SRB, was put down to a faulty batch of material. In fact, this had been the first flight of an uprated booster which, by being 4 per cent more powerful, allowed a 1,400-kg increase in payload. As with the upgrading of the SSMEs, this was part of the ongoing process of pushing the shuttle system closer to its designed limits. Erosion of the SRBs was to become a matter

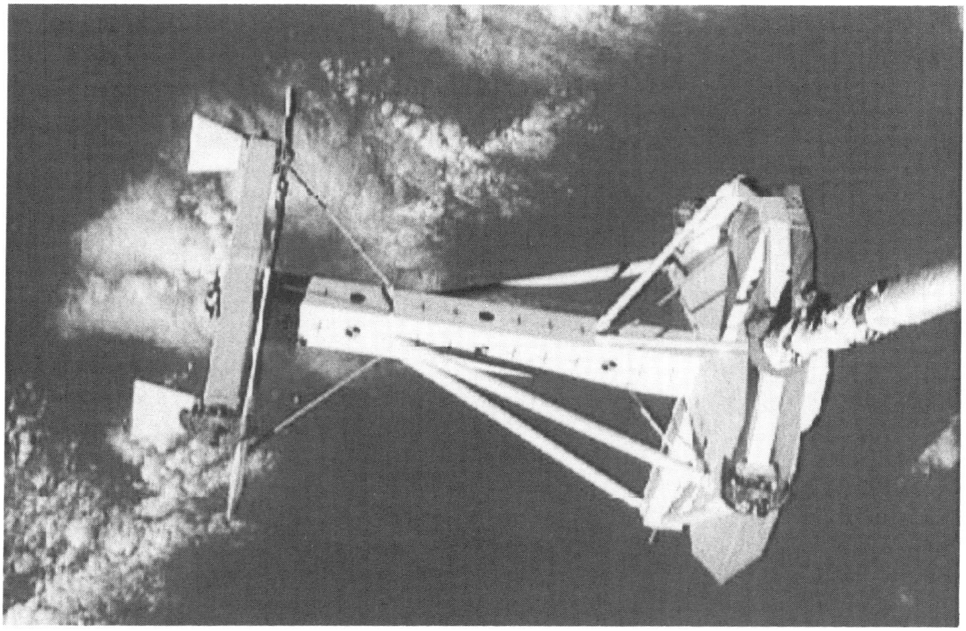

Fig. 2.17. The Payload Flight Test Article (PFTA), a low-fidelity mock-up of a large satellite, being manipulated by the RMS to test the ability of the arm to control massive objects.

Fig. 2.18. STS-8 followed its spectacular night launch by making the programme's first landing in darkness.

of some concern, not least because there seemed to be no clear pattern as to the nature of the problem.

As soon as it became clear that the solitary TDRS satellite would be on-line in October, John Young was given the go-ahead to fly Spacelab. However, Columbia's launch was put back a month while the SRB damage was studied.

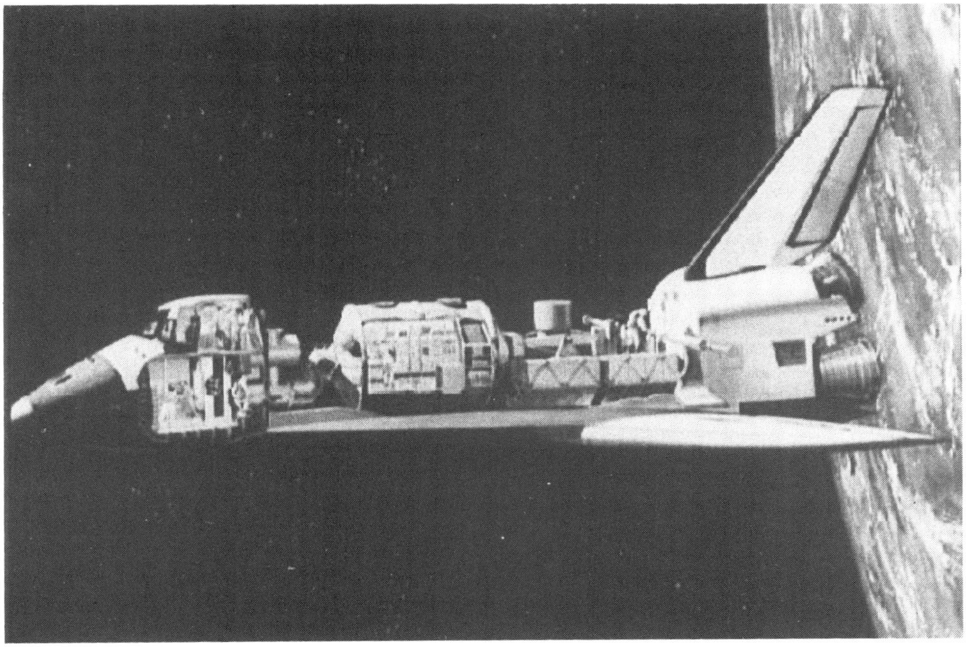

Fig. 2.19. A cutaway artist's impression of the orbiter's cabin and the double-length Spacelab used by STS-9.

The ESA-built pressurised module, with its external pallet of instruments, was the heaviest payload to date, and it took up the entire bay. It was a mission of *firsts*: the first to fly Payload Specialists, the first to fly a foreigner (Ulf Merbold, representing ESA, was a German), the first to use the 57-degree inclination to the equator for better latitude coverage, the first to operate a shift system to facilitate 24-hour operations, and the first to exploit the orbiter's maximum duration. In fact, the low rate at which the cryogenics were consumed meant that the flight was extended for a day. There was therefore very little margin when Young prepared Columbia for its deorbit burn. He subsequently admitted that he "turned to jelly" when, after he fired an RCS thruster to align the vehicle, a shock propagated through the orbiter's structure and the primary GPC dropped off-line. Another processor promptly took over, but it too shut down. This was the programme's first in-flight crisis. There was some concern that there might be a generic fault in the GPC system. It took seven hours to reinitialise the computers and reset for the deorbit burn. The radio blackout during the initial phase of the re-entry was the tensest moment in the programme since Young had flown Columbia back after its first test flight, and the Houston flight controllers were undoubtedly greatly relieved to see it make a textbook landing on the dry lake, not just because of the in-orbit problem but also because the laboratory made it the heaviest return-payload to date.

What nobody was aware of was that a hydrazine fire was raging in the aft compartment in which one of the APUs – the power units which drove the aerodynamic control surfaces – were located. In fact, the fire was not noticed until the recovery team moved in to vent the engine bay. Analysis revealed that the fire had flared up about two minutes prior to landing, and that it was due to microscopic debris which had contaminated circuitry. Although only one APU was needed, if an orbiter was denied power to move its control surfaces it would be doomed, so they had been physically isolated to minimise the risk of all three units being destroyed. It had been a narrow escape.

Fig. 2.20. Columbia lands after STS-9. Unknown to anyone, an Auxiliary Power Unit (APU) fire was raging in the orbiter's aft bay.

Even without the second TDRS relay, the Spacelab had downlinked more science data on its ten-day mission than Skylab had during its six months of habitation. More Spacelab missions were planned, and a complete TDRS network had been thought to be an essential prerequisite for their operation. However, it was evident that recertifying the IUS would be a protracted process, and the rest of the satellites would not be able to be added until 1985. This was also bad news for the Air Force because many of the satellites that it had waiting had been specifically designed for the shuttle and so were too large to be off-loaded onto its rapidly depleting stock of Titan rockets.

The long-term future for the shuttle programme received a boost in January 1984 when Ronald Reagan ordered NASA to build a space station "within a decade". This was *exactly* what NASA had been hoping for. It not only gave the shuttle a new strategic mission

Fig. 2.21. Touchdown in Florida. Contrast the narrow strip of the Shuttle Landing Facility (SLF) at the Kennedy Space Center with the vast expanse of the dry lake in California.

– the construction of the orbital facility – it also provided a *destination*, a space community to serve. The space station, however, was for the future; the overriding objective for 1984 was to increase the flight rate.

So many flights had been deleted from the shuttle's manifest that NASA abandoned the original straightforward numerical series for identifying flights. Instead, it introduced a scheme based on the Financial Year (FY) to which the cost of a mission was attributed. The flights in each year were identified by an alphabetic designator, with each year starting back at 'A'. The last digit of the FY was used to start the designator. To take into account flights launched from California, launches from the Kennedy Space Center were designated by a '1' and those from Vandenberg AFB by a '2'. This new scheme was introduced for FY84, which ran from October 1983 to September 1984. Although STS-9 had flown after the start of FY84, it fell within the FY83 budget. However, when the Air Force decided to cancel its flight rather than assign it a non-IUS payload, the 41A label was rendered void, so the new scheme got off to an unfortunate start, and not just by missing out its first designator; there were problems in space too.

In February 1984, STS-41B released two commercial satellites, but unfortunately both PAM stages fizzled out within seconds and stranded their payloads in low orbit. And the situation worsened. The balloon released to serve as a radar target to rehearse rendezvous procedures in preparation for the SolarMax repair inflated so rapidly that it ripped itself to shreds. Next, a fault in the RMS prevented the SPAS – the first satellite to be refurbished and reflown – from being released. However, the public imagination was caught when Bruce McCandless flew 100 metres from the orbiter to demonstrate the manoeuvring unit, the MMU; this was the first time an astronaut had formed an *independent* satellite.

Another first was achieved when Challenger landed at the Kennedy Space Center. This historic roll-out was marred only by a brake seizure when the orbiter tried to remain on the centreline of the 100-metre wide concrete strip. Returning to Florida rather than California shaved a week off the turnaround time.

The spectacular failure of the PAM rocket stages stalled the commercial satellite aspect of the manifest. Even though both the lightweight and heavyweight stages were now out of service for recertification, NASA advanced another strand of its multifarious manifest to the fore, and underlying the dynamic mission planning was the relentless drive to increase the flight rate.

In April, Bob Crippen took Challenger up to rendezvous with SolarMax to carry out the main feature of the early part of the programme: the repair of a satellite. To reach SolarMax, Challenger had to climb to 560 km, the highest operating altitude to date. Its launch marked the first time that a shuttle flew a *direct ascent* trajectory. In previous cases, the shuttle had settled into a low initial orbit, then used a series of OMS burns to raise its altitude to about 300 km. In this case, however, the OMS engines were fired immediately after the ET was released in order to yield a high initial apogee, marginally below SolarMax's orbit, at which point, around 45 minutes after launch, the OMS-2 burn was used to circularise at that level. This ascent profile both enabled LDEF to be deployed into a high orbit, safe from decay for the year or so that it was intended to leave it in space, and made rendezvous with SolarMax straightforward.

By now all the aspects of the SolarMax repair that could be tested had been rehearsed in space, so it was expected that the recovery would go smoothly. Unfortunately, the appa-

Fig. 2.22. The Long-Duration Exposure Facility (LDEF) is deployed. Its multifaceted surface was a test bed for materials intended to be used on future spacecraft.

ratus specifically built to mate with a grapple-fixture on the satellite refused to engage, and in trying to manoeuvre the satellite manually George Nelson induced a rotation that sent it spinning out of control. Without sunlight on its solar panels, the satellite switched over to its battery, but because this could support only a few hours of operation it began to look as if, far from being repaired, SolarMax had been killed off. Ground controllers were able to stabilise the satellite, however, so Challenger returned two days later. This time Terry Hart used the RMS to grab the satellite and set it down on a tilt-table at the rear of the bay. From this point, the repair operation went smoothly, and NASA's Public Affairs Office (PAO) slipped into overdrive.

Crippen had hoped, finally, to be able to land in Florida, but unacceptable weather there forced a diversion. His earlier enthusiastic remark, "What a way to come to California!" was haunting him.

Discovery's debut mission, as STS-41D, the twelfth flight of the programme, started as an exercise in frustration. In fact, it took *three* countdowns to get the new orbiter off the ground. After a perfect FRF on 2 June, NASA set the launch for 22 June, then postponed it to 25 June, but this launch attempt was scrubbed at T–9 minutes due to a GPC fault. The next day, the count ran smoothly right down to the point at which, at 120-millisecond intervals, the SSMEs were ordered to ignite. The hydraulically-activated fuel valve of the first engine failed to open. By the time this misfire was diagnosed and the controlling GPC

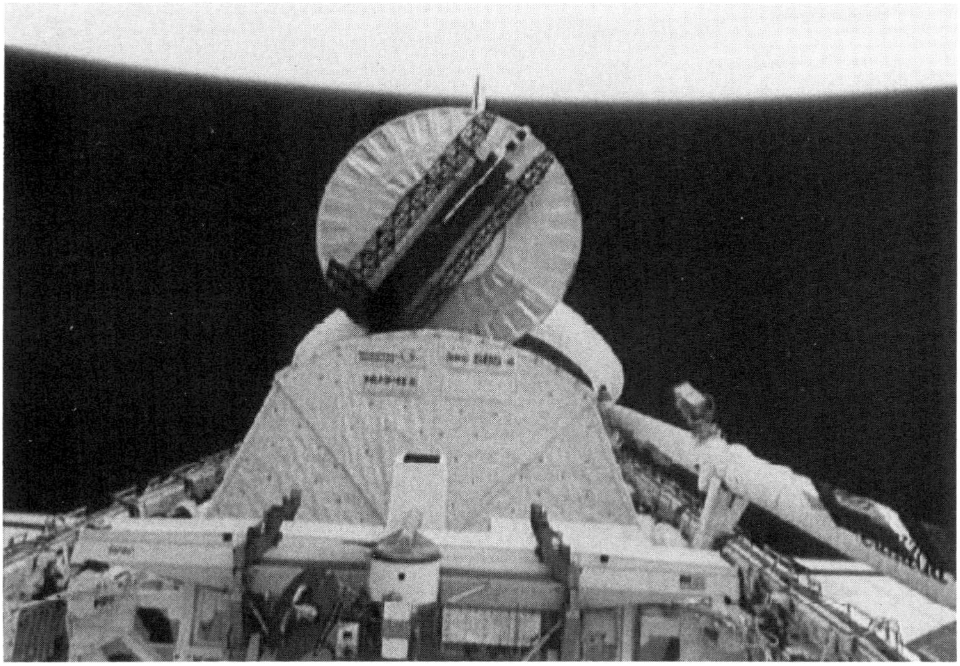

Fig. 2.23. The deployment of a Leasat. This type of communications satellite was one of the few that was designed to exploit the width of the shuttle orbiter's payload bay. Unlike the smaller HS-376, which was ejected longitudinally, the Leasat was rolled out of the bay, 'frisbee style'. The objects on the near side of the satellite, which is the top, are the antennas that will unfurl once the satellite reaches its operating station. This type of satellite remained the property of the manufacturer and was simply leased by the US Navy, hence its name 'Leasat'.

was able to intervene, the second engine had fired and was already up to 20 per cent thrust, so it was ordered to shut down and the third ignition was inhibited. The flame fizzled and the billowing cloud of steam rapidly dispersed, leaving the shuttle stranded on the pad. This was the first time that a launch had been scrubbed following engine-start (an event that NASA dubbed a post-ignition abort). Although spectacular, this was not an unprecedented situation, because the orbiter was in essentially the same state as after a FRF test. Forty-five minutes later, Hank Hartsfield's dejected crew disembarked. It subsequently emerged that a very hot colourless flame had persisted at the base of the stack for several minutes as free hydrogen had burned off.

Since the engines that had fired had to be refurbished, the launch had to be pushed back by a month. The FY84 schedule had slipped so far behind that NASA had already cancelled mission 41E; it now cancelled 41F and added one of that mission's commercial satellites to Discovery's manifest, and this made the revised 41D the first shuttle to be assigned a trio of satellites. The attempt to launch Discovery on 29 August was foiled by a timing fault in the orbiter's master event controller, but the next day, after a brief delay for a private aircraft to clear the restricted airspace around the Kennedy Space Center, 41D was finally dispatched. Once in orbit, everything went well. Two of the three satel-

Fig. 2.24. An artist's impression of the shuttle testing the solar panel intended to be used on a future spacecraft or space station.

lites deployed rode the recertified PAM stage and, to everyone's relief, were successfully boosted into geostationary transfer orbit. The third satellite was an HS-381 built specifically for the shuttle; it incorporated its own kick-motor. Immediately after the deorbit burn, the active APU developed a hydraulic leak, so the back-up, already running and standing by, was used to drive the aerodynamic surfaces during the re-entry.

The descent was routine until the nosewheel made contact with the Edwards dry lake, at which point the orbiter slewed to the right. Hartsfield applied the rudder to straighten up, to run parallel to the centreline. When he finally applied the brakes, the orbiter rocked back and forth with a frequency of several times a second, which led him to suspect that one of the brakes was repeatedly locking and releasing. It turned out that the shock absorber of the right gear had lost nitrogen pressure and the 'flat' mechanism had repeatedly bottomed out, causing that side of the vehicle to bounce. The fault had developed at precisely the moment that the wing passed through horizontal to lower the nose gear onto the runway, just as the pressure on the wing transitioned from up-lift to down-force. This roll-out, the trickiest yet, caused some concern, because although the runway at the Kennedy Space Center was more than long enough, it offered little room for lateral manoeuvring.

Fig. 2.25. Judy Resnik in a moment of reflection on the middeck during STS-41B.

The cancellation of 41E and 41F, and 41D's successful deployment of three satellites, meant that the thirteenth flight, 41G, effectively restored the year's schedule. And with two orbiters in service, it was possible to accelerate the pace: 36 days after Discovery's landing, Challenger was launched at the first attempt. For this science mission, the initial crew was augmented at the 'last minute' by two Payload Specialists (Garneau and Scully-Power) and their names had to be sewn on in an arc around the lower edge of the already prepared crew patch. Although Crippen had been assigned to 41G, this had been in the expectation that he would have a year following 41C to prepare his new crew, but 41G had been advanced by six months and much of the responsibility for preparing this mission fell to his rookie pilot, Jon McBride. The main event was the spacewalk during which the procedure that was to be used to replenish Landsat 4 was rehearsed. Much of the flight was spent on radar mapping, but this became an exercise in frustration because the orbiter's TDRS antenna, which was a dish on a short boom projecting over the starboard sill just aft of the cabin, ceased to follow the relay satellite. This meant that data could not be relayed in real time, so the only option was to lock the antenna and then repeatedly reorientate the orbiter, first to scan the surface and then to point the stuck antenna at the satellite to download each successive strip of data. As a result, instead of 50 hours, just 9 hours of data could be returned in the time assigned to the experiment. When Crippen finally managed to land in Florida, the Capcom teased him with the retort, "But Crip, the beer's been sent to Edwards!"

For the first flight of FY85, flight 51A, NASA set out on an ambitious rescue mission. After it had dispatched two more communications satellites, Discovery was to try to retrieve the two HS-376s which had been stranded in low orbit in February when their PAM stages had failed. This task would be far more difficult than had been SolarMax, not because there were two satellites, but because, unlike SolarMax, these satellites had not been designed to be manipulated in orbit. Nevertheless, after discarding a handling bar which would not fit, both satellites were securely stowed in the bay. It was a *tour de force* for the spacewalkers, Joe Allen and Dale Gardner.

The next flight was shrouded in mystery. This first Department of Defense mission was so classified that even the imminence of its launch was not announced until the clock picked up at the end of the T–9 minute hold; the secret had obviously been kept because, for once, there were no Soviet 'trawlers' offshore. With the exception of the 8-minute climb to orbit, all communications were encrypted. The Air Force reluctantly let NASA release occasional, rather bland, progress reports. No video downlink was released. Over the years, the media had grown accustomed to NASA's openness, so the 'closed door', which generated not a little resentment, became the focus of attention. In the revised FY85 manifest, 51B had been pushed back, so this military flight, the fifteenth of the programme, was 51C. Upon the recertification of the IUS, the Air Force had exercised its authority and requisitioned a schedule slot in the interest of national security. In effect, this was the 'STS-10' mission. The crew was the same as had originally been assigned, but with the late addition of Gary Payton, who became the first Spaceflight Engineer to fly; his role was to supervise the activities involving the classified payload. The Air Force intended to have one of its own officers accompany each classified flight so that it would not have to reveal too much about the payload to the NASA astronauts, even if they had previously been in the military.

Although the IUS' first-stage solid rocket motor underperformed, the controller burned the aft-pointing thrusters to make up the velocity shortfall so that the second stage was able to take over and deliver the secret satellite. With its satellite away, Discovery landed. At just three days, it had been the shortest flight since the shuttle had been made operational, but it reflected the Air Force ethos: get airborne, carry out the primary mission, and then return to base with the minimum of distractions; in other words, operational efficiency.

NASA, too, was keen to exploit the recertification of the IUS, so it assigned the second TDRS satellite to the next flight, designated STS-51E, which it scheduled for 7 March. But a week before launch was due, a thorough analysis of communications faults with TDRS-A prompted concern that there could be a flaw in the design, so the countdown was cancelled, and TDRS-B was unloaded so that it could be fully checked. The relentless launch process immediately switched its attention to STS-51D, which was set for April. The clock ran smoothly right to the point that the only issue holding up the launch was the weather; when Discovery finally lifted off, the window had only 55 seconds remaining. It deployed two communications satellites; however, one satellite, an HS-381 Leasat, failed

Fig. 2.26. The 'fly swatter' taped to the end of the RMS, ready to try to throw the switch that would activate a dormant Leasat.

to activate as it left the bay. Every effort to kick-start the mechanism using the RMS failed, so the slowly spinning hulk had to be abandoned in low orbit. But the real drama came at the very end of the mission. The landing on the SLF was the first in a crosswind. Development was underway to install nosewheel steering. The mechanism was in place, but the control system had not yet been certified, so Karol Bobko had to apply differential braking to stay on the centreline. The crosswind was just 8 knots, but the differential loading so stressed the main gear that one of the 275-psi tyres blew out, another was badly eroded, and one of the brakes seized. Luckily, Discovery rolled to a halt a few seconds later. If the blow-out had occurred earlier, the orbiter could all too easily have veered into the scrub adjacent to the strip. With hindsight, it was clear that expecting to land on such a narrow runway in a crosswind prior to introducing the nosewheel steering system had been inviting disaster.

The orbiter's brakes had been causing concern for some time because they had showed much greater wear than expected. Weight had been a key factor in their design. There were four brake units on each main gear. A carbon-coated beryllium stator disk was developed to combine low mass with high efficiency, but its size meant that it could not sustain heavy braking. NASA demoted the SLF to emergency-only status, and ordered future missions to land at Edwards, ideally on the dry lake where there was ample room to run wide.

Until the nosewheel steering could be introduced, no crosswinds would be allowed; this constraint was viable because there were a dozen runways on the dry lake, at all azimuths. As events transpired, however, it would be five years before another shuttle landed in Florida.

The countdown for the seventeenth shuttle flight – the postponed 51B mission – was held up for just two minutes to check that an oxygen drain valve was properly set, so Challenger launched a mere ten days after Discovery's return. It was the first flight to attract a group of protesters; the Spacelab carried two squirrel monkeys, which the Animal Rights movement considered to be cruel. Despite this added interest, 51B was the first shuttle launch *not* to be broadcast live by the television networks. The only excitement during the ascent was the shutdown of one of the three APUs. The crew operated two shifts to perform materials and life science experiments. The landing, at Edwards, was absolutely textbook. By mid-1985, the shuttle programme had reached the point where it was not likely to be reported unless something new and dramatic was on offer. By having become routine, the shuttle itself was no longer news. In fact, it is doubtful that anyone was aware that the missions were being flown out of their manifested order.

When STS-51G was launched in June it carried the outstanding satellite from the cancelled 41E mission, in addition to the two satellites with which it had started out. All three were sent on their way without difficulty. In addition, the SPARTAN astronomical free-flyer, which was deployed and retrieved by the RMS, had its first outing. The only excitement of the landing occurred towards the end of the roll-out when the left main gear sank into a soft patch of the dry lake and entrenched itself.

The attempt to launch Challenger as 51F on 12 July was scrubbed at T−3 seconds when a hydrogen coolant valve failed to close and the SSMEs, all three of which had ignited, had to be shut down. The launch two weeks later became the first mission to execute an abort during the ascent. After jettisoning its SRBs, Challenger continued to climb under its own power. The centre engine began to overheat, however. Controllers monitoring the situation throughout the 104 per cent phase were relieved to see that it did not stray into the red line before it throttled back to 65 per cent for the final phase of the ascent. However, it continued to overheat and, at T+350 seconds, the GPC shut it down. By this point, with two engines, the shuttle would have sufficient energy to achieve low orbit, so an Abort To Orbit (ATO) was ordered. This involved mission commander Gordon Fullerton turning the selector to 'ATO' and then pushing a button to enact his selection. The GPC raised the two remaining engines to 91 per cent, and added 70 seconds to the burn to compensate for the missing engine; this was feasible because all three engines drew propellants from the same tank. No sooner had the abort been executed, however, than a temperature rise was spotted in one of the remaining engines. At this point the controllers began to suspect a faulty sensor rather than a genuine problem with the engine, and they recommended that the crew intervene to inhibit the GPC from shutting down another engine. Unable to reach orbit on one engine, Challenger would have been forced to invoke a Transatlantic Abort Landing (TAL) which, in this case, was a landing at Zaragoza in Spain. After this extended two-engine burn, the OMS were fired to make up the velocity shortfall, and the orbiter limped into a 220-km circular orbit. It was not ideal for the observational programme planned for the telescopes on the palletised Spacelab, but any orbit was better than none, and the observing plan was hastily revised. NASA undoubtedly took

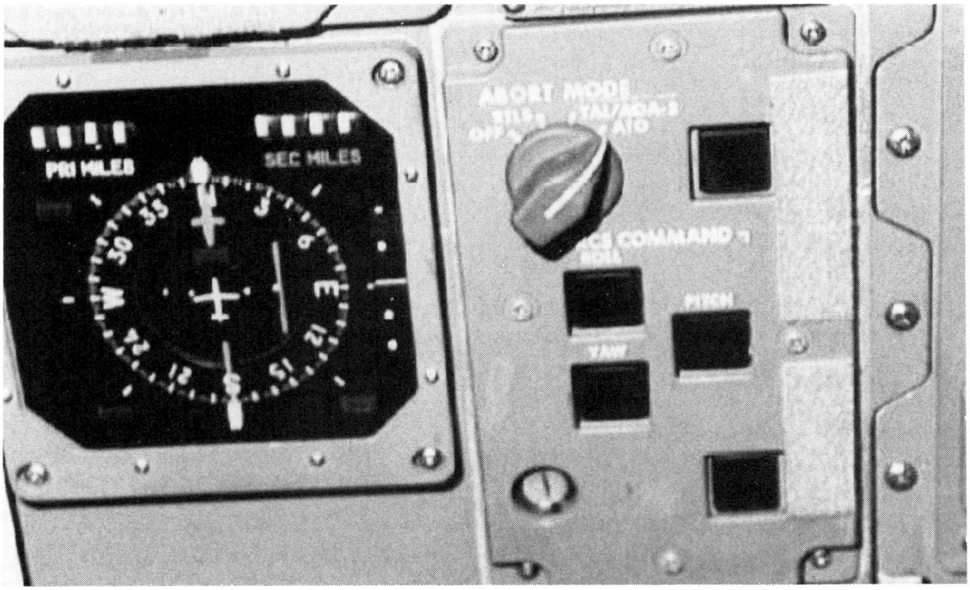

Fig. 2.27. Abort! Note the selector set to the 'ATO' option, indicating that the shuttle had performed an Abort to Orbit.

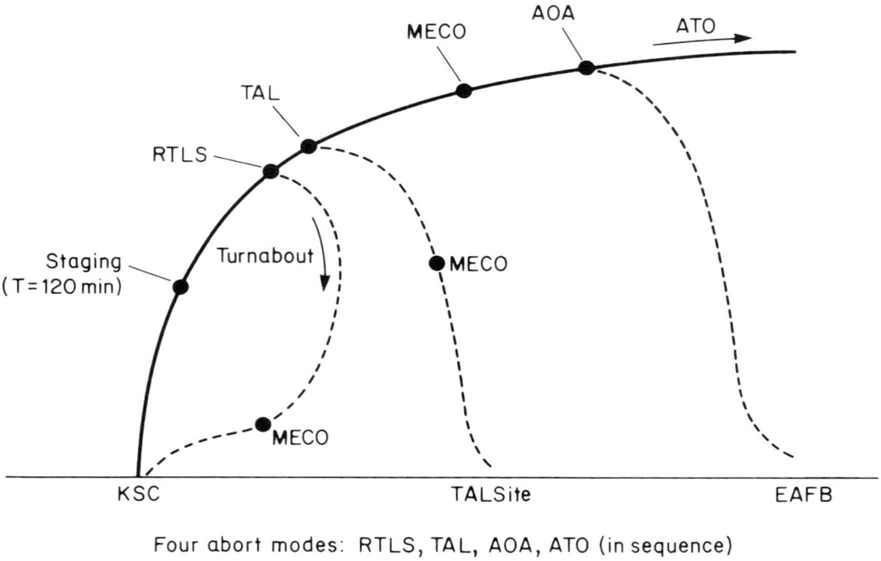

Fig. 2.28. A schematic of the abort modes which are available to a shuttle during its 8-minute ascent through the atmosphere.

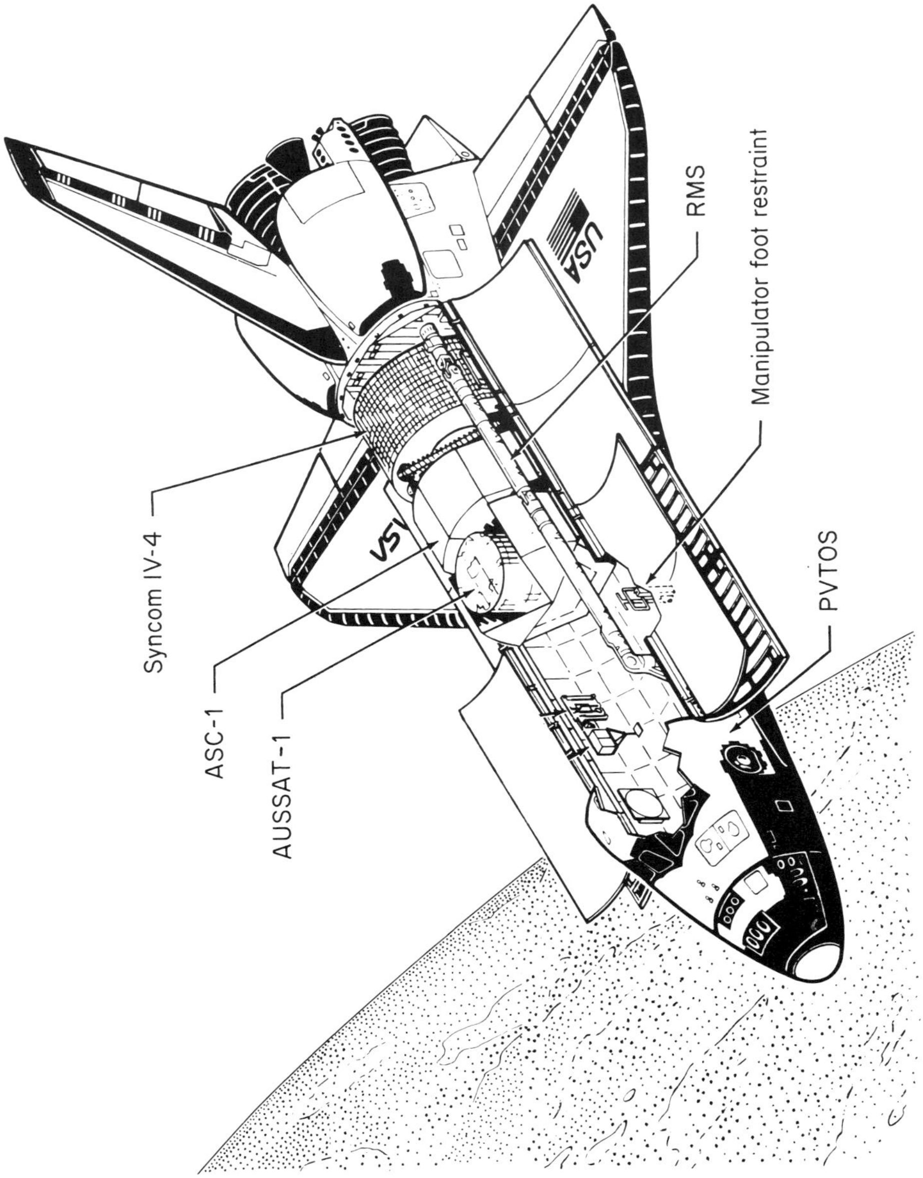

Fig. 2.29. A schematic of the payload bay for Flight 511, showing the loading of the three satellites.

comfort from the fact that the abort procedure had worked, even if in this case the term *abort* appeared to be overly dramatic.

A delay in integrating the telescope package for Flight 51F had led to Spacelab 2 being flown *after* the life sciences Spacelab 3 mission. To avoid a recurrence of this embarrassing situation, it was decided to split the Spacelab programme into a number of themes, name each to reflect its topic, and pursue these themes in parallel. Although ESA's preference was for multidisciplinary missions such as Spacelab 1, this change reflected NASA's wish to fly more focused missions so as to avoid the conflicting requirements of a mixed payload.

Precisely three weeks after Challenger landed, Discovery took off for the programme's twentieth mission, as STS-51I. In addition to deploying yet another trio of communications satellites, it rendezvoused with the Leasat that had been abandoned in April, and Bill Fisher and James van Hoften spacewalked to bypass its failed timer so that it could finally boost itself up to geostationary orbit. This rescue, undertaken *on the fly*, at the end of an already full satellite-deployment flight plan, demonstrated another aspect of the way that the shuttle programme was maturing: failed satellites abandoned by one flight could be retrieved, repaired and sent on their way by a later flight, an advance which provided a tremendous operational flexibility, and which vindicated those who had predicted that repair work would be one of the shuttle's roles. Impromptu spacewalks made crew training a dynamic process, required to accommodate substantial last-minute tasks. Overall, NASA was proving to be quite adept at what it had inappropriately dubbed *routine* operations.

STS-51J in October, which marked Atlantis' debut, was another exercise in frustration for the media because it was a classified mission, but evidently the deployment of an IUS carrying two strategic communications satellites for the Department of Defense was "highly successful", and Atlantis returned after four days with the minimum of publicity. It was not reported at the time, but the secure payload handing (classified satellites were not placed in the orbiter in NASA's Operations & Checkout Building; they were stored at the Air Force's Cape Canaveral Station, and were installed only after the shuttle had been checked out on the pad) and in-flight encrypted communications were costing the military dearly: possibly as much as $100 million per year. Three weeks later, Challenger launched with a record crew of eight astronauts and a Spacelab which had been co-sponsored by West Germany. This was flight STS-61A. Orbiter propellant had been traded for payload mass, but despite exploiting gravity-gradient stability and frugality with cryogenics there was insufficient consumables' margin to permit the hoped-for extension beyond the nominal mission, so Challenger came home after seven days. During the dry lake roll-out, the newly activated nosewheel steering was tested. In the absence of any crosswind, Hank Hartsfield steered Challenger off to one side of the centreline and then back again. As it would turn out, this would be the last time that this particular orbiter would return to Earth. The Spacelab flights had tested prototypes of the kind of scientific apparatus which could be used on the space station. The next flight, STS-61B, set off in November to test how that station should be assembled. After another trio of communications satellites had been deployed, Jerry Ross and Sherwood Spring built and stripped down sections of girders, and their activities were recorded stereoscopically so that a computer model of their working efficiency with different assembly techniques could be studied. This data was to re-

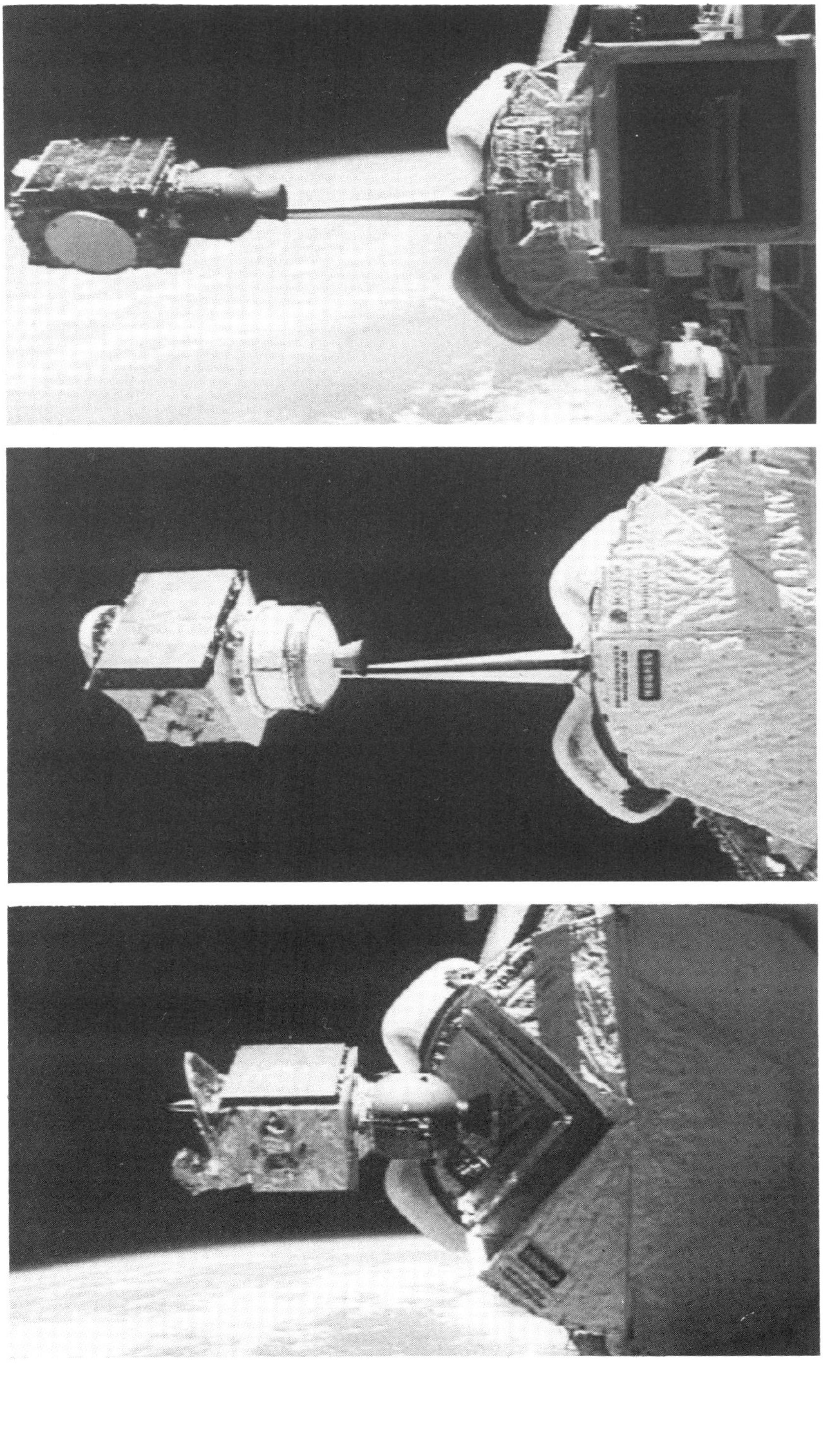

(a) (b) (c)

Fig. 2.30. One, two, three: (a) ASC, (b) Satcom-K, and (c) Arabsat.

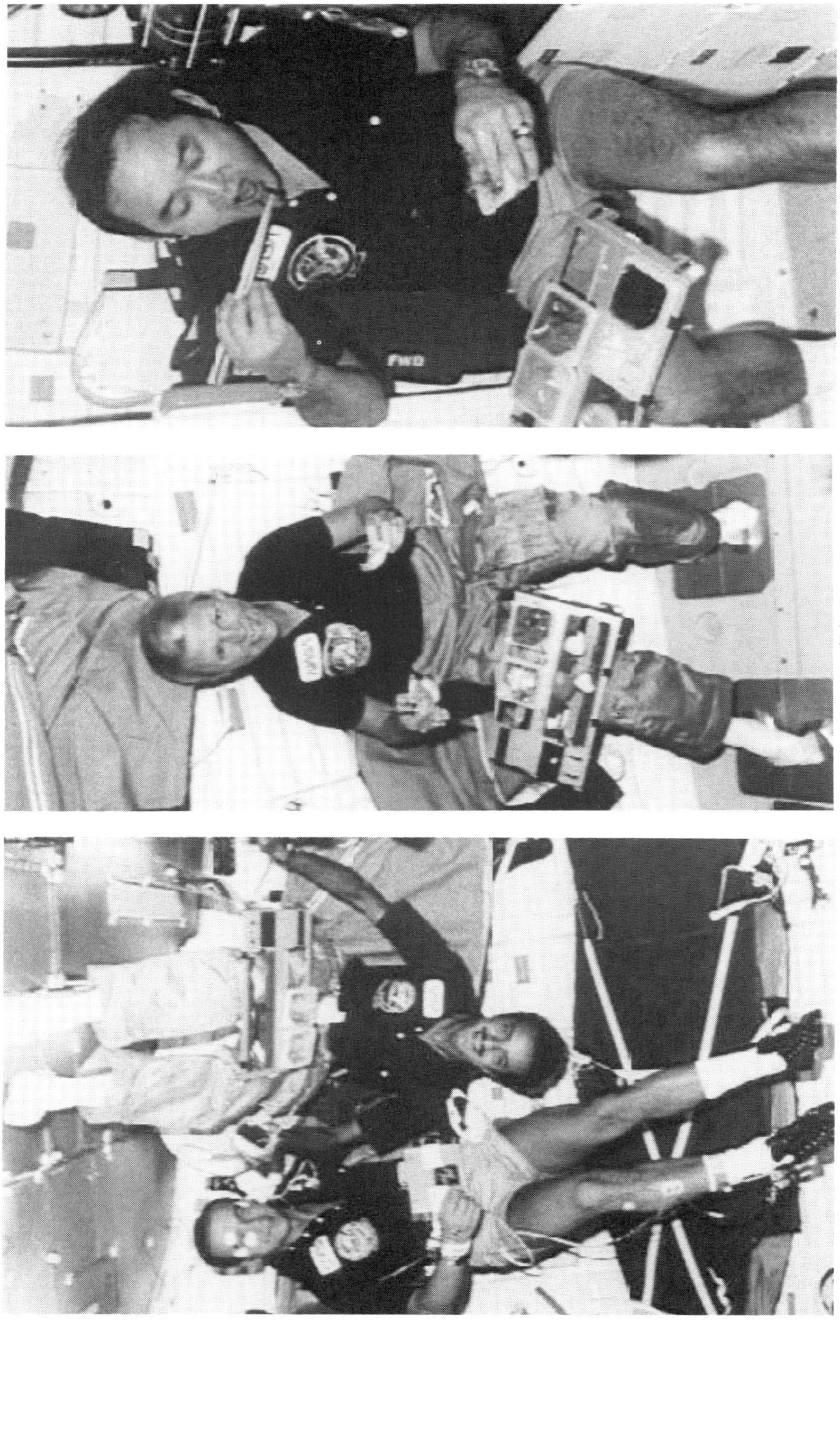

(a) (b) (c)

Fig. 2.31. Mealtime: (a) Patrick Baudry and Prince Sultan Salman Abdul Aziz al-Saud, (b) Rhea Seddon and (c) Ellison Onizuka. A food tray is strapped to the thigh to keep it in place. Onizuka demonstrates the use of chopsticks.

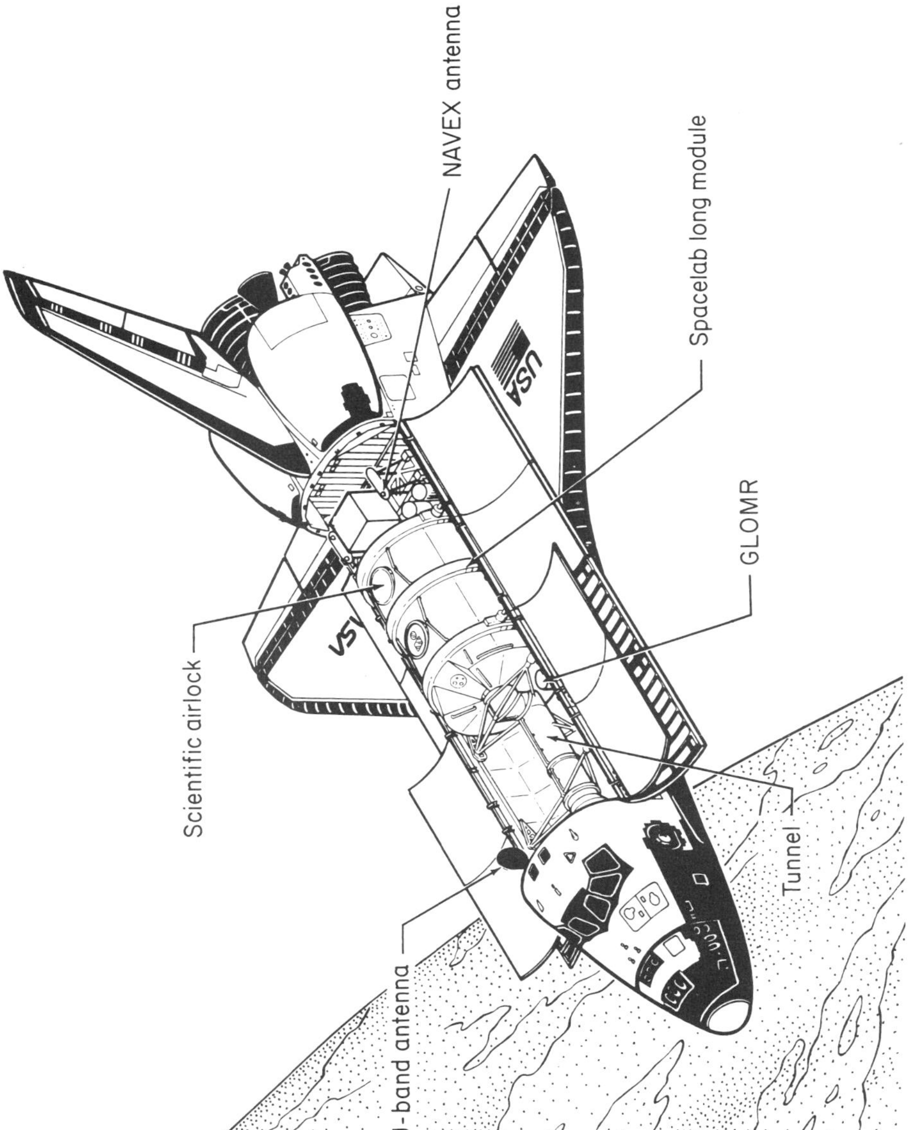

Fig. 2.32. A schematic of the payload bay layout for Flight 61A, showing the German-sponsored (D1) Spacelab and the GLOMR satellite, which is in the Get-Away-Special (GAS) can mounted on the sidewall.

solve a debate within NASA concerning whether it would be possible to use astronauts to assemble the station, or whether some semi-automated system would have to be developed. Thus, in addition to the ongoing deployment of satellites, the various strands crucial to the future of the programme were being drawn together.

Fig. 2.33. Challenger touches down to draw Flight 61A to a successful conclusion. In the event, this was to be the last time that this particular orbiter landed.

It had been hoped to mark Columbia's return to service by launching on 18 December, but a variety of problems early on consumed all the slack built into the countdown, and once it became clear that there was insufficient time left, the launch was postponed a day. Then, at T–14 seconds a hydraulics fault on one of the SRBs caused the attempt to be scrubbed, and the 6 January attempt ended at T–31 seconds when a valve in an SSME malfunctioned. A hold at the T–31 seconds point is very unforgiving because, under such circumstances, the APUs cannot be run for longer than five minutes, and if the fault cannot be rectified in that time the launch has to be scrubbed. A count the following day had to be abandoned due to unacceptable weather at the Spanish and West African TAL sites. The weather at the launch site precluded a launch on 10 January. Since the launch procedure did not call for the crew to enter the orbiter until after the ET had been loaded with cryogenics, once they were aboard it made sense to have them sit out unacceptable weather until the window closed. On a flight to rendezvous with an existing satellite, the window might be a few minutes, but on a less time-critical mission it could last several hours. With each launch postponement, Hoot Gibson's crew dutifully powered down the orbiter and disembarked. They finally left the ground, without a hitch, on 12 January.

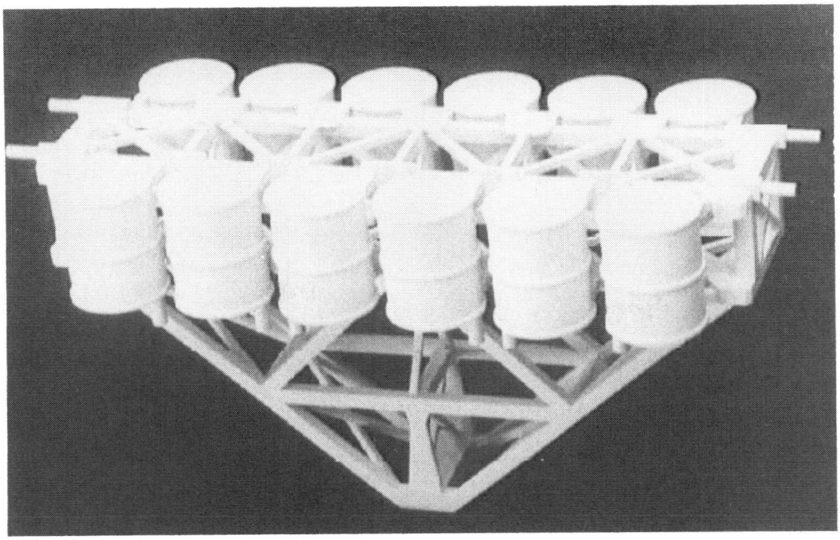

Fig. 2.34. An artist's impression of the Get-Away-Special (GAS) Bridge Assembly developed to enable more self-contained payloads to be flown.

As Columbia had just one satellite in its bay (the other, a Leasat, had been withdrawn for inspection after the loss of its predecessor, and replaced by the GAS Bridge Assembly), if it had been delayed *again* it is likely that this mission would have been cancelled and its payload added to a later flight, because Columbia had an appointment with Halley's Comet in March. As it was, 61C was recalled a day early so as to make a start on this turnaround, and so became the first flight to be cut short in the absence of any fault. With the nosewheel steering certified, Columbia had been assigned a SLF landing. However, this effort to save time was foiled by the weather not just on the recall day, but also on the originally assigned day, so, after a day's extension, Gibson landed at Edwards, and this automatically slipped the turnaround by a week.

As part of Columbia's refit, various sensors had been installed, and during this re-entry a camera mounted in a streamlined pod at the top of the tail took a series of images to record the evolution of the thermal stresses across the upper surface. Lightweight blankets were to be fitted in areas that could be shown not to require the full protection of the tiles. A sensor mounted in the nosewheel bay noted the chemical composition of the plasma that came into contact with the skin of the vehicle during the early phase of the re-entry, and another set of sensors recorded the stresses on the nose cap, which, together with the leading edge of the wing, bore the worst of the heating. Various manoeuvres were performed to further refine computer models of hypersonic flight. These sensors had been conceived when the shuttle was under development, but Columbia's refit had been the first opportunity to install them. Once the hypersonic characteristics of the orbiter were fully understood, and the turn-about manoeuvre required to execute a Return-To-Landing-Site (RTLS) was certified, it would be possible to relax the payload-mass constraints. This was all part of the ongoing process of stretching the envelope to make the shuttle more flexible.

Table 2.3. The first twenty operational shuttle flights

Flight	Orbiter	Launch	Return	Mission
STS-5	Columbia	11 Nov 1982	16 Nov	communications satellites
STS-6	Challenger	4 Apr 1983	9 Apr	TDRS satellite
STS-7	Challenger	18 Apr 1983	24 Apr	communications satellites
STS-8	Challenger	30 Aug 1983	5 Sep	communications satellite
STS-9	Columbia	28 Nov 1983	8 Dec	first Spacelab
STS-41B	Challenger	3 Feb 1984	11 Feb	communications satellites
STS-41C	Challenger	6 Apr 1984	13 Apr	deploy LDEF and fix SolarMax
STS-41D	Discovery	30 Aug 1984	5 Sep	communications satellites
STS-41G	Challenger	5 Oct 1984	13 Oct	environmental satellite
STS-51A	Discovery	8 Nov 1984	16 Nov	release and retrieve satellites
STS-51C	Discovery	24 Jan 1985	27 Jan	classified (DOD)
STS-51D	Discovery	12 Apr 1985	19 Apr	communications satellites
STS-51B	Challenger	29 Apr 1985	6 May	microgravity Spacelab
STS-51G	Discovery	17 Jun 1985	24 Jun	communications satellites
STS-51F	Challenger	29 Jul 1985	6 Aug	astronomical Spacelab
STS-51I	Discovery	27 Aug 1985	3 Sep	communications satellites
STS-51J	Atlantis	3 Oct 1985	7 Oct	classified (DOD)
STS-61A	Challenger	30 Oct 1985	6 Nov	German-sponsored Spacelab
STS-61B	Atlantis	27 Nov 1985	3 Dec	communications satellites
STS-61C	Columbia	12 Jan 1986	18 Jan	communications satellite

A MAJOR MALFUNCTION

The delay in launching 61C had undermined NASA's intention to take advantage of the fact that it now had a four-orbiter fleet to fly *fifteen* missions in FY86. There was therefore tremendous pressure to make up lost time. Challenger, the next in line as STS-51L, was to deploy the long-overdue second TDRS satellite. Its launch had been set for 22 January, but when it became clear that Columbia's return would be late, this was slipped to 26 January. When poor weather was predicted for that date, it was put back again, so it was 27 January before Dick Scobee's crew finally boarded Challenger. There was a frustrating delay while the handle on the hatch, which had jammed, was forcibly disengaged; then the count had to be held for crosswinds which would have jeopardised a RTLS landing, so the crew sat out the window until, finally, it closed. The countdown was recycled for a launch the next day, 28 January.

Although, at 28° latitude, the Kennedy Space Center is pretty much as far south as it is possible to go within the mainland boundary of the US, in winter the overnight temperature can plummet. On the night of 27/28 January 1986, the temperature on the pad (which, as it happened, was Pad 39B, the first time that this rarely-used pad had been used in the shuttle programme) fell so low as to threaten to crack water pipes on the fixed service structure, so the valves were opened so that the flow would prevent the pipes freezing.

Fig. 2.35. The crew of Flight 51L enjoy their breakfast. Left to right: Ellison Onizuka, Christa McAuliffe, Mike Smith, Dick Scobee, Judy Resnik, Ron McNair and Greg Jarvis.

This water froze on the gantry, covering its walkways with sheets of ice and adorning its structure with long icicles, but this was not deemed to be a problem because the ice was sure to melt when the Sun rose in the morning. The next morning was unusually chilly, however, and the ice was still very much in evidence when the pad-inspection team was asked to give its approval for the launch. The issue was not the cold; rather, it was that the sonic shockwave would loosen so much ice as to damage the orbiter's thermal protection tiles. It was decided to hold for two hours to allow the ice to melt, which it duly did.

There had been no problems with the shuttle, so it was launched at 1138 EST, at which time, even though it was rapidly warming, the temperature at the pad was 15 degrees colder than it had been for any previous launch. At Thiokol, the responsible group of engineers, led by Roger Boisjoly, had expressed doubts about the likely resilience of the rubber O-rings that would have to seal the joints of the segments that formed the SRBs, but Joseph Kilminster, the company's vice-president for boosters, was very aware of the imperative to build up the flight rate. Lawrence Mulloy, the manager at the Marshall SFC who managed the Thiokol interface, asked incredulously: "When *do* you want me to launch? Next April?" Kilminster overrode his engineers and recommended a launch. In the NASA way, a waiver was drawn up which recertified the SRBs for this new low temperature; it was seen as yet another step in the continuing process of stretching the shuttle's operating envelope. As the twin pillars of smoky fire lifted Challenger off the pad, the decision to launch was seemingly vindicated. However, although no-one was aware of it, the O-ring at the lowest field-joint of the right SRB had been so chilled that it had failed

Fig. 2.36. The crew of Flight 51L leave the Operations and Checkout (O&C) Building for the short walk to the Astrobus which will take them to the pad. Note the casual attire worn at this point in the programme.

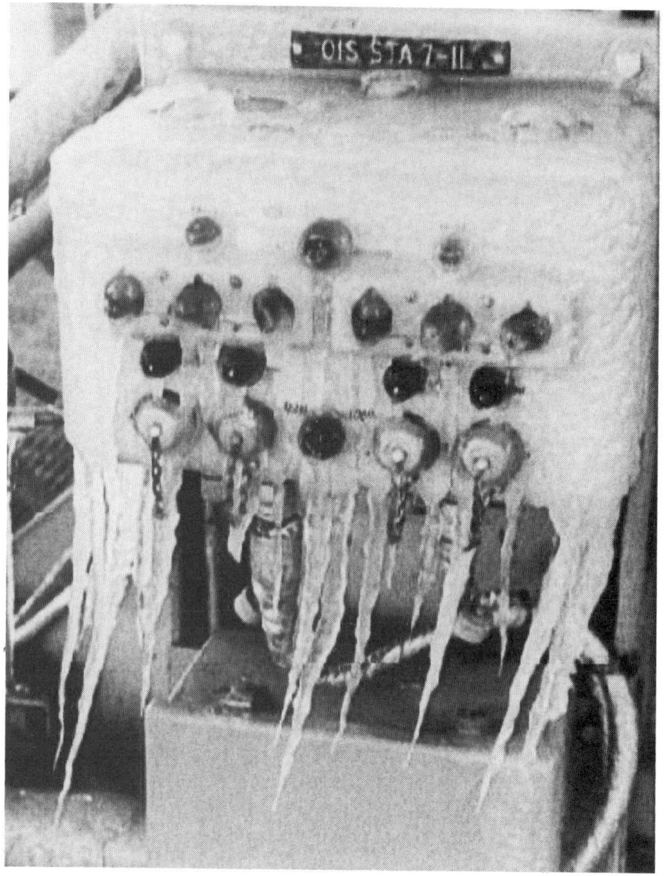

Fig. 2.37. The overnight freeze of 27/28 January 1986 is demonstrated by the presence of icicles on equipment on the pad.

to seat properly and had passed a blast of hot gas. The resulting puff of dense black smoke that leaked from the side of the casing was spotted only by the high-speed cameras that filmed each launch close-in for subsequent analysis. A rapid series of smaller puffs followed as the joint flexed in response to the casing absorbing the longitudinal stress of acceleration. Until the joint finally sealed, successive blasts of gas seared further into the O-rings. Controllers at the Kennedy Space Center's Launch Control Center and in Houston's Mission Control Center, and Dick Scobee and Mike Smith aboard Challenger, were all monitoring their displays, but there was no indication that the vehicle was in distress.

The thrust was throttled back so as not to overstress the stack as it went supersonic, and after the aerodynamic loads reached their peak the SSMEs were increased to 104 per cent and the geometry of the solid propellant in the SRBs increased their thrust for the long climb to orbit. The SRBs were due to burn out at T+128 seconds, so they were barely half-way into their cycle when the leak from the right-hand booster reappeared, now as a jet of flame. This blowtorch began to slice through the strut that held the the motor casing to the ET, seared the orbiter's wing, and burned through the insulation at the base of the tank itself. At T+73 seconds, almost instantaneously the tank ruptured and spilled its hydrogen into the roiling airflow, the GPC shut down the SSMEs, the strut gave way, and the booster, now attached only at its top end, pivoted, and its nose crushed the intertank, spilling the oxygen. Upon the structural collapse of the ET, the bolted-together stack disintegrated, and the asymmetric aerodynamic forces tore the orbiter apart. As the vehicle disintegrated, the hydrogen and the oxygen detonated in a massive fireball.

The visitors' stand at the Kennedy Space Center contained a few stalwarts of the Press, but most of the small crowd were friends of the astronauts and this was the first launch that they had seen. A few, believing that what they were seeing was the standard staging, began to cheer, but this petered out as those who recognised the evil cloud for what it was gasped in horror. The still-thrusting SRBs continued to ascend. Those observers familiar with the vernacular muttered "RTLS!" and waited for the orbiter to appear, but all that emerged was a hail of debris which splashed into the sea. After twenty-five years of launching astronauts on rockets, NASA had finally lost a crew.

CNN was the only major network to show the launch live; it used NASA's video feed. As the controllers stared in disbelief at their telemetry displays, the Agency's PAO, Stephen Nesbitt, explained to his television audience that it was, "obviously a major malfunction". He followed a few moments later with: "We have a report from the flight dynamics officer that the vehicle has exploded!" The loss was all the more shocking because there seemed to have been no indication of a problem.

Roger Boisjoly had raised a 'red flag' on the O-ring issue after inspection revealed that during a launch last April, gas had burnt completely through a narrow arc of the inner ring and had severely eroded the ring immediately beyond. Thiokol had set up a formal study in October, but the O-ring became just one of many pieces of hardware being considered for upgrade. The data was inadequate to characterise the fault. Given that the problem involved enduring stresses encountered in flight rather than on the horizontal test rig, developing this database required more shuttles to be launched.

This incremental process has been criticised for unnecessarily placing the shuttle at risk. But to argue that NASA should have grounded the shuttle so that the O-ring problem could be studied is naive.

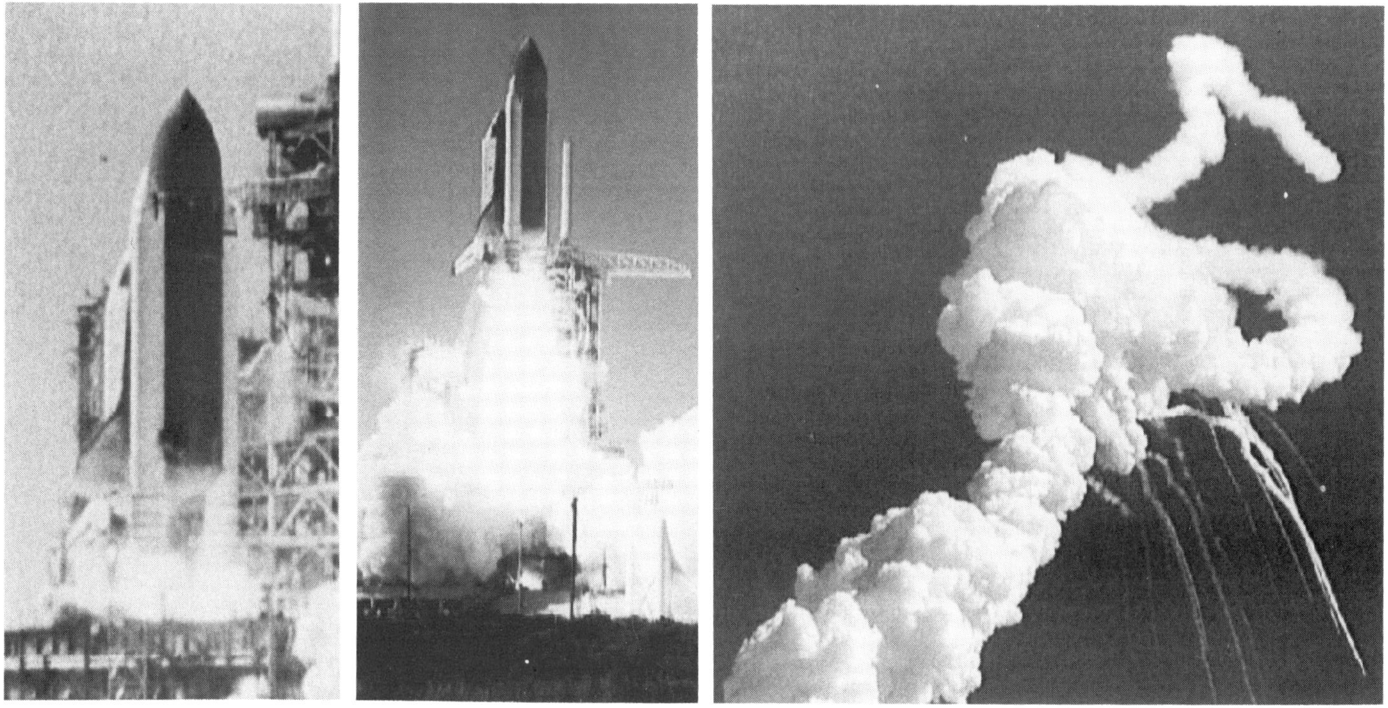

(a) (b) (c)

Fig. 2.38. Challenger's last flight: (a) Flight 51L's first moments - a puff of black smoke is seen leaking from the aft field-joint of the right-hand Solid Rocket Booster (SRB) at the moment of ignition. (b) Challenger clears the tower on 28 January 1986 on Flight 51L. Just over a minute later, high in the chilly Florida sky, the shuttle was destroyed by a massive explosion, and Dick Scobee, Mike Smith, Ellison Onizuka, Judy Resnik, Ron McNair, Greg Jarvis and Christa McAuliffe were all killed. (c) Flight 51L's final moments - 73 seconds after lift-off, the External Tank (ET) exploded. Even as the hail of debris emerged from the cloud of destruction and fell towards the Atlantic, the SRBs continued to climb, and they were subsequently destroyed by the Range Safety Officer (RSO).

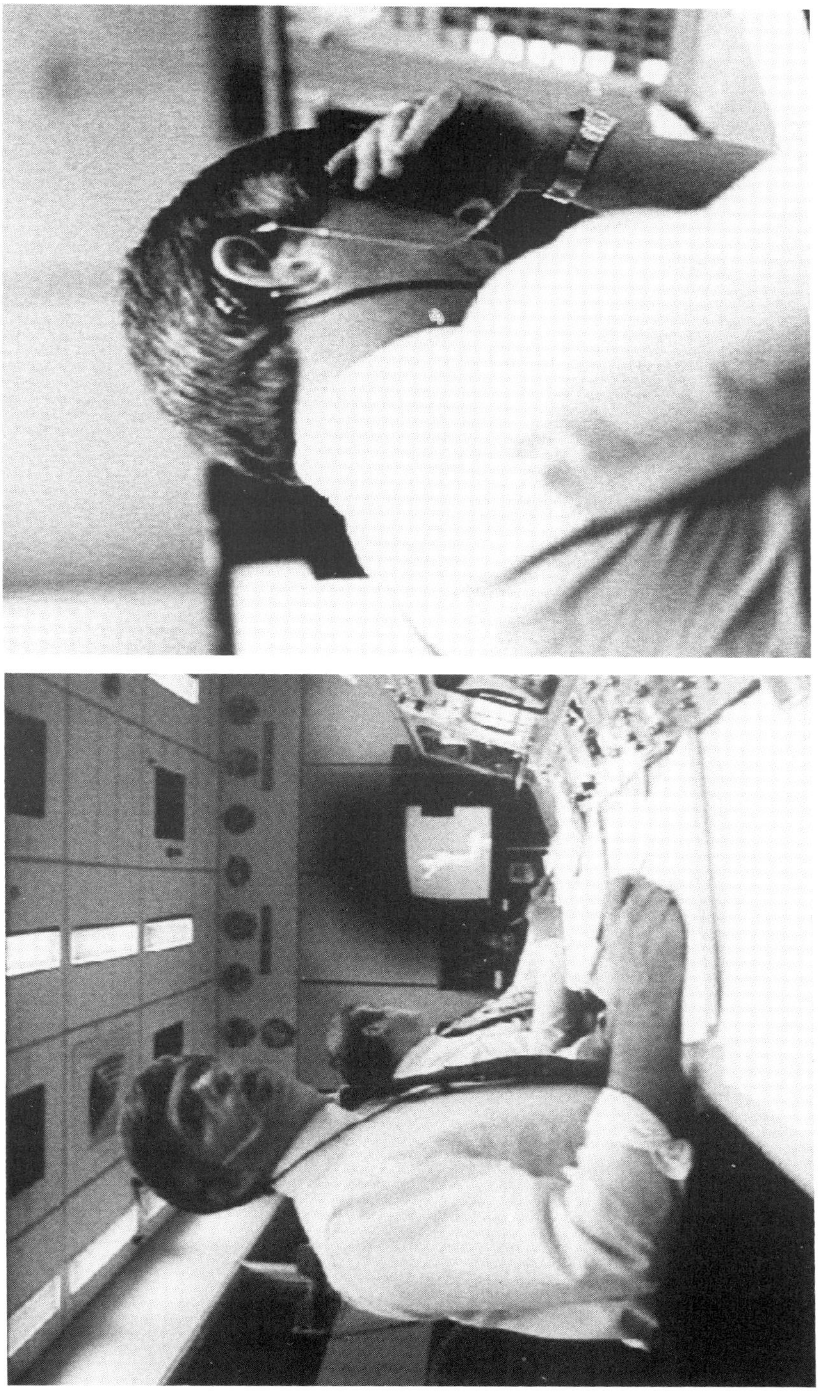

(a) (b)

Fig. 2.39. The scene in Mission Control as the TV monitor shows Flight 51L exploding: (a) the Flight Director first looks up and then (b) holds his head in disbelief.

Fig. 2.40. Wreckage of Challenger recovered from the Atlantic.

The process of incremental upgrading is standard practice in the aerospace industry. Production of airliners continues as 'fixes' for rare faults are developed. When it becomes available, a 'fix' is introduced to the production line. If practicable, it is retrofitted to existing aircraft. Only a catastrophic failure is permitted to ground an entire class of aircraft, and even then, to allow the aircraft to resume flying as soon as possible, the first step towards a solution is likely to be to impose an operational rule (to constrain the envelope) in order not to invite that particular type of failure. In the case of the shuttle, upon finding that there was a burn-through problem with its boosters, Thiokol had initiated this formal review process. However, this particular problem became just one more item on a long list of issues, and its significance, particularly its sensitivity to low temperature, was not widely recognised.

Table 2.4. Immediate post-Challenger manifest

Flight	Date	Orbiter	Mission
61E	Mar 1986	Columbia	ASTRO for Halley's Comet
61F	May 1986	Challenger	Ulysses deployment
61G	May 1986	Atlantis	Galileo deployment
61H	Jun 1986	Columbia	communications satellites
62A	Jul 1986	Discovery	classified
61M	Jul 1986	Challenger	TDRS satellite
61J	Aug 1986	Atlantis	Hubble Space Telescope
61K	Sep 1986	Columbia	Earth studies Spacelab
61I	Sep 1986	Challenger	retrieve LDEF
62B	Sep 1986	Discovery	classified
61L	Oct 1986	Atlantis	astrophysics Spacelab

In stretching the envelope to launch at the record low temperature, NASA had gambled, as it had gambled in the past, and as it would have to continue to do if it was to increase the performance of the shuttle. But this time it lost the bet, and Dick Scobee, Mike Smith, Judy Resnik, Ellison Onizuka, Ron McNair, Greg Jarvis and Christa McAuliffe paid the price.

As soon as it had absorbed the shock of the loss of Challenger, NASA's first imperative was to ensure that the flow of Columbia's turnaround continued, to ensure that it would be ready in the event that it proved feasible to proceed with 61E in March. To some outside the Agency this decision was met with disbelief, but it was perfectly in keeping with the NASA operating process of not closing off options. It was soon accepted, however, that the shuttle would have to be grounded while a Presidential Commission investigated what had come to be dubbed 'the Accident'.

Chaired by William Rogers, former Secretary of State in the Nixon administration, this commission included Neil Armstrong, the Apollo 11 moonwalker, Chuck Yeager, the pilot of the X-1 which had broken the sound barrier back in 1948, Sally Ride, a shuttle mission specialist, Robert Hotz, a former editor of the leading aerospace magazine *Aviation Week & Space Technology*, and Richard Feynman, the Nobel laureate in physics from Caltech. It was formed on 3 February, and after interviewing 160 individuals and examining 6,000 documents, and after instigating 35 panels to chase up specific issues in depth, on 6 June it announced its conclusions in *Report of the Presidential Commission on the Space Shuttle Challenger Accident.*

The media focused on the management process which had led to the decision to launch, the organisational changes designed to reduce the 'turf wars' within NASA concerning the shuttle, and a revision of reporting procedures to ensure that the views of the astronauts and engineers were fully represented in the decision-making process. All of these changes had merit, but they did not address the fundamental issue.

The real significance of the loss of Challenger is not that a rocket failed – it was inevitable that one day a component of this very complex system *would* fail – but that the onset of that failure was not detected in time to initiate the appropriate *abort*; in this case a contingency-RTLS.

The RTLS abort mode actually covered a spectrum of contingencies. If an SSME had to be shut down in the initial phase of the ascent, in the two minutes during which the SRBs fired, an orbiter would *have* to return to the Kennedy Space Center. In that case, no action would be taken until the SRBs had run through their cycle and been jettisoned in the normal manner. Ten seconds later, on its remaining engines, the orbiter would climb rather steeper than the normal trajectory, then level off at an altitude of about 150 km and fly out over the Atlantic. When several hundred kilometres out, it would pitch over to transition from flying on its back, with the ET above, to flying the right way up, *tail first*, with the ET beneath, a manoeuvre which has never been attempted in practice, but for which all shuttle pilots train. In this configuration, the orbiter would slow down and lose altitude, and follow a trajectory calculated to ensure that at the moment the ET ran dry and was jettisoned, the orbiter would be on track and on energy to join the nominal descent trajectory, in order to glide back to the Kennedy Space Center, just as if it had returned from orbit. The RTLS option remained open until about T+4 minutes 20 seconds; if an engine failed after that point, depending on the nature of the fault, a shuttle would either have to make a Transatlantic Abort Landing (TAL) or continue on to an Abort To Orbit (ATO), which the OMS engines could then either augment or abandon by performing an Abort Once Around (AOA).

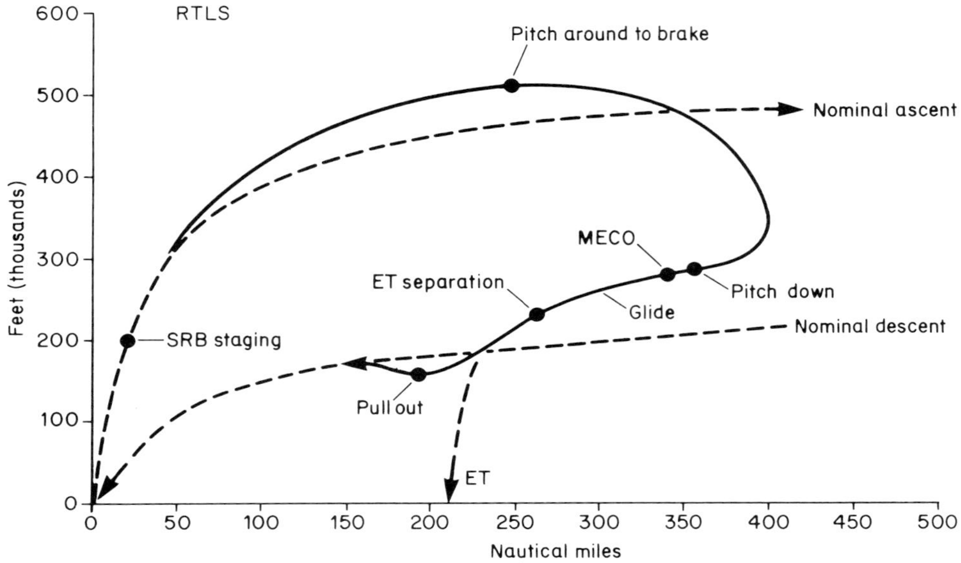

Fig. 2.41. A schematic of the standard Return to Launch Site (RTLS) abort procedure.

The progression of abort options is really meant to deal with SSME failures at various phases in the ascent, yet the *most* dangerous period is the SRB boost phase. Although the shuttle's design requirement did not explicitly specify that the orbiter must have a means of escape in the event of an SRB fault, this was not the same as indicating that there should be *no* chance. In addition to the above-mentioned abort modes, which were expected to have a high likelihood of success, the GPC had been programmed to make a 'contingency abort' during the SRB boost phase. In such an abort, the *best* that could be hoped for afterwards would be a ditching at sea, which would be risky, and would probably lead to the break-up of the orbiter, but some chance was better than no chance.

In the circumstances in which 51L found itself, the plan called for the still-firing SRBs to be jettisoned and for Challenger to continue its powered flight, to achieve an energy state suitable for a turn-about manoeuvre, to shed the ET, and then to attempt a 'normal' return. All of NASA's abort plans relied on there being sufficient warning of an impending failure to perform the necessary separation. But the instrumentation monitoring the performance of the vehicle did not show any evidence of the burn-through until the SRB's internal pressure began to fall. A RTLS *might* just possibly have been feasible. Jettisoning the SRBs would have been fraught with danger, because the separation might not have occurred cleanly; but by that point Challenger was already in extreme jeopardy. In such an organisation as NASA, renowned for its contingency planning, a successful abort, even if the orbiter had to ditch at sea, would have been an absolute vindication of the shuttle's operating procedure.

But how much time would have been available to invoke this most difficult of all the abort options? The abridged extract from the Commission's Report, in Table 2.5, replays the story in appalling clarity. (Some details of the nominal ascent following the termination of the smoke plume pulsations and prior to the onset of the flame plume have been edited.)

Table 2.5. The countdown to catastrophe

T+ seconds	
0	With all three SSMEs stable at 100 per cent of rated thrust, the two SRBs are ignited, and almost instantly build up to their peak output to lift the stack off the pad.
0.678	A strong puff of smoke spurts from the aft field-joint of the right SRB; further puffs occur with a rate of about four per second, which is the frequency of the field-joint's flexure in response to the loads that are inherent in the onset of flight; the thick dark smoke is suggestive of the O-ring passing hot propellant gas with resultant burning of the joint's grease and insulation; this O-ring failure was in the arc facing the ET.
2.733	The last smoke puff from the right SRB.
19	Smith: "Looks like we've got a lot of wind here today."
20	Scobee: "Yeah."
22	The internal pressure in the SRBs peaked, then began to decrease to accommodate increasing aerodynamic forces.
22	Scobee: "It's a little hard to see out my window here."
24	The computer commanded the SSMEs to throttle down to 95 per cent.
28	Smith: "There's 10,000 feet and Mach point five."
35	Scobee: "[Mach] point nine."
35.379	The computer commanded the SSMEs to throttle down to 65 per cent.
40	Smith: "There's Mach 1."
41	Scobee: "Going through 19,000 [feet]."
42	The SSMEs were now at 65 per cent.
43	Scobee: "Okay, we're throttled down."
50	Max-Q.
51.860	The computer commanded the SSMEs to throttle up to 104 per cent.
57	Scobee: "Throttling up."
58	Smith: "Throttle up [acknowledged]."
58.788	As the shuttle passed through the moment of maximum aerodynamic pressure and as the SRBs began to increase thrust, a flame appeared in the same position on the right SRB as smoke had previously been vented.
59	The SSMEs were now at 104 per cent.
59	Scobee: "Roger."
59.262	The flame was now a continuous well-defined plume; the supersonic slipstream played this flame across the base of the ET; as this flame grew, it impinged on the strut holding the SRB to the ET.
60	Smith [remarking on the acceleration]: "Feel that mother go."

Table 2.5. (continued)

T+ seconds	
60	Uncertain: "Woooohoooo."
60.004	Telemetry now began to show the two SRBs with different internal pressures, that of the right SRB being lower.
62	Smith: "35,000 [feet], going through [Mach] one point five."
62.484	Telemetry indicated that the right hand outer elevon control surface was in motion.
63.924	Telemetry showed a change in this elevon's actuator pressure.
64.660	The flame breached the ET and hydrogen began to vent, and burn in the impinging plume.
64.937	The computer started to gimbal the SSMEs with large pitch variations in order to overcome unwanted vehicle motions.
65	Scobee: "Reading 486 [the meaning of this is disputed] on mine."
65.524	Telemetry showed a pressure change in the actuator for the outboard elevon on the left side.
66	MCC: "Challenger, go at throttle up."
66.764	Telemetry showed a drop in the ET's hydrogen pressure.
67	Smith: "Yep, that's what I've got too."
70	Scobee [answering the ground's call]: "Roger, go at throttle up."
72.204	The strut holding the base of the right SRB to the ET came loose and the booster pivoted around its upper strut, its rear skirt impacting the orbiter's right wing; telemetry indicated divergent yaw rates between the two SRBs.
72.284	Telemetry indicated divergent pitch rates between the two SRBs; the loose SRB continued to thrash about.
72.497	The SSME gimbals were now rolling at 5 degrees per second in an attempt to damp out vehicle motions.
73	Smith: "Oh-oh."
73.124	The bottom cap of the ET failed, dumping pressurised hydrogen into its wake.
73.137	The nose of the still-thrashing right SRB struck the ET's intertank, fracturing the base of the oxygen tank, leading almost immediately to an explosive burn of mixed propellant; the asymmetric loading on the vehicle ripped the stack (and the orbiter itself) apart.
73.143	The SSMEs responded to the loss of propellant and the turbopumps began to overheat.
73.482	The SSMEs initiated a shut down.
73.618	Orbiter telemetry was terminated.
75	PAO: "1 minute, 15 seconds; velocity 29 hundred feet per second; altitude now 9 nautical miles; down range distance 7 nautical miles. Flight controllers here looking very carefully at the situation. Obviously a major malfunction. We have no downlink. We have a report from the flight dynamics officer that the vehicle has exploded!"

In addition to this 'official' timeline, in February 1997 Tim Furniss reported in *Flight International* that Ali Abu Taha, who had been given access to the NASA evidence and had carried out an independent analysis, had recovered a photograph which showed a 3.3-metre long plume of flame emerging from the SRB at about T+20 seconds, but this image has yet to be published. This coincides with the elbow in the SRB's internal pressure profile, so a flame leak at this point may have been caused by the casing relaxing as the pressure began rapidly to fall, in order to ease the stress on the vehicle in the run-up to Max-Q.

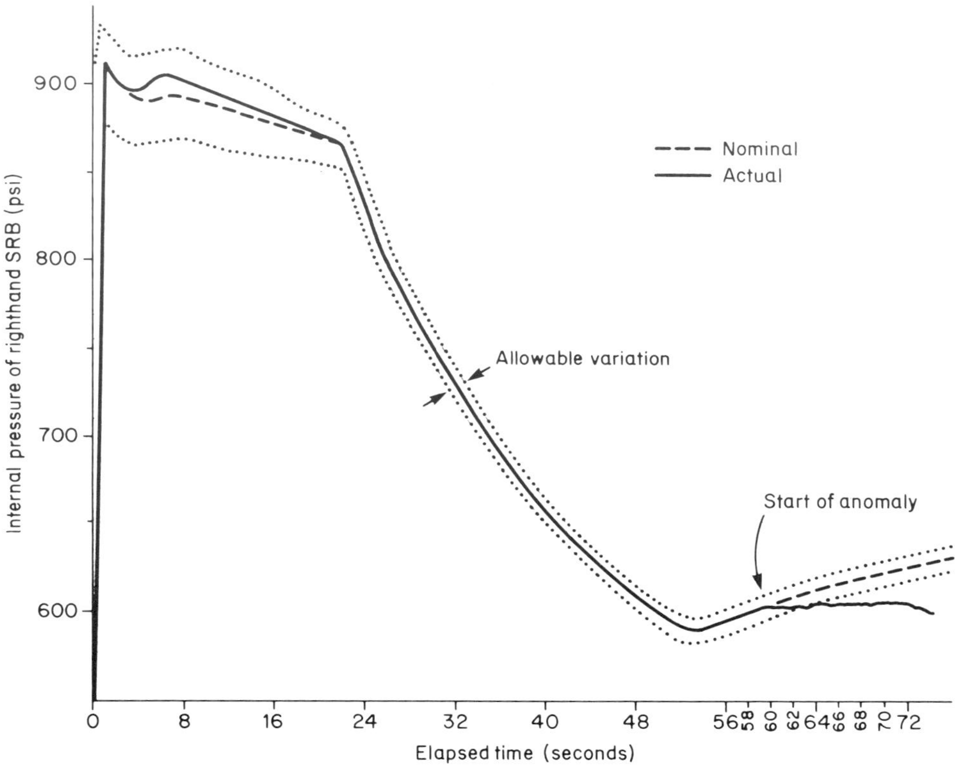

Fig. 2.42. A plot of the internal pressure in Flight 51L's right-hand Solid Rocket Booster (SRB). It clearly shows the timescale during which the booster's performance degraded.

The first indication in the telemetry that there was a fault in the right SRB was at T+60 seconds, when its internal pressure began to depart from nominal. Telemetry over the next 13 seconds confirmed and clarified the nature of the developing catastrophe. The issue for the flight controllers was whether the fall in internal pressure was a temporary fluctuation – in which case the focus of attention would move to how well the shuttle was able to correct its trajectory by swivelling the nominal engines – or whether it represented a divergent trend. The internal pressure could vary within a narrow band and still be acceptable, but at T+63 seconds it departed from this band; it was clearly under-performing, but the reason was not evident, at least not in the telemetry. In principle, a RTLS abort could have been invoked at this point. Putting it into effect would take three seconds. The situation was not yet critical, but at T+66 seconds, when the ET was fractured and started to vent hydrogen, the option of a powered-RTLS closed. From this point, Challenger was in dire peril. The SRBs could still have been dumped, but without hydrogen for the SSMEs, the likelihood of the orbiter being able to manoeuvre to a point suitable for an emergency landing was slim indeed. For the next six seconds the telemetry continued to track the falling pressures in both the right SRB and the ET's hydrogen tank. Time was running out fast, and when it came the end was essentially instantaneous.

The window for a powered-RTLS abort had lasted only six seconds, from the moment that the right SRB began to lose pressure to the time that the associated flame compromised the integrity of the hydrogen tank. This wasn't much, but it *was* a window of opportunity. Why wasn't a contingency abort ordered? Simply, because although the telemetry indicated that there was a fault in the SRB, and that this grew to include the ET, the telemetry did not reveal the cause. Although the flight controllers worked to understand the cause, they were denied the unambiguous indicator which would certainly have prompted them to call for an abort. Some have interpreted the terse exchanges between Scobee and Smith in the fateful final seconds as their having become aware of the fall-off in SRB pressure, but even if they had, it is unlikely that they would have initiated an abort without confirmation of this fault from their colleagues on the ground. If a window of only a few seconds seems to be asking too much of the controllers, how long *should* it have taken to figure out that disaster was looming?

The vital information that made the looming catastrophe clear was the imagery from the tracking camera which showed the plume of flame. However, the view fed to the television networks – the view that was available to the flight director – was a viewpoint from which the flame plume was obscured. Might events have turned out very differently if the output from the *other* camera had been shown live?

GROUNDED

The upshot of the Rogers Commission's report was that NASA must develop a safer SRB field-joint; that it must reassess issues of landing the orbiter safely, and pay particular attention to the tyre and brake failures; that it must upgrade contingency planning for aborts during the ascent phase; that it must stock sufficient spare parts not to need to cannibalise a recently-returned orbiter to enable another to launch; that it must resist the temptation to use the need to increase the flight rate to override established limits by the issuance of waivers; that it must precisely define the shuttle programme manager's authority and responsibility; that astronauts should participate in the management process; and that an independent panel should review safety issues, and report directly to the programme manager.

A few weeks later, Congress, keenly aware that the operational safety of the shuttle had assumed a higher profile, ordered NASA to cancel the Centaur upper stage because it could jeopardise an abort. Even if the orbiter made it to an emergency runway, the Centaur might well break free in a rough landing, spill its tonnes of cryogenic hydrogen and oxygen, then blow the orbiter apart in a devastating explosion. The cancellation of this powerful rocket threw the planetary programme into chaos.

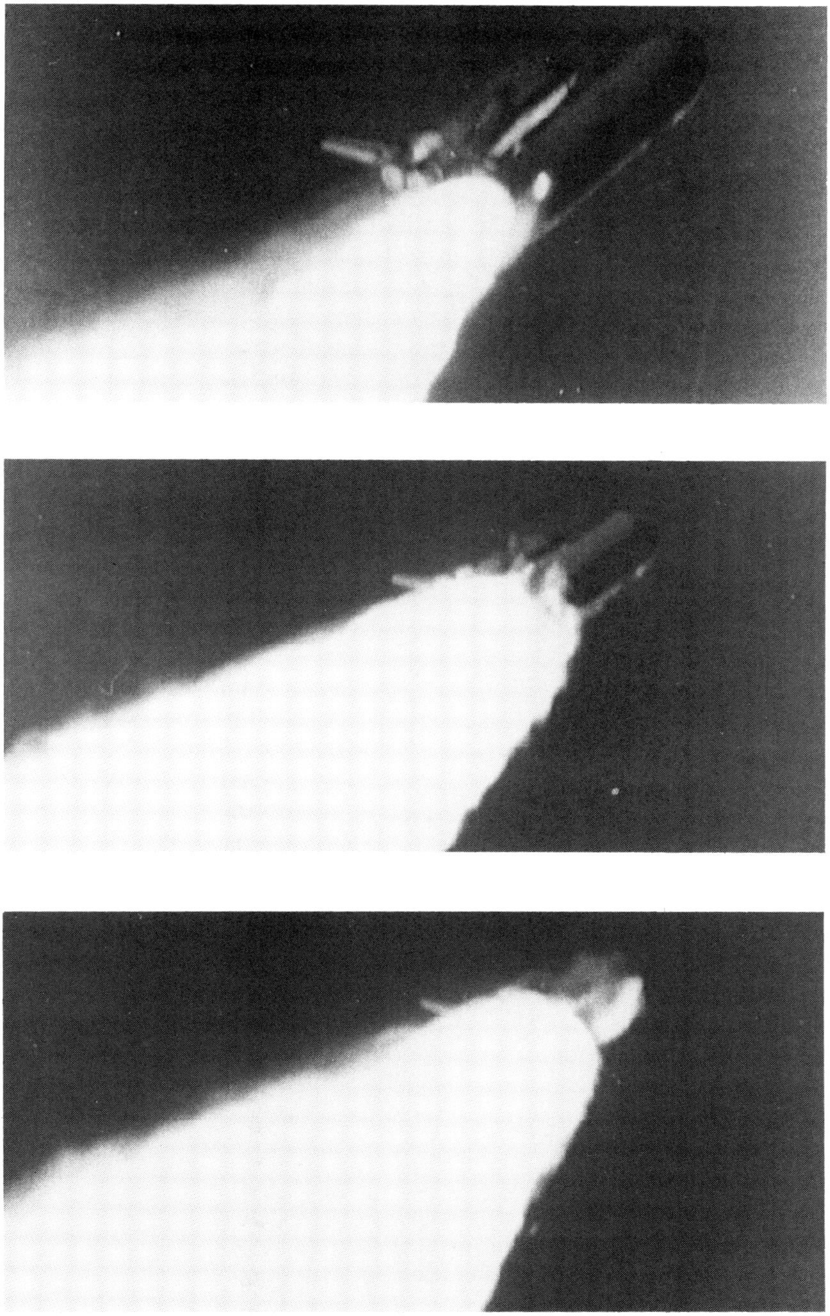

Fig. 2.43. The final few seconds of Flight 51L from a remotely-operated camera stationed north of the launch pad. In the first view (top), the 'anomalous' jet of flame can be seen venting from the aft field-joint of the right-hand Solid Rocket Booster (SRB).

However, there was some good news too. On 15 August, the White House ordered the construction of a replacement orbiter. However, it also instructed NASA to cease using the shuttle to launch commercial satellites, and not to use the shuttle to deploy any satellite that could be offloaded onto an expendable rocket. The adoption of this *mixed-fleet* policy was good news for the rocket makers. When Reagan had declared the shuttle operational, and it became *the* National Space Transportation System for government payloads, the makers of the rockets had been told that they could sell their services in the commercial sector, but the lower-than-cost fees for satellites booked on the shuttle had made this an empty gesture. As a result, the lightweight Delta had already ceased production, and the Atlas and Titan were to be scrapped as soon as the shuttle took the over the launching of polar-orbiting satellites. Now, with not only commercial, but also government payloads offloaded to expendables, there were rich pickings for the rocket manufacturers. And their only competition would be Arianespace. The production lines were reopened immediately, but until deliveries could be resumed the inventory of expendables would be extremely limited. In fact, there were only five Deltas left, and all of the dozen or so Atlas and Titan were reserved by the government.

Meanwhile, NASA worked on the shuttle. All but two of the SRBs had been recovered (both of STS-4's had been lost due to identical parachute faults), and evidence of hot gas flowing past the main O-ring had been found in *nine* casings. Although NASA redesigned

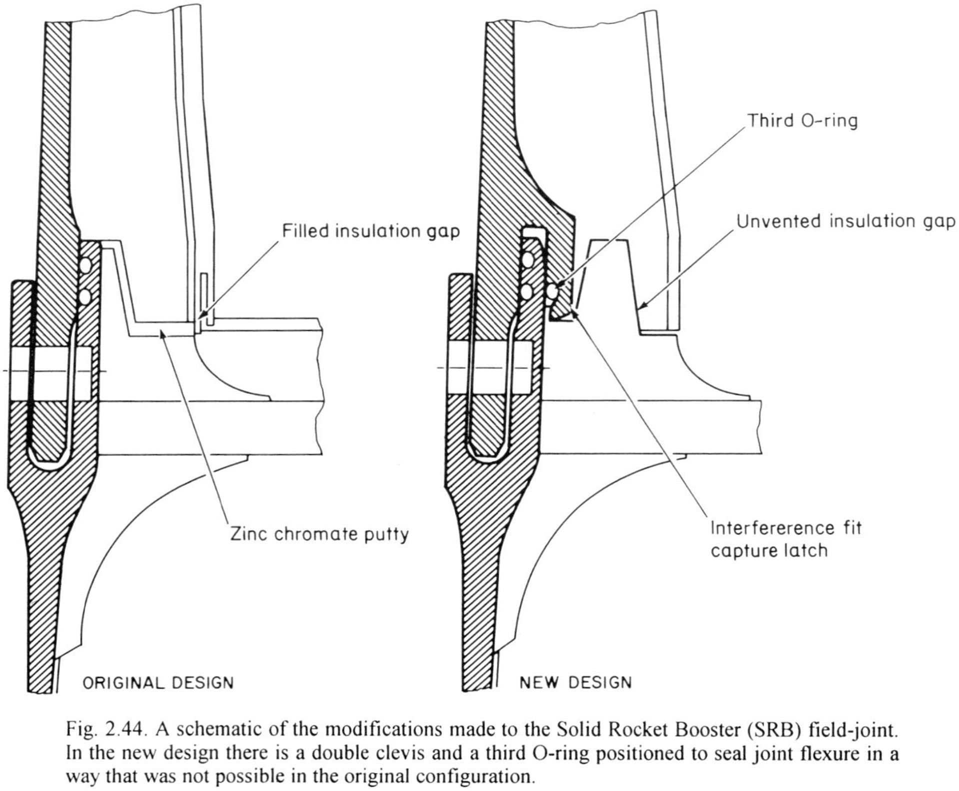

Fig. 2.44. A schematic of the modifications made to the Solid Rocket Booster (SRB) field-joint. In the new design there is a double clevis and a third O-ring positioned to seal joint flexure in a way that was not possible in the original configuration.

the tang-and-clevis joint to reduce flexure, it also minimised the modification, in order to be able to use the segments that it had in stock. In addition, an electrical heater was installed to protect the O-ring from low temperatures, an extra O-ring was added, the inside of the joint was sealed with extra putty, and, finally, a 'weather strip' was wrapped around the joint to keep out rain.

To follow up on another of the Rogers' Commission's recommendations, it was decided to develop a way for the crew to exit an orbiter during its subsonic glide, to deal with the contingency of a successful abort, either during the ascent or following re-entry, in

Fig. 2.45. The crew of STS-26 emerge from the Operations & Checkout (O&C) Building wearing the new pressure suits.

which it became clear that a landing would be impracticable. During the ALT and OFT phases of the programme, astronauts had worn full pressure suits and had ridden in ejection seats. Since then, the ejection seats had been replaced, and crews had worn lightweight blue flight suits, with helmets only to feed oxygen in the event of a cabin depressurisation. Future astronauts would wear pressure suits, and these were high-visibility orange to assist in rescue. The orbiter's side hatch was rigged with explosives so that it could be blown off. A telescoping pole was mounted on the ceiling of the middeck. This would be extended through the open hatch by the middeck crew, the orbiter would be placed on autopilot, then each astronaut in turn would loop a harness over the pole and bale out. The pole's job was to pull an evacuee through the slipstream and beneath the leading edge of the wing. Once clear, the astronaut would release the parachute integrated into the suit. Such an evacuation would clearly be a risky process, but it would offer a *better* chance of crew survival than an at-sea ditching.

NASA initially hoped to be able to complete its modifications within a year, and thus be able to resume operations in early 1987, but in the autumn of 1986 it acknowledged that it would not fly until 1988. The manifest published in July 1987 reflected both NASA's own priority and its obligation to the Air Force which, despite its disenchantment with the shuttle, had nevertheless exercised its right to commandeer flights in the interest of national security, because, until its Titan 4 could be introduced, the Air Force had no other way of launching its out-sized satellites. In resuming operations, NASA decided to revert to the sequential numbering scheme, even though it was likely that events would lead to missions being flown out of order.

Table 2.6. Shuttle manifest, *c*. mid-1987

Flight	Date	Orbiter	Mission
STS-26	Jun 1988	Discovery	TDRS-C
STS-27	Sep 1988	Atlantis	classified
STS-28	Nov 1988	Columbia	classified
STS-29	Feb 1989	Discovery	TDRS-D
STS-30	Apr 1989	Atlantis	Magellan
STS-31	Jun 1989	Columbia	HST

The first priority was the building up of the TDRS communications network. After two classified missions had been flown, and before it sent Magellan to Venus, NASA hoped to have time to complete its relay network. Magellan assumed priority over the Hubble Space Telescope because the launch window for the interplanetary mission was fixed by the dynamics of the Solar System. The assignment of the HST to Columbia was unfortunate; as the heaviest of the fleet, it was the least able to lift the 11-tonne telescope to the record 600-km altitude that it would need to assure minimal orbital decay. By early 1988, it was clear that Discovery would not make its June launch date, so the manifest was revised for a Return-To-Flight in August.

Table 2.7. Shuttle manifest, *c*. early 1988

Flight	Date	Orbiter	Mission
STS-26	Aug 1988	Discovery	TDRS-C
STS-27	Oct 1988	Atlantis	classified (DOD)
STS-29	Jan 1989	Discovery	TDRS-D
STS-28	Mar 1989	Columbia	classified (DOD)
STS-30	Apr 1989	Atlantis	Magellan
STS-31	Jun 1989	Discovery	HST
STS-32	Jul 1989	Columbia	Leasat-5; retrieve LDEF
STS-33	Aug 1989	Atlantis	classified (DOD)
STS-34	Oct 1989	Discovery	Galileo

Although the recycling of Discovery, Atlantis and Columbia was broken, the timescale was compressed in order to preserve the April and June dates, because the October window for launching the Galileo spacecraft was fixed. A fortunate result of swapping STS-28 and STS-29 was the reassignment of the HST to Discovery, which was far better suited to such a mission.

RETURN TO FLIGHT

By the time Discovery finally returned to space, on 29 September 1988, the manifest had been rearranged again to work within the constraints imposed by the fixed windows of the interplanetary spacecraft. The TDRS relays had retained their priority, but the Air Force's satellites had become bunched as a result of preserving STS-33's August launch. The loser was the HST; it had slipped eight months, to early 1990.

Table 2.8. Shuttle manifest, *c*. autumn 1988

Flight	Date	Orbiter	Mission
STS-26	Sep 1988	Discovery	TDRS-C
STS-27	Nov 1988	Atlantis	classified (DOD)
STS-29	Feb 1989	Discovery	TDRS-D
STS-30	Apr 1989	Atlantis	Magellan
STS-28	Jul 1989	Columbia	classified (DOD)
STS-33	Aug 1989	Discovery	classified (DOD)
STS-34	Oct 1989	Atlantis	Galileo
STS-32	Nov 1989	Columbia	Leasat-5; retrieve LDEF
STS-36	Dec 1989	Discovery	classified (DOD)
STS-31	Feb 1990	Atlantis	HST

The deployment of TDRS-C was completed on the fifth orbit, as was usual for an IUS payload. The only problem was that the TDRS satellite experienced difficulty deploying its twin primary antennas, but this was soon overcome and the new satellite took the load off TDRS-A, which had lost half of its K_u-Band capacity and was now put on standby. With two fully-operational satellites now in service – one above the Atlantic and the other over the Pacific – shuttles were in contact for 86 per cent of each orbit; the only dead zone was low over the Indian Ocean. The TDRS network was finally able to provide the service that it was to have offered within about six months of the shuttle becoming operational. Having achieved their primary mission, the crew settled down to a programme of microgravity experiments on the middeck. The nosewheel steering had been certified, but the test had been in the absence of a crosswind, so KSC was still rated as an emergency-only recovery site. Immediately upon touching down on the dry lake at Edwards, Discovery used differential braking to move off the centreline and back again to test the revised carbon–beryllium brakes which were later to be replaced by new carbon–carbon brakes. Overall, it was an excellent resumption of flight operations.

STS-27's launch slipped a day because of cloud that would have inhibited a RTLS, but it left on time the next day and successfully deployed its satellite which, despite the secrecy, was "reliably reported" to be the first Lacrosse all-weather radar imaging satellite. Unlike the earlier Department of Defense satellites, which had ridden IUS rockets to geostationary orbit, this reconnaissance package boosted itself up to about 700 km.

With two missions flown, NASA was confident that it would be able to fly seven more in 1989. However, routine post-flight inspection of STS-27's SSMEs revealed that one of the liquid oxygen turbopumps had a cracked bearing-race, so it was decided to exchange the pumps on STS-29, which had already been rolled out to the pad, delaying it by almost a month and marking a poor start to the new year's schedule. The only in-flight problem was the failure of the SHARE heat-pipe test, but the apparatus was in the payload bay and there was nothing that the crew could do to fix it.

The dispatch of the Magellan spacecraft was the first fixed point in the year's schedule. The window opened on 28 April, but the optimal date was 5 May because this would result in the best orbit around Venus; the energy of the transfer would thereafter deteriorate again until 22 May, when the window closed. If the spacecraft missed its window, it would have to wait 18 months. Rather than wait for the ideal date, and risk a delay, NASA scheduled STS-30 for the first day of the window, but the clock was stopped at T–31 seconds because of an electrical fault in a pump in one of the SSMEs. As it turned out, the first available date was 4 May so, in the end, Magellan was given the best possible start. The two stages of the IUS fired in succession to accelerate to escape velocity with such precision that the planned trajectory correction using the spacecraft's thrusters was cancelled. This was the first use of the IUS to dispatch an interplanetary spacecraft. After a 15-month cruise designed to enable it to approach Venus at a slow relative speed, the spacecraft fired its PAM motor to slow down sufficiently to enter orbit, released the spent motor, and then proceeded to spend four Venusian years (cycles) mapping the planet with its imaging radar before following up with a gravitational survey over a further two cycles, before it finally burned up in the planet's atmosphere.

Despite having dispatched Magellan in the first week of its window, NASA postponed one of the classified missions until after Galileo was also safely on its way, thereby further eroding the schedule. The effort to keep STS-28's payload secret was more successful than in the case of the 'open secret' of the Lacrosse; the only confirmed fact was that it did not use an IUS. Columbia still had the SILTS sensors in place, so further data were gathered on its re-entry characteristics as it made its way back to Edwards.

Table 2.9. Shuttle manifest, *c.* mid-1989

Flight	Date	Orbiter	Mission
STS-28	Aug 1989	Columbia	classified (DOD)
STS-34	Oct 1989	Atlantis	Galileo
STS-33	Nov 1989	Discovery	classified (DOD)
STS-32	Dec 1989	Columbia	Leasat-5; retrieve LDEF
STS-36	Feb 1990	Atlantis	classified (DOD)
STS-31	Mar 1990	Discovery	HST
STS-35	Apr 1990	Columbia	ASTRO-1
STS-37	Jun 1990	Discovery	GRO

The Galileo spacecraft, NASA's prestige planetary mission, was years late. Ironically, the five-year roundabout route to Jupiter started with a minimum-energy transfer to *Venus*. The window to Venus opened on 12 October, and lasted until 21 November. As before, the launch was set for the first available date, but an engine fault discovered early in the count forced a five-day postponement; then the clock was held for weather and finally cancelled. Everything went perfectly the next day, however, and Galileo was successfully dispatched. It was launched six months after Magellan, but as a result of the trajectories flown by the two spacecraft, it encountered Venus in February 1990, almost six months ahead of Magellan. Its close passage was designed to provide a gravitational-assist to accelerate the spacecraft because the IUS was not powerful enough to send such a massive payload directly to Jupiter. (Galileo subsequently received two further gravitational assists, both from the Earth, in December 1990 and December 1992.)

As for Atlantis, it was recalled two orbits early so that it could sneak into Edwards ahead of approaching bad weather. This required a 1,000-km crossrange manoeuvre which involved a modification to the hypersonic roll-reversals undertaken during re-entry to bleed off kinetic energy; more time than usual was spent in left-bank (in fact, most of the trajectory from the Mach 25 entry interface down to the Mach 13 point was flown in a left bank) to swing the ground track west. These dynamic soaring manoeuvres were performed by the computer. Unfortunately, this particular orbiter did not carry the SEADS and SUMS sensors to gather data on this first high-crossrange descent. The flight crew took over when Atlantis became subsonic, as usual, for the approach and landing. On this occasion, the nosewheel steering was subjected to a high-speed test immediately after touchdown.

With the fixed points in the schedule satisfied, NASA could switch to building up the flight rate. First in the queue was STS-33, the Department of Defense flight. Whatever the nature of its classified payload, orbital mechanics imposed the first night-time launch since operations had resumed. Another aspect of the orbiter's flexibility during the descent phase was demonstrated 35 minutes *after* the deorbit burn, when a change in winds prompted the southerly approach to the dry lake to be switched to a north-easterly landing on the concrete runway.

Accommodating the interplanetary spacecraft had resulted in only five missions of the seven missions planned for 1989 being launched, but hopes were running high for 1990, and the manifest was flexible right through to October, when the Ulysses spacecraft was to be dispatched to Jupiter.

All the shuttle flights since operations had resumed had been dispatched from Pad 39B. Columbia as STS-32 was to have been the first to use Pad 39A, in December, but the work on the modifications to the pad was running late, so this flight was postponed until early in the new year. In its own way, STS-32 was time-critical. It was to dispatch the final Leasat communications relay for the Navy and then retrieve the LDEF satellite, which had been in orbit since 1984. This materials-exposure satellite had not been meant to spend such a long time in space. Its orbit was rapidly decaying, and re-entry was predicted for March. NASA wanted LDEF back, so if STS-32 proved difficult to launch it could not be permitted to slip too far down the schedule. The situation was aggravated by the fact that Atlantis was to be launched soon, as STS-36. In the event, although Columbia was delayed a day because of bad weather, it was launched without difficulty on 9 January, the Leasat was deployed, and the LDEF was retrieved exactly as planned. In addition to the satellite operations, a full programme of middeck science was undertaken. The descent was put back a day by fog on the dry lake. To provide the longest possible launch window, Columbia had carried more propellant than usual so that, in the event that it launched late in the window, it would be able to effect the plane change needed to rendezvous with the LDEF. The launch had been perfect, however, and Dan Brandenstein had made a smooth rendezvous, so Columbia had a large amount of propellant in reserve. To dump this excess propellant, the deorbit burn was performed with the orbiter aligned 50 degrees off the velocity vector. This burned off 2 tonnes of propellant. With the 10-tonne satellite aboard, this was not only the heaviest orbiter yet to return, it also had the most-forward centre of mass, so the data from its aerodynamic sensors were particularly welcome. All of the flights thus far since Return-To-Flight had been brief – only four or five days. At ten days,

Table 2.10. Shuttle manifest, *c.*early 1990

Flight	Date	Orbiter	Mission
STS-36	Feb 1990	Atlantis	classified (DOD)
STS-31	Apr 1990	Discovery	HST
STS-35	May 1990	Columbia	ASTRO-1; BBXRT
STS-38	Jul 1990	Atlantis	classified (DOD)
STS-40	Aug 1990	Columbia	SLS-1
STS-41	Oct 1990	Discovery	Ulysses
STS-37	Nov 1990	Columbia	GRO
STS-42	Dec 1990	Atlantis	IML-1

STS-32 was to be the first full-duration flight, but after this wave-off it actually entered the record book as the longest flight yet. Although delayed through no fault of the shuttle itself, this textbook mission had not disrupted the new year's schedule. After a few days lost to weather and faulty ground systems, STS-36 launched on 28 February. It was the first shuttle to make a dog-leg as it flew out across the Atlantic in order to attain an inclination higher than the 57-degree limit that is imposed by the form of the Eastern Seaboard. In this case, it was to 62 degrees. This was undoubtedly done to enhance the mission of its classified satellite. With STS-31 next in line, at long last, the astronomical community saw its Hubble Space Telescope roll out to the pad.

The launch of the HST, which was not time-critical, had been set for 18 April, but, for once, NASA found itself ahead of schedule, so it brought the mission forward to 10 April. On that date, however, a fault developed in one of the APUs and, despite the T–31 seconds hold, the count had to be scrubbed. The auxiliary power unit was readily replaced, but the resulting delay required that the HST's battery be recharged, which took a week,

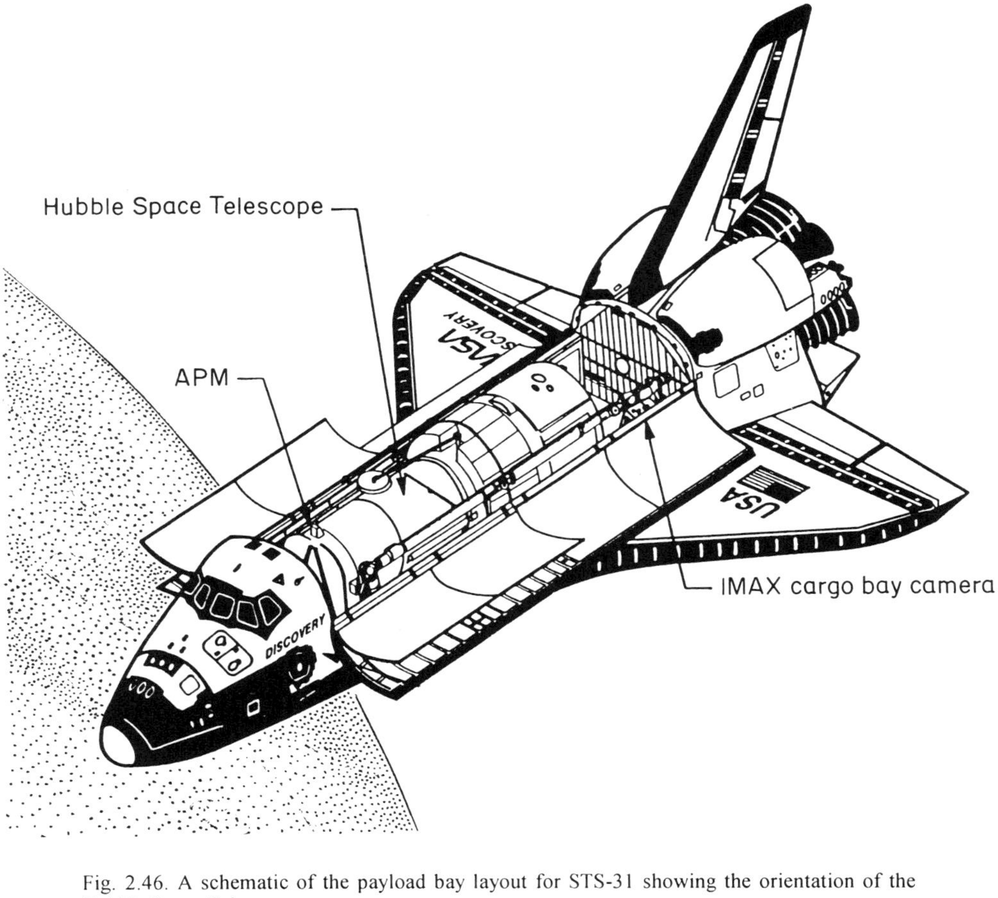

Fig. 2.46. A schematic of the payload bay layout for STS-31 showing the orientation of the Hubble Space Telescope.

so the effort to launch early proved fruitless. The countdown on 24 April, however, proceeded without incident. To ensure that the telescope's orbit did not suffer undue decay, it was to be deployed at an altitude of 600 km. To achieve this, Atlantis flew a direct ascent, with a steeper than usual climb through the atmosphere. At MECO, this ascent left the orbiter in a 50 × 600 km orbit. This was promptly circularised by the OMS-2 burn.

The following day, the telescope was deployed in a tricky operation that involved Steve Hawley, operating the RMS, first lifting the enormous bulk out of the bay, then rotating the 13-metre long stepped-cylinder so that he could see the indicators on its display panel from his work station on the aft flight deck. As the telescope was hoisted up out of the bay, it disconnected itself from the orbiter's power supply. The deployment process was now running against the clock. The spacecraft's battery was good for just eight hours. If the solar panels had not been deployed by that time, Bruce McCandless and Kathryn Sullivan would have to make a spacewalk to plug in an umbilical. While Goddard SFC methodically checked out the vehicle – for the HST was a *spacecraft*, not simply a telescope – the orbiter faced its belly towards the Sun in order to shade the HST until its thermal regulation system was brought on line. Finally, it was time to deploy the appendages. The first of the two 12-metre solar panels unrolled perfectly, but it required several attempts to complete the deployment of the second panel because the system that monitored the tension in the control wire overrode the motor; but when the astronauts reported that they could see no fault this safety feature was itself overridden and the panel rolled all the way out. While the problem with the panel was being studied, the clock had been ticking, so the orbiter was turned so that the solar panels could recharge the battery. Once the two TDRS antennas had been deployed, the spacecraft was released as planned. After so many years in development, and more years in storage waiting for the shuttle to be returned to service, the most expensive astronomical instrument ever built was finally in orbit.

Atlantis took up station 60 km away, and its crew performed middeck experiments for three days while Goddard continued to work through the HST's start-up procedure. It had once been intended simply to deploy the spacecraft 'clean' and leave it to deploy its various appendages, but the difficulty in deploying the solar panel had demonstrated the wisdom of not releasing the spacecraft until its power supply was secure. As Goddard neared the point of unlatching the telescope's aperture, Atlantis prepared to return, just in case the astronauts would be called upon to open it manually, but, once it was verified that the flap was open, the shuttle was cleared to withdraw. The descent from this unprecedented altitude required Atlantis to make the longest deorbit burn, and the longest coast down to the entry interface, of the programme to date. During the roll-out on the Edwards' concrete, Atlantis tested the new carbon–carbon brakes.

THE LONG HOT SUMMER

What was to have been a bumper year for the astronomical community turned out to be an unmitigated disaster. While Goddard methodically tested the HST instruments, STS-35 was to have flown the ultraviolet telescopes of the ASTRO-1 Spacelab. The count on 30 May had to be abandoned at T–6 hours, as the ET was being loaded, when a hydrogen leak was noted. When this was traced to the 17-inch disconnect valve, in the aft compartment of the orbiter, in the pipe that brought hydrogen from the ET, it became necessary to roll the stack back to the VAB. NASA decided to bring forward Atlantis' classified STS-38 mission to July, fly Columbia in mid-August as STS-35, and postpone its next mission, as STS-40, to the end of the year in order to allow sufficient time for the turnaround. The priority was not to jeopardise the dispatch of Ulysses in October, on Discovery. Meanwhile, NASA made the shocking discovery that the HST's main mirror was flawed!

Table 2.11. Shuttle manifest, *c.* June 1990

Flight	Date	Orbiter	Mission
STS-38	Jul 1990	Atlantis	classified (DOD)
STS-35	Aug 1990	Columbia	ASTRO-1; BBXRT
STS-41	Oct 1990	Discovery	Ulysses
STS-37	Nov 1990	Atlantis	GRO
STS-40	Dec 1990	Columbia	SLS-1
STS-42	Jan 1991	Discovery	IML-1

But worse was to come. At the end of June, a hydrogen leak was discovered on Atlantis, in the vicinity of its disconnect valve. It suddenly began to seem that there might be a *generic* fault in this crucial component. This appalling prospect forced NASA to ground all three of its orbiters, which threatened the chances of Discovery meeting the window for Ulysses. Fortunately, it turned out that Columbia's problem was a seal in the flapper valve, whereas Atlantis' was due to a leaky flange at the ET end of the pipe. It was *not* the same problem. The flange problem would take time to fix, but Columbia was soon repaired, and it was put back at the top of the queue in the hope that it would be practicable to launch it ahead of Discovery, but this hope was frustrated when a fault was found in one of the SSMEs, so at this point Discovery, with its fixed window, was assigned priority. When STS-41 was launched from Pad 39B on the set date, it drew this five-month hiatus to a close. It was a smooth mission. The two-stage IUS boosted the Ulysses spacecraft to escape velocity, and then the attached PAM motor fired to put it on a fast track to Jupiter, which it would reach in February 1992, almost three years ahead of Galileo. The dispatch of this last shuttle-deployed interplanetary probe restored flexibility to the manifest.

On the way back to Earth, Discovery demonstrated a procedure that would be needed in the event of a RTLS abort. Apart from when a Spacelab was carried, or in the case of when the LDEF had been retrieved, an orbiter touches down with a far-to-the-rear centre of mass. At launch, however, with a full bay, the centre of mass is generally much further forward, a fact which could jeopardise an already tricky emergency landing. The only way to slip the centre of mass towards the rear would be to dump propellant from the forward RCS tanks. The objective of Discovery's test was to demonstrate that the RCS thrusters could be safely fired deep in the atmosphere. The PTIs involved making yaw manoeuvres with the forward thrusters as the orbiter flew through Mach 13, Mach 6 and then Mach 4. This test was not just a step towards increasing confidence in the RTLS procedure, but also, by showing that an orbiter could adjust its centre of mass, it increased manifesting flexibility. These landing restrictions would apply to a TAL abort too, but in such an eventuality there would be more time to reconfigure the vehicle. In addition to testing the carbon–carbon brakes, Discovery exercised the nosewheel steering and thereby recertified the orbiter for a crosswind landing.

Table 2.12. Shuttle manifest, *c.* October 1990

Flight	Date	Orbiter	Mission
STS-38	Nov 1990	Atlantis	classified (DOD)
STS-35	Dec 1990	Columbia	ASTRO-1; BBXRT
STS-39	Mar 1991	Discovery	classified (DOD)
STS-37	Apr 1991	Atlantis	GRO
STS-40	May 1991	Columbia	SLS-1
STS-43	Jul 1991	Discovery	TDRS-E
STS-42	Aug 1991	Columbia	IML-1

The manifest could change significantly upon a scrubbed launch attempt because when an orbiter slipped down the manifest so too did all of its downstream assignments, unless the orbiter/payloads pairings were reassigned to maintain *mission* order. As a result, some missions slipped by up to six months at a time. Nevertheless, its hydrogen leaks overcome, NASA was confident that it would be able to make up for lost time.

CATCH-UP

It seems that shortly before the long-delayed STS-38 was finally launched, its classified payload was swapped to support Operation Desert Shield. The countdown for 9 November had to be abandoned early on, when a problem was found with the payload. Although this was soon rectified, by missing its slot on the Eastern Test Range schedule the shuttle was obliged to slip in behind an Air Force rocket that was due to leave from the Cape Canaveral Air Force Station on 13 November. This rocket, which was the first Titan 4, was launched successfully. It took two days to reconfigure the Test Range to handle a different launcher, so STS-38 followed on 15 November. After the secret satellite had been deployed, Atlantis prepared to return to Edwards even though the weather was unacceptable, in the hope that it would clear up, but, four minutes before the deorbit was due, the return was slipped a day. The weather was no better the next day, and White Sands was not available either, so rather than wait NASA recalled STS-38 to Florida. Thus, despite the fact that it still had the older carbon–beryllium brakes, Atlantis became the first orbiter to return to its launch site since Discovery had blown a tyre in 1985.

The long-delayed STS-35 was set for 2 December. ASTRO was the first Spacelab to be devoted to a single subject, in this case astronomy. The Instrument Pointing System (IPS), mounted on a pair of pallets at the front of the bay, carried three ultraviolet telescopes. These were controlled jointly by the astronauts and astronomers at the new Payload Operations Control Center (POCC) at the Marshall SFC, which was known by its radio call sign 'Huntsville'.

This mission was to have been flown as 61E immediately after the return of Challenger, on which occasion it was to have had the Large Format Camera in its aft bay to photograph Halley's Comet. The comet had long gone, but there was now a supernova, and the rear of the bay carried BBXRT, an X-ray telescope that was controlled by Goddard SFC. ASTRO was the first mission since Return-To-Flight to carry Payload Specialists; in this case two astronomers, Sam Durrance and Ron Parise. Unfortunately, the observational programme had to be curtailed when it was decided to land a day early so as to slip into Edwards ahead of bad weather. In order to pave the way for the Extended Duration Orbiter (EDO) system that would significantly increase the output from a Spacelab flight, a trash compactor, and a modified toilet (one that could be unloaded in space) were tested.

One aspect of the ASTRO-1 mission drew comment in the postflight debriefing: a 9-day flight with a crew of seven working on a 24-hour basis in the confines of the cabin was rather stressful. The off-shift crew members had to try to sleep on the middeck while the others worked 'upstairs' on the flight deck. Mike Lounge, one of the Mission Specialists, observed that it was "pretty stressful" operating the pallet-only Spacelab configuration.

By mounting STS-41, STS-38 and then STS-35 at monthly intervals, it finally began to seem that it would indeed be possible to increase the flight rate beyond the record of nine in 1985, so there was every expectation that the new year's schedule, which did not appear to be too demanding, would be met.

In retrospect, it is evident that in the immediate Return-To-Flight period, NASA was in no mood for dynamic scheduling. The missions were precisely defined and the flights were so straightforward that they embodied the military's philosophy: launch, complete the primary mission as soon as possible, and land. There was none of the vibrant spontaneity of the pre-Challenger days, and, very noticeably, no spacewalks.

Table 2.13. Shuttle manifest, *c.* early 1991

Flight	Date	Orbiter	Mission
STS-39	Feb 1991	Discovery	miscellaneous military projects
STS-37	Apr 1991	Atlantis	GRO
STS-40	May 1991	Columbia	SLS-1
STS-43	Jul 1991	Discovery	TDRS-E
STS-44	Aug 1991	Atlantis	DSP
STS-48	Nov 1991	Discovery	UARS
STS-42	Dec 1991	Atlantis	IML-1

STS-39 was duly rolled out to Pad 39A, but when cracks were found in a subassembly on the ET it was returned to the VAB; it was rolled out again on 1 April, for a target date of 23 April. While Discovery was being rolled back and forth, Atlantis, with the Gamma Ray Observatory (GRO) in its bay, had been set up on Pad 39B. The STS-37 count proceeded exceptionally smoothly and, exactly on time, it made a direct ascent into a 450-km circular orbit. At 16 tonnes, the GRO was the most sophisticated gamma-ray package ever to be put in space. Linda Godwin, operating the RMS, hoisted the satellite out of the bay and ran the pre-release checks. The solar panels unfolded as planned, but when the command was sent to unstow the solitary TDRS antenna, the 3-metre articulated boom remained tight against the underside of the satellite. Godwin moved the arm back and forth in an attempt to shake the boom free of its clamp, but it remained stuck fast; the only option was to send astronauts out to free it.

Two astronauts on each crew are trained to conduct a ‘contingency EVA’, to overcome any of a variety of problems which could jeopardise the orbiter, such as closing the payload bay doors. The first contingency was on STS-16, when David Griggs and Jeff Hoffman attached the ‘fly swatter’ to the RMS in order to try to activate the dormant Leasat. More recently, Bruce McCandless and Kathryn Sullivan were depressurising the airlock to go out and deploy the HST’s stuck solar panel when Goddard finally managed to coax it to unroll. As it happened, STS-37 was the first mission since Return-To-Flight to have a spacewalk assigned. Jerry Ross and Jay Apt promptly suited up and traversed the length of the arm to inspect the satellite. Ross confirmed that the clamp had indeed released the boom. It seemed that the boom’s hinge was frozen, so he gave the other end of the boom a hefty shove, then watched as it started its slow rotation towards its deployed position. It had taken only about twenty minutes, so it was decided to stay out to make an early start on the assigned external programme, which was to test a number of translation aids which were under consideration for use on the space station, and this work was continued the following day when Ross and Apt went out for the scheduled excursion.

Atlantis landed on 11 April. It had been hoped to dispatch STS-39 on 23 April, which was about as soon after Atlantis’ return as the rule book in effect at that time permitted, but that count had to be abandoned early on when an oxygen pump in one of the SSMEs was found to be running irregularly; the count was recycled, and then completed without further difficulty on 28 April.

Although STS-39 was a Department of Defense mission, it was the first such flight not to be completely classified; a few aspects of the mission remained secret, but most of the payload was announced. The *scientific* value of the observations that were to be made was explained, but it was evident that the sensors were really being evaluated to support military applications. In terms of orbiter manoeuvres, this was one of the most demanding missions to date; the plumes of the orbiter’s engines were studied both by a free-flyer released by the orbiter and by an SDI satellite. The most mysterious item was the small satellite which was ejected from a militarised GAS can dubbed the Multipurpose Experiment Canister (MPEC). The weather at Edwards was unacceptable, so Discovery was ordered down to KSC. When it arrived, there was a slight crosswind. However, Discovery had carbon–carbon brakes, so no problems were anticipated, and, indeed, it was viewed as an opportunity to test landing in a crosswind. In the event, the starboard main gear touched the concrete before the left and, as differential braking was used to line up on the centreline in advance of lowering the nosewheel, one of the tyres was shredded. Clearly, if high-speed differential braking was to be avoided, some form of stabilisation system would have to be devised to keep the orbiter lined up between the time that the main gear touched down and the nosegear touched down; the obvious solution was a drag chute, so work started to develop one. Despite the problem in this case, the status of the SLF was upgraded to be 'first equal' with Edwards.

Next in line was the long-delayed Spacelab for Life Sciences (SLS-1) mission assigned to STS-40. The count on 1 June had to be scrubbed at T–23 minutes when a fault was found in one of the IMUs, but it was completed four days later. For this, the first Spacelab to be devoted to life sciences, three of the astronauts were medical specialists. Their objective was to make the most systematic study so far of the first phase of the body's adaptation to microgravity. The primary programme comprised 18 experiments to monitor the response of the heart, lungs, kidneys, and endocrine glands, as well as muscle and bone degradation. These experiments formed an interrelated study, and the in-flight data was compared to post-flight tests of the readaptation to gravity. Columbia remained in space for the maximum nine days in order to provide the longest possible data set. Although this brief study could not possibly rival the year-long flights being undertaken by cosmonauts on the Mir space station, the fact that the 10-tonne Spacelab carried a comprehensive suite of apparatus meant that SLS-1 established the 'gold standard' for data on the first week of adaptation to the space environment. After landing at Edwards, Columbia was returned to Rockwell for its second refit, part of which would include the installation of the EDO facility.

The rest of the year's manifest was to be flown by Atlantis and Discovery. NASA was undoubtedly buoyed up by the fact that, despite minor problems, the manifest published at the start of the year was running on time. Misfortune struck again on 23 July, when the launch of STS-43 had to be halted due to a problem with the ET, and again the next day, when one of the SSMEs needed attention, but both of these problems were noted early on. Although the count on 1 August ran to the T–9 minute hold, the weather at the SLF was unacceptable for a RTLS, and the window expired before conditions improved. Everything went smoothly the next day, and the primary objective, the release of a new TDRS satellite, was achieved without incident.

In contrast to earlier straightforward satellite-deployment missions, in which the orbiter had landed after four or five days, STS-43 was to stay up for the maximum time permitted by Atlantis' consumables in order to obtain biomedical data for comparison with that from SLS-1. This required John Blaha's crew to maintain round-the-clock operations without a Spacelab module. The ASTRO mission, which had involved a palletised Spacelab, had demonstrated the difficulty of trying to run split shifts in a crew confined to the orbiter's cabin; the activities of those at work inhibited those supposedly sleeping, with the result that after several days the whole crew was exhausted. The situation had been exacerbated on STS-35 by the presence of two Payload Specialists, but, without passengers, Blaha's crew fared rather better. This aspect of STS-43 showed that operating the shuttle required a careful balance of conflicting requirements. The astronomers had been eager to operate the telescopes on a 24-hour basis so as to maximise the data from the limited time available, and they had wanted to fly the telescopes' designers in order to ensure that the best opportunity was taken of the situations that arose, but this had conflicted with long-term efficiency. The medics had wanted the best possible output from SLS, and the presence of the laboratory in the bay had made round-the-clock work sustainable. A satellite-deployment mission could not carry a laboratory as well, so, although such a flight could be extended and a shift cycle employed, it could not sustain the high-intensity science that was possible on a dedicated mission with a laboratory full of specialised apparatus. Making this balance would be more difficult once the EDO entered service, because 16 days was a long time for seven or eight people to be confined to the orbiter's cabin.

Atlantis had been scheduled to land at KSC, but upon reviewing Discovery's difficulty in landing in a crosswind, STS-43 was authorised to land at KSC only if crosswinds were negligible; which, in the event, they were, so Atlantis returned to the landing site. Edwards was to remain the prime landing site until the orbiters were fitted with tyres using a harder compound. This was yet another demonstration of the fact that, even after more than forty flights, the shuttle was still involved in a process of incremental development. In a very real sense, each mission was a *test* flight, and, as a result, as an operational system the shuttle was becoming progressively more effective.

The manifest had unfolded essentially as planned, but flying STS-39 after STS-37 had meant that Discovery's turnaround could not be guaranteed for STS-43, so that mission's TDRS had been switched to Atlantis, which had received a head start as a result of its having flown earlier. The consequence of this exchange was that STS-44 and STS-48 had also to be exchanged, and Discovery launched on the set day. This adjustment marginally affected the year's schedule, but the fact that this was running only a few weeks late was testimony to the fact that the orbiter flow at this flight rate was manageable.

The Upper Atmosphere Research Satellite (UARS) – the first element of the Mission To Planet Earth (MTPE) programme (or Earth Science Enterprise, as NASA recently renamed it) – was the largest environmental monitor to date. It carried such a suite of sensors that it required two days to check it out prior to deployment, but this was completed and the satellite was released on schedule. As with its predecessor, STS-48 was to land on the SLF if conditions were perfect; but they were not, and it was diverted to Edwards.

The launch of Atlantis on 24 November was delayed only by the need to carry out work on the mobile launch platform. Although it was a Department of Defense flight, the Defense Support Programme (DSP) satellite was not classified. The deployment, at the usual point in the flight for an IUS payload, was the first time that the release of a military satellite had been publicly displayed.

STS-44 was to have been a full-length flight, in order to enable Tom Hennen, an expert in interpreting reconnaissance imagery, to make experiments to test the ability of the human eye to resolve surface detail; to allow the LACE satellite to observe thruster plumes;

Table 2.14. Shuttle manifest, *c.* late 1991

Flight	Date	Orbiter	Mission
STS-42	Jan 1992	Discovery	IML-1
STS-45	Mar 1992	Atlantis	ATLAS-1
STS-49	May 1992	Endeavour	Intelsat repair
STS-50	Jun 1992	Columbia	USML-1
STS-46	Aug 1992	Atlantis	TSS-1; deploy EURECA
STS-47	Sep 1992	Endeavour	Spacelab J1 (SL-J)
STS-52	Nov 1992	Columbia	USMP-1; LAGEOS-2
STS-53	Dec 1992	Discovery	miscellaneous military projects

and to produce more biomedical data, but a fault in one of the three Inertial Measurement Units (IMU) required the flight to return three days early; and, as previously, unacceptable conditions in Florida prompted diverting the orbiter to Edwards.

The final mission assigned to 1991 slipped over the holiday season, but was dispatched on 22 January 1992 after an hour's hold for weather. As the first International Microgravity Laboratory (IML), this Spacelab was loaded for life sciences and materials-processing. Ulf Merbold, one of the two Payload Specialists, had flown on the first Spacelab. His presence reflected the fact that Germany had built much of the hardware. Roberta Bondar's inclusion resulted from NASA's long-term collaboration with Canada. Although a laboratory module was carried in the bay, and a two-shift cycle was invoked, only a seven-day flight had been assigned. By conserving the orbiter's consumables, however, the flight was able to be extended by a day to enhance the scientific yield, which was already high because the experiments had proceeded refreshingly flawlessly.

STS-45's count on 23 March was scrubbed during ET tanking, when sensors reported a concentration of hydrogen in the orbiter's engine bay. When an inspection failed to locate a leak, the launch was rescheduled for the following day, at which time Atlantis departed on time. In effect, ATLAS was a second strand of the MTPE programme. Its role was to complement the UARS. Like ASTRO, this was a full-length Spacelab mission in which the bay was full of pallets and the crew included two Payload Specialists, so the orbiter's cabin was crowded. The crew's spirits were raised, however, because the instruments were able to collect a vast amount of data on the state of the Earth's atmosphere and the way in which it is stimulated by the energy received from the Sun. Several of the experiments were flown because on their initial outing, on earlier Spacelab missions, their operation had been frustrated; unfortunately one of them (the SEPAC electron-beam) was foiled again, this time by a blown fuse, but not before it had returned a substantial amount of data. ATLAS was intended to be flown on a regular basis, to measure the solar–terrestrial relationship at different points in the 11-year sunspot cycle.

ENDEAVOUR

The most significant item on the 1992 manifest was the introduction of the new orbiter, Endeavour, which had been built to replace the lost Challenger. The count was held for 34 minutes to allow weather at the TAL site to clear but was otherwise uneventful, and the new orbiter launched on schedule.

Endeavour's task was to rescue Intelsat 603. There was a certain irony in this, because this type of satellite had been designed for carriage in the shuttle's bay, but, after the loss of Challenger, NASA had been told to cease its commercial operations. Actually there was nothing wrong with the satellite; it had simply been stranded in low orbit by a fault in the upper stage of its commercial Titan launcher.

The plan was for spacewalking astronauts to catch the satellite, mate it with a new kick-motor, and then release it so that it could resume its journey to geostationary orbit, where it would serve as a communications relay for at least a decade. Special tools had been made to handle the satellite, and training in the hydrotank had suggested that the operation should be able to be completed in two spacewalks. However, capturing the satellite proved to be more difficult than expected, with the result that it was still drifting alongside the orbiter when the time expired. In opting to continue with the Intelsat rescue, which was ultimately completed in triumph, NASA was obliged to cut short the secondary objective – testing procedures for assembling structures – and squeeze this programme into the one remaining spacewalk. This inaugural mission set several spacewalk records, including duration and the number of people involved, but the difficulty encountered in capturing the large satellite was a cause for concern.

Immediately after Endeavour's nosewheel settled, a parachute was deployed from the base of the tail. As soon as it had slowed the orbiter to 100 km/h, the chute was jettisoned to ensure that it did not snag the SSMEs, because that would have complicated the task of the recovery team. On this occasion, the drag chute had been used only to test its deployment mechanism. In service, it was to be deployed *prior* to nosewheel contact in order to slow the orbiter and keep it on the centreline, so that the brakes need not be used until later in the roll-out, just before wheel-stop. Finally, it seemed, the shuttle was capable of landing on a narrow strip in a crosswind.

Fig. 2.47. A returning orbiter uses a drag-chute to hold the centreline and to slow down.

STS-50 carried a Spacelab of microgravity experiments supplied by US establishments, including the various NASA-sponsored CCDS, so by analogy with IML this mission was designated USML. Columbia spent most of its time in gravity-gradient mode, which locked the vehicle stable without requiring thrusters to be fired, to present the materials-processing experiments with the best microgravity environment. The life sciences work was greatly enhanced by the fact that this was the first use of the EDO. Although the regenerative unit (the RCRS) failed, there were sufficient lithium hydroxide canisters for the back-up system to prevent a build-up of carbon dioxide. Two Payload Specialists were included in the crew, two shifts were operated, and, for the first time, NASA assigned each member of the crew a half-day off during which they could do whatever they liked, even if it was more work. After two weeks, Columbia returned to KSC because Edwards was 'socked-in'.

Flight operations were now proceeding smoothly. Not only was the manifest unfolding as planned, but NASA was able to dispatch STS-46 a few weeks *ahead* of schedule. This was not to be a straightforward flight. Radio interference delayed the deployment of EURECA by 24 hours. This ESA free-flyer carried experiments requiring an extended period of high-quality microgravity. It was to be left in space, and to be retrieved by another shuttle a year later. The US/Italian Tethered Satellite System (TSS-1) was to have unreeled a test satellite to a distance of 20 km, but it repeatedly jammed after just a few hundred metres and had to be reeled back in. A tethered payload offered significant scientific results, so once the fault in the deployment system was fixed it was virtually assured of a reflight.

Upon its return to space, Endeavour had a Spacelab in its bay. This was funded jointly by the US and Japan, and involved a large number of materials-processing, technology and life sciences experiments. It launched on time and, in contrast to its predecessor, caused no excitement; it managed to touch down at KSC between two periods of unacceptable weather. It was followed by STS-52. Located at the rear of the bay was the LAGEOS-2 satellite that had been supplied by Italy. This boosted itself to a high orbit in order to reflect lasers, not for weapons research but for geophysical research, such as measuring the rate of continental drift and of uplift in volcanoes. Apart from a few GAS cans, the only other item in the cavernous bay was a strikingly small pallet carrying three microgravity experiments, and the status of this payload was recognised in its designation – the US Microgravity Package (USMP). This Columbia mission carried such a lightweight payload that it was criticised for not being worth the cost of the launch.

The launch of STS-53 – the final mission of the year – had been planned for dawn, but it was delayed for 90 minutes to allow the ice that had built up overnight to melt, so that it would not be shaken loose and strike the orbiter's tiles. Although the temperature had dropped to 40 °F, which was only slightly warmer than that which Challenger had endured the night before its final launch, this was not seen as posing a problem for the redesigned SRBs because its O-rings were electrically heated.

On this Department of Defense mission, Discovery's bay was dominated by a classified satellite that was blandly designated 'DOD-1'. The only other external payloads were GAS cans, so this satellite, which was released a few hours after launch, was clearly substantial. However, other aspects of the flight were conducted openly, and many of these experiments continued earlier projects. The crew had to stay in the orbiter for two hours after landing at Edwards, so that the recovery team could deal with a leaking RCS thruster in the nose.

For once the *entire* year's manifest had been flown as planned. By this point, NASA had come to terms with the fact that it would never be able to fly a dozen missions a year, and although it had sometimes managed to mount missions within weeks of one another, it was evident that it would require a supreme effort to *sustain* a monthly flight rate. So, NASA settled down to an annual cycle of between six and eight flights, which experience had shown *could* be sustained.

Table 2.15. Shuttle manifest, *c.* late 1992

Flight	Date	Orbiter	Mission
STS-54	Jan 1993	Endeavour	TDRS-F
STS-55	Feb 1993	Columbia	Spacelab D2
STS-56	Mar 1993	Discovery	ATLAS-2; SPARTAN-201-1
STS-57	Apr 1993	Endeavour	Spacehab 1; EURECA retrieval
STS-51	Jul 1993	Discovery	ACTS; ORFEUS–SPAS-1
STS-58	Aug 1993	Columbia	SLS-2
STS-60	Nov 1993	Discovery	Spacehab 2; WSF-1
STS-61	Dec 1993	Endeavour	1st HST service

The count for STS-54 was uneventful, and apart from a short hold to enable high-level winds to be assessed, Endeavour was once again launched on schedule. The deployment of a TDRS satellite was now a matter of routine, but in this case, during the coast up to the top of the geostationary transfer orbit, the IUS' primary computer failed and the back-up had to step in to deliver the satellite.

This flight marked a significant change in NASA's attitude towards spacewalking. The capture of the Intelsat satellite had proven to be much more difficult than underwater training had suggested. Of course, masterful improvisation had overcome these problems, but it had taken significantly longer than the time assigned, and the price had been a curtailment of another task. NASA's immediate concern was that the training for the servicing of the Hubble Space Telescope, which would be far more demanding and more time-critical, might be inadequate. In the longer term, there was concern that it might be unwise to design the space station in such a way that it would *rely* on spacewalking for its assembly, so to build up experience in EVA, it was decided to add *generic* spacewalks to several already planned missions in order to run through basic procedures, and in so doing to assess the fidelity of the hydrotank training.

Another test was performed to reduce the technical risk in *operating* the space station. Standard operating procedure required a shuttle flight to be cut short in the event that a fuel cell had to be shut down, but because it is intended to power down an orbiter whilst it is docked with the space station, it had been decided to verify that a fuel cell could be restarted in space. On the last day of the flight, when it would not matter if the fuel cell could not be brought back on line, one of the three cells was shut down and successfully restarted. The planned descent was delayed for an orbit to allow the early morning Sun to burn off the fog masking the SLF.

Whilst the newest orbiter was proving to be an exceptionally lucky ship, gremlins were once again at work on Columbia, the oldest member of the fleet, and its February launch as STS-55 had to be postponed so that the high-pressure oxygen pumps of all three SSMEs could be serviced. The launch was initially set for 21 March, but the shuttle had to compete with other rockets for the tracking facilities of the Eastern Test Range, and it had to be put back a day when a Delta 2 was delayed. Then the count had to be aborted at T–3 seconds when an incorrectly set valve inhibited the start of one of the SSMEs. It was the first post-ignition abort since Return-To-Flight. The rule called for engines which had undergone a prestart to be serviced, so this imposed yet another delay.

With Columbia stuck on the ground, attention switched to Discovery's STS-56, which had already slipped several weeks from the planned late-March date. The count on 6 April was scrubbed at T–11 seconds, when it seemed that a valve in one of the main engines had not cycled correctly. Frustratingly, it turned out to be a faulty sensor; the valve had operated normally. The launch two days later was perfect, however.

On this occasion, a reduced set of ATLAS instruments had been squeezed onto a single pallet, so there was room at the rear of Discovery's bay for SPARTAN-201-1, a free-flyer which studied the Sun for two days. This ATLAS flight, following precisely a year after the first, provided a corroborative check on the state of the atmosphere immediately after a northern winter.

Meanwhile, Columbia had been serviced on the pad. NASA wanted to launch it as soon as possible after Discovery's return so as to minimise the disruption to the overall schedule. In the event, Discovery was waved off for an extra day to await better weather at KSC, so it did not land until 17 April. Nevertheless, NASA set out to launch Columbia on the first available date, on 24 April. This valiant effort was foiled at T–6 hours when one of the IMUs failed to start properly. Two days later, the countdown was flawless, and Columbia launched with the second German-sponsored Spacelab. This represented a turning point for the Spacelab programme, because the German Space Research Agency (DLR) had decided that this would be its final 'national' mission. NASA's inability to reduce the cost of flying the shuttle to anything approaching that of the original hype had made mounting a Spacelab mission so expensive that, from now on, Germany would fly its experiments in the context of the international Spacelabs. Although the D2 Spacelab was on Columbia, the EDO wafer was not carried. However, because the orbiter spent most of its time in the gravity gradient, the rate at which it consumed resources was low enough to allow an extra day for the crew – which included two German Payload Specialists – to carry out the 88 experiments of the multidisciplinary programme. Much of the laboratory apparatus was being evaluated for use on the Columbus module which ESA planned to add to NASA's space station.

The manifest was now running significantly late. The factor dictating the launch date of Endeavour's STS-57 mission was the orbit of EURECA, which was to be retrieved. It would have been possible to launch in mid-May, but this would have meant both launching and landing in darkness; so, despite the consequences for the schedule, NASA delayed this flight to 3 June in order to guarantee a daylight landing. However, when it was found that a spring had been damaged in an engine that was being serviced, it was decided to replace the part on Endeavour's engines, as a precaution, because if the spring fractured in flight it would cause an explosion. On 20 June the launch window expired before rain over the SLF lifted sufficiently to permit a RTLS abort, but the next day conditions were perfect, and Endeavour launched on time.

This was a landmark mission for NASA, because it carried the first Spacehab 'middeck augmentation' module. The module had been designed to accommodate the same lockers as the orbiter's middeck; its carriage doubled the usable volume of the orbiter's cabin without taking up too much of the bay. This facility had been developed by a private company, with NASA's encouragement, to perform microgravity experiments on a commercial basis. For this inaugural flight, NASA had booked half of the module's lockers for experiments from its CCDS; the other experiments were supplied by commercial customers.

The retrieval of the EURECA free-flyer turned into a race against time. To supply power to the satellite in the short time between it retracting its solar panels and being deposited on its cradle, at which time it would be able to draw power from the orbiter, a socket had been installed in the end-effector of the RMS; but it had been set in reverse, and the satellite was forced to live off its battery. This would not have been a problem if the satellite could have been stowed immediately, but its whip antennas refused to retract. The rule book prohibited stowing EURECA until its antennas had retracted, but, with time running out, it was decided to carefully lower the satellite onto its cradle. Another generic spacewalk had been assigned to this mission, so, before they began testing the tools that were to be used to work on the HST, David Low and Jeff Wisoff retracted the whips so that they would not thrash about during the landing and damage the GAS Bridge Assembly stationed immediately in front of EURECA's cradle.

With the manifest slipping ever further behind, NASA's priority was to mount the HST servicing mission on time in December. Although the telescope was returning amazing views, its flawed mirror was still a supreme embarrassment. The successful refurbishment of the telescope was expected to ease the way for the space station budget through a hostile Congress. If the astronauts damaged the telescope, NASA would probably be forced to abandon its plans for a space station. Given the delays, it was hoped to launch two of the three missions in the queue ahead of the HST service, but STS-60 was slipped into the new year.

The launch of STS-51 Discovery had been set for two weeks after Endeavour's return, but the countdown on 17 July was foiled when a fault was spotted in the system which armed the explosive bolts which would have to fire at T=0 to release the stack. On 24 July, the count was scrubbed at T−19 seconds because of a faulty power unit in one of the SRBs. NASA then decided to wait until after the peak of the Perseid meteor shower (and any possible attendant hazard from meteoroid impacts), which occurred on 11 August. The next day produced a post-ignition abort at T–3 seconds, which required that the engines be refurbished. The new orbiter's run of luck had evidently taken a turn for the worse, and the schedule was being ripped to shreds. Endeavour did not finally launch until 12 September.

The main item was the release of the Advanced Communications Technology Satellite (ACTS), which had been developed by NASA's Lewis Research Center to demonstrate the utility of the high-capacity K_a-Band. This 1,500-kg satellite was to be lofted to geostationary orbit by the new Transfer Orbit Stage (TOS), which was being used for the first time on the shuttle. When the command was sent to detonate the pyrotechnics to eject the stage from its cradle, a wiring fault fired both the primary and the secondary systems. Although it did not affect the payload, debris was blasted out and some punched straight through the aft wall of the bay, but luckily the APUs in the compartment beyond were not damaged. The next order of business was to deploy the ORFEUS–SPAS free-flyer, which carried ultraviolet telescopes and was to make astronomical observations while the crew ran middeck experiments and made another spacewalk to test procedures for the HST service. In the final phase of the rendezvous with ORFEUS–SPAS, a bay-mounted laser rangefinder (the TCS) was evaluated as part of the preparations for the newly announced

Shuttle–Mir programme. Unacceptable weather prompted a day's extension; then, on 22 September, Endeavour made the first ever night landing at the SLF, a descent which showed the need for improved runway lighting.

The second life sciences Spacelab that was loaded into Columbia's bay for STS-58 was to exploit the EDO to significantly extend the database on human adaptation. The count on 14 October was scrubbed at T–31 seconds when a problem was discovered with the ground equipment of the ETR. The next day, a fault in the orbiter's S-Band communications link ended the count at T–9 minutes. On 18 October, however, the only problem was an aircraft which flew into the restricted airspace over the Cape. It was hardly credible that a private sightseeing aircraft could cause a shuttle to miss its window, but as soon as this plane was shooed away, Columbia was dispatched.

The long-term objective of the biomedical programme was to develop countermeasures against the effects of weightlessness, particularly the bone mass loss that appeared not to stabilise and was slow to recover following return to Earth. Cellular activity is slowed in the absence of gravity, and the terrestrial equilibrium in the relative rates of cell production and destruction is upset in space; the result is a set of imbalances which affect body chemistry. The effects were best studied in rodents. Previously, rats had been dissected upon their return to Earth, but by that time their organs had already begun to readapt to gravity, so on STS-58 the rats were killed and dissected in space to preserve the state of their organs in their space-adapted state. Martin Fettman, a veterinarian, flew as a Payload Specialist specifically to attend to this delicate task. A new experiment, dubbed PILOT, was a programme running in a laptop computer rigged to accept steering inputs from the orbiter's command stick. This enabled a member of the flight crew to simulate a re-entry, approach and landing, so as to refine flying skills. Under normal circumstances, astronauts regularly flew T-38 jets, so two weeks of not-flying represented an aberration. This experiment not only maintained 'proficiency'; it also produced quantitative data on how flying skills degraded as a result of being in space.

As this was the longest flight to date, it was decided that Columbia should return to Edwards in order to take advantage of the flexibility of the multiplicity of runways out on the dry lake, but in the event it landed on the concrete strip.

THE BIG TEST

With STS-58 safely back on the ground, attention switched to the attempt to service the Hubble Space Telescope. This was fully expected to be the most demanding mission of the programme to date. Although the first countdown on 1 December had to be scrubbed because the weather over the SLF would have inhibited a RTLS, there were no problems the next day, and the launch was exactly on time; which was fortunate, because NASA was eager to complete this tricky mission before the Christmas break, to keep costs down.

After a two-day rendezvous, Endeavour approached the HST from directly below so as to employ differential gravity as a brake, and thereby reduce the degree to which its thruster plumes would contaminate the telescope's optics. Claude Nicollier then grappled the HST with the RMS and settled it onto a rotating tilt-table at the rear of the bay.

The activity plan was for a-spacewalk-a-day for the next five days, with each excursion taking advantage of the maximum capacity of the EMU life support system. Pairs of astronauts were to go out on alternating days in order to spread the load, and to provide a day's rest between excursions. It was to be a carefully phased programme of work. Story Musgrave, Jeff Hoffman, Tom Akers and Kathryn Thornton had spent *hundreds of hours* immersed in the WETF, training for this mission. They had rehearsed each individual task, and had undertaken integrated simulations in which the tasks were performed in sequence to make sure that they did not conflict. Working on the telescope was fully expected to be demanding. The individual tasks called for delicate, precise manipulation of bulky objects. Seven hours was a long time to sustain concentration, and a single misjudged act could disable the telescope. It would be important to pace the work, but it was recognised that an unforeseen problem might force a spacewalk behind schedule. If the work proceeded well, but slower than planned, the astronauts were to focus on the primary tasks and leave other operations for a later mission. In fact, in more than 35 hours of external work, *every* item on the task list was completed more or less within the expected timeline.

Correcting the telescope's flawed optics was a triumph; a magnificent demonstration of the ability of astronauts to work in space. With this success, and the September agreement to work with Russia to build an *International* Space Station, the Agency's long-term future seemed assured.

Table 2.16. Shuttle manifest, *c.* late 1993

Flight	Date	Orbiter	Mission
STS-60	Jan 1994	Discovery	Spacehab 2; WSF-1
STS-62	Mar 1994	Columbia	USMP-2, OAST-2
STS-59	Apr 1994	Endeavour	SRL-1
STS-63	Jun 1994	Discovery	Spacehab 3; SPARTAN-201-2
STS-65	Jul 1994	Columbia	IML-2
STS-66	Sep 1994	Endeavour	ATLAS-3; CRISTA–SPAS-1
STS-64	Sep 1994	Discovery	LITE; SPARTAN-204
STS-68	Dec 1994	Atlantis	SRL-2
STS-67	Dec 1994	Columbia	ASTRO-2

The new year began with the postponed STS-60. It was set for 27 January, but delayed so that a leak in one of the aft RCS thrusters could be investigated. It was duly launched on 3 February, however. Although Discovery had a Spacehab and a novel free-flyer in its bay, the media focused on the fact that cosmonaut Sergei Krikalev was aboard. His inclusion in the crew was the result of the agreement signed by President George Bush and President Boris Yeltsin on 17 June 1992.

As previously, NASA had booked half of the lockers on Spacehab for the experiments supplied by its CCDS. The Wake Shield Facility (WSF) was also important for NASA because it had been privately developed to demonstrate the commercial potential of making semiconductors in space. The intention had been to release this free-flyer for two days, but an assortment of technical problems intervened and prevented it from being deployed.

This was a frustrating outcome for Ron Sega, a member of the development team who had since become an astronaut and who was flying to supervise the test.

Conditions at KSC were marginal, but after a one-orbit wave-off permission was given for the deorbit burn. However, the weather closed in, and when the orbiter arrived it found 80 per cent cloud cover. Nevertheless, upon emerging from the final turn of the heading alignment cylinder, mission commander Charlie Bolden reported that he could see the runway through a small clearing, and he proceeded to make a perfect landing.

STS-62 Columbia was put back a day because poor weather was forecast, and it was launched on 4 March. In addition to the USMP pallet, it carried a Hitchhiker Bridge with an assortment of apparatus on trial for use on future spacecraft. As previously, the lightweight payload attracted criticism. This time, the microgravity experiments were able to exploit the presence of the EDO wafer, and in order to provide the best possible conditions, the orbiter adopted gravity-gradient stability for most of that time. In effect, this flight served as a trial run for the kind of research that was to be undertaken on the International Space Station.

Although delayed a day by weather, Endeavour's launch on 9 April marked another step in unfolding the planned manifest. STS-59's bay was dominated by the Shuttle Radar Laboratory (SRL). This combined the L/C-Band radars, flown twice before, with a new X-Band radar developed jointly by Germany and Italy. At 10 tonnes, the new SIR-C configuration was the largest instrument ever designed to work in the bay. Between them, the radars scanned a swath centred on the ground track for over 100 hours, and mapped 25 per cent of the Earth's land surface with a linear resolution of better than 25 metres, and 10 metres under ideal conditions. To perform this extraordinarily productive mission, and to facilitate stereoscopic imaging, Columbia had to make a record 400 thruster firings to refine its orbit. The microwave radars proved particularly adept at penetrating sand in arid regions to reveal the topography of the rock beneath. They detected several ancient river beds in terrain which is now desert. The flight was extended by a day to await better weather at KSC, but when this was not forthcoming Columbia was ordered down to Edwards.

It had become clear that, apart from the lockers booked by NASA, the Spacehab was so under-subscribed that its next flight could not be justified, so Discovery's STS-63 mission was postponed until the company managed to secure commercial customers. To the distress of all concerned in the venture, the market for commercial microgravity applications had not developed as rapidly as had been predicted. There was significant industrial participation in the semi-academic CCDS, but this market was being directly serviced by NASA's block bookings. There were too many such payloads for the middeck, but insufficient to justify a Spacehab every six months. Columbia's turnaround from STS-62 could not be accelerated, so STS-65 could not be advanced. It launched on time on 8 July. This record 15-day flight of the second IML Spacelab tested a new level of interactivity between the apparatus on the orbiter and the scientists who had built it. This 'telescience' method of working was to be used on the International Space Station. Upon landing, Columbia was returned to Rockwell for a refit. Meanwhile, NASA had swapped orbiter/mission assignments. Atlantis (just out of refit) became STS-66 ATLAS. Endeavour took STS-68 SRL-2, and was advanced to 18 August. However, this launch was aborted at T–2 seconds because one of the oxygen turbopumps overheated. Instead of waiting for these engines to be refurbished, the engines in Atlantis were commandeered, and, while Endeavour was being prepared, Discovery was launched on time, on 9 September.

STS-64 was a busy mission. In addition to deploying the SPARTAN-201 free-flyer for the second time, it included a spacewalk to test a new manoeuvring unit and the demonstration of a new instrument designed to monitor the atmosphere. The Lidar-In-space Technology Experiment (LITE), which was carried on a pallet in the bay, was a laser with a boresighted telescope. The pulsed laser illuminated a *column* of the atmosphere and a matched detector in the telescope noted the way that the light was scattered. By slewing the laser back and forth across the ground track, it was possible to make a fine-resolution survey of the three-dimensional distribution of particulate matter (aerosols) in the atmosphere. Although lidars were standard equipment on meteorological research aircraft, this was the first time such an instrument had been flown in space. A faulty tape unit interfered with the trial, but 50 hours of data were collected, much of it in concert with aircraft whose data were to enable the space sensor to be calibrated. Once refined, such an instrument may be installed on an automated satellite as part of the MTPE programme. To determine the degree to which the orbiter's thrusters would impinge upon Mir in the forthcoming rendezvous, the RMS retrieved the 10-metre SPIFEX boom and used the sensors at its end to sample the plumes directly. The visual highlight of Discovery's mission, however, was the spacewalk by Mark Lee and Carl Meade to test the backpack-augmentation thruster unit developed for astronauts on the International Space Station. This was mounted on the base of the life-support backpack. Its nitrogen-gas jets would enable an astronaut with a loose tether to recover to the station's structure.

Discovery was ready to descend, but Florida weather forced first a day's extension and then diversion to Edwards. Ten days later, on 30 September, Endeavour lifted off on time for the second flight of the SRL. Its slippage from August to October meant that Endeavour would not be able to be turned around in time for the STS-67 mission that it had inherited from Columbia as part of the remanifesting earlier in the year, and which had been set for January. So this ASTRO reflight was postponed to March 1995. As it happened, this left a timely gap in which to reschedule the long-delayed STS-63. But the immediate priority was to fly STS-66. It had taken longer than expected to rustle up new SSMEs for Atlantis, so it was delayed a week to 3 November, on which date it was

Table 2.17. Shuttle manifest, *c.* late 1994

Flight	Date	Orbiter	Mission
STS-63	Jan 1995	Discovery	Spacehab 3; SPARTAN-201-2
STS-67	Mar 1995	Endeavour	ASTRO-2
STS-71	May 1995	Atlantis	Mir 1
STS-70	Jun 1995	Discovery	TDRS-G
STS-69	Jul 1995	Endeavour	WSF-2; SPARTAN-201-3
STS-73	Sep 1995	Columbia	USML-2
STS-74	Oct 1995	Atlantis	Mir 2
STS-72	Nov 1995	Endeavour	SPARTAN-206; SFU retrieval

launched without difficulty. Both of its payloads – the ATLAS pallet and the CRISTA-SPAS free-flyer – monitored the atmosphere, in this case just before the onset of the northern winter. Atlantis became the fourth flight of the year to be diverted to California due to inclement Florida weather.

The various science programmes were now unfolding nicely, and flights were receiving less and less publicity; orbital operations had become routine. But political events spawned an entirely new programme, one which would repeatedly raise the visibility of the shuttle to the lead item on the nightly news shows. America was not the only space-faring nation.

VISITING RIGHTS

For STS-63, Discovery was still to have the Spacehab and SPARTAN-201-2 payloads, but it had acquired an extra crewman and an important new objective: cosmonaut Vladimir Titov was to supervise a rendezvous with the Mir space station. The greater the interval between the time that Mir's orbital plane intersected the launch site, and the moment that the shuttle took off, the more complex would be the rendezvous. The window was particularly narrow due to the Russian requirement that the final part of the rendezvous take place within range of the Russian communications network. In the event that the shuttle could *not* achieve the desired initial orbit, the rendezvous with Mir would be cancelled because, its historic nature notwithstanding, it was only the *secondary* objective and its achievement could not be at the expense of the deployment and retrieval of the SPARTAN free-flyer. In fact, after a day's delay to replace a faulty IMU, Discovery lifted off at the optimum moment on 3 February. It was flown by Jim Wetherbee and Eileen Collins, the first female shuttle pilot (and later to be the first female shuttle commander).

Despite concern over leaking RCS thrusters which might contaminate the instruments on Mir's surface, the rendezvous three days later was flawless and the orbiter drew to a halt just ten metres from the docking port at the end of Mir's Kristall module. For ten minutes the two 100-tonne spacecraft maintained formation, then Discovery pulled back 100 metres to perform a photographic fly-around prior to withdrawing to conduct its primary mission.

Lacking commercial payloads, Spacehab Incorporated had renegotiated its contract with NASA to enable it to use all of the lockers on the next two modules instead of half of the lockers over the next four flights. One of the experiments flown this time was a robot called Charlotte. It had been built to tend to other experiments.

Towards the end of this flight, in a spacewalk to assess procedures for the International Space Station, Bernard Harris and Michael Foale hid in the orbiter's shadow to test how cold they became, and when Foale reported that it felt as if his fingers were in an icebox the test was curtailed. Clearly, the spacesuit would need some modification.

The repeatedly delayed STS-67 was finally launched without incident on 2 March. As Endeavour was an EDO-capable orbiter, it was able to undertake the full-length ASTRO-2 mission that had originally been assigned to Columbia. However, this was an ordeal for the crew, which once again included Sam Durrance and Ron Parise, operating around the clock in the confines of the cabin. Indeed, by being extended a day, Endeavour set a new shuttle endurance record.

Fig. 2.48. Eileen Collins, the first female shuttle pilot, works through a checklist during Discovery's rendezvous with the Mir space station.

On 14 March, as the ASTRO-2 mission drew to a close, astronaut Norman Thagard took off on a Soyuz rocket from the Baikonour Cosmodrome in Kazakhstan, and, two days later, boarded the Mir space station. This flight fulfilled the other side of the agreement signed with the Russians in 1992, in which, in return for flying a cosmonaut aboard the shuttle, an astronaut would be allowed to spend a tour aboard Mir. With the subsequent decision to merge the two national programmes, it had been decided to conduct the next crew handover as part of the Atlantis docking mission. Although this STS-71 flight dominated the summer schedule, the date could not be set until the Spektr module had docked with the Mir complex, and the launch of that module, which was to deliver most of Thagard's scientific apparatus, had been repeatedly delayed. As STS-71 slipped, so did the next mission in line, STS-70, but in April, when it became clear that Mir would not be ready to receive Atlantis until early July, NASA decided to launch Discovery on 8 June, as originally scheduled, but for a truncated five-day mission. If STS-70 ran into substantial delays, it would be pushed back to August so that STS-69 could fly in late July, after Atlantis. In effect, the rest of the year's manifest had, by virtue of the need to turn the orbiters around for their future assignments well into the following year, been made secondary to the first Mir visit. On 2 June, STS-70 had to be rolled back into the VAB so that the insulation on its ET could be refurbished to fill in holes in the foam made by woodpeckers trying to make nests. The mission to symbolically end the Cold War was thus assigned to NASA's 100th crew.

The count on 23 June had to be abandoned early due to a nearby storm. The next day, it was scrubbed when the hold at T–9 minutes, which was extended awaiting an improvement in the weather, encroached on the 10-minute window. On 27 June, however, Atlantis lifted off precisely on time. It was a significant occasion; for once, a shuttle had had somewhere to go. Two days later, mission commander Hoot Gibson eased his vehicle into the Kristall module's docking ring. When the hatches were opened an hour or so later, he and Vladimir Dezhurov, the Mir commander, shook hands to formalise the union. In the control centre at Kaliningrad, Dan Goldin, the NASA Administrator, and Yuri Koptev, the head of the RSA, did likewise. When the crews congregated in Mir's base block, it became evident that the station's designers had never anticipated that it might one day host *ten* people. For Bonnie Dunbar, this was a bitter-sweet mission. As Thagard's backup, Dunbar had initially hoped to swap places with him and serve on with Anatoli Solovyov and Nikolai Budarin, the new residents who had flown up on Atlantis as passengers, but Mir's Soyuz escape capsule was limited to three people, and ESA had booked a tour of duty aboard the complex for Thomas Reiter. In addition to the ceremony of the event, there were scientific data to be gleaned from subjecting the departing Mir crew to a full battery of biomedical tests with apparatus such as was carried on the SLS missions. This work, supervised by Thagard and Ellen Baker – both of them physicians – was done in the Spacelab carried in Atlantis' bay, and these data vastly enhanced NASA's database on the adaptation of the body to the space environment. On 4 July, after a miscellany of cargo had been transferred, Atlantis undocked.

So that Thagard, Dezhurov and Strekalov would not have to sit upright for the return to Earth, they used reclined couches installed on the middeck. Upon landing on 7 July, Thagard staggered off the orbiter, but his colleagues obeyed instructions and allowed themselves to be carried off recumbent so that they would not jeopardise the follow-on biomedical tests.

Despite having finally postponed STS-70, it was not, as had been announced, put back after STS-69; instead STS-69 was slipped so that STS-70, which by then was ready, could be launched in record time after Atlantis' return, and when Discovery lifted off on 13 July this was achieved. After the new TDRS satellite was released, this mission was continued to its originally planned length so that a variety of middeck experiments could be conducted.

At this point, an alarming discovery was made. Erosion of the inner lining of the SRBs was not uncommon, but post-flight inspection of STS-71's boosters found the most severe O-ring scorching to date. The nozzle-joint of an STS-60 SRB had also been damaged, so it was decided to augment the joints of the already-mated boosters, which meant that STS-69 missed its 5 August count. A fuel cell that would not start properly prompted a T–8 hours scrub on 31 August, but Endeavour was launched on 7 September without difficulty. In the bay were *two* free-flyers – the SPARTAN-201 and the WSF. On this occasion, the WSF was able to be released, but it suffered attitude-control problems.

The spent SRBs from STS-69 were inspected as a matter of urgency, to clear the way for Columbia, whose boosters had also had their nozzle-joint O-ring insulation augmented. One of the three SSMEs used by Discovery had tested a higher performance oxygen pump, and because this had performed well, Columbia had been fitted with two of these upgraded motors. Once this new oxidiser pump was being flown as standard, effort would switch to improving the hydrogen pump, in order to increase the capacity of the shuttle to carry cargo to the 51-degree inclination orbit in which the International Space

Station was to be assembled. In the event, STS-73's count on 28 September had to be abandoned when the hydrogen pump on one of these new engines sprung a leak. Starting on 5 October, three successive counts were scrubbed due to a storm, a hydraulics fault and an engine fault. After the weather pre-empted the 15 October count, the shuttle was obliged to yield ETR priority to an Atlas 2 rocket, but it was finally launched on 20 October. By relying on telescience, the USML-2 operations served as a high-fidelity rehearsal for the way in which science would be undertaken on the International Space Station. Just before the deorbit burn, a small piece of orbital debris struck one of the recently closed payload bay doors.

Upon Columbia's safe return on 5 November, attention switched to Atlantis. When the year's manifest had been drawn up, it had assigned STS-74's launch to 26 October, but the summer hiatus had prompted the *compression* of the post-STS-71 schedule, with missions being flown back to back. The progressive delays in dispatching Columbia had eaten up the slack and then begun to push back Atlantis. Having demonstrated that it could launch just a week after a landing, NASA assigned STS-74 to the first available window. Unacceptable weather at the TAL site pre-empted the 11 November attempt, but Atlantis lifted off at the optimum point in the all-too-brief Mir window the following day. STS-74's mission was notable for the fact that this was the first time that a shuttle was to add a module to a station. In keeping with the refreshing new spirit of international cooperation, this five-metre long airlock had been constructed by the Russians. Mounting the Docking Module (DM) on Kristall would, in future, guarantee the orbiter clearance from Mir's solar panels, thereby eliminating the need to swing Kristall onto the complex's axis, as had been done for the STS-71 docking, because repeatedly relocating Kristall would interfere with Mir's ongoing work. Although an assortment of cargo was transferred in both directions, Atlantis carried no astronaut to be left aboard the station, because ESA's Thomas Reiter was serving his tour of duty. Atlantis' landing on 20 November brought flight operations for the year to a conclusion. Endeavour was still being turned around after its much-delayed STS-69, so it was not possible to launch it as STS-72 in December, and it slipped into the new year.

Table 2.18. Shuttle manifest, *c.* late 1995

Flight	Date	Orbiter	Mission
STS-72	Jan 1996	Endeavour	SPARTAN-206/OAST; SFU retrieval
STS-75	Feb 1996	Columbia	USMP-3; TSS-2
STS-76	Mar 1996	Atlantis	Mir 3
STS-77	May 1996	Endeavour	Spacehab 4; SPARTAN-207/IAE
STS-78	Jun 1996	Columbia	LMS-1
STS-79	Aug 1996	Atlantis	Mir 4
STS-80	Nov 1996	Columbia	WSF-3; ORFEUS–SPAS-2
STS-81	Dec 1996	Atlantis	Mir 5

Its customary luck having returned, Endeavour lifted off for STS-72 on 11 January in the centre of the window for the rendezvous with the Space Flyer Unit (SFU). Japan had launched this satellite on its new H-2 rocket the previous March, on the understanding that NASA would retrieve it, and Koichi Wakata was flying to snatch it with the RMS. The retrieval was marred only by the failure of the two long solar panels to retract, so they were jettisoned and then the 3-tonne package of experiments was captured and placed on a cradle in the bay. NASA was also considering the retrieval of an errant Chinese satellite. The next day, the SPARTAN was deployed, and while this was performing its experiment programme for NASA's Office of Aeronautics and Space Technology (OAST), two more spacewalks were conducted to rehearse procedures intended for the International Space Station.

A month later, Columbia launched on time as STS-75. Its primary task was to rerun the TSS experiment, and on this occasion, although the tether unreeled smoothly, it broke just before it reached its 20-km limit, and the satellite was drawn away by differential gravity. Nevertheless, just before this failure the act of trailing the conducting tether in the Earth's

Fig. 2.49. Italy's Tethered Satellite System in the process of being unreeled. The purpose of the experiment was three-fold: to demonstrate the deployment system; to observe the dynamics of a long tether; and to assess the prospects for generating electrical power by trailing a 20 km wire through the Earth's magnetic field. Unfortunately, the deployment system malfunctioned.

magnetic field had generated a low-power current with a potential difference of 3,500 volts, satisfying the scientific objective of the test. The rest of this cabin-confined EDO mission was devoted to the USMP experiments.

THE MILK RUN

Atlantis' count for STS-76 was slipped a day to 22 March, due to excessive winds, but on that date it launched on time. NASA then entered a new phase in its operations. Over the next two years, a succession of astronauts were to serve tours aboard Mir, with Atlantis routinely running handover and resupply missions. Atlantis was the only orbiter in the fleet equipped to carry the docking unit. The Shuttle–Mir flights had become the main events on NASA's manifest, and dispatching Atlantis on schedule was to take first priority. What is more, dedicating an orbiter to Mir for so long meant that all the payloads which it would otherwise have carried had either to be deleted or be reassigned to other orbiters, because by then it was clear that the flight rate could not be increased beyond the eight per year that was achieved during a good year.

Rather than continue to fly a Spacelab – as had been done on the first docking in order to use its medical instruments – NASA had negotiated a contract with Spacehab Incorporated in which the company would bolt together two of its modules to form a bulk cargo hauler. As this was not yet ready, for STS-76 Atlantis carried a standard module in its bay. Although, previously, the Spacehab had been carried forward, so as not to prevent the rear of the bay being assigned to free-flyers, the fact that the ODS had to be up-front meant that the squat module had to be set further back. But placing Spacehab nearer the orbiter's centre of mass enabled it to carry a heavier payload than would otherwise have been feasible.

Rich Clifford and Linda Godwin spacewalked to remove superfluous apparatus from the DM, and to install a package of experiments. They wore the SAFER backpacks, in case their tethers broke. When the time came for Atlantis to depart, Shannon Lucid remained on the Mir side of the hatch. In an emergency, she would return to Earth with Yuri Onufrienko and Yuri Usachyov in the Soyuz lifeboat, but if everything went as planned she would be retrieved by Atlantis when it returned four months later.

Endeavour's count for STS-77 was on schedule. In addition to the final Spacehab with NASA-funded lockers, the bay contained a SPARTAN free-flyer with the innovative inflatable antenna experiment. It comprised a 15-metre diameter dish and a 28-metre tripod, on which to mount a transponder. The deployment was considerably more *dynamic* than had been expected, and the 1.2-tonne SPARTAN carrier tossed and turned as the antenna inflated, and it was still pitching end over end when the entire structure snapped erect. It was a technology test for the practicality of inflatable structures. The antenna was subsequently jettisoned so that the carrier could be retrieved. Upon landing, Endeavour was returned to Rockwell for a refit in which it would be upgraded to fly assembly missions for the International Space Station.

The year's manifest had been unfolding as planned, and Columbia's launch as STS-78 on 20 June was right on time. For the first time, a video camera recorded activity on the flight deck throughout the ascent. This Life and Microgravity Sciences (LMS) Spacelab took advantage of the EDO wafer and was another exercise in round-the-clock science with broad international participation to set the scene for the International Space Station.

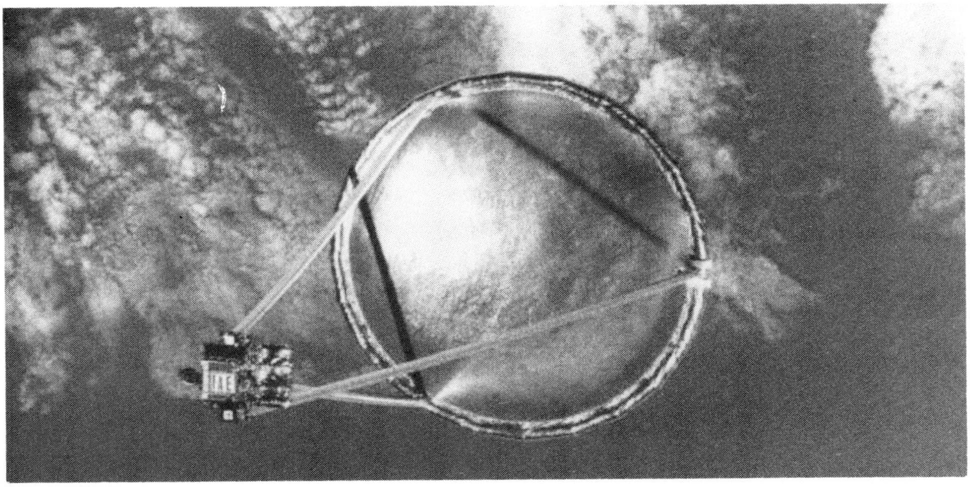

Fig. 2.50. The Inflatable Antenna Experiment (IAE) following its deployment by a SPARTAN carrier. Its objective was to assess the scope for using inflatables to construct large framework structures in space.

Post-flight inspection of STS-78's SRBs found a significant degree of hot-gas erosion. It had not damaged the O-rings, but it represented an unacceptable risk. Tests showed that the new sealant that had recently been introduced was at fault, so it was decided to revert to the older material. The new water-based type had been substituted for the old methyl-based putty which had been banned by the Environmental Protection Agency. STS-79's boosters had already been fabricated, so the only option was to strip them down. In order not to delay Atlantis for longer than necessary, the segments intended for STS-80 were ordered to be mated using the old sealant and reassigned to STS-79. Even so, this pushed the mission back by six weeks to 12 September. On Mir, Shannon Lucid, who was enjoying her tour, took this news in her stride. On 4 September, Atlantis had to be rolled back into the VAB to avoid a hurricane, and this delayed the launch to 16 September, but the launch on that date was faultless.

Having spent six months aboard Mir, which the RSA now regarded as a normal tour of duty for residents, Lucid handed over to John Blaha. This was the first time that NASA had had occasion to perform this ritual. It marked a significant milestone in the programme. The double-sized Spacehab in Atlantis' bay carried the heaviest load of cargo to date, and it took several days to perform the transfers to and from the station. In addition, after Atlantis had undocked, a variety of experiments were performed in the module, to test apparatus for later use on the International Space Station. Upon landing, Lucid expressed her surprise at how readily she readapted to the inexorable tug of gravity.

Despite having reverted to the original sealant, it turned out that STS-79's boosters had suffered significant erosion. In one case, the damage took the form of a series of 6-cm deep grooves in the carbon cloth liner 'downstream' of small depressions carved in the throat of the nozzle. Although in no case had this erosion reached the nozzle's casing, if the 3,100 °C gas had come into contact with the metal the resulting breach could all too easily have led to a catastrophic failure. Grooving had been observed earlier, but not on this scale. The effort to catch up on the schedule by dispatching STS-80 on 31 October was made contingent on a review of the recent changes to the booster's manufacturing process and, once the fault had been traced, the count was recycled for a 15 November launch. Weather and competing calls on the ETR put STS-80 back to 19 November. On that date, after the count was held for a few minutes in order to allow a potential hydrogen leak to be investigated, and found to be spurious, Columbia was finally dispatched. The WSF and the ORFEUS-SPAS free-flyers were deployed and retrieved, but a series of spacewalks to test tools for the International Space Station had to be abandoned when it was found that the outer hatch of the airlock would not open because a screw had come loose and jammed the ratchet. The rest of the mission passed off smoothly, until it was time to return; after two days of wave-offs for weather, Columbia was ordered back to Edwards. At 18 days, therefore, this flight stretched the EDO facility to its absolute limit. Apart from anything else, STS-80 was notable because it was veteran astronaut Story Musgrave's final shuttle mission; his sixth.

With Atlantis committed to Shuttle–Mir, its schedule revolved around visits to Mir every four or five months, so the delay in launching STS-79 was propagated to STS-81, pushing it into early 1997. It lifted off precisely on time, on 12 January. The sceptics who had said that the shuttle would not be able to meet the tight launch windows for Mir had so far been proven unduly pessimistic. The Shuttle–Mir programme had settled into a routine, with its own rituals. Blaha's handover to Jerry Linenger was uneventful.

Table 2.19. Shuttle manifest, *c.* late 1996

Flight	Date	Orbiter	Mission
STS-81	Jan 1997	Atlantis	Mir 5
STS-82	Feb 1997	Discovery	HST service
STS-83	Apr 1997	Columbia	MSL-1
STS-84	May 1997	Atlantis	Mir 6
STS-85	Jul 1997	Discovery	CRISTA-SPAS-2
STS-86	Sep 1997	Atlantis	Mir 7
STS-87	Oct 1997	Columbia	USMP-4; SPARTAN-201-4
STS-88	Dec 1997	Endeavour	ISSAF-1

There was nothing like success for fostering a *can-do* attitude, so when Discovery took off to service the Hubble Space Telescope for the second time it all seemed very routine. The count had actually been advanced two days, to 11 February, in order to plan for minor slippage (because the ETR was booked by various rockets for the week after 14 February) but in the event there were no delays. The only surprise in space was that some of the HST's thermal insulation had cracked and peeled, so an extra spacewalk was made to stick on makeshift protection over the most damaged sections. The landing at KSC was in darkness, so newly installed halogen lamps on the centreline of the SLF were illuminated to help Ken Bowersox line up for the final approach. One day, using GPS navigation, orbiters should be able to fly *blind* landings.

After lifting off a day late, on 4 April, STS-83 rapidly deteriorated as a fuel cell that had never really settled down finally had to be switched off. Columbia was recalled only four days into its planned 16-day flight. The Materials Sciences Laboratory (MSL) was to have been the last in a series of eight Spacelabs, and was to have undertaken the most extensive investigation to date of combustion in space. This basic research was considered to be so important that, for the first time, NASA decided to turnaround an orbiter and refly it with the same payload and the same crew. In the event, there was a convenient gap in the manifest in the summer, caused by the postponement of the initial assembly flight of the International Space Station, so Columbia was refurbished (a process that was eased by the fact that the payload did not have to be reintegrated), redesignated STS-94 and slipped into the schedule behind the next Mir mission. STS-84 Atlantis was launched on time on 15 May, retrieved Jerry Linenger, and dropped off Michael Foale. Columbia's count on 1 July was uneventful, and the 16-day MSL programme was completed.

Slowly but surely, NASA was readopting dynamic scheduling as its standard operating procedure. This was just as well, because in April it had to postpone STS-88 Endeavour by eight months, to mid-1998, because the Russians were behind schedule in constructing the Service Module that was to form the International Space Station's habitat. The manifest for the rest of the year, and on into 1998, was therefore revised.

Table 2.20. Shuttle manifest, *c*. mid-1997

Flight	Date	Orbiter	Mission
STS-85	Aug 1997	Discovery	CRISTA-SPAS-1
STS-86	Sep 1997	Atlantis	Mir 7
STS-87	Nov 1997	Columbia	USMP-4; SPARTAN-201-4
STS-89	Jan 1998	Discovery	Mir 8
STS-90	Apr 1998	Columbia	Neurolab
STS-91	May 1998	Discovery	Mir 9
STS-88	Jul 1998	Endeavour	ISSAF-1

STS-85 Discovery lifted off on 7 August, as planned. It was an international mission. In addition to the US/German CRISTA-SPAS free-flyer, which was left to monitor solar-terrestrial relationships for a week, the payload bay included a Hitchhiker Bridge, GAS canisters, and an MPESS carrying a prototype of the robotic arm that is to be used to service experiments on the outside of the Japanese laboratory of the International Space Station. About the only unplanned moment of the mission occurred when the PAM motor which had failed to boost Westar-6 into geostationary transfer orbit, so many years ago, made a close pass by the free-flyer, but there was no danger of collision.

The next mission to Mir became a cliff-hanger even before it left the ground. At issue was whether Mir had become too dangerous to risk another four-month tour of duty. NASA had weathered the storm during Linenger's tour without too much criticism, but the recent collision while a discarded Progress cargo ferry was being manoeuvred around the

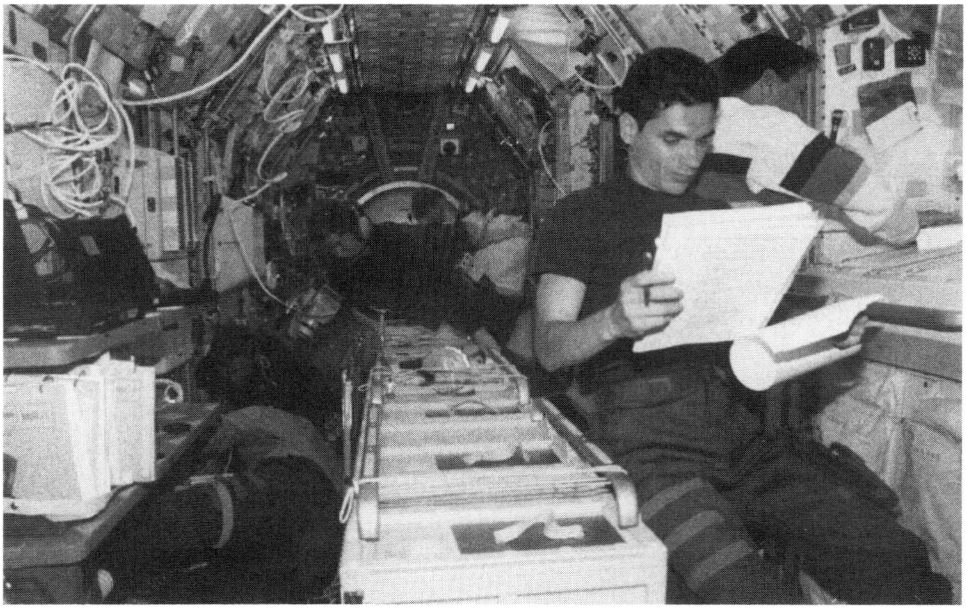

Fig. 2.51. Almost the entire crew is seen at work in this Spacelab set up as the Materials Science Laboratory (MSL).

station, and the crisis that this had precipitated, had led some members of Congress to call for the joint programme to be terminated, arguing that it had already served its purpose. Dan Goldin waited until a few hours before launch before he announced his decision: NASA would honour its commitments. In fact, a change had already been made. Wendy Lawrence was to have replaced Foale, but when it was realised that she was too short to fit the Russian EVA suit, it had been decided to send up David Wolf instead. He had been training for the next mission, and so would be able to help out in an emergency. Lawrence would still fly, because of her detailed knowledge of the cargo transfer procedures, but she would remain aboard the shuttle when it undocked. Once again, Atlantis blasted off at precisely the optimum moment in the narrow launch window for a rendezvous with what was now effectively an international, rather than a Russian space station.

The STS-86 crew included Vladimir Titov and Jean-Loup Chrétien, the Frenchman who had already visited two of the Salyut stations. Scott Parazynski made a spacewalk with Titov to retrieve experiments which had been left on the DM by STS-76, and then tested apparatus to be used on the International Space Station.

After landing, Atlantis was flown to Rockwell for a well-earned refit. The final two flights of the Shuttle–Mir programme, which had been added to the initial seven-flight sequence so as to help the Russians in extending Mir's utility until assembly of the International Space Station was well underway, were to be flown by Discovery, but with the start of the ISS delayed, the newly refurbished Endeavour was grounded without a mission, so it was decided that it should fly the first of these missions.

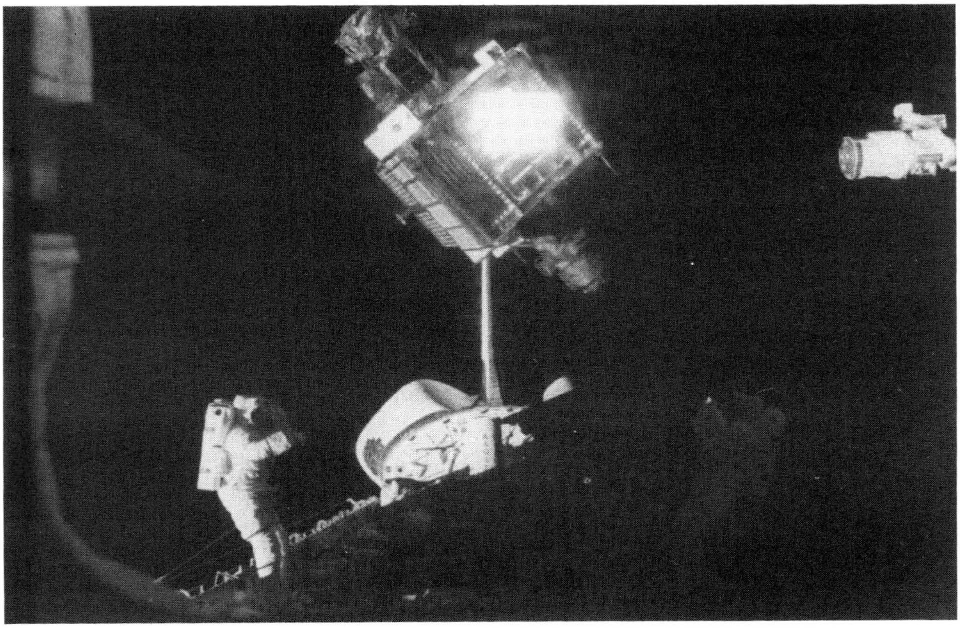

Fig. 2.52. Vladimir Titov and Scott Parazynski spacewalked during STS-86, Atlantis' final visit to Mir, to retrieve the Mir Environmental Effects Package (MEEP) which had been set up on the Docking Module (DM).

The run of luck came to an end on STS-87. After the SPARTAN-201 free-flyer was released, it failed to start its sequencer and, because it did not achieve attitude-hold, it did not cancel out the slight spin that it acquired when the RMS was withdrawn; it simply drifted in space. This put NASA to the test. It had rescued errant satellites before, but it had always done so by training a subsequent crew to perform the operation with specially-designed tools. In this case, there was no convenient way of tacking a recovery onto a future mission because once assembly of the International Space Station had begun, shuttles would fly in higher-inclination orbits. This satellite would have to be recovered now. As it happened, Winston Scott and Takao Doi had planned to make a spacewalk to test procedures for the International Space Station, so they set themselves up on either side of

Table 2.21. Shuttle manifest, *c.* end-1997

Flight	Date	Orbiter	Mission
STS-89	Jan 1998	Endeavour	Mir 8
STS-90	Apr 1998	Columbia	Neurolab
STS-91	May 1998	Discovery	Mir 9
STS-88	Jul 1998	Endeavour	ISSAF-1

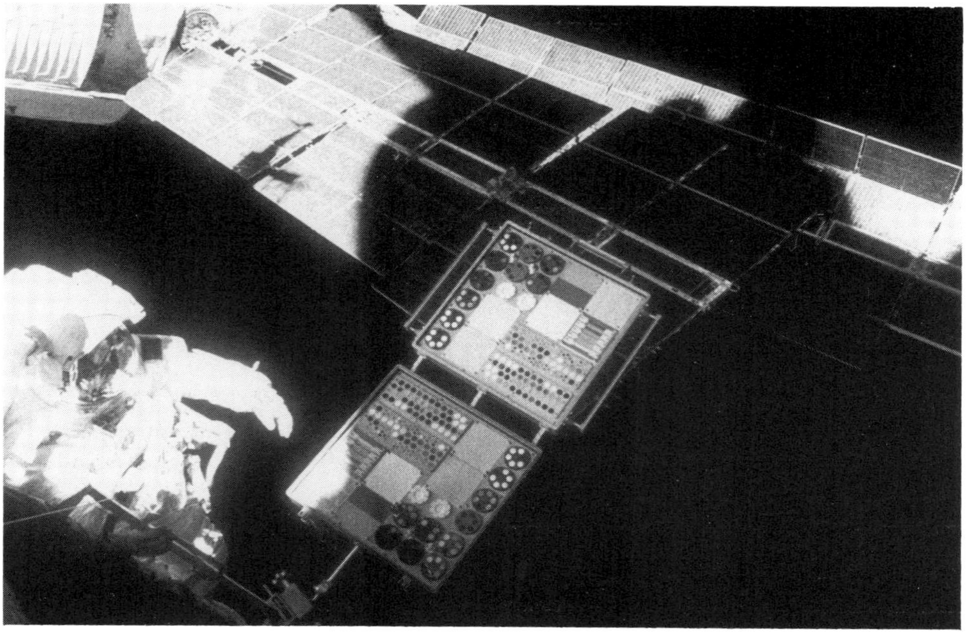

Fig. 2.53. Winston Scott and Takao Doi stand in Portable Foot Restraints (PFR) mounted on either side of the payload bay (one in shadow in this view), poised to take hold of the drifting Spartan 201 free-flyer.

the bay and waited while Kevin Kregel manoeuvred Columbia to place the satellite between them, at which time, using the 'universal tool', the human hand, they grabbed it, and then gently lowered it into its cradle. The casual banter that they kept up throughout, made it all seem so straightforward.

The penultimate visit to Mir went smoothly, and Wolf, who had served a refreshingly quiet tour, was replaced by Andy Thomas for the final phase of the two-year continuous American presence aboard the station. A month later, if all goes well, Endeavour will start the process of assembling the International Space Station. For the next five years, Atlantis, Discovery and Endeavour will be dedicated to this effort. Only Columbia will carry out independent missions, such as the deployment of the third of the Great Observatories, the Advanced X-ray Astrophysics Facility (AXAF). Perhaps the only certainty concerning the station assembly manifest is that its timeline *will* soon be disrupted. But as this history has hopefully shown, NASA will cope, and fulfil this ambitious orbital construction contract.

3

Communications satellites

It was Arthur C. Clarke who first pointed out that a satellite orbiting 36,000 km above the equator would remain at a fixed point in the sky, due to the fact that the period of its orbit would be 24 hours. Clarke, a fervent advocate of spaceflight, published his analysis of the geostationary orbit (GSO) in *Journal of the British Interplanetary Society* in May 1945. Later that year, in the October issue of *Wireless World*, he proposed that if three satellites were stationed 120 degrees apart they would be able to relay radio signals between any two points on the planet.

A space-based communications system was beyond the technological state of the art of the late 1940s, of course, not just in terms of rocketry but also in electronics; a decade later, and even after the first satellite had been launched, the development of the communications technology still represented a considerable challenge.

The first commercial communications satellite, launched in 1962, was Telstar. This was built by American Telephone & Telegraph (AT&T) as a private venture. Although it carried the first live TV across the Atlantic, its low orbit restricted its use as a long-distance relay to brief periods. AT&T intended to launch a constellation of Telstars, and so capture the entire market for itself, but the US Government passed the Communications Satellite Act and then ordered that a national consortium be established to develop and operate a system with the capacity to serve the domestic telecommunications industry. The Communications Satellite Corporation (Comsat) was created in February 1963. Rather than pursue the low-orbiting Telstar, it decided to use geostationary relays because these would provide *continuity* of service. A geostationary satellite represented a considerable technical challenge, but NASA was in the last stage of developing Syncom as a technology demonstrator. This was placed over the International Date Line in 1964, just in time to relay live coverage of the Olympic Games in Japan to a fascinated US audience, and the banner 'Live By Satellite' immediately became the defining symbol of that time.

The International Telecommunications Satellite Consortium (Intelsat) was established in August 1964 to deploy a network of geostationary relays to provide telecommunications on a worldwide basis. It went to Hughes, which had built the Syncom, and asked it to modify its design. When the Early Bird was stationed over the Atlantic in 1965, it became the first commercial geostationary relay. Of the 11 founder members, the largest was Comsat. Over the years, as the benefits became manifest, Intelsat membership soared ten-fold. Each member built its own ground stations, and contributed funding for satellite development in proportion to its take from the revenues. Intelsat soon had the majority of the international telephone traffic and almost all the transoceanic TV-relay business. The exponential growth in demand ensured that it made financial sense to continuously develop new generations of ever more capable satellites.

Having supplied Syncom, Early Bird and the Comstars with which Comsat served the domestic market, Hughes established a clear lead in the satellite-building business. But to *maintain* its lead it had to exploit every advance in microminiaturised electronics to build ever-better satellites. In the early 1970s it came up with a lightweight, compact bus that was capable of providing a high-capacity service. The 300-kg HS-333 was able to relay either a colour TV channel or 1,000 voice channels. It was a spin-stabilised drum with a conformal array of solar cells and a despun antenna on top. The electronics shaping this beam to cover the customer's target area, and *only* that area, was the key. Shaping the beam delivered all the radiated power directly to the customer, rather than distributing it over a wider area, and it allowed neighbouring customers to radiate on the same frequencies without incurring any interference. Frequencies in the C-Band were rationed to avoid interference, so this ability to double-up was a significant marketing advantage for Hughes.

The Telesat consortium was formed in 1971 by the Canadian government in partnership with the domestic telecommunications companies to develop an integrated national service. It was responsible for voice and data communications across the country, and it was the main broadcast service for the widely separated communities in the vast northern territories. In deciding to operate its own communications satellites, Telesat rode the crest of the wave made possible by the introduction of the HS-333 configurable-footprint system.

Telesat had a requirement for only two satellites, but it ordered another as a ready spare in case one was lost. The system was called 'Anik', which means 'brother' in Inuit, the language of the Eskimo of northern Canada. Since the satellites were to be replaced by a more advanced variant at some point, the first generation was marked by the letter 'A'. Anik-A1 was launched in November 1972 on a Delta rocket, and A2 followed in 1973. Having proved themselves, they attracted business, and to satisfy this increased demand, the spare was added in 1975 as A3. Although Telesat hoped that the HS-333 satellites would survive for the advertised seven years, it knew that it would have to start launching replacements by the early 1980s, so it set out to explore the limits of the technology, to set the requirements for its follow-on system. The communications unit for Project SCORE in December 1958 – in which the Department of Defense had an Atlas rocket place itself in low orbit for a pioneering experiment in space communications – had been built by the Radio Corporation of America (RCA). This replayed a tape-recorded message by President Eisenhower. In the mid-1970s, Telesat asked RCA to design an experimental communications satellite to carry both C-Band and K_u-Band transponders. Launched in 1978, Anik-B1 was the first satellite to try broadcasting TV directly to customers. It was not DBS as it is today, however; the antennas were two metres wide, so were convenient only for hotel complexes and similar placements. The experiment was so successful that Telesat decided to set up *two* networks of satellites in parallel: one to continue the C-Band facilities of its then fully-operational system, and the other to exploit the increased capacity of the K_u-Band. For this Anik-C, Telesat ordered the latest satellite produced by Hughes, the HS-376, which was able to accommodate customer requirements. And to develop a domestic manufacturing capability, Spar Aerospace bought a license to build an HS-376 clone as the Anik-D model, to sustain the C-Band service well into the 1980s.

Satellite Business Systems (SBS) was set up in 1975. The majority of the shares in this consortium were owned by Comsat and IBM. Comsat leased transponders on its Comstar satellites to the telecommunications industry. By the mid-1970s, its big clients were AT&T and General Telephone & Electronics (GTE), which ran a lucrative business relaying voice for a diverse community of users. SBS was established to sell communications services *directly* to end-user businesses with widely distributed facilities on the mainland. It offered an all-digital service integrating voice, data, video and E-mail. Boeing, General Motors and Westinghouse were early customers. Spare capacity was leased to the telecommunications companies as before, for voice traffic. To implement this, SBS ordered a fleet of HS-376s fitted with K_u-Band transponders.

As the National Space Transportation System, the shuttle's introduction was extremely timely from the viewpoint of commercial satellites because the market for such satellites had just opened up. For the first time, it was financially justifiable for individual corporations to buy communications satellites for private use, instead of buying services indirectly through Comsat's satellites. In addition, members of Intelsat were buying satellites to provide local telecommunications services. All of these satellites would need to be launched. On the basis of production orders, it looked as if the most popular type would be the Hughes HS-376. At 4.5 metres wide and 18 metres long, the shuttle's bay was long enough to accommodate five such satellites, and because it had only to ferry the satellites into low orbit and deploy them, using the same support equipment repeatedly, its turnaround time and operating costs would be minimised. Launching commercial satellites was evidently going to be a lucrative operation.

Although the shuttle was given a monopoly in the domestic market, NASA hoped to fly it on a weekly basis. To ensure that the shuttle had sufficient work, NASA also set out to attract foreign business. Long before the first shuttle test flight, several governments had taken up a special introductory deal whereby a number of satellites could be launched for a bargain-rate fixed fee during the first five years of shuttle operations. The shuttle would not be without competition in the international market, however. The European Space Agency (ESA) was developing the Ariane rocket specifically to satisfy this need for launchers, and this too was to be operated as a business, in competition with the shuttle.

SBS was seduced by NASA's package, and it booked all of its satellites on the shuttle. However, it could not afford to wait-out the shuttle's ongoing development delays, so the first two satellites were offloaded and launched on Delta rockets. SBS-1, launched in 1980, became the first HS-376 to enter service. Telesat also booked all of its new satellites on the shuttle. However, although its HS-333 satellites ran for longer than Hughes had predicted, Telesat was obliged to transfer Anik-D1 to a Delta so as to prevent its C-Band system from degrading. The shuttle's entry to service, therefore, was eagerly awaited by the new service providers of the telecommunications industry.

THE ACES

After four test flights the shuttle was declared operational, so when Columbia lifted off as STS-5 in November 1982 it carried two HS-376s in its payload bay. They were mounted vertically in separate cradles towards the rear of the bay. Each satellite was a drum coated with solar cells. On station it would be spinning for stability, and this would smooth out the thermal stresses. But sitting in the bay, it would be baked on one side and frozen on the other, so, as soon as the bay doors swung open, clamshell covers were closed over the satellites to protect them, and heaters kept them warm. The covers were not opened until just before the satellites were to be deployed. Columbia faced its payload bay forward and tilted its right wing towards the ground, so that as a satellite emerged it would be aligned with its velocity vector, as it would have been if it had been delivered by an expendable rocket. The final stage of the preparation was to stabilise the satellite by spinning it on a turntable in the cradle. Eight hours into the mission, SBS-3 became the first satellite to be released from the shuttle. During the deployment, the orbiter placed its K_u-Band TDRS link on standby so as to preclude the possibility of its beam damaging the departing satellite's electronics. The release was videotaped and relayed on the downlink immediately afterwards. It showed the spinner being spring-ejected from its cradle, pass the shuttle's tail, and slowly drift towards the horizon. It had been a long time coming, but the National Space Transportation System was finally in the commercial satellite business.

The satellite had a perigee-kick motor affixed to its base. McDonnell Douglas had made its Payload-Assist Module (PAM) compatible with both the shuttle and its own Delta in order to ease its customers' transition onto the shuttle. The 2,200-kg solid rocket could accelerate a 1,250-kg satellite into geostationary transfer orbit (GTO). After forty-five minutes, while over the equator, the PAM ignited. It was due to fire for about a minute and a half. Because a solid motor tends to accrete a residue of spent propellant in the throat of its nozzle, which deflects the thrust, the rocket would have gradually diverged from the desired trajectory if it had not been set spinning. The spin induced a cork-screwing motion which wasted energy, but this 'coning' did not seriously jeopardise its short cruise. As soon as the motor shut down, it was jettisoned. Six hours later, upon reaching apogee high over the Eastern Pacific, SBS-3 fired its own thrusters to circularise for GSO. If it slowed down marginally, it would descend and speed up, which would cause it to drift east; if it climbed slightly, it would slow down and drift west. Over the next few days it refined its position and settled into its assigned station at 94° W. The HS-376 was an ingenious design, with a double drum of solar cells, one inside the other. This simple feature meant that the 6.6-metre configuration could be carried in a 3.7-metre enclosure, and extending the outer drum from the base doubled the satellite's power. The second HS-376, called Anik-C3, was dispatched in identical fashion on the second day of the mission, heading for its station at 115° W.

The astronauts celebrated their successful double deployment by displaying a placard to the video downlink; it identified their employer as the 'Ace Moving Company', and cited the corporate slogan, 'We Deliver'.

SATELLITES GALORE

Over the next three years the shuttle released a total of 20 commercial satellites, *all* of which were intended to be boosted up into GSO to serve as communications relays.

Table 3.1. Commercial satellites launched by the shuttle

Satellite	Mission	Manufacturer	Bus	Operator
SBS-3	STS-5	Hughes	HS-376	SBS, USA
Anik-C3	STS-5	Hughes	HS-376	Telesat, Canada
Anik-C2	STS-7	Hughes	HS-376	Telesat, Canada
Palapa-B1	STS-7	Hughes	HS-376	Perumtel, Indonesia
Insat-1B	STS-8	Ford	—	ISRO, India
Westar-6	STS-10	Hughes	HS-376	WU, USA
Palapa-B2	STS-10	Hughes	HS-376	Perumtel, Indonesia
SBS-4	STS-12	Hughes	HS-376	SBS, USA
Telstar-3C	STS-12	Hughes	HS-376	AT&T, USA
Anik-D2	STS-14	Hughes	HS-376	Telesat, Canada
Anik-C1	STS-16	Hughes	HS-376	Telesat, Canada
Morelos-1	STS-18	Hughes	HS-376	SCT, Mexico
Arabsat-1B	STS-18	Aerospatiale	—	ASCO, Arab League
Telstar-3D	STS-18	Hughes	HS-376	AT&T, USA
Aussat-A1	STS-20	Hughes	HS-376	Aussat, Australia
ASC-1	STS-20	RCA	S-3000	ASC, USA
Morelos-2	STS-23	Hughes	HS-376	SCT, Mexico
Aussat-A2	STS-23	Hughes	HS-376	Aussat, Australia
Satcom-K2	STS-23	RCA	S-4000	RCA, USA
Satcom-K1	STS-24	RCA	S-4000	RCA, USA

With so much of the geostationary relay business concentrated on the North American continent, the problem of overcrowding in the arc of the equator covering this region grew progressively worse. The satellites had to be distributed several degrees apart, so that they would not interfere with one another. Interference was more likely at lower frequencies, so the switch to the K_u-Band not only eased the problem by allowing satellites operating in the two bands to be interleaved; it also enabled the satellites to be spaced closer together. It was nevertheless often necessary to rearrange several of the existing population to accommodate newcomers.

All three of Telesat's Anik-Cs and the second of its Anik-D satellites were deployed by the shuttle. All reached their stations, and operated successfully for many years. Although two of the K_u-Band relays provided sufficient capacity to carry its load, Telesat exploited NASA's multiple-satellite deal to have the third sent up early to serve as an *in situ* spare. Anik-C1 took up its station and awaited the command to start relaying traffic. In light of the scramble for launchers after the loss of the Challenger, its early installation proved to have been a wise decision. In fact, Anik-C2 was initially spare capacity too, and until May 1984, when Telesat brought it on line, its transponders were leased to market-test pay-to-view satellite-to-home broadcasting, and in so doing it set the scene for the direct-broadcast systems (DBS) which followed.

The Morelos satellites were operated by Mexico's Secretariat of Communications and Telecommunications (SCT). Some 25 per cent of the 75 million population lived in the immediate environs of Mexico City. The rest had poor communications and no TV. The satellites were to remedy this. Like Telesat, SCT opted for the HS-376. It named them after Jose Maria de Morelos y Pavon, the hero of the revolution, and booked them on the shuttle. The first was dispatched in June 1985, and even though a severe earthquake in September soaked up the funds intended for the construction of ground stations, it was decided to proceed with the launch of the second in November. Morelos-1 proved its value in coordinating the recovery in the aftermath of the earthquake. When the second satellite was finally brought on line, in 1989, Telesat had to move Anik-C3 out of its way, so Anik-C1 was activated and the older satellite put in reserve.

When American Telephone & Telegraph became the first private company to design and operate its own communications satellites, it had called its system Telstar. When it decided to purchase HS-376s in order to operate independently of Comsat, the company revived its famous brand name. Telstar-301 was sent up on a Delta in July 1983, but the network was completed by the shuttle, with two deployments in 1984 and 1985. For the next 10 years, these satellites relayed C-Band for the company's domestic customer base.

The American Satellite Company (ASC) was an early player in offering fully-integrated communications, and it focused on the various agencies of the US government. It had been formed in 1972 by Fairchild Industries, so was able to design its own satellites. In 1974, in order to build a customer base, it leased transponders on Westar satellites, and it later bought a 20 per cent stake in that company. ASC built satellites with C/K_u-Band transponders on an RCA three-axis bus. It booked two on the shuttle, but as soon as ASC-1 was in service in 1985, the company was sold to the Continental Telephone Corporation (Contel). ASC-2 was to have gone up in 1986, but it could not be launched until 1991, at which time it rode a Delta. A few months later, Contel merged with GTE, which operated the Spacenet series which used the same RCA bus, and ASC-2 was renamed Spacenet-4.

In addition to creating Astro Electronics to sell satellites as a business, in 1976 RCA set up RCA Americom to beam TV to hotels, hospitals, schools and residential complexes with two-metre dish antennas. In 1972 the Federal Communications Commission had opened up the market to satellite broadcasting. The cable-TV operators had been eager to expand into this new area because it would enable them to reach a vastly enlarged audience without the expense of laying cable. Satcom-1 was launched in 1975, and by 1982 there were four such satellites. Home Box Office was set up specifically to beam new-release movies to pay-to-view customers via the Satcom satellites. The first generation used C-Band, but because the K_u-Band could be picked up by a much smaller receiver, which would further expand the market to individual households, the follow-on model used the K_u-Band, and four of these new Satcom-K satellites were booked on the shuttle. The Satcom-K was too heavy for the standard perigee-kick motor, so the uprated PAM-D2 was used. Following the earlier PAM failures, insurance premiums had risen dramatically. The rate for the first to use this new motor was excessive, so RCA opted to dispense with insurance – a gamble which paid off. The first two were successfully deployed on the flights immediately prior to the loss of the Challenger. Satcom-K2 was effectively leased to NBC TV News, and Satcom-K3 was leased to Home Box Office. With no prospect of launching the two remaining satellites in the near future, RCA put them up for sale. One of them became a one-of-a-kind Intelsat TV-relay, and the other eventually appeared as Astra 1B. In 1986, RCA sold Astro Electronics to GE Spacecraft Operations, which built satellites as GE Astro Space. As part of this deal, RCA Americom became GE Americom, and continued to run the Satcom satellites until GE Americom launched a new generation of C-Band satellites in the early 1990s.

Comsat sold its holding in Satellite Business Systems in 1984. After SBS-3 had become the first satellite to be deployed by the shuttle, the company had planned to send up two more HS-376s and then start a new series of the more powerful HS-393. Only SBS-4 made it before the shuttle had to be grounded. In 1987, SBS was bought by The Satellite Transponder Leasing Company. The new HS-393 – the largest commercial communications satellite ever built – had been designed to exploit the shuttle's wide payload bay, and had twice the mass and triple the capacity of the HS-376. When SBS launched its first HS-393, in 1990, it was on an Ariane rocket. The shuttle did not encounter one of these giants until some years later, and it was an occasion on which spacewalking astronauts pulled off yet another triumphant repair operation.

The shuttle entered service too late to help Western Union, which was actually the first telecommunications company to use private geostationary satellites to integrate voice, data, video and fax for US customers. Its first Westars were HS-333, two of which were sent up in 1974, with a third following in 1979. Their successors were HS-376. Westars-4 and -5 went up on Delta rockets in 1982. Westar-6, intended to be the final satellite in the series, was put on the shuttle, but its PAM misfired and it was stranded in low orbit. Although the shuttle released the satellite perfectly, its loss dented NASA's image. However, the video of spacewalking astronauts retrieving it a few months later perfectly projected the agency's 'can-do' attitude; for NASA, in these early 'golden years', it seemed as if every cloud had a silver lining. The HS-376 was refurbished and sold to the Asian Satellite Telecommunications Company of Hong Kong. Renamed Asiasat-1, in April 1990 it became the first US-built satellite to be launched on a Chinese Long March rocket. By this point though, Western Union had been bought by Hughes, which had replaced the Westars in the busy American arc, on a one-to-one basis, with its own Galaxy satellites, which were also derived from the HS-376.

This overcrowding was in marked contrast to the situation in the Far East. Australia and Indonesia had no problems finding convenient locations for their satellites. Although both countries were members of Intelsat for international links, neither had operated satellites for local services. As with Canada, the satellites were to overcome geographical adversity, and consortia were formed to undertake the task.

In 1981 the Australian government formed Aussat with Telecom Australia, the domestic telecommunications company. Up to this point, only the five main cities had state-of-the-art local communications, and they were weakly interconnected. So three HS-376 satellites were ordered and booked on the shuttle. In addition to business communications, the K_u-Band transponders were to relay the national TV network. However, the largest single user was the Homestead & Community Broadcasting Service. This produced TV for one million people in settlements too remote to be serviced by conventional transmitters. An L-Band transponder relayed air and maritime traffic. The footprint covered Australia, a large part of the archipelago to the north, and New Zealand, so it represented an enormous advance. The first two satellites were deployed before the shuttle was grounded, and the final satellite went up on an Ariane rocket in 1987. When the government deregulated the market in 1991 it sold its majority share of Aussat to Optus, which gave its name to the second-generation system. The HS-376s were to have been successively retired as the new series (using the HS-601, the first three-axis bus built by Hughes) became available in the early 1990s, but when the first satellite suffered problems and the second was lost, the Aussats, all of which were still functional, were retained to keep the by-now vital service running.

Indonesia faced an even worse geographical problem. Its 200 million population lived on more than 13,000 islands running almost 5,000 km along the equator, and even substantial communities had little direct contact. A geostationary relay was the *only* cost-effective way of linking the nation together.

The Perumtel consortium was created specifically to develop the service. In 1975 it placed a contract with Hughes for an HS-333, and in 1976 this was launched as Palapa ('the fruit of great endeavour'). To follow up, Perumtel ordered the HS-376. As in the case of Telesat, upon which it had modelled its operations, Perumtel ordered two satellites, plus a spare, and booked them all on the shuttle. Unfortunately, Perumtel was not to have Telesat's luck. Although the first was delivered to its operating position without incident, Palapa-B2 became stranded in low orbit when its PAM motor misfired. The spare had been purchased for such a contingency, so its launch was brought forward. However, the shuttle was grounded before this satellite could be deployed, and it was finally launched on a Delta in 1987. The retrieved satellite was refurbished and relaunched on a Delta in 1990, to supplement the network.

India had had an active space programme for some time. It had developed its own satellite-launching capability, and, as a participant in Intercosmos, it had sent a cosmonaut to one of the Soviet Union's Salyut space stations. Its problem, however, was not so much a scattered population as such a large population that it was impracticable to develop a telecommunications infrastructure by conventional means. India had a long-standing interest in acquiring a geostationary capability, but it preferred a multipurpose satellite for a range of applications. The Indian Space Research Organisation (ISRO) defined requirements for carrying cameras for weather and Earth-resources imagery in addition to transponders for C-Band telephone relays and for S-Band TV broadcast. The contract for this customised platform was awarded to Ford Aerospace, a company which specialised in three-axis systems. A fully stabilised bus offered the advantage over a spinner in that it could expose a larger array of solar cells on flat panels that could be kept facing the Sun, and so run higher-power transmitters. However, it required additional systems to steer the panels and to control the orientation of the bus sufficiently accurately to keep the narrow beam properly aligned, and if either system failed the spacecraft became useless as a relay. Although the size of a spinner limited its power, because only a small fraction of its solar cell array could face the Sun at any given moment, it offered the advantage that its angular momentum made it inherently stable. Its design was complicated, however, by the fact that the antenna had to be mounted on a despun platform. Insat-1A went up on a Delta in April 1982, but it had to be abandoned a few months later, after it expended all its propellant in an attempt to recover from an attitude-control fault. The rest of the satellites in this series had been booked aboard the shuttle.

Insat-1B left its cradle with all of its various projections tightly folded against its boxy bus. As soon as it jettisoned its PAM stage, the satellite used its thrusters to cancel the spin. Upon assuming its station, at 74° E, it had difficulty in unfolding one of its solar panels, and it actually took several weeks for the panel to flatten out. The astronauts had reported hearing a 'clunk' as the satellite had been spun up in its container, which prompted the suspicion that it might have been damaged, but a replay of the deployment video failed to show any specific sign of an impact. Insat-1C had been scheduled for launch in 1986 but, upon the grounding of the shuttle, it was offloaded. Competition for rockets was so intense that it did not reach space until July 1988, when it rode an Ariane. Unfortunately, it lost control after 18 months and had to be written off. Insat-1D followed on a Delta in mid-1990, and Insat-1B was retired a year later. However, a few months later a power failure disabled half of 1D's transponders, and the venerable old 1B had to be reactivated until the first of the follow-on series could be launched. For this series, ISRO combined C/S/K_u-Band transponders and, despite having lost two satellites to instability, it retained three-axis control. But this time it developed its own bus. The first of the new satellites, Insat-2A, was launched in 1992 and proceeded to function satisfactorily.

Of the troubled series, only Insat-1B had been sufficiently robust to deliver the intended service. Virtually single-handedly, it had delivered social and educational TV to 75 per cent of the vast population, and returned hourly cloud surveys spanning the hemisphere from Egypt to the China Sea. In doing so it confirmed India's need for a geostationary capability.

The Arab League created the Arab Satellite Communications Organisation (ASCO) in 1976. It set the specification for a system to link the whole of North Africa and the Arabian Peninsula using a single C-Band footprint. A trio of three-axis satellites was ordered from Aerospatiale in 1981. Arabsat-1A went up on an Ariane in February 1985, and Arabsat-1B was deployed by the shuttle a few months later. When these were retired in 1992, the third satellite, bought as a spare, was launched. By this time, however, demand had outstripped the capacity of a single satellite, and more advanced satellites were not due until the mid-1990s. There was only one way to add capacity. Even after a satellite has exceeded its anticipated service life, if it is still operating reliably it can be sold on the second-hand market, and the price is often little more than the insurance premium for the launch of a new satellite. This provides a convenient way for newcomers to establish an infrastructure before investing in brand new satellites. Argentina, for example, took over the aging Anik-C1 and Anik-C2 in 1993 when Telesat (by this time Allouette Communications, because the government had sold its share of Telesat) introduced its C/K_u-Band Anik-E series. And when Telesat retired Anik-D2 in 1991, it leased it to GE Americom, which used it to fill a gap caused by the loss of Satcom-4. In 1993, Anik-D2 was sold on to ASCO, which operated it as Arabsat-1D. Similarly, when Telstar-301 was superseded in 1994, AT&T sold it to Telesat, which used it to cover for the ailing Anik-E2; and when this satellite returned to service several months later the Telstar was sold to ASCO, which retired Anik-D2/Arabsat-1D and replaced it with the Telstar, which it renamed Arabsat-1E. So this succession of old satellites supplemented Arabsat-1C's capacity and covered the gap until the next generation could be introduced. A communications satellite is rarely deactivated until it is completely worn out, or until it is no longer cost-effective in terms of the state of the art.

EVER MORE CAPACITY

In a move strongly reminiscent of the pioneering spirit of the Syncom, Earth Resources Technology Satellite and Applications Technology Satellite programmes, NASA developed the Advanced Communications Technology Satellite (ACTS) to demonstrate the advantages of high capacity K_a-Band communications, in the hope that by cutting the technological risk it would stimulate the commercial sector. ACTS was sponsored by the Agency's Office of Commercial Programmes and managed by the Lewis Research Center.

The K_a-Band offered the prospect of an unprecedentedly high degree of integration in global communications, but to realise it would take the development of new technology. At the heart of NASA's demonstration was a computer processor to sort and route traffic, and a microwave switching matrix to facilitate routing at much higher data rates than was possible using conventional satellites. In particular, ACTS used steerable pencil-beams which could be aimed at specific localised sites on Earth. Because the higher frequency band supported such a high data-rate, it was possible to adopt a 'burst' mode of operations in which a beam would 'hop' between several sites and transmit so much data in each time-slice that it could serve each site as well as if it had a dedicated link. The processor's task was to exploit this capability. ACTS was to be capable of linking supercomputers and relaying high-definition TV transmissions.

NASA had conceived the project in the early 1980s and had issued the contract to RCA Astro Electronics in 1984. Congress had backed the idea because it felt that the move to this higher frequency band was inevitable, and it was concerned that domestic service providers might be reluctant to risk rendering their expensive K_u-Band systems prematurely obsolete. But development suffered repeated technical and budgetary crises, and it fell far behind the originally intended 1989 launch.

At 2,540 kg, ACTS was too large for the PAM-D2 perigee-kick motor, so NASA used the Transfer Orbit Stage (TOS) built by Orbital Sciences Corporation. This had acted as an escape stage for the Mars Observer spacecraft, but the September 1993 ACTS mission was its first ride on the shuttle. The payload was carried in an annular support cradle. When the command was issued to detonate its pyrotechnics, the primary and the secondary charges both fired. A shower of debris peppered the bay's aft bulkhead, but there was no serious damage. The TOS successfully boosted the satellite into GTO, which manoeuvred into its slot at 105° W without incident.

After initial indifference, industry used ACTS extensively. Although it had been built to operate for five years, it was still running in 1997 and its services were in demand 24 hours a day. The Federal Communications Commission received several early applications to run K_a-Band systems. Motorola has used ACTS-like technology for inter-satellite linkage in its Iridium global telephone system. Hughes is using it in its Spaceway videophone satellites. Teledesic – a joint venture involving Microsoft and McCaw Cellular Communications (a part of AT&T) – is to build a constellation of K_a-Band satellites to provide an Internet-in-the-sky. In addition, a K_a-Band technology developed independently by Matra Marconi is to be used for the Wideband European Satellite Telecommunications system.

Critics have argued that there was no requirement for NASA to develop ACTS because the technology was already imminent. However, the fact remains that even though it was delayed, ACTS was the *first* on the scene, and its presence *did* eliminate the technological risk from the subsequent commercialisation.

TRACKING AND DATA RELAY

When the Skylab programme ended, NASA shut down many of the stations in its tracking network. To maintain contact with the shuttle it planned to build a network of geostationary satellites as a Tracking & Data Relay System (TDRS). Although NASA had planned a straightforward design that emphasised throughput, the Department of Defense insisted that the specifications be modified to provide secure communications for *its* shuttle missions. It had been NASA's intention to launch the satellites on Atlas–Centaur rockets, and to do so prior to the shuttle's first test flight; but in their upgraded form they were too big. As a result, when John Young and Bob Crippen took Columbia up for its shake-down mission, they were in contact with the flight controllers for at most 20 per cent of each 90-minute orbit, and that time was badly fragmented. Once the shuttle became operational, one of its prime tasks was to deploy these satellites.

The satellites were built by TRW for Space Communications Corporation, a consortium of Contel and Fairchild Industries formed specifically to serve NASA's communications needs. Although NASA defined the specifications, it would initially lease the satellites; it would not actually take ownership for six years. If the satellites operated far beyond their ten-year design-life, NASA would reap the benefit, but if they failed early it would not be unbeneficial. Spacecom hoped to build similar satellites for commercial leasing. At least, that was the theory. Contel bought out Fairchild's interest in 1985, and was itself taken over by GTE in 1991.

The 2,250-kg three-axis stabilised bus carried a comprehensive suite of S/C/K_u-Band transponders. Each had two redundant primary systems involving a 2-metre dish dedicated to the ground link, and a steerable 5-metre dish for single-access S/K_u-Band which tracked and relayed signals to and from individual spacecraft. A set of S/C-Band transponders was mounted on the Earth-facing side of the bus for parallel use on a multi-user basis. Electrical power was provided by a pair of solar panels. Each transponder was required only to act as a *bent pipe*, simply relaying a transmission without processing. The system was meant not only to relay voice and telemetry for shuttle missions, but also data streaming from satellites at rates as high as 300 megabits per second. The Great Observatories, and satellites such as the follow-on Landsats, were being designed to employ high capacity real-time downlinks. Each TDRS satellite could simultaneously serve two K_u-Band users at the highest rate, and two dozen S/C-Band users at 50 kilobits per second.

The plan was to deploy three satellites immediately, and to add another later to serve as an in-orbit spare. A ground station was built at White Sands, New Mexico. As viewed from White Sands, the two TDRS stations were designated 'East' (at around 45° W) and 'West' (at around 172° W). Arthur C. Clarke's concept had envisaged a third satellite on the far side of the planet, but NASA did not initially intend to exploit this relay point, because it would require this third satellite to relay *through* one of the others to maintain contact with White Sands. NASA was content to have the at-least-85 per cent coverage that a pair of satellites could provide.

Although NASA booked three satellites for immediate use, another as an in-orbit spare, and one as a ground spare, Spacecom signed up Westar to lease a satellite, so it ordered six satellites from TRW; but in 1982, when NASA requested this extra satellite, Spacecom had to cancel this commercial contract.

The first – TDRS-A – went up on STS-6 in April 1983. It was too big for the PAM motor that had dispatched the HS-376 commercial satellites on the preceding flight, and, besides, it was not intended to perform its own circularisation burn. The TDRS was to be deposited in geostationary orbit by the Inertial Upper Stage (IUS) which Boeing had built for the Air Force. This had two motors: one for the GTO-insertion, the other for GSO-circularisation. The satellite had its own propulsion system, but this was designed to enable it to *adjust* its station, not to perform radical orbital manoeuvres.

The payload was carried on an annular cradle at the rear of the bay. The 17-metre stack completely filled the bay. As soon as the bay doors were opened, the orbiter turned its belly to the Sun so that the satellite would be shaded, to prevent it from overheating. It could not be exposed to the Sun until its environmental system was activated. The first step in bringing the stack to life was to elevate the cradle to 30 degrees. The IUS was heavily instrumented. Each stage's redundant control systems were verified. The astronauts used a camera showing the first stage's nozzle, to verify that it gimballed properly. The satellite itself was tested whilst in contact with the control centre, and once the stack had been checked out, the power umbilical was disconnected, and the cradle was elevated to 60 degrees. At the appointed time, the ring-clamp holding the IUS in place was released, so that a spring could eject it. This was the first time that the mechanism had been used, and it worked flawlessly. The 18-tonne vehicle – the heaviest payload yet released by the shuttle – drifted lazily over the top of the orbiter's cabin, then off into the distance. The shuttle's part of the process was over.

Ten minutes later, by which time Columbia had withdrawn, the IUS activated its flight control system. Half an hour after that, it took a series of star sightings to update its inertial platform. As it crossed the equator, it ignited its first stage motor. This delivered 18,500 kg of thrust for two-and-a-half minutes, then shut down. Now in GTO, the first stage used its small thrusters to reorientate the stack and set up a slow roll (dubbed the 'barbecue' roll) to even out the thermal stresses during the six-hour climb. Several minutes before reaching apogee, the first stage reorientated the stack for the circularisation burn, and then released its payload. The second stage was to have delivered 2,750 kg of thrust for 103 seconds but, near the of the burn, the oil-filled seal of the gimbal deflated, the nozzle slewed off-axis, and the offset thrust induced a 30-rpm end-over-end tumble. As soon as the controllers noted this in their telemetry, they commanded the second stage to release the satellite. Although TDRS-A was soon stabilised, it was in a 21,700 × 35,550 km orbit which had little to recommend it as a communications relay. Not only was it non-synchronous, because the IUS's burn had not yet cancelled the 28.5-degree inclination of the shuttle's orbit, it was nodding 3 degrees either side of the equator. The IUS had redundant systems to overcome many faults, but there was nothing it could have done to recover from the mechanical failure in the manifold. It was immediately withdrawn from service.

NASA had set up a carefully interleaved schedule. The next TDRS satellite was to have been deployed within months, so that Spacelab, requiring a continuous high-capacity data downlink, could fly on STS-9.

The TDRS network would obviously have to await the recertification of the IUS, which could easily take a year. Since it would be NASA's only asset for the foreseeable future, it was vital that TDRS-A be manoeuvred up to its assigned station by using its thrusters. In fact, it took *two months* to nudge the satellite into GSO, and it did not reach its assigned 'East' slot, at 41° W, until early October, by which time it had used half of its propellant, a loss which would seriously reduce the time it could remain on station. As soon as it was clear that the relay would be on line in the autumn, NASA authorised the Spacelab mission, which would have to make do as best it could with a single relay. However, many

Table 3.2. Intended TDRS schedule, *c.* late 1982

TDRS	Date	Mission	Orbiter
TDRS-A	Jan 1983	STS-6	Challenger
TDRS-B	Jul 1983	STS-8	Challenger
TDRS-C	Jan 1984	STS-12	Discovery
TDRS-D	May 1984	STS-15	Discovery

planned payloads could not be sent up until the network was operational, so an early recertification of the IUS was imperative.

When the IUS flew next, in January 1985, the first stage's performance fell short. This time though, it was able to correct the problem and successfully deliver its military satellite. With the IUS back in service, TDRS-B was assigned to flight 51E, which was to launch in March. However, engineers analysing timing errors on TDRS-A feared that there might be a generic fault, so it was decided to offload TDRS-B for tests, and 51E was cancelled just a week before it was to have launched. Once the satellite was verified, NASA reassigned it to 51L, but it was lost along with the rest of the Challenger stack on 28 January 1986. When the shuttle resumed flying, building up the TDRS network was given precedence even over the backlog of military satellites. The manifest drawn up in late 1986 envisaged resuming operations in early 1988. It put a TDRS on the first flight, Department of Defense payloads on the next two, then the final TDRS on the fourth, but, as the dates progressively slipped, the third and fourth flights were transposed.

The situation had degraded somewhat while the shuttle had been grounded. In late 1986 one of TDRS-A's steerable antennas developed a problem, which denied the satellite half of its K_u-Band capacity. When TDRS-C took up the West station, at 171° W, it encountered difficulty deploying its antennas, but once this was overcome NASA had, for the first time, satellites in both primary stations, although with one satellite operating at reduced capacity. In March 1989, TDRS-D replaced TDRS-A in the East and the older satellite was moved to 79° W to act as an in-orbit spare. Unfortunately, TDRS-C lost part of its K_u-Band capacity in January 1990. In May, it was shunted to 174° W and TDRS-A took up the 171° W slot to augment it. It had been hoped to add TDRS-E in the autumn, but events conspired to slip it to August 1991. It replaced TDRS-C at 174° W, and the older satellite was moved to 62° W, as a spare. By this point, NASA had decided to run the satellites in pairs. As a result, when TDRS-F was launched in January 1993, it was placed at 46° W, to supplement TDRS-D at 41° W, in the East. This new operating policy was to ensure that neither station would be left uncovered by the total loss of a satellite. With the satellites proving less robust than hoped, NASA had ordered a replacement for the unit lost on the Challenger. When this last satellite, TDRS-G, was launched in July 1995, it was placed at 171° W, to serve alongside TDRS-E at 174° W, in the West.

Meanwhile, in late 1993 TDRS-A had been moved to 85° E to assist the Gamma Ray Observatory which, having lost its tape recorder, was forced to transmit data in real time. It required only an S-Band link, so TDRS-A was fully capable of handling it. A new ground station was set up at Tidbinbilla in Australia specifically to receive the data and relay it on to White Sands via the Intelsat network. When TDRS-C replaced it in this role in mid-1995, TDRS-A was finally retired.

Table 3.3. TDRS relay network, *c.* late 1995

Satellite	Launched	Status
TDRS-A	Apr 1983	retired
TDRS-B	Jan 1986	lost
TDRS-C	Sep 1988	85° E
TDRS-D	Mar 1989	41° W
TDRS-E	Aug 1991	174° W
TDRS-F	Jan 1993	46° W
TDRS-G	Jul 1995	171° W

It took much longer than expected to build up an effective network, and in a sense it has yet to be completed, because even though the farside station was finally occupied, it was by the degraded satellites. The primary stations can cover only 85 per cent of the shuttle's orbit. In addition, because it proved necessary to place two satellites in each station to guard against losses, many more satellites than planned had to be launched. As a result of the delays and the uncertainties over the system's reliability, the ground station network was not actually decommissioned until the early 1990s.

In space, the orbiter swung a small steerable dish over the starboard sill, just aft of the cabin. This was deployed as soon as the payload bay doors were opened, and retracted just before they were finally closed. A benefit of having a relay over the Pacific Ocean was that during re-entry, when the orbiter was immersed in ionised plasma, it could still communicate using its omni antenna. It was not all that important for the crew, but it removed a source of anxiety for the ground controllers during this critical phase of the flight.

The plan to develop an Advanced TDRS satellite was abandoned in 1993 when funding was withdrawn, but in February 1995, by which time the form of the International Space Station had become clear, NASA was authorised to build three satellites to extend the life of the existing system and, much to TRW's dismay, NASA awarded this contract to Hughes. The second-generation type will use the HS-601 bus, but will be TDRS-like in its mode of operation whilst extending the system's range into the K_a-Band and doubling the maximum data rate. This expanded network is to be in place by the time the space station is complete.

MARKET FACTORS

Three conclusions are immediately clear: communications was the commercial satellite business; the HS-376 was the satellite-of-choice; and the *shuttle* was the carrier-of-choice for this type of satellite. This status was a significant achievement for NASA, because half of the HS-376s that it released were for overseas clients. These contracts were won against competition from ESA's Ariane rocket. In a sense, NASA had succeeded in demonstrating that there was a market for the shuttle's services.

Appearances can be deceptive, however. NASA's fee for deploying satellites had been calculated from early projections for flight rates, turnaround time and operating overheads. By the end of 1985 it was evident that these expectations were unrealistic. Even operating at its best flight rate, in 1985, NASA managed only nine flights. At this lower rate of flights, and with each flight costing *far* more than predicted, the fee did not even cover the cost. It was this artificially low fee that was attracting clients.

In fact, the situation was becoming worse, because the terms of its overly-generous deal committed NASA to a similarly unrealistic fee for follow-on satellites throughout the period of the introductory offer, and with each passing year of escalating costs the satellites were being ferried into space at an increasing operating loss. The shuttle was busy, but it was not a viable business. For the first five years of its operations, NASA was actually trapped; and, worst of all, the US tax-payer was actually *subsidising* foreign satellites.

Intelsat's first geostationary satellite, Early Bird, could relay 240 telephone circuits or a single monochrome TV channel. The technological state-of-the-art advanced at such a pace that the relay capacity increased dramatically with each new generation of satellites.

Table 3.4. Generations of Intelsat satellites

Series	First	Voice	Supplier
1/2	1965	240	Hughes
3	1968	1,500	TRW
4	1971	4,000	Hughes
5	1980	12,000	Ford
6	1989	120,000	Hughes

The communications industry could satisfy this soaring demand only by continuously investing in new technologies. It was a *constructive* cycle in which the forecasted increase in demand funded the development of the technology, which not only immediately began to repay its investment but also further stimulated demand. It was the growth rate that opened up geostationary communications satellites to private operators.

The HS-376 dominated the shuttle's commercial manifest, so it is worth examining this satellite further. Not only could its antenna footprint be shaped by the customer's territory, but its bus was sufficiently versatile to accommodate a varied package of transponders in a mix designed to suit the customer. Another crucial factor was that the technology had achieved the point of being able to support a 10-year operating life.

The life expectancy of the early geostationary communications satellites had been three years. This had imposed an ongoing replacement strategy that only consortia could afford. By doubling this time, the HS-333 had allowed time to recoup the investment, which made this satellite a realistic option for a major telecommunications company to use to service its clients. The HS-376's operating life was designed to turn in a healthy profit. Furthermore, it enabled individual companies to link their scattered sites by their own satellite. But it was not simply a matter of developing ever-more durable satellites. With technological advance, there was a switch to higher frequencies. In the latest case, this was from the C-Band to the K_u-Band. A satellite with a life much longer than 10 years was likely to turn into a liability towards the end, because it would be obsolete. The HS-376 was therefore a rather effective compromise. In a static market, extending the life of the product would compromise future sales. But telecommunications was not a static market. It had always been – and looked as if it would remain – an expanding market. The HS-376's longevity did not serve to limit sales. Quite the contrary. Its low cost let individual telecommunications companies run their own satellites. It stimulated the creation of specialist leasing companies to compete with Comsat in the domestic market, and it led to the creation of Comsat-style consortia which provided national services where there had previously not been integrated communications systems. It also later served as the basis for the early Galaxy class, which Hughes itself set up as the world's largest fleet of privately-owned commercial communications satellites. As a result, the HS-376 production line expanded with each passing year, which cut the unit cost and thus made it an even more attractive product. It appeared at just the right time to create the critical mass of orders that ensured that it became the *satellite of choice*. Put more bluntly, it was a product whose time had come.

The pace of technological development in this market is well illustrated by Telesat's A-series and D-series Aniks, both of which had C-Band transponders for relaying either voice or colour TV.

Table 3.5. Comparison of Hughes' satellites

First	Type	TV	Voice
1972	HS-333	1	1,000
1982	HS-376	24	20,000

By integrating voice, video and data, the telecommunications companies with such satellites were able to support video-conferencing, an innovation which benefited customers by enabling them to make more effective use of their time. Businesses buying services on such satellites found themselves faced with an unprecedented degree of interconnectivity.

Communications satellite technology exploited microminiaturisation in electronics. Each generation of solid state electronics delivered more power at less cost then its predecessor. A satellite operator could pass on a saving to its customers without undermining revenues. When Early Bird entered service in 1965, it cost \$9 to make a 3-minute telephone call across the Atlantic. By the time the shuttle deployed its first satellite in 1982, this had fallen to \$3. If the cost had kept pace with inflation, it would have been \$30. Over that same time, the traffic increased from 2 million calls to 200 million calls. Thus, while the cost had fallen by an order of magnitude, usage increased by *two* orders of magnitude.

Direct-dialling and the World Wide Web would not be practicable without a massively-interconnected low-cost global communications system.

Although the most advanced modern satellites are capable of a fully-integrated all-digital high-throughput K_a-Band service, the lower frequencies continue to form an essential part of everyday communications. Motorola's Iridium satellites, the modern telephone relay, not only use L-Band to communicate with portable phones, they relay signals around the world by K_a-Band, and, so as to be able to pick up the weak transmissions from a hand-held unit, they have been deployed in low orbit rather than in geostationary orbit.

In various forms, the HS-376 is still being manufactured. In 1997, Russia ordered its first US-built communications satellite, and it was a customised HS-376. Over the years, in excess of fifty HS-376 have been placed in geostationary orbit, and even the retired ones are still there. Early Bird is also there, and when someone suggested that it be briefly reactivated to celebrate Intelsat's twentieth Anniversary, it came back on line. At only 38 kg, Early Bird was a truly remarkable device for its day. How many Early Birds could a single shuttle have deployed? The payload bay could have accommodated five HS-376 cradles, so the shuttle *could* have deployed a flotilla of satellites; but it never did, because the resulting forward centre-of-mass could have jeopardised the orbiter in an emergency landing. The largest number of satellites the shuttle ever carried was three. It did so on four occasions. In each case, only two were HS-376s. During Columbia's refit in 1985 it was equipped with instrumentation to measure the orbiter's hypersonic flight characteristics. Once this was understood, the payload was to be increased. The Department of Defense intended to exploit this. It had a constellation of 24 NAVSTAR satellites to launch, and it wanted to deploy them four at a time.

Under the shuttle-only policy, the prospects for the rocket-builders had been bleak, but NASA rode the wave of commercialisation in triumph because, although it was not running profitably, it was at least able to demonstrate that the shuttle had successfully met one of its primary technical objectives. But its opportunity to profit from commercial satellites ended in a cloud of destruction in the chilly Florida sky on 28 January 1986.

During the years that the shuttle was active in the commercial satellite business, eight Ariane rockets deployed 13 communications satellites, one rocket lost two others, and one dispatched the Giotto probe to Halley's Comet. It is noteworthy that Ariane attracted only one HS-376: Brazil's first domestic relay satellite.

Despite the intention to phase out the Atlas, it also was very busy during these years. In the time the shuttle was actively deploying commercial satellites, 17 out of 18 Atlas rockets were successful. However, their payloads were mostly for the Department of Defense and the National Oceanic and Atmospheric Administration, and only Intelsat placed communications satellites on the Atlas. The 13 Titan 3 launches each carried a single Department of Defense satellite, one of which was lost. The Delta had long been the rocket-of-choice for the small satellites sponsored by the many government agencies, but it had attracted the earliest Anik, Westar, Satcom and SBS satellites as well, and during the time that the shuttle was active six of the 12 launches carried commercial communications satellites.

Overall, therefore, even discounting the Palapa and Westar satellites which were left in low orbit by faulty kick-motors, the shuttle was able to carry as many commercial satellites as the Delta and Ariane rockets combined, and considering that it did this in the context of a wider programme, it really did rather well.

Table 3.6. Satellite launches and rocket losses while the shuttle was operational

Rocket	Launches		Satellites		
type	Total	Lost	Total[1]	Lost	Commercial[2]
Shuttle	25	1	33	1	20
Ariane	10	1	16	2	13
Delta	12	0	13	0	6
Atlas	18	1	18	1	5
Titan	13	1	<classified>		0

Notes:

1. Satellites of all types but, in the case of the shuttle, not free-flyers.
2. Commercial communications satellites.

The shuttle's grounding led to a complete reversal of the policy of operating the single National Space Transportation System. Immediately after receiving the report of the inquiry into the loss of the Challenger, the Reagan administration reverted to a *mixed-fleet* policy. The Department of Defense was authorised to buy more Atlas and Titan 3 launchers, and to order improvement programmes to update the rockets. The manufacturers eagerly restarted their production lines.

If the Challenger had *not* been lost, and if the shuttle had continued releasing commercial satellites, NASA would not have been able to lift its unprofitable fee until its five-year term expired in 1988. It is debatable how much business it would have attracted thereafter, if it had charged customers the economic rate. In the event, while the shuttle was grounded, the Reagan administration ordered NASA out of the commercial satellite business, so the US market was thrown wide open. Free of the shuttle-only policy, McDonnell Douglas with its low-end Delta, General Dynamics with the medium-lift Atlas, and Martin Marietta with the heavyweight Titans were determined to seize what were clearly large profits.

Whereas previously the rocket manufacturers had sold their goods to the government, whose agencies had worked with either NASA or the Air Force to prepare and launch their payloads, they could now offer a full service encompassing payload integration and launch, a service for which a genuine fee would be charged. It was not only the commercial satellite business that was up for grabs. All the satellites that the shuttle was to have carried for the government, including the 24 NAVSTARs, were also in need of rockets. Over the years, each rocket had been tailored to a given payload mass. Competition had been squeezed out during the lean years, so, in this new world of rocket supremacy, there seemed to be a convenient niche for each of the survivors. It would take time, however, for production to match the demand, because it meant reversing the process of decay that had accompanied the phasing out of expendable rockets. All the remaining Atlas and Titan 3 rockets were reserved by the government.

The Delta was an effective rival for Ariane but, in contrast to the shuttle, it would have to be run economically. Unfortunately, production had terminated when the shuttle entered service. There were only five left, and of these, two went to the Strategic Defense Initiative Office, two to NOAA (one of which was actually lost), and the last was used to compensate Indonesia for the embarrassing stranding of its shuttle-deployed Palapa.

There were simply not enough rockets to meet the sudden demand. Only Arianespace was in position to increase its operations. It immediately won three commercial satellites withdrawn from the shuttle: Aussat-A3, Insat-1C and SBS-5. The nine Ariane launches during the 32 months that the shuttle was grounded sent 16 satellites successfully on their way and wrote off another. In effect, therefore, Arianespace actually had the commercial business all to itself.

Table 3.7. Satellite launches and rocket losses while the shuttle was grounded

Rocket	Launches		Satellites		
type	Total	Lost	Total[1]	Lost	Commercial[2]
Ariane	9	1	17	1	13
Delta	5	1	5	1	1
Atlas	8	1	8	1	0
Titan	5	1	<classified>		0

Notes:
1. Satellites of all types.
2. Commercial communications satellites.

One final point deserves to be explicitly stated: the shuttle is a *rocket*; no rocket is infallible, and even the most reliable rockets fail occasionally. The Atlas and the Titan 3 were considered to be well-proven systems yet, in the short time that the shuttle was grounded, one of eight Atlas rockets failed, as did one of five Titan 3 rockets. In the light of these loss rates, it is clear that during its brief period of commercial satellite operations the shuttle was a remarkably productive transportation system.

4

No mission too difficult

One of the roles advocated for the shuttle was that it would be able to rendezvous with a defective satellite so that spacewalking astronauts could attempt to repair it; if an in-orbit repair proved to be impracticable, the satellite would be able to be retrieved.

When Ed White left Gemini 4 for 20 minutes in June 1965, spacewalking seemed easy. It was not until the next excursion, on Gemini 9 a year later, that the difficulty of extravehicular activity (EVA) became evident. Whereas White enjoyed the sensation of weightlessness by simply floating on the end of a long umbilical, Eugene Cernan was to undertake *work* outside. Cernan found it so difficult to maintain any given position that he floundered about. His suit was unable to cope with his exertions, and he was soon drenched in sweat. When the suit's visor fogged over, he abandoned the plan to make his way to the rear of the spacecraft, extract a rocket-backpack, and assess its utility. By the time he had felt his way blindly back to the hatch, he was exhausted. It was quite evident that evaluating the Astronaut Manoeuvring Unit (AMU) was asking too much at this early stage in the programme.

As it happens, Gemini 9 was the one flight of the series on which a spacewalker could possibly have overcome a problem that led to part of the plan having to be abandoned. An important objectives of the Gemini programme was to perfect rendezvous and docking. The Agena upper stage had been fitted with a special collar so a Gemini could mate with it. The Agena would then fire its motor for orbital manoeuvres. The first to try it was Gemini 8. It docked easily, but when a thruster misfired sending the spacecraft cartwheeling through space, it was forced to withdraw and make an emergency landing. Gemini 9 was to have repeated this experiment, but the rocket carrying its Agena target exploded, and a secondary target (the ATDA), built for engineering tests, was hastily sent up. It did not have an engine, but would serve as a docking target for Gemini 9. Unfortunately, the ATDA failed to shed the shrouds protecting the docking collar, rendering docking impracticable. The frustrated astronauts took pictures of the fouled shroud – whose appearance they likened to an 'angry alligator' – and then withdrew.

If Ed White's EVA had been more demanding, it might have alerted NASA to the difficulties of spacewalking. Forewarned, and better trained, Cernan might have been able to don the AMU, cross to the Agena, and slip off its shroud to clear the way for a docking. This would have been a triumph. However, it is unlikely he would have been given permission to try. Yet-to-fly astronaut Buzz Aldrin, a member of the support crew, was castigated when he suggested the possibility. It was even considered too dangerous to attempt to nudge off the shroud with the nose of Gemini 9. Instead, this part of the mission was cancelled.

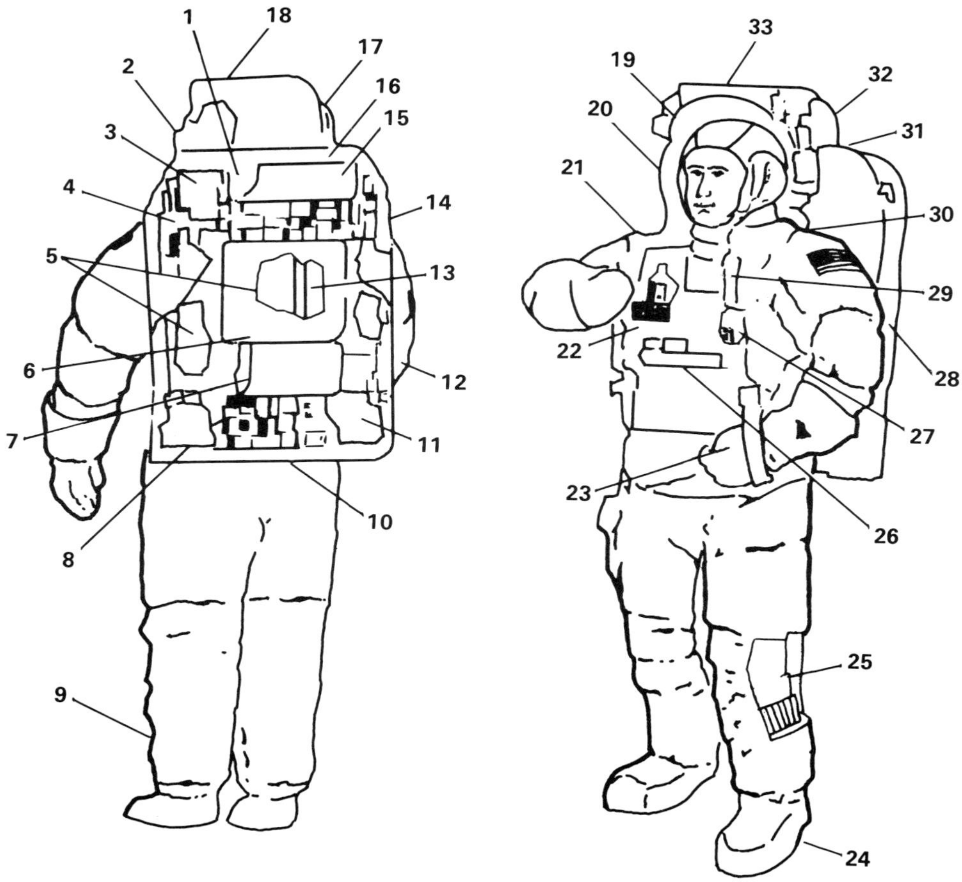

1. Radio
2. Lights
3. Caution and Warning Computer
4. Sublimator
5. Primary oxygen tanks
6. Contamination control cartridge
7. Battery
8. Secondary oxygen pack
9. Lower Torso Assembly (LTA)
10. Oxygen regulators
11. Secondary oxygen tanks
12. Primary life-support system
13. Water tank
14. Fan/separator/pump/motor assembly
15. Muffler
16. Antenna
17. Lights
18. TV camera
19. Visor Assembly
20. Helmet
21. Upper Torso Assembly (UTA)
22. Display and control module
23. Gloves
24. Boots
25. Liquid cooling and ventilation garment (within outer garment) (LCVG)
26. Oxygen control activator
27. Temperature control valve
28. MMU mount
29. Connection for service and cooling umbilical
30. In-suit drink bag
31. Communications carrier assembly
32. Lights
33. TV camera

Fig. 4.1. A schematic of the Extravehicular Mobility Unit (EMU) worn by astronauts making spacewalks.

It was decided not to fly the AMU again until basic spacewalking techniques had been perfected. Although much simpler tasks were assigned to Michael Collins and Richard Gordon on the next two flights, Geminis 10 and 11, and both achieved their main objectives, they wore themselves out in trying to reach and then to maintain their position at the worksite. The need for handholds had not been fully appreciated. Spacewalking was one thing; working in space was quite another. Aldrin spent many hours rehearsing procedures for using both handholds and footholds, using neutral-buoyancy in a water tank to simulate weightlessness. The result was that on the final mission of the programme, Gemini 12 in November 1966, he was not only able to complete *all* of his assigned tasks, but was able to do so relatively easily. He used strategically placed handholds to reach a worksite, and then used the footholds to lock himself in position, which left both hands free to work. And because he did not have to struggle to maintain his position, he was able to pace himself. As a result of this success, the Weightless Environment Training Facility (WETF) was built at the Johnson Space Center. The experience of spacewalking on Gemini revealed that a foot restraint is crucial to efficiently *working* in space, and on the shuttle this function is magnificently provided by the Remote Manipulator System (RMS).

A number of demanding spacewalks were needed to overcome the damage sustained by Skylab during its ascent through the atmosphere, and to maintain its solar telescope. These proved conclusively that astronauts could indeed work effectively outside, so NASA was in a confident mood as the shuttle neared its long-awaited flight test.

ANTICIPATION AND TRIUMPH

The first flight was kept simple. Its objective was simply to demonstrate that the shuttle could reach orbit and return to Earth, and the payload bay contained only instruments to record the status of the orbiter. But on STS-2, Joe Engle was to don the new two-piece spacesuit– or Extravehicular Mobility Unit (EMU) – seal himself into the middeck airlock, and conduct a depressurisation test. However, one of the fuel cells flooded and had to be switched off, so the test was cancelled and the flight curtailed. The depressurisation test was reassigned to STS-4. As it happened, the cover of the classified CIRRIS sensor system in Columbia's bay had refused to open, and Ken Mattingly, who was to test the suit, proposed that he go out and open it manually, but permission was denied. The rules stated that astronauts only make spacewalks in pairs, so they would be able to help one another out. Spacewalking was not planned for the test-flight phase; EVA was just a contingency in case the payload bay doors failed to close.

As soon as the National Space Transportation System was declared operational, two of STS-5's four-man crew were assigned the first excursion into the bay. This was postponed for a day because Bill Lenoir suffered a dose of 'space sickness'. The next day, however, he and Joe Allen spent four hours prebreathing pure oxygen to purge nitrogen from their blood streams (such a long time was necessary because the orbiter had yet to be tested with partial-pressure air mix). Problems arose even before they entered the airlock. In the process of testing the suit components, Allen found that a fan in his life-support backpack was running hot. Even though the excursion had to be cancelled, Lenoir went ahead and donned his suit to test it, but unfortunately, its regulator would not maintain the necessary 4.3-psi pure oxygen pressure, so even the depressurisation test had to be cancelled. It was not a very promising start.

The first real objective for spacewalking was to capture, repair, and release SolarMax, a satellite crippled by a malfunction in its attitude-control system. Although this was ambitious, if it could be successfully achieved it would be a convincing demonstration of the shuttle's unprecedented capability. On Challenger's first mission, Don Peterson and Story Musgrave became the first astronauts to venture outside, but only after the TDRS/IUS stack had been inserted into geostationary transfer orbit (GTO), *en route* to geostationary orbit (GSO). This deployment was accomplished on the first day. As initially manifested, Challenger was to have returned to Earth the next day, but, in the light of the earlier problems, the flight plan had been augmented to accommodate the evaluation of the spacesuit.

Fig. 4.2. Story Musgrave wearing the undergarment of the Extravehicular Mobility Unit (EMU). The undergarment contains a network of fine tubing through which water is pumped to heat or cool the astronaut. Note the wrist-loops used to ensure that the sleeves do not bunch up when the arm is pushed into the sleeve of the Upper Torso Assembly (UTA). By the time Musgrave finally retired in 1997, he had flown on the shuttle six times.

On 7 April 1983, as before, Peterson and Musgrave began their preparations with a full prebreathing. The suits raised no issues, so they put them on and sealed themselves into the airlock. After opening the outer hatch, they used the handholds on the other side of the cabin wall and attached safety lines to a bar just above the circular opening. With this vital line in place, they each moved to a sidewall and hooked a tether with a slider to a cable running along each sill; then they released the hatch tethers. Each spacewalker could now traverse the full length of the bay without becoming entangled in the other's tether. If a tether broke, the orbiter would manoeuvre to retrieve the drifting astronaut.

The primary objective on this first outing was to demonstrate a contingency procedure. A large annular cradle was mounted at the rear of the payload bay. It had held the IUS, and it had been rotated up to about 60 degrees for payload deployment. It was meant to turn back again afterwards, and it did, but because it might have jammed, possibly even with the payload still in place, there was a procedure for rotating it manually. As with the procedure to close stuck payload bay doors, this was a contingency which NASA hoped never to have to use, but it was wise to verify that it was feasible. Peterson and Musgrave made

Fig. 4.3. During the first spacewalk from the shuttle, Story Musgrave examines the Airborne Support Equipment (ASE) cradle which had just despatched the first Tracking & Data Relay System (TDRS) satellite.

their way to the aft wall of the payload bay, set up a winch, ran a cable from it through a block on the tilting part of the cradle, drew it vertical, and locked it into position. It was a fair afternoon's work. About 3.5 of the 4.3 hours they spent on their suits' environmental systems was time in the bay. There was much more to be done though, before it would be practicable to tackle a job as complex as repairing a satellite.

On the next mission, the environmental unit was reset in orbit for the first time to adjust the standard 14.7-psi cabin atmosphere down to 10.2 psi while maintaining the mix of 20 per cent oxygen and 80 per cent nitrogen. Operating at this reduced pressure would slash the prebreathing requirement in advance of a spacewalk. No external activity was assigned, but STS-7 was to carry out an important task in preparing for SolarMax. The

SPAS retrievable satellite had been deployed soon after reaching orbit. As Challenger returned to recover it, Bob Crippen flew the orbiter from the aft flight-deck station, looking 'up' through the overhead window at the satellite. The immediate objective was to use the opportunity of the first-ever recovery operation to assess the orbiter's manoeuvrability in close proximity to another object and to determine the extent to which it was disturbed by the Reaction-Control System (RCS) efflux. In the event, the orbiter proved to be an extremely controllable vehicle, and its thrusters did not disturb its much smaller companion.

The next step in preparing for SolarMax was assigned to STS-10, in February 1984. A manoeuvring unit had not been required for Apollo, but a redesigned form of the backpack that Gene Cernan had struggled to reach had been flown *inside* Skylab's cavernous orbital workshop because (according to the belief at that time) an astronaut setting off to recover an errant satellite would require independent manoeuvrability. The shuttle variant was dubbed the Manned Manoeuvring Unit (MMU). By this time, NASA was loading spares for critical items. Only two spacesuits were required, but a third upper-torso (the part with the systems in it) was carried so that a spacewalk would not have to be cancelled if one was found to be faulty. Two MMUs were carried just in case one proved unworkable. They were stowed in Flight Support Structures (FSSs) mounted on the bay sidewalls, to either side of the airlock.

On this occasion, two spacewalks were planned. Both were to be conducted by Bruce McCandless and Bob Stewart, so a day of rest was scheduled in between. McCandless was first out. Because he had been the astronaut most involved in the MMU's development, he had been given the task of putting it through its paces. It was actually a pallet with a recess. Once McCandless had reached the port FSS, he turned, and eased the integrated pack of his suit into the recess. Two arm rests were then swung down. Each had a controller with

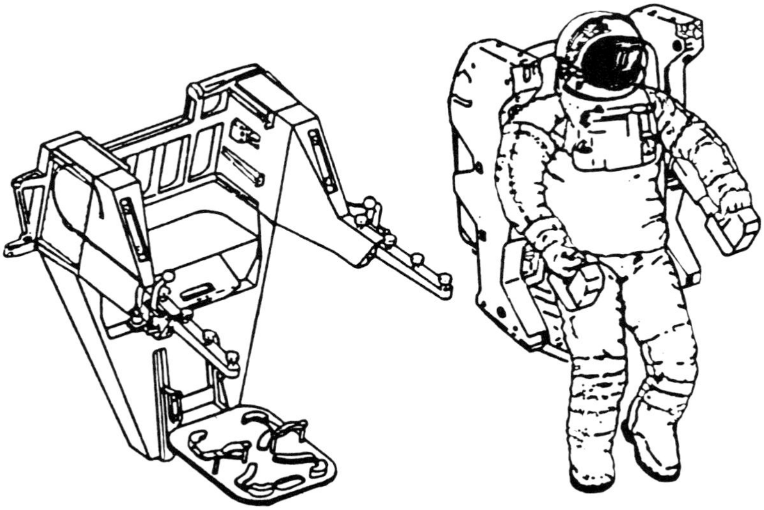

Fig. 4.4. A schematic of an astronaut who has just donned the Manned Manoeuvring Unit (MMU) and disengaged from the Flight Support Structure (FSS) that is mounted on the payload bay's sidewall.

push buttons and a joystick. The right arm specified rotation; the left specified translation. In all, the *pilot* had control of six degrees of freedom. Once he had verified the MMU's systems, McCandless released his tether and swung the levers of the FSS's locking mechanism, and in so doing he became the first human to fly in space as an independent spacecraft.

There must have been a temptation to zoom off and fly loops around the orbiter but, test pilot that he was, McCandless worked methodically through a series of thruster firings to assess the manoeuvrability and stability of the MMU in each of its modes. With Stewart observing from the sidewall, McCandless moved up and down, left and right, forward and back, and then pitched, yawed and rolled, all well within the payload bay. Once each of the degrees of freedom had been verified, he invoked a multi-axis rotation and then activated the automatic stabiliser to verify that it could halt the motion and then hold him stable. With this established, he began to move straight up out of the bay and did not stop until he was about 50 metres from the orbiter. After a few minutes, he returned. The MMU's nitrogen gas jets (there were 24 in two independent sets of 12) delivered only 0.75-kg thrust. It had a mass of 150 kg, as did the suit, so the combined mass of an astronaut, his suit and the unit made all movement slow and deliberate. In fact, the trip out and back had lasted about 15 minutes. To follow up, McCandless went out again, this time to 100 metres. Upon finally returning, he positioned himself in front of the FSS, eased the MMU into its receptacle, and raised the levers to engage the lock.

The MMU had performed magnificently. What could the Gemini astronauts have done with such a device? Could one of them have repaired the crippled Orbiting Astronomical Observatory satellite? Probably not, for the simple fact that the spacecraft had not been built to be serviced. SolarMax, in contrast, was the first satellite ever designed to be worked on by astronauts, and it was this that made it so attractive as NASA's first candidate for repair in space.

While McCandless had been playing Buck Rogers for the TV cameras, the downlink of which was broadcast live by the US networks, Stewart had been testing another piece of kit that would be crucial in a satellite repair operation. The Manipulator Foot Restraint (MFR) was a platform incorporating a pair of 'golden slippers'. It was mounted on the end of the Remote Manipulator System (RMS) mechanical arm. Eugene Cernan's Gemini spacewalk had demonstrated that a free-floating astronaut cannot do anything useful if he needs both hands just to maintain position. Buzz Aldrin had shown that the most valuable tool a spacewalker could have was a well placed foot-rest, because it leaves both hands free for work. Stewart retrieved the MFR from its sidewall stowage point and Ron McNair swung the end-effector of the RMS arm down to him so that it could be attached. With the restraint securely on the arm, Stewart climbed aboard and allowed McNair to swing him around as if on the end of a cherry-picker to determine how accurately the arm-operator on the flight-deck's aft-station could position an astronaut within the bay. Each time Stewart reached a notional work-site, he would jiggle around to assess the extent to which body movement would shake the arm, thereby disturbing his carefully arranged position.

By the time Stewart finished the MFR trial McCandless had returned with the MMU, so they exchanged roles. While Stewart flew the MMU, McCandless inserted a different foot-restraint, the Portable Foot Restraint (PFR), into the starboard sidewall. This placed him in front of a Stowage Assembly Box (SAB) in which were various tools and a mock-

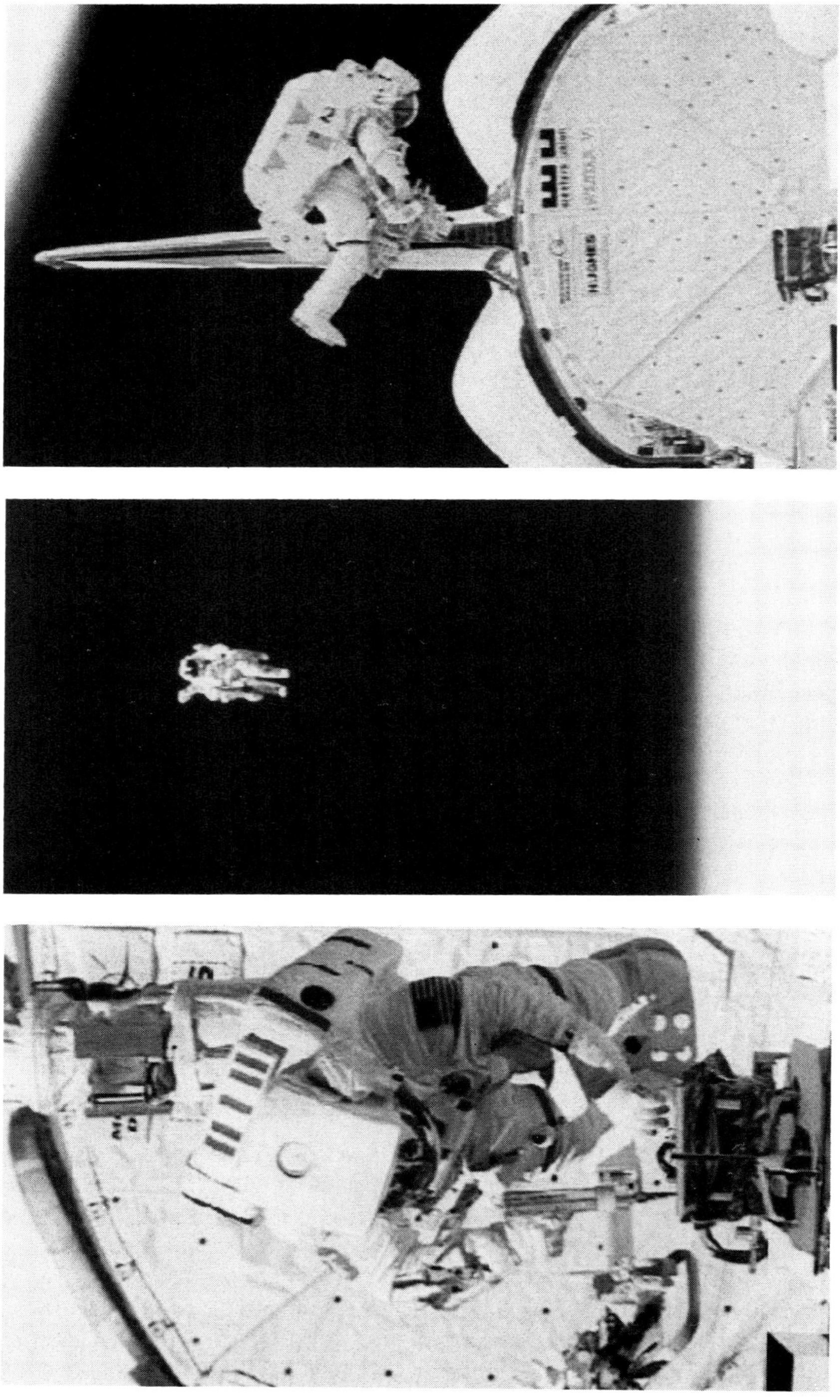

(a) (b) (c)

Fig. 4.5. On Flight 41B: (a) Bob Stewart stands in the Manipulator Foot Restraint (MFR), on the end of the Remote Manipulator System (RMS), to test a power tool designed to work in space, (b) Bruce McCandless pilots the Manned Manoeuvring Unit (MMU) away from the orbiter and (c) with the Trunnion Pin Attachment Device (TPAD) in position across the MMU, approaches the Mission Peculiar Experiment Support Structure (MPESS) in the centre of the bay to simulate the capture of the SolarMax satellite.

up of the electronics unit on the SolarMax satellite. His task was to establish whether the tools really worked, because sometimes apparatus designed for use in space failed. It was not simply a matter of making tools that would work in weightlessness; they also had to function in vacuum and survive the thermal stresses. Testing the SolarMax tools now would reduce the risk of a nasty surprise later on. Stewart had allowed the MMU to build up much higher speed than had McCandless, so he was soon back. It was just as well, because at just over six hours, they were at the limit of their suits' nominal life support capability. Having achieved all their objectives, they returned, highly satisfied, to the airlock.

Their second excursion was marred by a fault in the RMS. This ruled out a repetition of the cherry-picker tests, but the first test had been sufficient to demonstrate that the arm was an accurate and stable platform. While McCandless donned the second MMU, his colleague unstowed the Trunnion-Pin Attachment Device (TPAD) and then installed it on the front of the MMU. Projecting from the front of the TPAD was a stubby hollow cylinder that was a copy of the arm's end-effector. The plan had been to hoist a SPAS deployable pallet out of the bay and slowly turn it on the end of the arm so that McCandless could attempt to dock with it by sliding the TPAD over a pin on the satellite, as another astronaut would hopefully do with the slowly turning SolarMax. Unfortunately, the electrical fault in the arm affected its wrist, the very joint needed to rotate McCandless' target, so this part of the test had to be cancelled. It was still possible to verify the TPAD's locking mechanism with the SPAS still mounted in the bay by positioning himself over the pin, and this he proceeded to do several times.

Stewart's main task this time was to work at another SAB – one which contained a replica of SolarMax's propellant system. He coupled hoses to valves, then pumped fluid from one tank to another. SolarMax was not in need of refuelling, but the basic spacecraft bus was of a multipurpose design which had been exploited by several satellites, and one, Landsat 4, was in trouble; part of its communications system had failed. If the SolarMax repair proved successful, Landsat was likely to be the next to receive the astronauts' attention. However, its orbit was far too high for the shuttle, so it was planned to have the satellite lower itself. But it would expend most of its propellant in doing so, and if it was to be repaired it would also have to be replenished. This Orbital Refuelling System (ORS) test was a step towards determining whether this would be practicable. The fluid Stewart pumped was not hydrazine rocket fuel; it was just a dye. The valve and the pump worked, so a later flight would try it out for real. If it proved possible to pump fluid into satellites, then it would make sense to design orbital telescopes using coolant to be capable of being replenished, so that their service life could be extended. Clearly, there was a multiplicity of ways in which astronauts could make themselves useful in space.

Once McCandless returned with the TPAD, the two astronauts exchanged roles again. As Stewart was returning to stow the MMU, he spotted that the PFR had come loose and was drifting out of the bay. Even before Stewart could reverse his undoffing, McCandless made his way hand-over-hand down the sill of the starboard sidewall. By the time he reached the position of the PFR, however, it was too high for him to reach whilst retaining hold of the orbiter, so he called steering commands to Vance Brand, the commander on the flight-deck, who manoeuvred the orbiter to place McCandless' outstretched hand on the loose item. It was a remarkable demonstration of teamwork. By the time the spacewalkers returned to the airlock, they had each accumulated 12 hours outside. Events had gone very well. The only aspect of the SolarMax capture that they had not been able to simulate was its rotation, but apart from that, the prospects of success were looking good. The activities had been caught on 35-mm film by cameras equipped with ultra-wide-angle lenses, in a format designed for projection on a planetarium dome to give a 360-degree wraparound view. An IMAX camera was also carried for the first time; its 70-mm film combined a large format with high resolution.

The astronauts' satisfaction was tempered by the fact that the other goal of the mission, the deployment of a pair of commercial satellites, had turned sour by motor failures that had prevented them from reaching their assigned orbits. Nevertheless, they had cleared the way for the attempt to fix SolarMax two months later.

In November 1980, only a few months after it had been launched, SolarMax had lost its attitude-control system; a power spike in the electronics had blown the fuses. Its mission was to follow up the Orbiting Solar Observatory satellites by monitoring the Sun during the most active part of the 11-year sunspot cycle. Most of its seven instruments had to be aimed precisely at the Sun. The satellite was put into a spin so that the few instruments which did not require a steady view could carry out a restricted programme. The Sun had passed its peak and was nearing its minimum level of activity when STS-11 finally drew up alongside the stricken satellite. In addition to restoring its attitude-control capability, the astronauts' list of tasks included repairing an instrument that had been a little flakey from the very start. If SolarMax could be made fully functional, it stood every chance of surviving long enough to follow the Sun through its minimum and up to the next maximum, and thereby carry out its original mission.

The rendezvous with SolarMax required Challenger to climb to 560 km – higher than the shuttle had been before. On the way, it dropped off its primary payload, the Long-Duration Exposure Facility (LDEF). By the time the orbiter assumed station about 100 metres away, the controllers at the Goddard SFC had virtually cancelled SolarMax's spin and powered down most of its systems. Its fate rested with George 'Pinky' Nelson and James 'Ox' van Hoften.

Nelson donned the MMU and collected the TPAD, and then set off towards SolarMax. As had his predecessors, he flew very slowly, so it was almost 10 minutes before he drew up to study his target. SolarMax was actually two units bolted together. At the base was the spacecraft bus, and above it was the scientific package. A pair of broad solar panels projected either side of the interface ring. SolarMax was not actually static. It had a stable axis, but it was rotating at about one revolution per minute. The pin on which the TPAD was to lock was on the bus, so Nelson had to manoeuvre under the level of the solar panels and make sure that they did not clip the top of the MMU as they passed slowly over his head. Everything looked good, so when the trunnion pin came into view Nelson moved in to drive the collar of the TPAD over it. Apart from the spin, this was exactly what McCandless had done. On this occasion, however, when Nelson activated the mechanism the TPAD refused to engage the pin. He withdrew, waited for SolarMax to present the pin again, then redocked with the same result: the end-effector in the TPAD refused to lock on the pin it had been specifically designed to grip. In frustration, Nelson withdrew a few metres to consider what to do. But what, indeed, could he do?

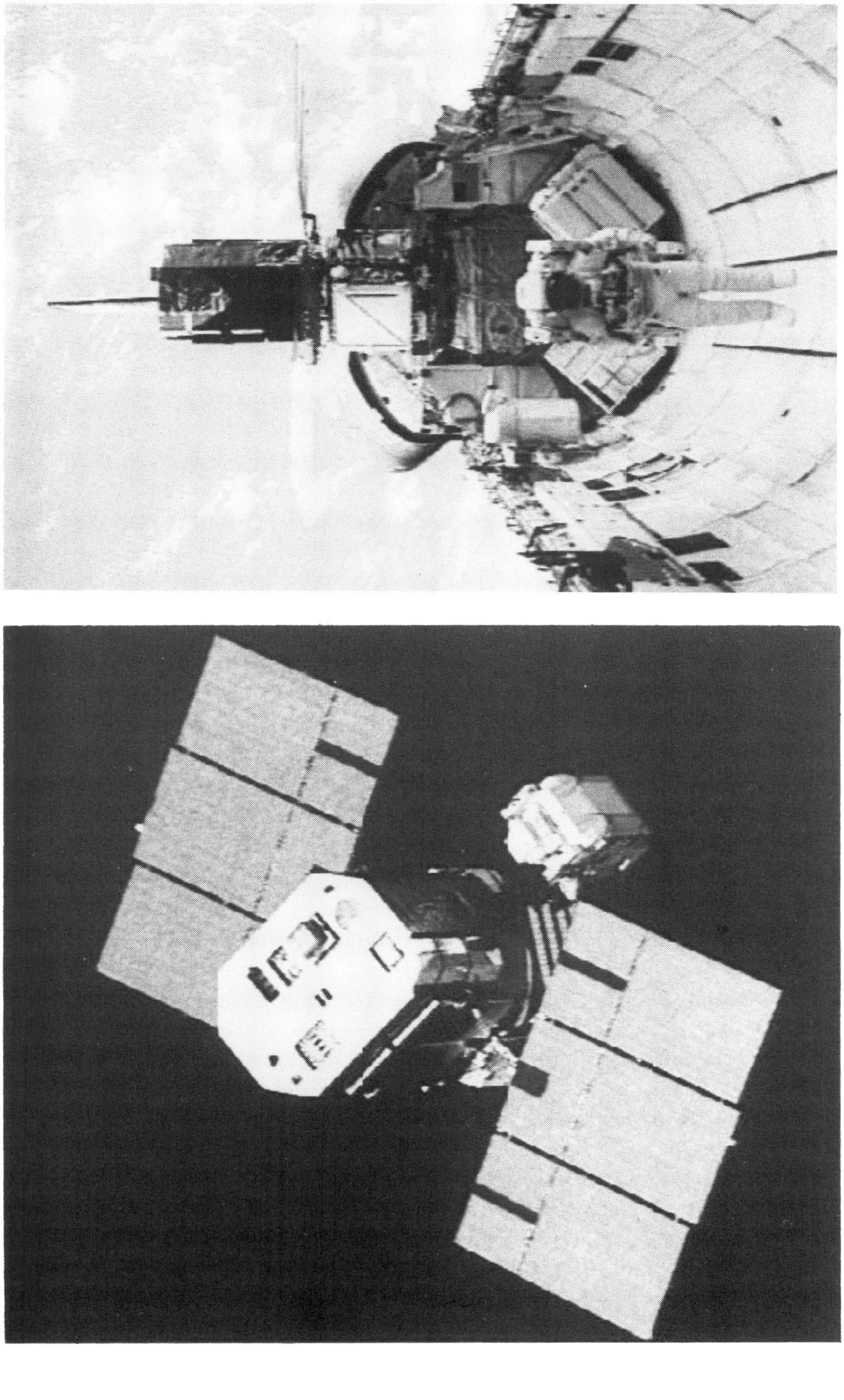

(a) (b)

Fig. 4.6. Flight 41C: (a) George Nelson has piloted the Manned Manoeuvring Unit (MMU) over to the SolarMax satellite and is about to attempt to dock using the Trunnion Pin Attachment Device (TPAD), and (b) the satellite mounted on the Flight Support Structure (FSS) at the rear of the bay.

Almost as an afterthought, Nelson reached out and grabbed hold of the tip of one of the solar panels as it drifted by. If he could cancel the residual spin, Challenger's RMS might be able to do what his TPAD had not. Unfortunately, before the Goddard controllers could intervene to warn Nelson off, his action had transformed the axial spin into a crazy pitching roll that threatened to incapacitate SolarMax. No longer able to keep its solar panels facing the Sun, it began to draw on its battery. Under normal circumstances, this chemical battery was used only during the half-hour or so that SolarMax was in the Earth's shadow, and it would then be immediately recharged when the spacecraft passed back into sunlight. Even with most of its systems powered down, SolarMax would rapidly exhaust its battery. Nelson attempted to overcome the tumbling motion, but the conservation of angular momentum prevented him from doing this, and he only made matters worse. With his MMU running low on nitrogen propellant, Nelson was recalled to the orbiter. With shocking suddenness, it seemed that the satellite that he had set out to rescue had been written off! Challenger manoeuvred clear, and the astronauts did their best to sleep. It was a close call, but Goddard managed to orientate SolarMax so that its solar panels provided enough power to run the core systems, which provided the time needed to stabilise the spacecraft; so Challenger returned two days later.

This time it was decided to try to snatch SolarMax with the RMS. This had not been the favoured means of capture, because it required the arm to absorb the energy of any residual spin, which risked damaging the arm's delicate joints. In the event, it was trivially easy for Terry Hart to station the end-effector and then slide it over the pin on SolarMax's bus when it rotated into view, and the arm absorbed the energy without difficulty. It was immediately deposited onto the FSS at the rear of the payload bay so that it could draw power from the orbiter via an umbilical. This structure incorporated a large U-shaped cradle and a tilt-table. It had been built to enable the shuttle to carry into orbit new satellites using the same type of Fairchild multipurpose bus, so if it proved impracticable to fix SolarMax it was to be returned to Earth. Glad to have recovered the situation, the astronauts took the rest of the day off so that they would be refreshed for the repair operation the following day.

While Nelson retrieved the PFR, van Hoften attached the MFR to the arm and mounted it, and then, with Nelson tagging along, Hart transferred them to the rear of the bay, where Nelson set his foot-rest into a socket near the sill of the sidewall. The tilt-table could turn as well as tilt, so it was positioned to give the repairmen access to SolarMax's attitude-control system. The bus comprised three boxes, each about 1-metre square, set 120 degrees apart. In fact, because the nature of the fault was not understood, it had been decided to replace the entire controller. In practice, this meant replacing one of these boxes. The bus had been designed to be serviced, so the astronauts had only to undo a set of bolts to release the box. The new controller was stowed in a container on the FSS, and there was a container ready to take the old unit. It was a simple matter for Hart to swing van Hoften with his bulky package down for the swap. The heavy lifting work would have been feasible without using the RMS, but the arm made it easier, and faster, and *time* was the limiting factor in a spacewalk.

Installing the new controller fulfilled the main objective, because it restored SolarMax's ability to aim its instruments at the Sun. If they had been running late, the astronauts might have called it a day, but they were ahead of schedule, so they set out to fix the

(a) (b)

Fig. 4.7. Flight 41C: (a) George Nelson and James van Hoften work on SolarMax, and (b) the repaired satellite following its release.

coronagraph, which had never really worked properly. This was a much trickier task, because they could not replace it, and they had to take it apart to repair it. First though, they had to gain access to the instrument. The satellite was repositioned on the FSS to present the instrument section. The first task was to slice through the thermal insulation blanket with a pair of heavy-duty scissors, and then the loose flaps were folded back and held in place by duct tape. Six screws had to be undone to release the underlying hatch. The screws could damage the instrument if they found their way into it, so they had to be retrieved – an awkward procedure for gloved hands. It would be impracticable to reuse the screws, so van Hoften set another strip of tape to act as a hinge. The astronauts were making excellent progress, which was just as well because to disconnect the instrument's controller required undoing an additional 22 screws! The task of wiring up the new controller fell to Nelson. Fast-action spring-clips were used to re-establish the 11 circuits. This done, the hatch was closed and taped in place, and the insulation was flattened and taped in place. Although this tape would alter the thermal properties of the instrument unit, once the spacecraft was operating this surface would not face the Sun. Having completed both their primary and secondary tasks, van Hoften and Nelson retreated to the airlock. Their excursion had lasted a record 7.3 hours.

SolarMax was released the next day, 12 April, the third anniversary of the first shuttle's launch. Once more fully able to adjust its orientation, SolarMax aimed its instruments at the Sun, and the coronagraph proved to be operating better than ever. This was a well-timed repair, because a fortnight later, even though the Sun was well past its peak, the largest flare since 1978 was seen. SolarMax recorded it, in all its violence. As the solar cycle slowly worked up to its new peak, it excited the upper layer of the Earth's atmosphere, which increased the rate of decay of the satellite's orbit. A propellant replenishment was tentatively assigned as a secondary objective on the 1990 mission which was to retrieve LDEF (which was also suffering from this increased drag) but, as it turned out, SolarMax re-entered the atmosphere in December 1989, having maintained its watch on the Sun right to the end. The retrieved attitude-control system was returned to Fairchild, which repaired it and then used it for the Upper Atmosphere Research Satellite (UARS); so a piece of SolarMax is still hard at work.

To the critics who argued that it would have been cheaper to build another SolarMax, and send it up on a Delta rocket, NASA responded by pointing out that this repair had been undertaken *on the fly* as a secondary task after Challenger had deployed its main payload. Some argued that a dexterous robot could have performed the repair, so such autonomous robotic spacecraft should be developed, but NASA maintained that a human presence made the vital difference between overcoming the unexpected and being overcome by it. It was a long-standing difference of opinion.

In any event, even as the repair of SolarMax had been underway, other astronauts had been rehearsing a procedure to recover the satellites lost by STS-10. NASA was like any sales organisation in that no matter what the current target for its operatives, it had to go one better once this had been achieved, for the simple reason that it could not afford to stand still. So it was that while Joe Allen and Dale Gardner trained in the WETF for the next ambitious spacewalk, two manifests were merged, and STS-12 became the first mission to carry three communications satellites. Although no spacewalk had been planned for this flight, a build-up of ice outside the waste water vent on the port side of the orbiter created an icicle half a metre long. A similar icicle had formed on STS-10 and been ignored, and was later believed to have snapped off just before re-entry and collided with the bulging OMS pod, damaging the thermal insulation. Therefore, it was decided to knock this one off. The water vent was in the sidewall, about level with and to the aft of the crew hatch. Henry 'Hank' Hartsfield was able to contort the RMS to reach the side of the orbiter, and nudged the icicle several times, but it refused to budge. He had to take care not to damage the thermal insulation on the leading edge of the orbiter's wing, which was just beyond the icicle. Each crew includes two EVA-trained members whose role is to go out and perform exigent tasks. It was decided that such a 'contingency' excursion was called for in this case. When Hartsfield finally announced that he had finally snapped the icicle off at its root, Richard Mullane and Steven Hawley were in the airlock and almost ready to enter the bay. Once the vent's nozzle had been redesigned, it gave no further trouble.

The next planned spacewalk was on STS-13 in October 1984, when Kathryn Sullivan and David Leestma took the ORS a stage further. This time, it was on an MPESS installed at the rear of the payload bay to hold a large format terrain-mapping camera. Their objective was to use seven special tools to insert a valve into the plumbing mock-up. It could be done only once, so they worked as a team. Once they had left the bay, 90 kg of hydrazine was pumped through the valve for a high-fidelity simulation of replenishing a satellite. This had not been performed while the astronauts were outside, in case the highly toxic fluid leaked; there was no simple way to decontaminate suits in the airlock. Bob Crippen, commander of this flight, had initially been reluctant to use hydrazine, but the flow was completed without incident. This led to the replenishment of Landsat 4 being approved. However, because the satellite's polar orbit was unreachable from Cape Canaveral, the flight would have to depart from Vandenberg AFB. It was assigned to a mission then scheduled for 1987.

TWO UP AND TWO DOWN

The next spacewalking drama – the recovery of the two lost satellites – took place in November. Again, it was done on the fly, because STS-14's primary objective was to deploy a pair of communications satellites. Better yet, the insurer had paid NASA a fee to bring back the old satellites so that they could be refurbished. The solid-rocket motor of a PAM was supposed to burn for 83 seconds, but a common fault had caused both of these motors to fail after just 10 seconds, sufficient only to raise the apogees to 1,000 km. After the useless motors had been jettisoned, the satellites had lowered themselves using their own small thrusters. The two satellites were now in similar orbits a few hundred kilometres apart, so this recovery mission involved far more manoeuvring than any previous flight.

Discovery drew to a halt barely 10 metres away from Palapa. Its 60-rpm spin had been virtually cancelled; its slow residual roll ensured axial stability. There was no time to waste, so Joe Allen and Dale Gardner were soon in the bay preparing their tools. The fact that the Palapa and Westar satellites were both HS-376s was lucky, because it meant the same tools could be used in each case.

Because these satellites had been designed to operate in GSO, nobody had expected that they would be seen again after deployment, so they did not have a trunnion pin. It was therefore not possible for the RMS simply to reach out and snatch them as it had Solar-Max; this time the participation of the spacewalkers was crucial. The absence of a pin also meant that the TPAD could not be used. A new capture procedure had had to be devised, and necessary equipment manufactured. To capture the satellites without damaging the solar cells wrapped around their sides, it had been decided to develop a device which would hook onto the ring at the base that had mated with the PAM stage. Although its designers referred to this as the capture device, its shape had inevitably resulted in the astronauts dubbing it the 'stinger'. It was really several devices in one. Like the TPAD, it had a clamp to attach to the arms of the MMU. A long rod (the stinger itself) projecting in front was to be inserted into the throat of the satellite's apogee kick-motor. A ring at the base of the rod was to mate with the adapter on the satellite. Lastly, there was a trunnion pin on the right-hand side to enable the RMS to take the satellite once it had been stabilised. This was ingenious improvisation, but would it work?

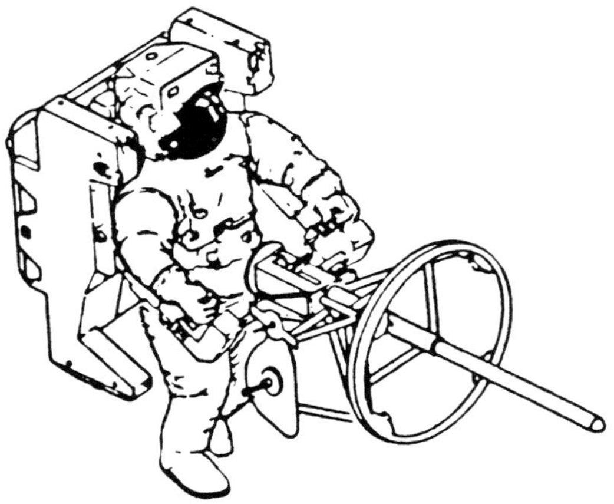

Fig. 4.8. A schematic of the 'stinger' designed to capture an HS-376 satellite. Because the design of this type of satellite did not incorporate a trunnion pin attachment, it was decided that the best way to recover it would be to poke the stinger through the nozzle of its motor to align a circular clamp which could then engage the support ring which had been used by the now discarded Payload Assist Module (PAM) perigee kick-motor.

Allen manoeuvred into position directly behind Palapa, waited a few minutes, and then eased forward. The stinger's success was evident even before Allen reported it, because his audience saw him suddenly adopt the satellite's spin. The MMU soon cancelled this. Once Allen had manoeuvred into a convenient orientation, Anna Fisher brought the RMS in over his shoulder to snare the stinger's trunnion pin. With Allen still attached, Fisher pitched the satellite vertical, then eased it down into the bay where Gardner was waiting with shears to cut off the omni antenna rod that projected out from the top of the satellite. The only way to stow the satellite in the bay was to turn it upside down. A bracket had been built to run over the top of the satellite, but it could not be attached. It had been tested on similar satellites at Hughes, but the HS-376 was tailored to suit its customer's requirements, and the feed horn on the folded antenna on top of Palapa projected a little further than the tool's designers had been led to believe, and this prevented the clamps from engaging. The plan had been to have Gardner attach the bracket, for Fisher to raise Allen back out of the bay and release him, for him to then pitch the satellite over, and for Fisher to snatch the grapple fixture on the bridge that Gardner had installed. Allen would then disengaged the stinger and Fisher would rotate the satellite and lower it onto the awaiting pallet. But the bridge would not fit. It was time to improvise.

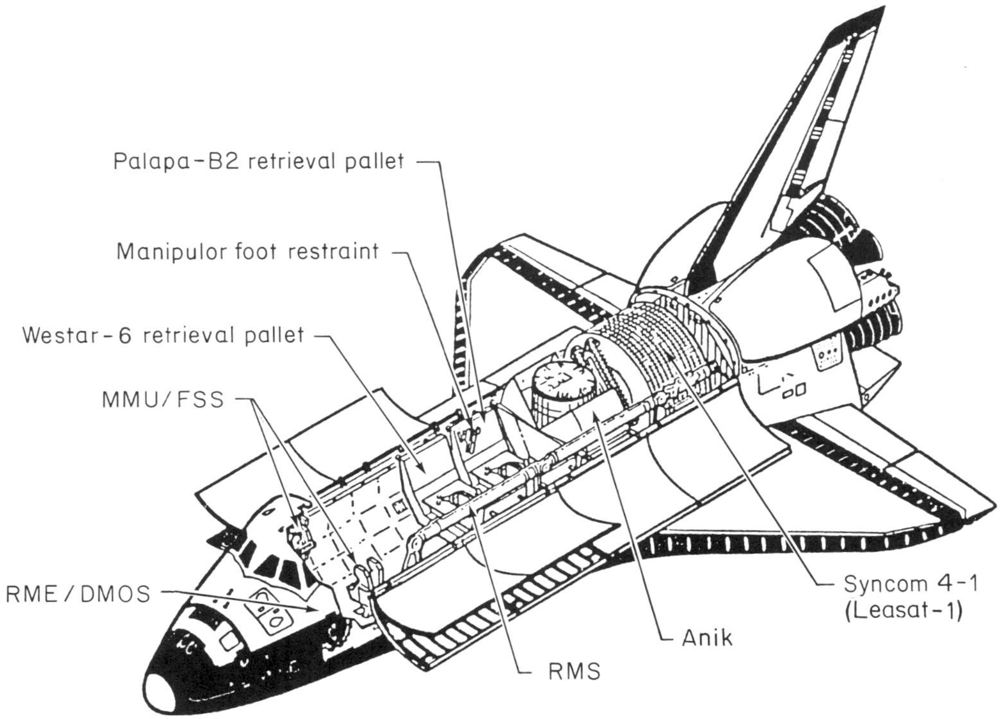

Fig. 4.9. A schematic of the payload bay layout for Flight 51A showing, at the rear the satellites which were to be deployed and, further forward, the pallets on which the two HS-376 satellites were to be mounted, if they could be captured.

In the revised plan, Allen disengaged from the stinger, which remained on Palapa, with the arm holding it. While Allen stowed the MMU, Gardner moved the PFR to the sidewall. Allen then stood on the PFR, and took hold of the satellite by its still-folded main antenna. When Fisher withdrew the arm, Allen did his best to hold the satellite steady while Gardner disengaged the stinger. Gardner placed a 'shower cap' over the rocket nozzle so that flakes of solid propellant did not fall out into the bay when the orbiter landed. Gardner then lifted a mount off the floor of the bay. This was an A-frame with a ring matching that on the base of the satellite; mating the rings fastened the frame to the satellite. All this time (well over an hour) Allen held the satellite steady. Now Allen and Gardner together pitched Palapa down into the bay and then Gardner fastened the A-frame to its floor mount. Despite the failure of the common bracket clamp, as the top bracket was called, the

(a)

(b)

Fig. 4.10. Flight 51A: (a) Joe Allen closes in with an HS-376 to insert the stinger into its nozzle, and (b) Allen manoeuvres the captured satellite so that Dale Gardner, standing on the Manipulator Foot Restraint (MFR) on the Remote Manipulator System (RMS), can take hold of its far end; (c) the plan to use the RMS to manipulate the satellite in the bay had to be abandoned because a bracket would not fit, so Allen and Gardner manhandled the satellite on the circular clamp on the bay floor; (d) with the quarry safely stowed, the astronauts pose on the RMS and present their 'FOR SALE' sign. Both of these satellites were refurbished and relaunched.

(c)

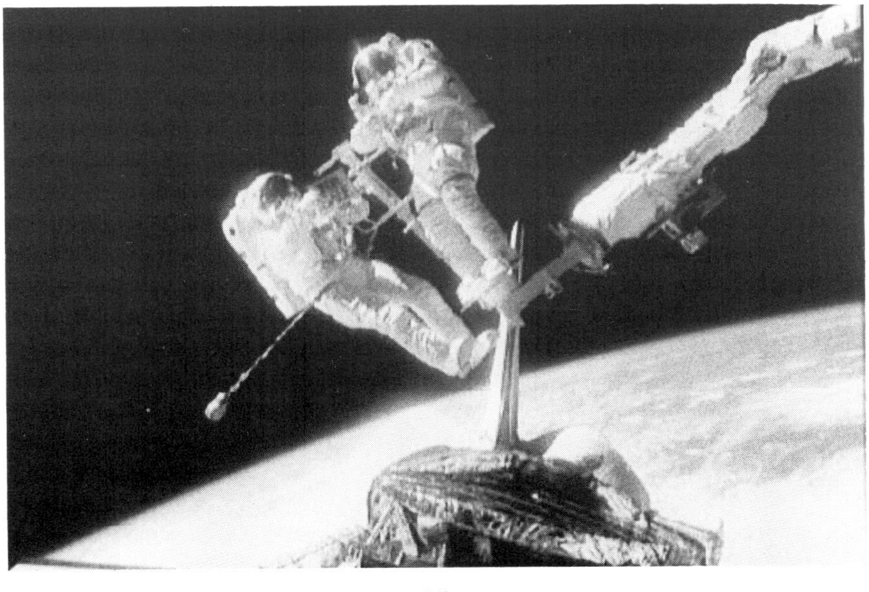

(d)

two men had managed to lock the errant satellite into the bay. By the time they closed the airlock, they had been out for six hours.

Two days later, Discovery pulled up alongside Westar, and Allen and Gardner went out to retrieve it. This time Gardner used the MMU. Rather than waste time testing to determine whether the bracket would fit this satellite, it had been decided to adopt a revised version of the established procedure. This time the arm carried the MFR and Allen rode it. The satellite was captured in the same way, and Gardner flew it over to Allen, who took hold of it using the antenna, as previously. The stinger was immediately disengaged and stowed, together with the MMU. With Gardner on the bay floor, Fisher slowly brought the arm down. The fact that Allen had Westar by its antenna meant that the satellite arrived the right way up, so Gardner was able rather more easily to fit the A-frame and mount it on the bay floor. The final act was to sever the omni antenna so that it would not interfere with the closing of the bay doors. Before they came in, they perched on the arm and displayed a placard to the cameras announcing that they had two satellites for sale. Later, they displayed another card, echoing the card shown by the crew that deployed the shuttle's first commercial satellites, this one identifying the shuttle as operating for the *Ace Repo Company – The Sky's No Limit*. The crew of Discovery were subsequently awarded the Silver Medal by Lloyds of London, the insurers. This award was issued to persons whose "extraordinary exertions contributed to the preservation of property". Both satellites were subsequently sold, and relaunched under new names.

Once again, the astronauts had overcome equipment failure and used improvisation and muscle to complete their objective. It is also important to note that the satellites had run into trouble only nine months earlier, and although NASA had almost immediately agreed to try to recover Palapa, Westar had been added to the plan only two months before the flight; so in contrast to the SolarMax repair, into which years of preparation had gone, this double repossession mission had posed a different organisational challenge. NASA had eased into a dynamic planning mode in which subsequent flights rectified the faults of earlier ones, *in addition* to achieving their own objectives. Considering the complexity of mounting each flight, this dynamic planning was extremely impressive.

HOT-WIRING A SATELLITE

As the Leasat-3 satellite emerged from Discovery's bay, its release was meant to trigger the switch that activated it. The deployment had seemed to go well, so Discovery withdrew to leave the satellite to run through its start-up sequence. When the satellite did not send any telemetry to its ground station, it was realised that the infallible switch had somehow failed.

Rather than leave the dormant satellite for a later flight to deal with, it was decided to extend the flight, and rendezvous for a contingency-EVA. As it happened, STS-16 had an influential passenger. Senator Jake Garn was 'observing' in his capacity as chairman of the subcommittee that oversaw NASA's budget. Surely all that was required was to throw the switch! This was a great opportunity to show that the most effective back-up system was a human presence.

The spacewalkers were David Griggs and Jeff Hoffman. The initial plan called for them to mount the RMS and gingerly throw the switch. Ideally, that would start the timer of the sequencer which controlled the engine firings during the GTO manoeuvres. But what if the satellite started up in a disturbed state and fired its motor immediately!? It was reluctantly decided that having the astronauts throw the switch manually would be too risky. A method would have to be devised which would not require an astronaut to be near the satellite when it came to life. The crew was told to get some sleep, while the support staff on the ground considered the matter, but the astronauts stayed up half the night devising a scheme of their own.

The next day, a plastic document cover was cut to construct a flexible 'fly swatter' with a narrow slit near its tip. Griggs and Hoffman took this out into the bay and taped it to the 40-cm wide cylindrical end-effector. Once the spacewalkers had retreated to the airlock for safety, Rhea Seddon brushed the tip of this masterpiece of make-do against the solar cells coating the surface of the slowly rotating drum, then waited for the switch to come around. After several near-misses, the fly swatter's slit caught the switch and threw it, just as it was meant to. Discovery withdrew 100 metres to observe the satellite's activation, but when the rod of the omni antenna did not swing up on time it was acknowledged that the satellite was lifeless, so it had to be abandoned. It was evident that the satellite had not been disabled by a simple switch fault; if it had, the fly swatter should have worked. If a way could be devised to jump-start it, the derelict satellite might still be rescued. A detailed analysis of the likely failure mode revealed just such a possibility, and several months later, after deploying its trio of brand new satellites, Discovery drew up alongside Leasat-3.

James van Hoften was waiting on the RMS, and Mike Lounge eased him out to the satellite, which was still axially stable due to its slow roll. Van Hoften had taken a specially-built bar to span the gap between two of the sockets used to hoist the satellite on the ground. He waited until the attachment points rotated into view, then pushed the capture bar into position. When it came around the next time, he grabbed it and manually cancelled the spin, using his arms to absorb the energy. This had been achieved despite a fault which impaired the movement of the RMS.

Meanwhile, Bill Fisher had set up station on a PFR, on the starboard sidewall. The arm slowly moved van Hoften down towards the bay, and he dragged the 7-tonne satellite with him. Once it was in place, just over the bay, Fisher attached a second handling bar and took the satellite. Van Hoften replaced his bar with a trunnion pin assembly and then dismounted the arm and removed the MFR from the end-effector. The RMS then relieved Fisher of the satellite. It was teamwork all the way.

With the Leasat in place directly above the bay, both men were free for the really tricky work – the installation of the Spun-Bypass Unit (SBU). This was another piece of apparatus built for this mission, and contained the timers required to activate the satellite. All they had to do was to install it so that it would bypass the faulty switch. It was to be attached to a point on the side of the drum, and cables were run out to the circuitry. The system had been tested on a satellite in the Hughes factory, but doing it in space was not expected to be easy. What made it so tricky was that the HS-381 had most definitely *not* been designed to be worked on by astronauts.

The immediate task was to safe the satellite by disabling its pyrotechnics, so that there was no danger of triggering something while disabling the seemingly dead Post-Ejection Sequencer and installing SBU. All of this tricky rewiring was completed without incident.

(a)

(b)

Fig. 4.11. Flight 51I: James van Hoften, on the Manipulator Foot Restraint (MFR) high above the payload bay, waits for the orbiter to draw alongside a dormant Leasat; (b) once the satellite had been repaired, and released, van Hoften had to manually spin the massive satellite back up for stability.

The SBU diagnostic display gave the welcome news that the satellite's battery was in good shape; there had been concern that it might have degraded. The next big task was to install a Relay Power Unit (RPU). This directly closed the relay in the satellite which deployed the omni antenna and, to everyone's delight, this swung up. It also started the telemetry, and enabled the satellite to take commands. By this point, the astronauts had effectively restored Leasat-3 to the state equivalent to just after a successful deployment. The two men had been out for almost seven hours, so there was no time for redeployment. It was decided to leave this to another excursion, which meant extending the flight. Unlike previous practice, these spacewalks took place back-to-back because there were consumables for only one extra day in space; there would be no day of rest, therefore. Leaving the satellite on the arm overnight was not really desirable with the arm in its partially degraded state, but breaking the repair had left no option.

One of the uncertainties in reactivating the Leasat after so long in its dormant state, was that if the solid propellant in its integrated perigee-kick motor had grown so cold that cracks had formed, it was likely to explode upon ignition. On the second excursion, a probe was inserted into the nozzle of the engine so that the temperature could be determined. This was equipped with a telemetry link so that it could monitor later attempts to warm the propellant. Once this was verified to be functioning, the arming pins were pulled from the SBU to start the various timers. Fisher resumed his station on the sidewall and held the satellite, then the arm disengaged and van Hoften removed the trunnion pin. After slipping the MFR back on the arm, van Hoften mounted it. He then retrieved a third handling bar. Rather smaller than the capture bar, this was the spin-up bar. Once van Hoften had the satellite, Fisher removed his bar. Lounge drew the arm up out of the bay, and van Hoften dragged the satellite with him. It could not just be released. As it had first emerged from the bay, six months ago, it had picked up a 2-rpm roll, but this had been cancelled when van Hoften had grabbed it. Now he had to restore it.

The HS-381 was the first satellite built by Hughes specifically to exploit the great width of the shuttle's payload bay. This 3-metre high, 4-metre wide drum was the largest satellite so far encountered by a spacewalking astronaut. Even a mighty heave barely made it spin. Two further boosts were required to build up to the required rate. The task was complicated by the fact that the initial impulse also set the satellite drifting away, so the arm had to chase it so that van Hoften was in place to grab the bar each time it came around. The sight of one of its astronauts manually spinning up this enormous satellite was tailor-made for NASA's Public Affairs Office. The bypass performed flawlessly. The solid motor was left to warm for six weeks before being commanded to fire, and it propelled the satellite up to GSO as intended.

This Leasat repair is an excellent case study of how NASA added activities to planned flights. STS-16 had been forced to abandon the satellite, and withdraw, but it had later become apparent that it should be possible to perform a rescue. STS-17 was a Spacelab flight. This module took up the entire bay, so there was no room to work on a satellite. In any case, the RMS was usually not carried on such a flight. STS-18 was a satellite-deployment mission, and it could have been assigned the job, but this would have allowed very little time for the crew to prepare. STS-19 was another Spacelab. Thus, by a process of elimination, STS-20 was the first available mission to offer both a clear bay in which to work and sufficient time to train. The dynamic planning cycle-time enabled mistakes to be rectified about six months later. In fact, it is amazing just how flexible the shuttle was proving to be. Deploying three satellites and hot-wiring another was pretty good for a single flight by any standard, but the fact that it had been pulled off at short notice indicates that as an operational system, the shuttle had matured rapidly. Spacewalkers were being assigned ever more complicated tasks, and were succeeding. And with each success, they became more adventurous. Although they were always careful to employ a step-by-step methodology in developing the state of their art, they were definitely taking increasingly 'longer strides'. If a placard were to have been shown after the Leasat rescue, surely it would have read: *No Mission Too Difficult.*

ASTRONAUT HARD-HATS

In early 1984, President Reagan authorised NASA to build a space station. Back at the start of the 1970s, when the shuttle had first been proposed, it had been as *part* of a project to build a space station; the shuttle was to facilitate assembly, and the station was to provide the shuttle with a community in space to serve. But the projected cost had been so astronomical that NASA had been obliged to delete the station. The cost depended not only on *what* was to be built, but also on *how* it would be constructed. In particular, it depended on whether complex semi-automatic deployment systems would have to be developed, or whether most of the assembly work could to be undertaken by astronauts.

The fact that spacewalkers were not only performing so much work, but were doing so *by hand*, encouraged NASA to believe that it would be able to rely on spacewalkers to put together the structures that would form the space station, and surely that would cut costs. It was important to assess fabrication techniques as soon as possible. With the results injected into the design process, it would be possible to select between different station strategies, and so refine time and cost projections. An experiment to test two means of erecting trusses was assigned to STS-23. This was no 'fly off' to choose between competing techniques; the objective was to ensure that the *problem* was understood. No tools were required, as both structures were to be built with rods and individually-shaped nodes which snapped together by hand.

On the first of two excursions, Jerry Ross and Sherwood Spring began by going to the MPESS situated conveniently at the front of the bay. They elevated a platform up alongside the sidewall sill. On this platform, they were to erect a 15-metre tall triangular cross-section truss. For this Assembly Concept for Construction of Erectable Space Structures (ACCESS) test, the MPESS was loaded with 100 tubular rods and a stock of tongue-in-groove nodes using sliding-sleeve locks. It would take only a few seconds to make each joint, although the astronauts had to fetch, transport, orientate and attach each element in turn. At first, Ross took up station at the base and observed while Spring built the structure, doing all his own fetching. At the half way point they exchanged places, and Spring passed successive pieces up to Ross, so that they worked as a team, with each performing his own task. It took only half an hour to build the 10 frames required for the tower's full length. It was immediately disassembled, so that they could move on to the other apparatus, known as Experimental Assembly of Structures in EVA (EASE) which tested a different construction strategy. This used rods twice as long as ACCESS, so it was impracticable to construct a full truss. In fact, only one 2-metre triangular segment was erected.

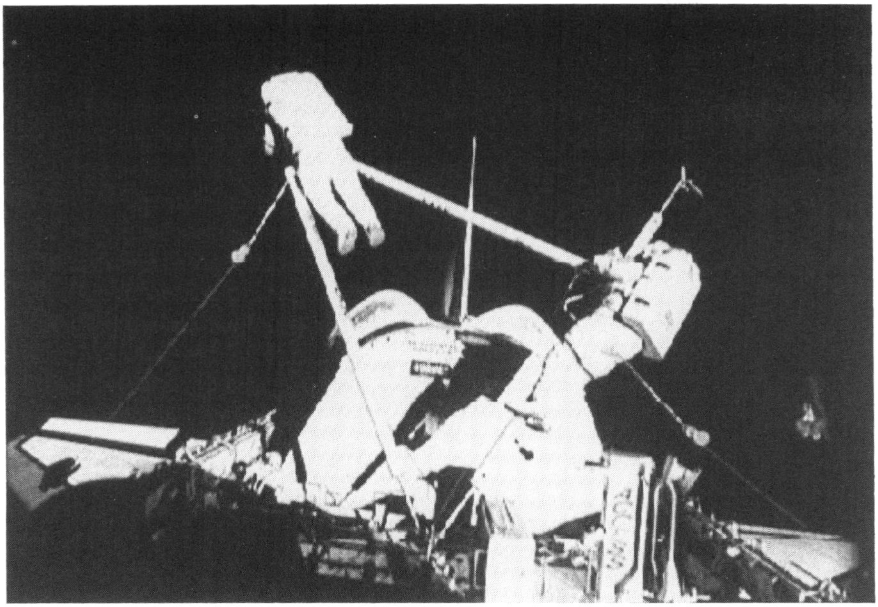

(a)

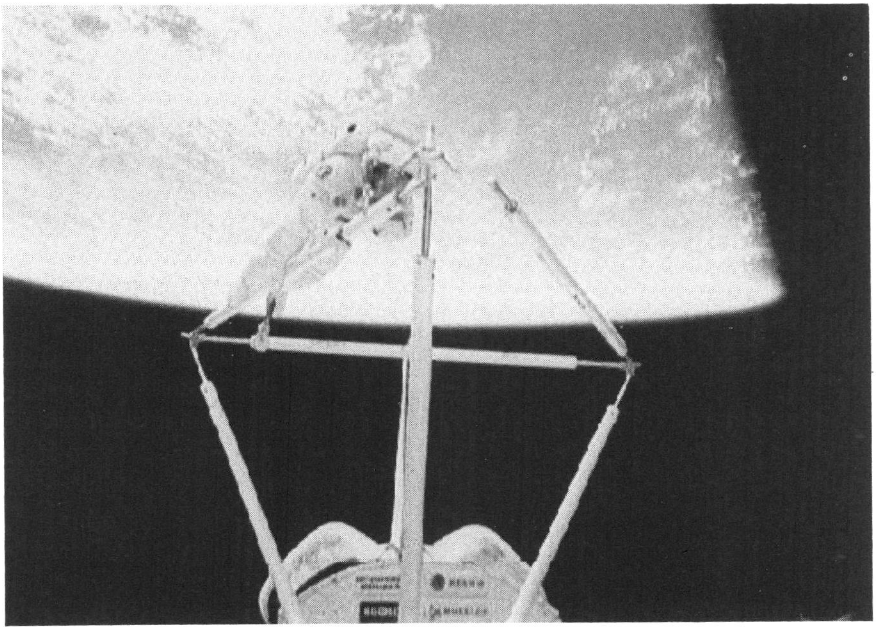

(b)

Fig. 4.12. Flight 61B: (a) and (b) Jerry Ross and Sherwood Spring assemble a triangular segment of the Experimental Assembly of Structures in EVA (EASE), to assess their ability to construct a truss for the space station. All the work was done while freely floating.

Many such structures would have to be joined together to create a truss. It was built on top of the MPESS, in inverted form. Ross and Spring worked as a team by coordinating to deal with *opposite ends* of each element in turn. Because they had to hold onto the structure with one hand, they had to locate a rod for connection and engage it using only the other hand. This was an acquired art. As soon as it was finished, the structure was disassembled. After the two structures had been built and stripped several times, the astronauts called it a day. Although at 5.5 hours this was not the longest spacewalk, it had been the most energetic yet. It had also been the most dexterous. Both men reported that their fingers ached.

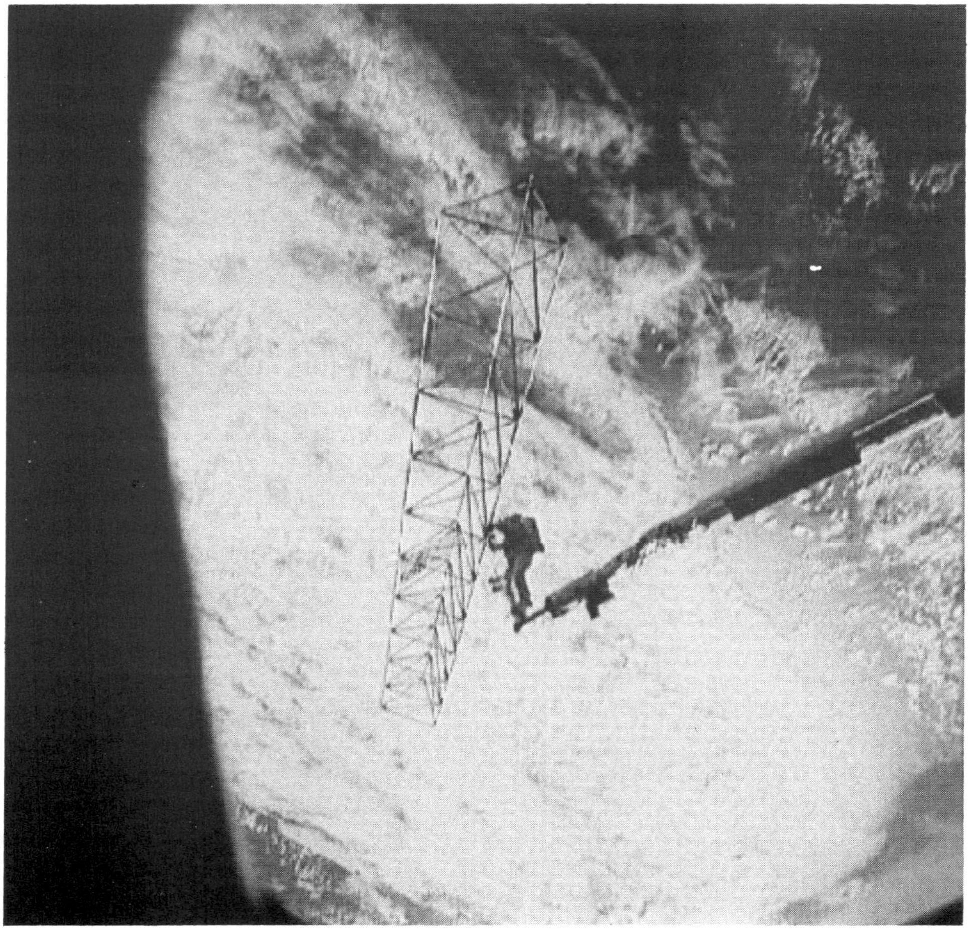

Fig. 4.13. Flight 61B: Jerry Ross, perched high above the bay on the Manipulator Foot Restraint (MFR), manipulates the truss that he had erected to test the Assembly Concept for Construction of Erectable Space Structures (ACCESS).

Nevertheless, after a day's rest, they went back out to do it all again. This time, Ross stood on the MFR to build the tower, which required him to coordinate with Mary Cleave, the RMS operator. Once it was complete, Ross took a cable from within the MPESS and ran it up the length of the tower, clipping it into place as might later have to be done to embed an umbilical in the space station's truss. Once this had been stripped out again, Spring released the tower from its base and Ross took it with him when Cleave lifted him high above the bay. Once he was clear, Ross swung the truss level to test the ability of an astronaut to manipulate a large object. After he had rotated the structure several times, he reorientated it vertically, and Cleave lowered him so that Spring could reattach the truss to the base. As a final test before stripping it down, Spring replaced an element half way up the tower, to simulate repairing damage to an erect structure. As a variation on the EASE test, the triangular structure was erected with one astronaut on the RMS and the other on a PFR, in order to compare their efficiency with that when they had been free. Not unexpectedly, having both hands free made inserting the rods into the nodes trivial, but it required close coordination with the arm operator to move around to fetch the elements, and this was another factor that had to be taken into account in assessing working efficiency.

The objective had been to establish whether it would be practicable to rely on astronauts or whether it would be necessary to develop self-deploying structure. Stereoscopic cameras recorded the entire exercise so that the activity could be modelled in 3-D and then subjected to a detailed time-and-motion analysis. By comparing work efficiency with that achieved in the WETF, the fidelity of training methods would be able to be improved. It was clear that although the WETF offered effective preparation for experiencing weightlessness, the drag of the water significantly inhibited movement. Because this was absent in space the pace at which the astronauts were able to work had been underestimated; in some tasks they were *twice as fast* as in the water tank. This reality check also showed that they could maintain a set pace without becoming exhausted. Lest this seem insignificant, compare the degree of dexterity with that undertaken on the Gemini spacewalks, when simpler tasks had involved a major effort which had soon led to virtual exhaustion.

Concerning the issue of how the space station should be assembled, this simple test was sufficient to confirm that astronauts could erect structures, so there was no need to develop self-deploying structures. NASA's long-standing advocacy for the role of the human being in space seemed to have been vindicated. With a space station to be built, it seemed that there would soon be plenty of work for astronaut hard-hats.

Unfortunately, Challenger was lost a few months later, and while the remaining shuttles were grounded, operating procedures were reassessed. Heightened concern for crew safety prompted the decision that crews would be assigned their objectives well in advance; there was to be no more last-minute building up of the activity plan to include tricky spacewalks. Spacewalking was to be minimised. Astronauts would be restricted to the payload bay and be tethered, and the MMU was grounded. Overnight, NASA's confidence had been sapped, its *gung-ho* spirit reversed. Now *any* mission was difficult if it involved unprecedented risk. Buck Rogers had evidently perished along with the Challenger Seven.

The shuttle resumed flying in September 1988, but it was April 1991 before astronauts next ventured into the bay. Atlantis was to deploy the Gamma-Ray Observatory (GRO), but the satellite's antenna had stuck in its stowed position. Jay Apt and Jerry Ross already had a spacewalk scheduled for the following day, but they went out a day early to fix the GRO. This did not take long, as the long boom was released from its clamp by a single shake by a spacegloved hand.

Since they were out, it was decided to make a start on the activities set for the following day which, as it happened, were to evaluate tools designed for use aboard the space station. The GRO remained poised on the arm high above the heads of the two men as they worked in the bay, so they could pursue only tasks which did not involve the use of the arm. Ross ran a tether across the bay and proceeded to assess how easy it was to pull himself along it. Although this may sound trivial, the focus of their spacewalking assignment was to assess moving about *rapidly*, because even after the space station is assembled, spacewalkers will have to maintain it, and on a structure that large it will be vital to move around rapidly so as not to waste valuable EVA time. While Ross scooted back and forth, Apt stood on the PFR of the Crew Loads Instrumented Pallet (CLIP) and acted out various actions so that sensors could record the forces he transferred through his feet. Then, as was standard practice, they exchanged places and repeated the experiments. Underlying such tests was a determination to make sure that every problem was thoroughly understood, to make sure there were no little surprises. After about four hours, they ended the spacewalk. The GRO was dispatched as soon as the bay was clear.

The next day, Apt and Ross returned for their planned excursion. Continuing the theme of rapid deployment, the Crew & Equipment Translation Aids (CETA) experiment assessed a number of prototype rail-carts under consideration for enabling astronauts to move both themselves and their equipment inside the space station's truss. The evaluation rail ran the full length of a sidewall. In each type of cart, the astronaut stood in a PFR. It was the mode of locomotion that differed. The simplest device was fully manual. The astronaut lay prone, and used a hand-over-hand action to pull himself along; he stood upright on the mechanical cart and pumped a lever in much the same way as on a manual railroad buggy, and operated a rotary dynamo to drive the electric cart. After testing how a solitary astronaut could travel, everything was repeated with one astronaut serving as the bulky equipment in need of rapid transfer. Once the cart evaluation was finished, Apt mounted the arm and was swung about at much higher speed than usual to assess the loads on the arm's structure. This kept them fully occupied for six hours. Compared with the antics of earlier excursions, it was rather dull, but the objective was to acquire engineering data. The favoured translation technique proved to be the straightforward manual hand-over-hand cart. In an operational system, the cargo will be strapped to a mount behind the foot restraint. It seemed that in weightlessness there was simply no need for the mechanical advantage provided by the more sophisticated devices. Some of the tests were purely passive: Ross, for example, tried new gloves designed to facilitate greater dexterity. Having participated in the ACCESS/EASE evaluation, he was well qualified to judge their effectiveness, but getting the gloves just right would prove to be a major headache.

TRY, TRY AND TRY AGAIN

STS-49 in May 1992, which was Endeavour's first mission, marked NASA's return to satellite-rescue. This time though, it was not a shuttle that had lost one of its satellites, and it was no last-minute affair. In fact, NASA had played no part whatsoever in losing the

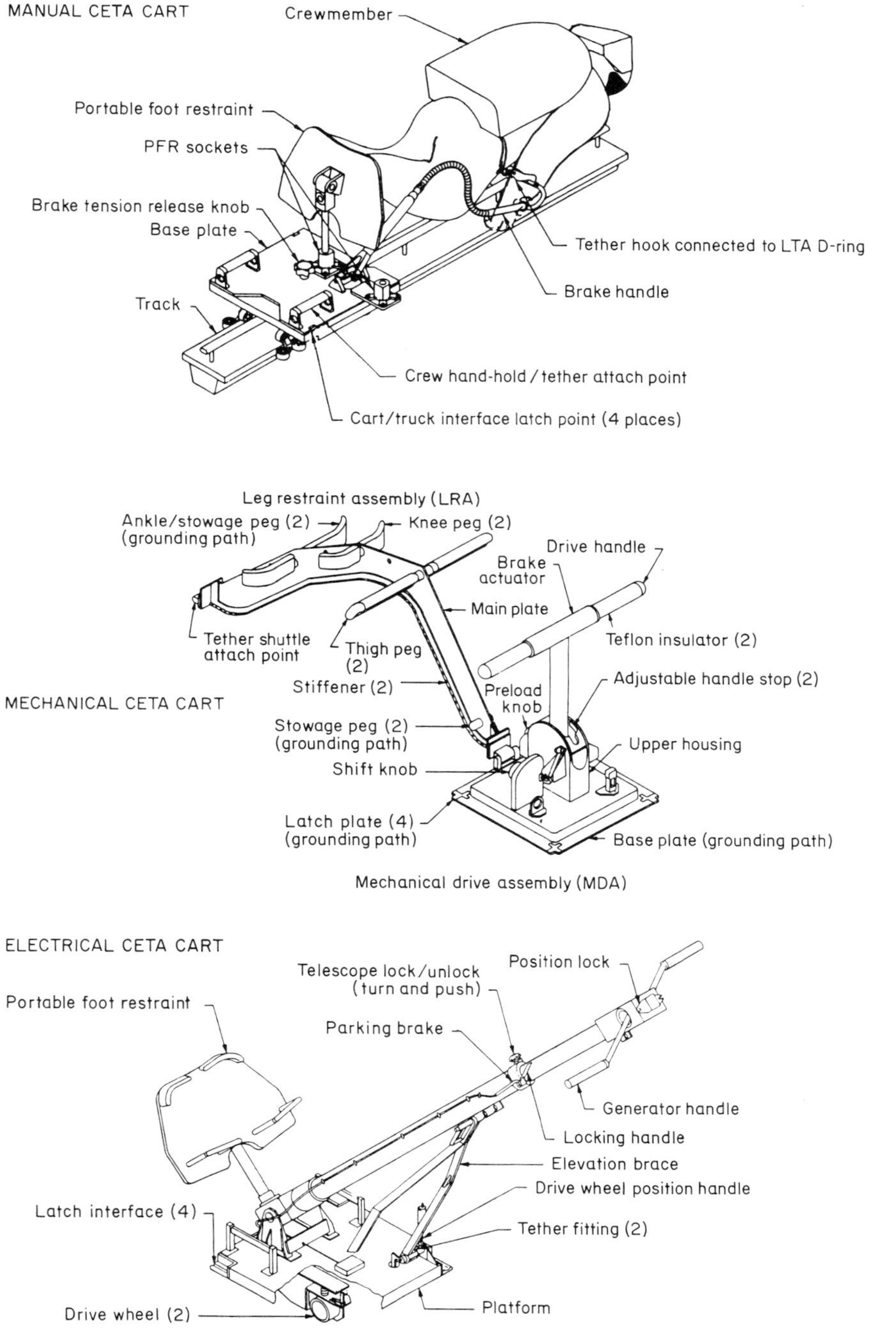

Fig. 4.14. Schematics of the carts proposed as Crew & Equipment Translation Aids (CETA) which was under development for space station use.

satellite; the agency had become involved only when it was asked to attempt a rescue, and it had spent *two years* assessing the feasibility of the task and working out how best to go about it. For NASA, there was a rich irony in the fact that it had been called upon to rescue this particular satellite.

As with the Leasat series, the HS-393 had been designed specifically to take advantage of the width of the shuttle's payload bay. On the outside, therefore, it was a taller version of the HS-381 drum. Its upgraded HS-376-type transponders could relay 120,000 telephone calls simultaneously. Intelsat had ordered half a dozen and only just booked them on the shuttle when the Challenger was lost, so it was left to compete for the limited stock of rockets. The problem was that because it was the largest commercial communications satellite ever built, only a heavyweight rocket with an outsized aerodynamic shroud could accommodate it, and, in the scramble for launchers, the government agencies took priority.

Intelsat had been forced to wait for either the upgraded Ariane or for the introduction of the Commercial Titan, so to cover itself it booked one of each. The Ariane became available first, so the first satellite (Intelsat 602; designations can be arcane) went up in October 1989, and the second (Intelsat 603) followed six months later on a Titan. It was only the second flight of the Commercial Titan; the first had carried two smaller satellites, and the deployment system was new. It was supposed to be able to accommodate variable numbers of payloads, but it was improperly wired and so the payload did not separate from the upper stage of the rocket. To prevent the satellite from being dragged back into the atmosphere along with the spent stage, it was ordered to separate from its perigee-kick motor and then to fire its own thrusters to raise its orbit a little. Although stranded in low orbit, at least the satellite was in space, and if a shuttle crew could fit a new kick-motor it would be able to continue up to GSO. So it was that NASA was invited to rescue a satellite which it had originally contracted to deploy.

The task had been assigned to Endeavour by a process of elimination. There had to be time to train; it had to be a flight on which the bay would be clear for retrieval work; and it could not be a flight with a massive satellite for deployment, because the new kick-motor would require a support cradle. In the event, Endeavour was the first mission to satisfy the criteria. During the two years following the mishap, a dozen flights had carried the HST, the GRO, the UARS, a TDRS, four Spacelabs, three military payloads, and the Ulysses spacecraft. Endeavour's bay was already given over to spacewalking, so it was the obvious choice.

Several motors were considered. Since the satellite had used a substantial fraction of its propellant to stave off orbital decay, which would restrict its operating life if ever it reached GSO (cutting it from 15 years to 12 years), an IUS made an attractive option. Being a two-stage vehicle, it would be able to perform the circularisation burn in addition to GTO insertion. However, bolting an HS-393 onto an IUS would not be easy. The second option was to use Orbital Science Corporation's Transfer Orbit Stage (TOS). This had been built specifically to boost heavy shuttle payloads into GTO. It had never been used, but it was rated for flight. The third option was to fit the same type of motor – an Orbus – as was to have been used originally. In fact, the TOS and the first stage of the IUS used variants of this same motor. It was decided to fit a new Orbus, and carry it in a modified form of the TOS cradle.

The challenge was to figure out a way to capture Intelsat 603, which was far larger than anything astronauts had ever worked with before. The only option was to develop a capture bar for an astronaut to attach to the support ring at the base of the satellite. This would have a trunnion pin on it, so that the RMS could take the captive satellite and manoeuvre it into position directly above the motor. If the motor had a suitable adapter, the bar could connect the satellite to the motor; the RMS could be withdrawn, the trunnion pin removed, then the payload released. The tricky part would be the capture. The support ring was three metres wide; a stinger configuration was impracticable, so a diametric bar was devised. The HS-393 was a spinner, so an astronaut attempting to lock both ends of the bar simultaneously would have to deal with a moving target. It was only *half* a revolution per minute, though, and WETF trials had indicated that it *could* be done.

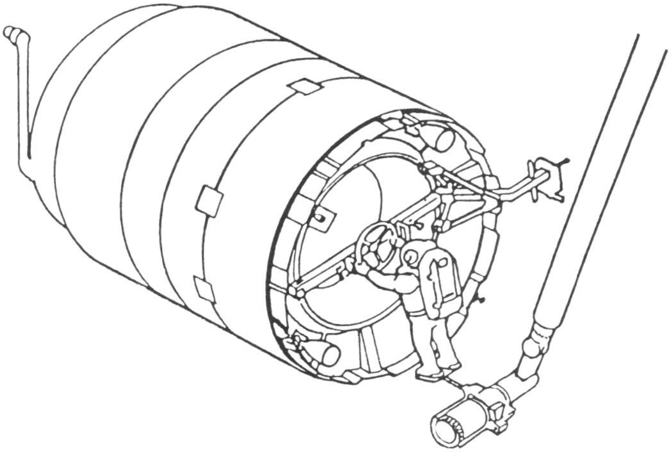

Fig. 4.15. A schematic of how Pierre Thuot was intended to ride the Remote Manipulator System (RMS) and use a capture bar to retrieve the Intelsat satellite.

So it was that Bruce Melnick swung Pierre Thuot on the RMS up to the slowly rotating Intelsat 603. Rick Hieb remained in the bay. After carefully aligning the capture bar against the ring on the tail end of the enormous satellite, Thuot edged the bar forward until it was in contact. An automatic mechanism was supposed to trigger latches to secure both ends at the same time, but Thuot had not struck the target with enough force to activate the mechanism. He withdrew, and then shoved the bar in harder. This time the latches fired. Unfortunately, when Thuot attempted to use the bar to cancel the residual spin, it slipped off, and as soon as the satellite was free it reacquired its spin. Furthermore, as a result of Thuot's actions its rotation had been disturbed, and it had adopted a coning motion. By precessing as well as spinning, it became more difficult to align the bar. Not only did further attempts to secure the satellite fail, each time Thuot touched it he made the pitching motion more pronounced. After three hours he finally admitted defeat, and Endeavour withdrew, leaving Intelsat 603 with a slow 50-degree nutation.

One fact was evident. The WETF simulations had not accurately represented the degree to which the satellite would be disturbed by light contact. The clue was the fact that it had reacquired its spin as soon as the capture bar had disengaged. It was a sure sign that the liquid propellant was retaining sufficient momentum to spin the satellite up as soon as it was free. This also explained the precession; the liquid was sloshing around. If he ever managed to capture it, Thuot would have to hold the satellite steady until the liquid had lost its kinetic energy, and then watch out every time he moved it. Despite training in a water tank, they had been caught out by the physical properties of fluids in weightlessness.

Overnight, the satellite's controllers eliminated both the precession and the residual spin, so when Endeavour returned the next day the satellite was completely stable. Thuot went out again, confident that he would be able to capture it. This time he eased the bar against the ring and activated the latches manually, but they failed to engage. He persevered for *five hours*, but was obliged to abandon the attempt to capture the satellite. This time he left it in a flat spin. After Endeavour withdrew, the controllers once again set out to stabilise the satellite, but it was beginning to look as if *homo spacewalker* had finally met his match, and that Intelsat 603 would have to be abandoned. Just one excursion had been allocated for its repair; the second attempt had used the first of two spacewalks intended for other activities. Should this other programme be sacrificed for a third attempt on the satellite?

While the crew slept, their colleagues on the ground, in the support team backroom and in the WETF, tried to work out an alternative way of capturing the rogue satellite. It was all too evident that the capture bar would have to be abandoned. To design, build and certify it for flight had cost in excess of $2 million, and it almost certainly met its specifications. But there were cases where practising something on the ground and doing it in space were somehow significantly different, and this was one such case. But how could it be done? How could an astronaut capture a 4-metre wide, 6-metre tall drum *without* using the bar? The bar had to be in place for the RMS to manoeuvre the satellite over the motor, and it was essential if the satellite was to be attached to the motor. Could Thuot capture the satellite *by hand*? Could he hold it steady while Hieb worked on it? If Thuot stood on the arm to hold the satellite, Hieb would not be able to reach it unless the orbiter manoeuvred in really close. If Thuot lost control of the satellite, it might strike the orbiter. It could easily damage the payload bay doors, or the vertical stabiliser. In space, the crew of Endeavour had also come to the conclusion that the ideal tool for grappling a satellite was the human hand, but they added a twist of their own. The shuttle carried a spare suit, if Tom Akers, who was scheduled for a later excursion, were to come out too, then Thuot could grab the satellite and slowly swing it down into the bay, where Hieb and Akers would be waiting for it; then they could hold it steady while he attached the bar, free of any concern for disturbing the satellite's motion. As soon as the bar was on, they could revert to the original plan. When this proposal was put to Houston, doubts were expressed whether three people would fit in the airlock. A WETF test showed it to be a tight fit, but it was feasible, so the plan was approved.

Endeavour had propellant for just this one rendezvous. If Intelsat 603 eluded them this time it would have to be written off even though it was perfectly healthy. If they failed, it would not be for lack of experience. Dan Brandenstein was Chief of the Astronaut Office. He manoeuvred the orbiter to put the satellite a mere two metres over the bay, with its base facing down. Hieb had taken up station on a PFR on the starboard sill, and Akers was on top of an MPESS positioned right in the centre of the bay. It was just above their heads. Melnick, who had already demonstrated his prowess on the arm, positioned Thuot around the far side. For 10 minutes they just observed the satellite's motion. It had a slight nutation with a 3-minute cycle. They waited until it was vertical with respect to the bay, and then six hands simultaneously grasped the base ring. They held the satellite still for several minutes to give the fluid time to settle, but although they had captured it in a vertical alignment, it was inconveniently orientated for fitting the capture bar (which Hieb had left to the sidewall below his station) so, working in concert, and a little at a time, they gingerly spun the drum through 120 degrees.

When it was aligned, and stable once more, Hieb let go, retrieved the bar, and shoved it across just below the ring. From his position on the sill he could not reach the controller

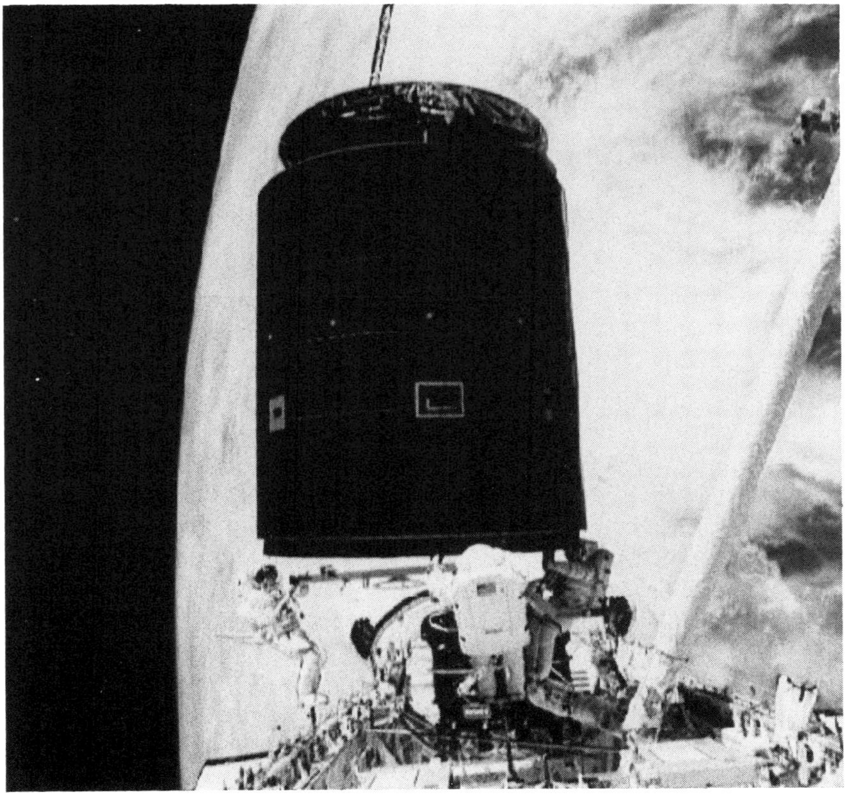

Fig. 4.16. Rick Hieb is on a Portable Foot Restraint (PFR) on the starboard sill, Tom Akers is on a PFR on an Mission Peculiar Experiment Support Structure (MPESS) in the centre of the bay, and Pierre Thuot is on the Manipulator Foot Restraint (MFR), riding on the Remote Manipulator System (RMS). It was the first spacewalk with three astronauts. After they grabbed hold of the enormous Intelsat satellite, Hieb retrieved the capture bar and slid it into position beneath the satellite so that Thuot could finally attach it.

in the centre of the bar to engage the latches, so he held the bar in position with one hand and the satellite with the other. Akers served as a steady anchor. Thuot then manoeuvred beneath the satellite, indicating his instructions to Melnick by way of hand movements. The bar proved simple to fit now that the target was unable to move. With the latches engaged, Thuot tightened a set of bolts to ensure that the bar was firmly attached, then he dismounted the arm. Melnick dispensed with the MFR and then took hold of the satellite, and Akers, who had held the satellite in a vice-like grip for an hour and a half, was finally able to let go. It was now a straightforward matter to reconfigure the bar by deleting various appendages used during the capture phase, and installing a clamp at each end. Once the arm had positioned the satellite on the motor, the clamps were engaged. The job was done.

The surprise came when Kathryn Thornton sent the command to eject the stack from its cradle and nothing happened, but the back-up circuit worked and Intelsat 603 set off on the first stage of its journey to GSO.

BACK TO BASICS

The spacewalkers had triumphed again, but only after having switched strategy. Maybe the most important conclusion to be drawn from it was that the human hand is the best tool, and ought to be used as the *tool of choice*, not in the last resort. Clearly, if they had set out to do it the way that they finally did, they could have fully dealt with Intelsat's lost satellite on the first day. In the event, the experiment known as Assembly of Station by EVA Methods (ASEM) was considered sufficiently important to extend the flight to accommodate a single spacewalk for Akers and Thornton to carry out a truncated version of this programme.

The first task was to construct a truss segment, with both astronauts floating freely. In fact, as in the case of EASE, only one section was built, but it represented half of a 14-metre cube that would form one element of a truss. As with EASE too, it was found that the best way to handle a rod was to set it free a few centimetres from a node, rapidly swap hands on the structure, prepare the node with the now-free supporting hand, and then retrieve the rod and pop it in. This worked even when the far end of the rod was not anchored, but it took some time to become accustomed to letting the rod go, because the WETF training had not lent itself to this technique. It was therefore rather easier to assemble the joints in space than it had been in training. A key objective of the test was to uncover such empirical data. When they were finished, the MPESS on which the rods had been stowed, and on which Akers had stood the day before, was hoisted by the RMS until it was just above the bay. The spacewalkers extended legs from the triangular structure, and then, using only voice commands, directed Melnick to lower it into position for mating to the assembled truss segment. This tested the team's ability to coordinate to the extent needed for the arm-operator to work 'in the blind' on construction tasks. The entire assembly was immediately stripped down to leave time to test the Crew Propulsive Device (CPD).

When Ed White floated out of Gemini 4, he took with him a simple handheld gas-jet to control his movements, but he was barely able to familiarise himself with it before it ran out of nitrogen propellant. Michael Collins tried it on Gemini 10 and found it difficult to use, and an improved version had been evaluated inside Skylab. Now another such gas jet

Fig. 4.17. Tom Akers and Kathryn Thornton test erecting the Assembly of Station by EVA Methods (ASEM) structure.

had been developed, and Akers briefly tested it in the front of the bay. The problem with each of the hand-held devices was that it was difficult to aim the impulse through the centre of mass, so it tended to impart an unwanted rotation in addition to a desired translation. The MMU was able to isolate the individual degrees of freedom, and this made it simple to use. It had been deemed impracticable to give MMUs to astronauts venturing outside the space station. The gas jet was one of several devices being considered as a means of enabling an astronaut with a broken tether to make a hasty return to the nearest part of the structure, to await rescue. It did not seem likely that a hand-held jet would be that device. Akers and Thornton crammed as much of the ASEM programme into their 7-hour walk as they could, but some tests – such as an inflatable rod – intended for their second excursion – had to be deleted.

Endeavour's maiden voyage had set EVA records. In particular, it saw the first three-person spacewalk. This had been tough on Thuot and Hieb. The flight plan had called for a total of three excursions, with Thuot and Hieb making the first and the last, and Akers and Thornton taking the middle one, so as to give the others a day off. In the event, Thuot and Hieb had made three walks, two of which had been back-to-back. However, this provided additional data on human endurance. NASA had learnt a lot about spacewalking from this mission. The ASEM rods had been easier to assemble than expected, which was good news, but the sensitivity of the big satellite had come as an unpleasant surprise. Evidently, the WETF was excellent for preparing astronauts for the sensation of weightlessness, but the water-drag and the residual action of gravity in the tank was impairing training for manipulating objects. This prompted concern for the practicality of relying on spacewalkers to build the space station. To minimise this technical risk, it was decided that the contingency-EVA astronauts on forthcoming missions should venture into the bay to rehearse basic procedures. No activity was to be taken for granted. In the event, however, the big problem turned out to be *inactivity*.

Endeavour flew the first of these generic spacewalks as STS-54. After it had deployed its TDRS satellite, the bay was clear for the tests to begin. They started in the airlock, with Mario Runco and Greg Harbaugh taking turns to move one another so that their backpacks slotted into the storage brackets on the wall. This may sound trivial, but the astronaut doing the work could not see the bracket, and in the bay they slipped into and out of foot restraints, without bending down to see their feet. Even on the ground, many everyday tasks become difficult if direct vision is denied. It was prudent to establish now that such tasks could be achieved in space, to preclude the possibility of finding out later that they couldn't. Another deceptively trivial assignment was the attempt to move *purposefully* in the bay. This was the first time that a spacewalk did not have a specific imperative; it was an exercise in data acquisition. The goal was to demonstrate the feasibility of tasks, not to assess how rapidly they could be achieved, so Runco and Harbaugh paced themselves.

Certain items located outside the space station's pressurised modules will have a finite operational life, so they will need to be replaced from time to time. Chemical storage batteries, for example, are to be based at the far end of the truss, near the solar panels, and astronauts will have to carry fresh batteries out and return with the old ones. To simulate manipulating bulky Orbital Replacement Units (ORUs), Runco and Harbaugh dragged one another along the sill, employing a single-handed translation technique. They found this rather tiring. The inefficiency of handling a bulky object whilst floating freely argued that it would be wise to fit foot restraints at all sites on the station where astronauts were intended to carry out such work.

When Endeavour flew next, as STS-57, David Low and Jeff Wisoff rehearsed specific procedures for the repair of the HST. To rectify the flawed optics, a package as bulky as a telephone booth would have to be accurately inserted into the telescope's instrument bay, and it was crucial that this be done without causing any damage. Their first task, however, was to stow the antennas of the just-retrieved EURECA satellite (its retraction mechanism had failed). This proved more difficult than expected, so it was two hours before they were able to turn to their assigned programme. Wisoff played the role of a bulky package, Low rode the RMS, and Nancy Sherlock operated it. Low picked up Wisoff, then instructed Sherlock to move him to a specific location. Such tests had been performed previously to evaluate the stresses on the arm when making large-scale movements. This time, it was the ability of the arm to make *fine* adjustments that was on trial. After retrieval, the HST was to be stationed on a FSS at the rear of the bay, so these tests were conducted with the 15-metre RMS fully extended. After the arm trial, Wisoff assessed a new torque-wrench. Working on bolts had been awkward in the WETF, but they posed no difficulty in space.

A few months later, Jim Newman and Carl Walz continued the tool tests in Discovery's bay. Story Musgrave was to head the HST repair operation. While rehearsing in a vacuum chamber he had suffered mild frostbite in his fingers. Newman held the palm of his hand against one of the high-powered floodlamps which illuminated the bay while it was in shadow, to determine whether an astronaut could overcome chilled fingers, and thus preclude having to curtail an excursion. Although it transpired that the heat radiated by the lamp was able to penetrate the fingers of his glove, this fact demonstrated that the glove was not as effective a thermal barrier as had been thought. This realisation prompted concern that it was likely to lose heat if subjected to prolonged shadow, as it had in the vacuum chamber. This directly affected the plan for working on the HST; while it was held captive by the orbiter, it would be without power, and to prevent it overheating the orbiter would face its belly to the Sun. Each spacewalk was expected to last at least seven hours, and would be in shadow for the whole time. Several spacewalks were assigned, but time would be tight, and if it turned out that the astronauts had to curtail the work because their fingers were too cold to work with the required precision, then the repair would falter. A better glove was ordered, but because this would not be available in time, an over-glove was devised as an interim measure in the hope that the additional layer would reduce heat loss. However, care had to be taken not to degrade the glove's already marginal dexterity.

The equipment tested this time included a power ratchet, a semi-rigid tether designed to stabilise the operator of a torque-inducing tool, and a new PFR that incorporated a platform that could be raised and tilted by activating pedals with the feet. The only issue was that the SAB steadfastly refused to shut. That this took 45 minutes was a timely reminder that even a familiar object could set back a schedule. Sensors monitored the environment in the bay while the astronauts were out, to assess the risk to the HST from any contaminants emitted by a spacesuit; absolutely nothing was being left to chance. The rendezvous with the HST two days into STS-61 was to be followed by an unprecedented *five consecutive days* of spacewalks. Story Musgrave and Jeff Hoffman were to alternate with Kathryn Thornton and Tom Akers. Further excursions could be made if appropriate, but Endeavour's maximum endurance was 11 days; at that time, outstanding tasks would have to be left. The orbiter carried 7 tonnes of ancillary apparatus in its bay. In 400 hours of WETF-time, Musgrave *et al* had choreographed each action that they were to make in space. After mastering each component task, they had made a series of 10-hour simulations to check that the individual tasks followed on from one another seamlessly, and that there were no omissions. No crew had ever trained more singlemindedly for a spacewalk. They *had* to succeed. There was far more resting on the outcome than the future of the telescope; their actions would also determine the prospects for the space station.

Claude Nicollier snatched the HST with the arm, and then set it down on the FSS at the rear of the bay. This could rotate and tilt the HST so that the astronauts could work on any part of it and remain directly visible to the arm operator. Power was supplied by umbilical, so the orbiter immediately reorientated itself to put its captive in shadow.

Three of the six gyroscopes in the HST's attitude-control system had already failed, and another was showing intermittent signs of wear. If (or rather, when) this failed too, the

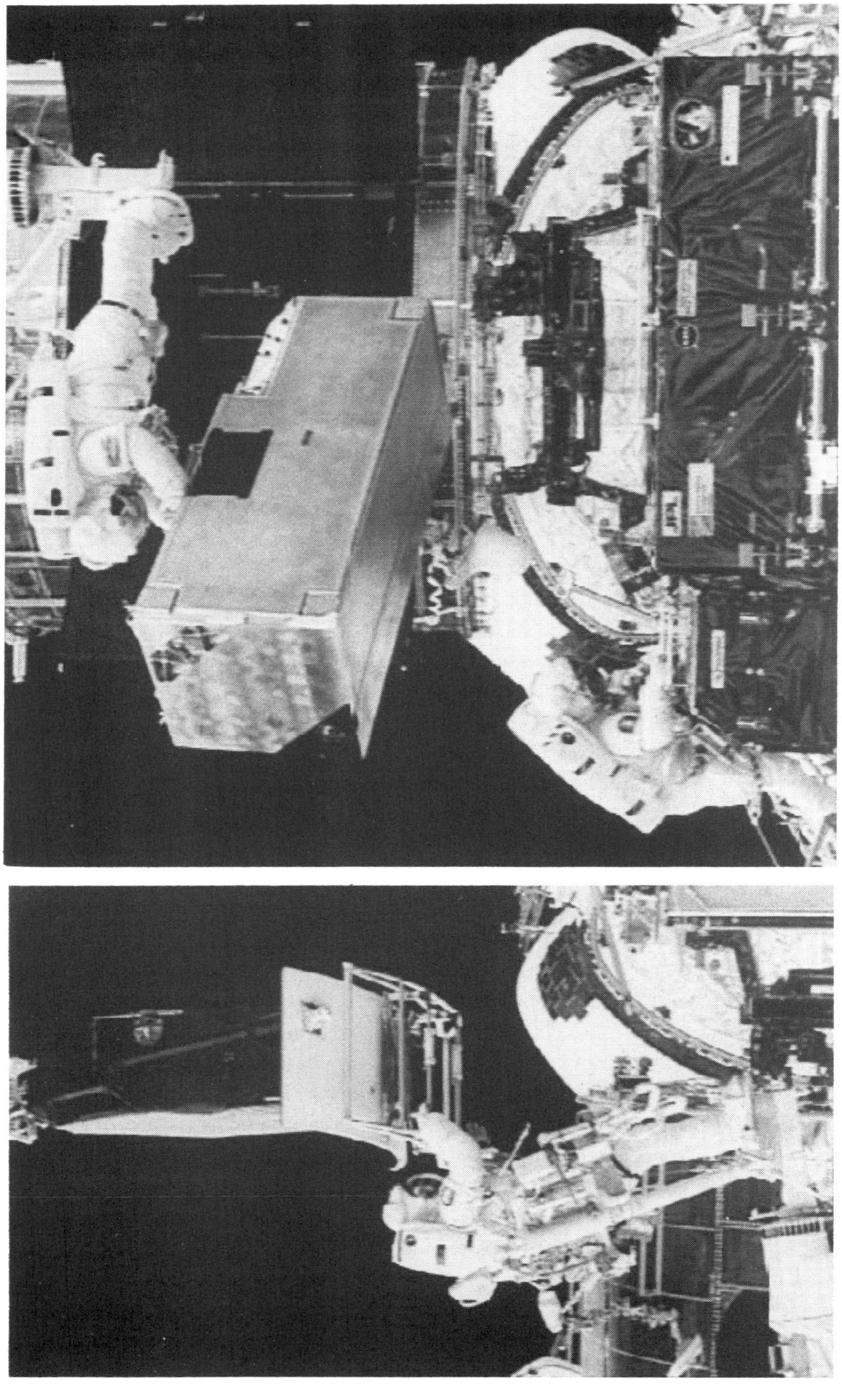

(a) (b)

Fig. 4.18. Servicing the Hubble Space Telescope (HST): transferring (a) the Wide Field/Planetary Camera (WF/PC), and (b) the telephone-booth-sized Corrective Optics Space Telescope Axial Replacement (COSTAR) package.

telescope would have to be withdrawn from service. Musgrave's and Hoffman's primary task on the first outing was to restore this capacity. The only problem they encountered was at the end, when two of the four bolts proved reluctant to engage. It took longer than expected to close the cover of the compartment, and the outing was extended to nearly eight hours. The following day, the Goddard SFC controllers commanded the solar arrays to roll up so that they could be dismounted and replaced. The starboard array was badly distorted, and it stuck, but this did not really pose a problem. Akers and Thornton retrieved the other array and stowed it in a container in the bay for carriage back to Earth where it could be carefully examined. After Thornton had detached the stuck array, Nicollier raised her well clear of the HST on the end of the arm so that she could cast the twisted array adrift. The replacement arrays were affixed to the HST, but were not extended.

Fig. 4.19. A wide view of the Hubble Space Telescope (HST) set on the Flight Support Structure (FSS) at the rear of the payload bay. The coast of Australia is visible in the background. The telescope's new solar arrays have yet to be unrolled. The orbiter's cabin is reflected in the telescope's shiny thermal insulation. For a sense of scale, note the tiny figures of Jeff Hoffman and Story Musgrave riding on the Remote Manipulator System (RMS) at the top of the telescope.

For their second outing, Musgrave and Hoffman extracted the enormous wedge-shaped WF/PC-2 instrument from the side of the HST, then slid in the new one. They had expected that they might encounter difficulties aligning the new unit to fit on the guide rails, but it slipped straight in; the HST had been designed for in-space servicing. With the main task complete, they moved onto the magnetometers, two of which had failed. These units were on the skin of the tube, but it was impracticable to detach them, so the new units were mounted on top of the old ones and bypasses fitted. Akers and Thornton now had to remove the HSP instrument and replace it with COSTAR, the unit that was to provide the corrective optics for the telescope's three remaining on-axis instruments. The HSP and COSTAR were each the size of a telephone booth. An astronaut perched on the arm had to hold the 300-kg box of COSTAR and position it within a few millimetres of the guide rail that was not directly visible. It required a great deal of teamwork, but when it was properly positioned COSTAR slipped straight in. For the follow-up task of this outing, the telescope's computer system was upgraded by installing a 386 coprocessor. On the final excursion, Musgrave and Hoffman replaced the controller for the solar array motors in order to overcome a fault that had limited the rate at which the arrays could be rotated, and the new arrays were then unrolled. All that remained was to install a crossover in the power supply of the GHRS, so as to restore a failed data channel. And that was it! The HST, now in better condition than when it was initially deployed, was released the next day.

To everyone's relief, all of the tasks had been accomplished more or less in the allotted times. Words can hardly express the concentration that must have been required to work on the world's most expensive telescope for up to eight hours at a time, in the knowledge that the slightest mistake could write it off. Nor can words readily convey the intricacy of the operation. Every iota of experience gained from previous spacewalks, and from the WETF training, had been exploited in devising the individual operations, estimating their durations and integrating them into a viable sequence for each spacewalk. In practice, the planned 30 hours of EVA time had stretched to 36 hours, but this was to overcome problems such as a cover that would not shut, not because the astronauts had fallen behind schedule with one of the main tasks. In fact, the outcome proved that the repair operation had been thoroughly understood, although the telescope had been designed for in-orbit servicing. All the tools had worked as intended, and the frostbite issue also appeared to be resolved.

That the training regime had been sound was now self-evident, but Musgrave *et al* had virtually monopolised the WETF for almost a year. Training at this level of intensity could not be made available to the astronauts who would perform the spacewalks to assemble and maintain the space station; these tasks would require several hundred hours of EVA effort per year. The supreme effort to fix the HST had paid off, but assembling the space station would require making such intense activity routine. Any apparatus that would reduce this workload was to be welcomed. STS-62 tested the Dexterous End-Effector (DEE) that was under development to enable the RMS on the space station to undertake delicate operations. It was able to sense the dynamic loads that it imparted on an object as it picked it up, to give its operator a degree of feedback that the orbiter's arm could not offer. No spacewalk was necessary, however, because the RMS was able to retrieve the DEE from its stowage rack, test it and put it back. When the DEE enters service on the station, it should reduce the need for astronauts to go outside to undertake basic installation and maintenance tasks. The next spacewalk, on STS-64, returned to the topic of how an astronaut with a loose tether could effect a hasty recovery. The new device was a compromise between a handheld gas jet and a fully fledged backpack. This was a compact box which could be fitted to the base of the standard life-support backpack. Given its purpose, it was dubbed the Simplified Aid For EVA Rescue (SAFER).

The excursion undertaken by Carl Meade and Mark Lee was specifically to test this new propulsive device, so they were able to subject it to a thorough evaluation. In doing so, Lee made NASA's first untethered spacewalk since 1984. The control unit was on an umbilical stowed on the side of the pack. He had only to reach down to retrieve it. SAFER had been designed to serve four functions in sequence: firstly to stabilise a drifting astronaut, then to turn and head for a site of safety, and finally to halt. The 38-kg pack was essentially a tank of nitrogen, with 24 jets positioned to deliver impulses to isolate all six degrees of freedom. Although the 1.4 kg of nitrogen in its tank was suitable for its role as an emergency device, it did not offer much scope for a comprehensive trial, so a tank had been fitted in the bay in order that the astronauts could top-up with gas.

The evaluation began with Lee methodically testing the SAFER's ability to control each degree of freedom. Then Meade, riding the arm, manhandled Lee to impart rolling, pitching and yawing rotations in order to assess the ability of the attitude-hold function to overcome them. They then swapped places, and Meade performed translation trials, moving out along the arm and back again. Finally, he headed straight for the airlock to demonstrate that the SAFER provided sufficient directional control for him to reach a specific point without the need to waste gas making mid-course corrections. A data unit recorded its performance, and later analysis confirmed the astronauts' enthusiastic recommendation that the backpack augmentation unit be adopted. Significantly, most of the training for this EVA had been conducted with a virtual reality system. Computer simulation already provided the basis for training to use the RMS. Training for spacewalking was more demanding, but as the fidelity of virtual reality simulations improved, it was hoped that they would reduce the demand for the WETF.

Whereas to date astronauts had tackled one spacewalking assignment at a time, and had been able to devote months to learning what they had to know to undertake a small number of specific tasks, on the space station they would have to apply themselves to a much wider range of tasks. It would clearly be impracticable for every spacewalker to know everything about every system they would encounter, so an Electronic Cuff-Checklist (ECC) had been devised. Worn on the forearm, this 1-kg system had an 8 × 10-cm screen with a 2-megabyte memory. In effect, it would enable each astronaut to look up procedures and checklists in a 500-page manual, the contents of which could be loaded from a laptop immediately prior to venturing out. Although Meade and Lee reported it to be quite effective, a problem arose on STS-63, a few months later, when Michael Foale and Bernard Harris found that it failed when subjected to the intense cold of a prolonged period in shadow. Finding out that the ECC did not like the −100 °C chill was a bonus of their excursion.

Their main task had been to assess the revised gloves. When astronauts are in sunlight the suit has to work hard to keep them cool, and in shadow it has to keep them warm. It had turned out that Musgrave *et al* had been so active during the HST servicing that their own metabolism had kept them warm during the extended time in shadow. To exacerbate the effect of heat-loss through the gloves, Foale and Harris tethered themselves to the end of the RMS high above the bay and proceeded to do *absolutely nothing*. The idea of having astronauts remain idle for several hours was anathema to planners accustomed to ensuring that each and every five-minute block on a flight plan was put to good use; but in this case, doing nothing was essential. After two hours, both men reported that their fingers felt as if they were touching ice, so they went on to a follow-on experiment in which they were to assess their ability to manipulate, and accurately position, the 1,200-kg SPARTAN free-flyer. It was cold, too; in fact, *so* cold that it was painful to grip, so they were recalled. Work sites within the space station's truss would be in semi-permanent shadow. Jim Voss and Michael Gernhardt continued this trial on STS-69. The undergarment had been modified so that its coolant loop could be switched off without disabling the loop which kept the suit's electronics cool. This had been a single loop. This meant that when an astronaut spent any time in shadow the cooling system could be deactivated. Voss and Gernhardt found that their body heat was sufficient to keep warm, even while they were idle, perched on the end of the RMS. To overcome the original glove problem, tiny electrical heaters were installed at the finger tips. The results indicated that the problem of working in shadow had been solved. Even the improved ECC survived the cold this time. The back-to-basics strategy had paid a rich dividend. The prospects for the space station improved with every flight. Voss and Gernhardt went on to undertake generic space station tasks such as removing protective blankets from simulated ORUs, using tools to unplug their umbilicals, swapping them, and then reconnecting and shielding them.

Two spacewalks on STS-72 were conducted by Dan Barry, Winston Scott and Leroy Chiao (who went out twice) for "the most extensive hardware evaluation ever". The 20 items that they tested for the space station included utility boxes, fluid connectors and a rigid umbilical to be used to run bundled electrical and fluid lines along the truss. To make the tests as realistic as possible, so that the ergonomics of the tasks could be studied, parts of the bay had been rigged to resemble the relevant work sites on the station. One of Scott's assignments was to stand idle on an MPESS throughout a period of orbital darkness in order to assess his suit's performance. It is important to note that all these excursions intended to reduce the technical risk facing the assembly of the space station were secondary objectives on already planned flights, and this vital experience was purchased at minimal cost.

The next time astronauts ventured out, it was from Atlantis, docked with the Mir space station. Rich Clifford and Linda Godwin worked on the interface module that the shuttle had attached to the station on its previous visit. They removed items such as a video camera which was no longer required, and then installed the four exposure cassettes of the Mir Environmental Effects Package (MEEP). As the orbiter would not be able to undock to chase after them if a tether broke, both carried the SAFER as a precaution. A year later, Jerry Linenger joined Vasili Tsibliev in spacewalking in the Russian Orlan spacesuit. He was swung on the end of a crane so that he could install another experiment on Mir. A few months later, Michael Foale and Anatoli Solovyov went out to inspect the damage to the Spektr module which had resulted from a collision with a Progress cargo ferry. And a month after that, Vladimir Titov joined Scott Parazynski in spacewalking from Atlantis to retrieve most of these instruments.

The plan for STS-80 to assess a crane designed for moving bulky ORUs was frustrated when Columbia's outer hatch could not be opened. When the mechanism was stripped after the mission, a tiny screw was found to have worked loose and fouled the ratchet.

Tammy Jernigan and Tom Jones, clearly frustrated, had had to repressurise the airlock and re-enter the cabin. The real concern over the stuck hatch was that if the TDRS antenna ever jammed, preventing the payload bay doors from closing, or if the doors themselves failed to close, astronauts would have to go out and rectify the fault. But if a stuck hatch blocked their way into the bay, the orbiter would be stranded in space. NASA promptly installed a set of tools so that, if necessary, the ratchet mechanism could be dismantled and the hatch released manually. This was another valuable lesson learnt.

THE TASK AHEAD

Cosmonauts working outside the Mir space station have demonstrated that construction in orbit is feasible. Their achievements are manifest in pictures of Mir taken at various times during its long life. Yet the scale of the task facing NASA over the next few years is an order of magnitude greater.

With the knowledge that its WETF is a high-fidelity training facility for spacewalking, and with proven procedures and tools, NASA is well set to start a task that will be far more demanding in terms of workload, complexity and importance, than it has ever undertaken before in space. Even if everything goes exactly according to plan, assembling the International Space Station will involve at least 600 hours of external activity – fully three times all of its spacewalking experience to date. *Homo spacewalker's* big test is therefore still to come.

Table 4.1. Spacewalks during the shuttle era

#	Mission	Date	Hours	Astronauts	Objective
1	STS-6	7 Apr 1983	3.5	Peterson and Musgrave	demonstration
2	STS-10	7 Feb 1984	6	McCandless and Stewart	SolarMax rehearsal
3	STS-10	9 Feb 1984	6	McCandless and Stewart	SolarMax rehearsal
4	STS-11	8 Apr 1984	6	Nelson and van Hoften	SolarMax capture
5	STS-11	11 Apr 1984	7.3	Nelson and van Hoften	SolarMax repair
6	STS-13	11 Oct 1984	3.5	Sullivan and Leestma	ORS test
7	STS-14	12 Nov 1984	6	Allen and Gardner	Palapa-B2 retrieval
8	STS-14	14 Nov 1984	6	Allen and Gardner	Westar-6 retrieval
9	STS-16	17 Apr 1985	–	Hoffman and Griggs	to attach fly-swatter
10	STS-20	1 Sep 1985	7.2	Fisher and van Hoften	Leasat-3 capture
11	STS-20	2 Sep 1985	4.5	Fisher and van Hoften	Leasat-3 release
12	STS-23	– Nov 1985	5.5	Ross and Spring	ACCESS test
13	STS-23	– Nov 1985	6.6	Ross and Spring	EASE test
14	STS-37	7 Apr 1991	4.6	Ross and Apt	CETA test
15	STS-37	8 Apr 1991	6.2	Ross and Apt	CETA test
16	STS-49	10 May 1992	3.7	Thuot and Hieb	Intelsat capture
17	STS-49	11 May 1992	5.5	Thuot and Hieb	Intelsat capture
18	STS-49	13 May 1992	8.5	Thuot, Hieb and Akers	Intelsat capture/repair
19	STS-49	14 May 1992	7.7	Akers and Thornton	ASEM test

Table 4.1. (continued)

#	Mission	Date	Hours	Astronauts	Objective
20	STS-54	17 Jan 1993	4.5	Runco and Harbaugh	ISS preparation
21	STS-57	25 Jun 1993	5.8	Low and Wisoff	HST preparation
22	STS-51	16 Sep 1993	7.1	Newman and Walz	HST preparation
23	STS-61	5 Dec 1993	7.9	Musgrave and Hoffman	HST service
24	STS-61	6 Dec 1993	6.6	Akers and Thornton	HST service
25	STS-61	7 Dec 1993	6.8	Musgrave and Hoffman	HST service
26	STS-61	8 Dec 1993	6.9	Akers and Thornton	HST service
27	STS-61	9 Dec 1993	7.3	Musgrave and Hoffman	HST service
28	STS-64	16 Sep 1994	6.8	Lee and Meade	SAFER test
29	STS-63	9 Feb 1995	4.6	Foale and Harris	ISS preparation
30	STS-69	16 Sep 1995	6.8	Voss and Gernhardt	ISS preparation
31	STS-72	15 Jan 1996	6.2	Chiao and Barry	ISS preparation
32	STS-72	17 Jan 1996	6.9	Chiao and Scott	ISS preparation
33	STS-76	27 Mar 1996	6.1	Clifford and Godwin	Mir experiments
34	STS-82	13 Feb 1997	6.7	Smith and Lee	HST service
35	STS-82	14 Feb 1997	7.5	Harbaugh and Tanner	HST service
36	STS-82	15 Feb 1997	7.2	Smith and Lee	HST service
37	STS-82	16 Feb 1997	6.6	Harbaugh and Tanner	HST service
38	STS-82	17 Feb 1997	5.3	Smith and Lee	HST service
39	Mir	29 Apr 1997	5.0	Tsibliev and Linenger	Place/retrieve experiments
40	Mir	6 Sep 1997	6	Solovyov and Foale	Spektr inspection
41	STS-86	1 Oct 1997	5	Parazynski and Titov	Retrieve experiments
42	STS-87	24 Nov 1997	7.5	Scott and Doi	Rescue SPARTAN
43	STS-87	3 Dec 1997	5	Scott and Doi	ISS preparation
44	Mir	14 Jan 1998	3.8	Solovyov and Wolf	Kvant 2 inspection

5

The darker side

In 1970, when NASA decided what kind of a shuttle it would need, it opted for a fully-reusable configuration of two piloted winged vehicles bolted together for a vertical launch. In the upper atmosphere, the big carrier would release its small companion, which would continue on to orbit. The carrier would fly back and land on a runway at the launch site, just as would the orbiter upon conclusion of its mission.

NASA's main requirement of the orbiter was that its payload bay be 4.5 metres wide so that it would be able to accommodate a module for the space station that it hoped eventually to build. It opted for a bay 12 metres long and a payload capacity of 12 tonnes, but it was otherwise fairly flexible.

'PACT WITH THE DEVIL'

Because this innovative configuration was so different from the enormous and powerful Saturn 5 rocket built for Apollo, development would be expensive. In fact, if NASA was to be the shuttle's only user, as it had been for the mighty Saturn, the cost would be prohibitive. NASA set out to broaden the shuttle's utility, to build up the political support required to talk Congress into awarding appropriate funding. The first step in this process was to refer to the shuttle as the Space Transportation System, to show that it was a utility, not an end in itself. The second step was to carry out a cost-benefit analysis to indicate that if the shuttle flew sufficiently often (on a weekly basis, in fact), the cost-per-kilogramme of payload would fall below that of expendable rockets. This meant that these rockets could be scrapped, that the shuttle could become the *National* Space Transportation System, and that the high cost of development could be offset against the savings accrued over a 25-year period of operations. The most crucial step, however, was NASA's attempt to secure the Department of Defense's backing for the shuttle.

In effect, the military's space programme was run by the Air Force, which had a fleet of proven launchers derived from the Atlas and Titan missiles, so it was not very receptive to NASA's overtures. In fact, Robert Seamans, then Secretary of the Air Force, informed Congress that the shuttle was "not essential" for national security. The only way that

James Fletcher, the newly-appointed NASA Administrator, could secure Seamans' support was to let the Air Force set the shuttle's performance characteristics. In return, however, to ensure that the shuttle became an *essential military requirement*, Fletcher demanded that the Air Force phase out its rockets and adopt the shuttle-only policy. Rather reluctantly, the Air Force agreed, but it insisted upon the right to requisition shuttles in the interests of national security, to restore the scheduling flexibility of response it had enjoyed when using its own rockets.

The most powerful rocket in the Air Force's inventory at that time was the Titan 3. The Air Force insisted that the shuttle provide twice the Titan's payload mass and three times its payload volume. Specifically, the Air Force demanded that the shuttle's payload capacity be increased to 30 tonnes (half that into polar orbit), and that the payload bay be lengthened to 18 metres. Scaling up the orbiter to accommodate these requirements effectively ruled out a fly-back carrier. A partially reusable system would be cheaper to develop, although slightly more expensive to operate, but there was no option. To make the larger orbiter more manageable, it was decided to carry its propellant in an external tank; and a pair of massive strap-on solid rockets would be developed to provide the initial lift. It was at this point that the shuttle's stack assumed its final form.

The orbiter's shape also had to be revised. NASA had planned to launch from, and land at, the Kennedy Space Center, Florida. It had envisaged a vehicle that would re-enter the atmosphere at a high angle-of-attack, using its exposed belly as a brake, and lower its nose only once it had slowed down, in order to minimise thermal stresses. It was to have stubby straight wings for its subsonic approach and landing. The Air Force, however, wanted the flexibility to land at any convenient airbase. It demanded that the orbiter be able to lower its nose early, in order to diverge from its straight-in path. This *hypersonic soaring* capability dictated a delta wing for a better lift-to-drag ratio, and it also substantially increased thermal stress. In effect, the Air Force saw the shuttle as a way to revive the role of the X-20, a project which it had been forced to abandon a decade previously, in which a winged spaceplane was to be launched on a rocket, make a surprise reconnaissance pass over a target on the far side of the world, and land 90 minutes later. Since the Earth would have rotated about 23 degrees during that time, the landing site would be some 2,500 km east of the re-entry track, so this became the orbiter's *crossrange* requirement. So the Air Force was not simply specifying a shuttle that would supersede its rockets; it was also *expanding its mission*.

On the other hand, NASA saw an opportunity for a trade. The Air Force needed to send the shuttle into polar orbit. This was not feasible from the Kennedy Space Center because it would involve making the initial phase of the ascent over land, and if anything went wrong the debris would rain down on an angry population. The highest inclination attainable from Florida was 57 degrees. At Vandenberg AFB in California it was possible to launch south out over the ocean, so payloads destined for polar orbit departed from there. It was evident that the Air Force would have to build a shuttle complex at Vandenberg. NASA suggested that if the Air Force let it fly from there, it would assign one of its orbiters to the Air Force.

This process of horse-trading led to the shuttle's configuration being dominated by the requirements of the various government agencies within the Department of Defense.

AN INAUSPICIOUS START

Many of the satellites which the Air Force launched were not only heavy and bulky; they were also sent up to geostationary orbit (GSO). Having agreed to use the shuttle, the Air Force set the requirements for the two-stage manoeuvring unit which was to fly its satellites to their final destinations after the shuttle had released them into low orbit. Boeing won the contract for this Inertial Upper Stage (IUS). Just in case the shuttle's development was delayed, the Air Force required the IUS to be Titan-compatible. In the event, the IUS suffered from political infighting and it did not become available until the shuttle was already flying. The first IUS actually rode a Titan into orbit in October 1982, and went on to take a pair of military communications satellites up to GSO. After the shuttle's first commercial flight, a month later, the Air Force requisitioned a flight in 1983 so that it could start to clear its backlog of secret satellites, many of which had been built specifically to exploit the orbiter's cavernous bay, and so had been in storage awaiting the shuttle's entry into service. NASA planned to use the IUS to launch the TDRS communications satellites. It had assigned them to STS-6, STS-8, STS-12 and STS-15. As soon as the first two relays were in place, STS-9 was to carry the first Spacelab. The Air Force, therefore, booked STS-10.

After initiating commercial operations, Columbia was returned to Rockwell to have the development-phase systems removed. Although Challenger had been scheduled to make its maiden flight, as STS-6, in January 1983, problems with its engines forced a slip to April. Alarmed by this three-month delay, the Air Force put in a late order for a dozen expendable rockets in order to guarantee that it would have a reserve of rockets with which to dispatch any satellites which NASA could not accommodate at short notice.

Table 5.1. Military shuttle assignments, *c.* mid-1982

Mission	Date	Orbiter
STS-10	Nov 1983	Challenger
STS-13	Mar 1984	Challenger
STS-22	Nov 1984	Discovery
STS-25	Feb 1985	Discovery
STS-27	Apr 1985	Columbia
STS-30	Jul 1985	Columbia

Challenger released the TDRS/IUS stack without incident. The IUS' first motor burned successfully, and when the stack reached the top of the geostationary transfer orbit (GTO), the second motor ignited for the circularisation burn. Towards the end of this burn, the actuator for the thrust-vectoring system failed and threw the gimballed nozzle off axis, which in turn put the stack into a tumble. Ironically, the specifications for the IUS required it to be the most reliable transfer-stage ever built. It had been fitted with redundant avionics, but it could not recover from this mechanical fault. As soon as the controllers at the Air Force's operations facility in Sunnyvale, California realised what had happened, they released the satellite, which was eventually able to reach its operating station by firing its own thrusters.

Pending the resolution of the IUS problem, all the shuttle payloads that required it were cancelled, and this seriously jeopardised the Air Force's ability to support the many agencies operating within the orbit of the Department of Defense which relied upon it to run their various secret satellite programmes.

RECONNAISSANCE

When the IUS flew again, in January 1985, it was the first stage that caused concern. Although the solid motor fell short of its nominal thrust, the inertial flight control system noted this and fired its thrusters to make up the 16 metres per second shortfall in velocity, just as it was meant to. The circularisation burn was perfect, and the payload was deposited on its assigned station. If it had not been for the IUS problem in 1983, this satellite would have been deployed by the first classified shuttle. Most of the crew were the same as for that cancelled mission.

Although eavesdropping *ferret* satellites were tried in the 1960s, signal processing was not up to the task until solid-state microelectronics became practicable in the 1970s. It seems that the 300-kg Rhyolite that first appeared in 1973 was set up to listen in on telemetry from weapons systems, and that the 1,200-kg Chalet that joined it in 1978 monitored voice links, but it is impossible to be sure because these systems rely on their capabilities being secret. The new satellite's role was meant to be a secret too, but the *Washington Post* reported that it was an electronic intelligence gathering platform. Called Magnum, it was ordered by the National Security Agency (NSA) when it lost its listening post in Iran following the overthrow of the Shah in 1979. The satellite's role was to listen to the entire radio spectrum emanating from the Warsaw Pact and relay its 'take' to Washington.

Since the IUS/Magnum stack was 17.5 tonnes, it was not practicable in the aftermath of the loss of Challenger to offload the second satellite to an existing rocket, so this had to wait until the shuttle resumed flying; it was deployed by STS-33, in November 1989. The Titan 4 was developed to match the shuttle's capacity in terms of volume and mass, and to employ the Centaur as an upper stage. It is likely that further Magnums will be launched on this configuration. The shuttle took Magnum in its stride; it was simply a big satellite on an IUS in need of a ride into low orbit.

Another classified reconnaissance satellite was exposed in the press long before it could be launched. In mid-1987, with the shuttle grounded, *Aviation Week & Space Technology* noted that an imaging radar satellite with 1-metre resolution would not only be able to track armoured vehicles in the field, it would be able to classify them too. This would provide an order of magnitude improvement in battlefield reconnaissance capability. No such radar had yet been demonstrated. The Navy's SeaSat, which operated for a few months in 1978, and the Soviet ocean-surveillance satellites carried radars with resolution on the 25-metre scale. To track a vehicle, a radar operating at a shorter microwave frequency would be necessary. In addition to monitoring armoured forces, such a radar could easily identify mobile ICBM launchers, and so find targets for follow-up inspection using imaging satellites, and thereby police the treaty limiting the deployment of these missiles. This high-resolution radar would also enable the Defense Mapping Agency to survey routes through air defences, in order to produce safe routes for terrain-following cruise missiles. And, of course, it would enable the Navy to take a covert look at the weapons installed on its rivals' ships. The multifaceted benefit to be derived from possessing such a radar guaranteed that it would be built as soon as it was technologically feasible.

The first of these satellites, named Lacrosse, was launched in December 1988. The fact that it was assigned to the first military flight after the shuttle re-entered service indicated its importance. Because radar reconnaissance was impracticable from geostationary orbit, this satellite had no need of the IUS. For the greatest coverage of the Soviet Union, it needed a high inclination. Atlantis employed 57 degrees, which was the highest inclination it could fly from Cape Canaveral without flying a dog-leg over the North Atlantic. The satellite was lifted out of the bay by the RMS and checked out. The 23-metre solar panels were difficult to deploy, but once everything was verified the satellite was released. When the orbiter had cleared the scene, the satellite boosted itself up to about 680 km.

If Challenger had not been lost, Lacrosse would probably have been launched in 1987. Because it had been designed to exploit the payload bay, it could not be offloaded; it simply had to wait for the shuttle to resume flying. It is likely that the second Lacrosse was launched on a Titan 4 in March 1981.

At the end of the Second World War, the United States had felt invulnerable to attack; it alone had the atomic bomb. It thought it had a decade's lead on the Soviet Union, so was shocked when the Soviets detonated their own bomb in 1949. President Truman immediately ordered development of the more powerful thermonuclear fusion bomb. The Strategic Air Command's mission was to retaliate if the United States was attacked. During the 1950s, it progressively modernised its fleet of strategic bombers to be ready to carry out the policy of massive retaliation, but the problem was that it had little knowledge of its targets. At first, it issued pre-war commercial maps to its pilots. Whilst this was sufficient to find and bomb a city, it did not help with military targets. It was not a matter of navigating to known targets; without up-to-date photographic intelligence it could not even compile a meaningful target list. The U2 was built in the mid-1950s specifically to search for targets. When this super-high-flying subsonic aircraft proved vulnerable to a newly deployed surface-to-air missile, the multisonic SR-71 took over; but even this was regarded as a stop-gap solution, because what was *really* needed was a camera in orbit. As soon as the Air Force converted its Thor missile into a launcher it made this programme – which was appropriately named Discovery – its immediate priority.

Several cameras, collectively called Key Hole (KH), were evaluated in the early 1960s. The KH-4 Corona had 3-metre resolution from an altitude of 200 km, and was used to find interesting targets for the KH-8 Gambit, the *close-look* counterpart that skimmed the upper atmosphere at perigee to take detailed pictures with 15-cm resolution. Both recorded their imagery on film and returned it to Earth for processing, so their individual missions were of short duration. The Corona was superseded in the 1970s by the KH-9 Hexagon, which to some extent combined the roles by providing wide-swath surveys at medium resolution, but the Gambit was still flown for follow-up studies. In order to make photo-interpretation more straightforward, the Hexagon used a sun-synchronous polar orbit, and to increase its useful life it carried several film-return canisters. The operating procedure was transformed in 1976 by the introduction of the KH-11 Kennan. This CCD unit produced a digital image that could be radioed to Earth, so it could provide *real-time* intelligence. Since there was no need to make film drops, a satellite could be used almost indefinitely. These reconnaissance satellites had soon outgrown the Thor. The KH-9 model – appropriately known as Big Bird – rode a Titan 3D. By the time the KH-11 had grown to require a Titan 34D, it was a 20-metre long cylinder weighing in at over 13 tonnes. In effect, each of these later satellites was like a 'Hubble Space Telescope', only aimed down at the Earth instead of out into space. Although the Air Force ran them, many of the satellites were flown on behalf of the Central Intelligence Agency (CIA).

It was generally believed that an even larger reconnaissance satellite had been built to be deployed by the shuttle and it was referred to as the 'KH-12' in anticipation. The classified satellite deployed by STS-28 in August 1989 managed to retain its cover, but the fact that it was dropped off in a 57-degree orbit and did not require an IUS strongly suggested that it was the long-awaited newcomer. If it was, and if it had been built to exploit the shuttle's capacity, it must have been a real heavyweight! But what was it capable of? All speculation was greeted by official silence. What is certain is that when Columbia was clear, the satellite raised its orbit to 460 km and then disappeared. This prompted the suggestion that it had manoeuvred into a highly eccentric high-inclination orbit, as used by the Satellite Data System (SDS). These satellites spent most of their time high above northern mid-latitudes, and served as relays for the real time imagery produced by KH-11s in low orbit. The SDS made its appearance in 1976, the same year as the KH-11. Given the secrecy surrounding the shuttle's payload, this seemed to be as good a guess as to its nature as any. Whatever it was, however, the fact that it took such a high priority immediately after the shuttle returned to service implied that it must have been considered vital. Intriguingly, in December 1992 STS-53 flew a virtual replay of this mission, so it may well have delivered a second such satellite. In July 1996, a Titan 4 with no upper stage delivered a satellite into a 300-km orbit at inclination 55 degrees, and within days this began to manoeuvre into such an orbit at 63 degrees. Was this related?

Perhaps the satellite released by STS-38 in November 1990 was the much-talked-about next-generation imaging satellite. At first, the 28-degree orbit suggested that it was to be flown up to geostationary orbit, but Boeing has not claimed this as an IUS flight. In fact, upon being released at 240 km, the satellite climbed to 750 km, which was consistent with an imaging reconnaissance satellite except for the fact that at its low inclination it did not overfly Soviet territory. In late 1990, however, the Department of Defense was heavily involved in the Middle East, and this satellite was ideally placed to make favourable passes to spy on Iraq. It has also been claimed that this satellite was a last-minute substitution. The flight had long ago been classified, and its July launch had been delayed because of hydrogen leaks. Classified payloads are not loaded until several days before launch, so it is certainly possible that this flight was reassigned. In fact, its launch was set for 9 November, but "payload problems" had caused it to miss that date. It takes two days to reconfigure the Eastern Test Range for a different type of vehicle on a different azimuth. Having missed its slot, Atlantis yielded to a Titan 4, which left on 13 November. The shuttle launched two days later. The Titan 4 had an IUS which delivered to GSO a satellite designed to detect ballistic missile launches. The Air Force had already used its final Titan

34D, and Titan 4 production was still in low gear, so whatever the nature of its payload, on this occasion, when the Air Force was in real need of a launcher, the shuttle was on hand.

But the satellite released by STS-36 in February 1990 caused the most speculation. The shuttle needed to fly an inefficient dog-leg over the North Atlantic for a 62-degree inclination, so this must have been *very* important for the payload. The Stabilised Payload Deployment System (SPDS), used for the first time, deployed the 17-tonne object by rolling it over the port sill.

But what was the payload? What is known is that three days later, a tracking radar observed four distinct objects in its orbit. This prompted speculation that its manoeuvring system may have failed, or that its battery may have exploded. The overall result was that in the Press it was another 'wasted' shuttle flight. However, in September it emerged that the satellite was operating in an 800-km circular orbit at the slightly increased inclination of 65 degrees, an orbit with excellent coverage of the Soviet Union. It seemed that the 'debris' had been covers ejected during its check-out sequence. This satellite was later acknowledged to have incorporated both digital imaging and signals-intelligence receivers for the CIA and NSA.

The shuttle, therefore, introduced three new reconnaissance satellites: Lacrosse was a radar system providing all-weather vehicle-tracking capability; Magnum was an advanced geostationary listening post; and the anonymous third category may well have been the eagerly anticipated successor to the KH-11, or maybe it was not. In addition, a pair of support satellites appear to have been released. Whatever they were, from the shuttle's viewpoint they were missions successfully accomplished.

COMMUNICATIONS

The Department of Defense had been very impressed by Syncom, the first relay satellite to be placed in geostationary orbit. It was eager to integrate its communications into a single global system, so it ordered development of a series of satellites that would provide secure communications. This required a significant advancement of the technological state-of-the-art because each satellite was to provide 1,300 channels, with transponders operating in the X-Band. Whereas Syncom had been a mere 65 kg, this new satellite was to be a 600-kg monster which would require the Titan 3C, then the most powerful rocket in the inventory. As development was expected to be protracted, a compromise was ordered as a stop-gap. A single-channel relay would be small enough for a Titan 3C to send them up eight at a time. A 24-satellite constellation was to be established in low orbit, which, although not providing a global uninterrupted network, would be a major step towards integrated communications. The operational network was to be the Defense Satellite Communications System (DSCS). Deployment of the Phase-I stop-gap was completed in mid-1968. Although the design life of the initial satellite was two years, most of the system was still operational when the first of the Phase II geostationary relays was introduced in 1971. Built by TRW, it was a 3-metre diameter spin-stabilised drum with a despun antenna, and was powered by peripheral solar cells. Each satellite was expected to last five years. The first of the Phase III satellites was launched in 1982, on the maiden flight of the IUS. Although the 850-kg new model offered only a 50 per cent increase in capacity, its electronics were better *hardened* and its service was considerably more secure. Built by GE Astro Space, it employed a three-axis stabilised bus and was powered by large solar panels.

Shortly after the shuttle's first test flight in 1981, the Air Force announced that building up this new DSCS network would be a high priority for the shuttle, but the grounding of the IUS in 1982 forced it to continue to use the Phase-II satellites. After the successful IUS flight in early 1985, the Air Force booked a shuttle to build up its degraded DSCS network. When Atlantis flew this mission in early October 1985, it climbed to 500 km – slightly higher than usual – so that the IUS could carry two of the heavyweight Phase-IIIs. Another pair was scheduled for November 1986 to bring the network up to full strength, but in the aftermath of the loss of Challenger, as part of its withdrawal from the shuttle, the Air Force offloaded the DSCS onto the Titan 34D. It was this rocket that had carried the IUS on its maiden flight, and it was unable to carry two of the Phase-IIIs. This meant that, as before, a Phase-II and a Phase-III would have to be carried, so it would take longer to build up the new part of the network. Unfortunately, a Titan 34D had failed a few weeks after STS-21, and a second was lost in April 1986. Both of these carried reconnaissance satellites. Having suffered two successive catastrophic failures, the Titan 34D was grounded. Since it had been in the process of phasing this rocket out to comply with the shuttle-only policy, the Air Force had only a few such rockets left. The Titan 34D did not fly again until February 1987, and reconnaissance took priority over communications, so the DSCSs were not launched until September 1989, on the very last rocket of this type. A few of the TRW satellites confounded their supplier by running for 15 years. The first Phase-III, however, showed signs of premature degradation, and had to be shut down. Starting in 1992, the Air Force launched the Phase-III satellites individually, on the new Atlas 2 rocket. This did not require the IUS because it injected its payload directly into GTO, and a new motor made the circularisation burn. Two satellites were added to the network every two years.

The shuttle's role in this programme, although curtailed by the loss of Challenger, had at least had the advantage of being able to dispatch the satellites two at a time. Clearly, if Challenger had been able to deploy the second pair of Phase-IIIs in late 1986, that system would have become operational at that time. As a result of the scramble for rockets after the shuttle's grounding, and the general offloading of Air Force payloads, the network did not achieve this status until mid-1993.

The DSCS network was designed for *strategic* communications between the Pentagon and its various Commands around the world, and it used super-high-frequency links which needed sophisticated processing facilities; it took a giant C-5 Galaxy to airlift a Phase-III ground terminal. Between 1978 and 1980, a simpler UHF system was set up to communicate with nuclear forces. The Navy sponsored the development of this geostationary network and, in keeping with its penchant for compressing functional descriptions to construct acronyms, it named this system Fltsatcom. The TRW-built satellites also carried Air Force transponders, so the Air Force referred to the network as Afsatcom. Although used primarily for secure voice communication at the *tactical* level of command and control, the low data-rate could relay authorisation for nuclear weapons release.

In 1978, when the Navy specified the system that would replace Fltsatcom, it was decided not only to launch the new satellites on the shuttle, but to design them to exploit the capacious payload bay. Hughes won the contract and came up with the HS-381. In contrast to the HS-376 – which it built for the commercial sector, and which had been constructed as a tall thin drum to suit the narrow payload shroud of a rocket such as the Delta – the drum of the HS-381 was over 4 metres wide and 3 metres tall. Whereas the HS-376 had been mounted on a perigee-kick motor and set upright in the bay, the HS-381 incorporated its own motor and was carried on its side, and it was *rolled* out of the bay, 'frisbee-style'. It was a monster in every sense. However, although it weighed 7 tonnes, half of that was propellant for the trip up to GSO, so it did not require the services of an IUS. As a result, this series was not held up by the grounding of the IUS. The full network required four satellites, but a fifth was built as a spare. In the commercial spirit of the time, the satellites remained the property of the manufacturer, and were *leased* to the Navy which, fittingly, called them Leasats.

Between August 1984 and August 1985, Discovery deployed four satellites, the first on its maiden trip, a flight that was memorable for being the first to release three satellites. The spring-loaded cradle not only set the satellite spinning at 2 rpm for stability; it also triggered the switch that activated the sequencer. The satellite's first act was to raise its omni antenna so that Hughes could check it out before it set off for geostationary orbit. With two Leasats in place, it came as a shock when the third rolled out of the bay and failed to activate. It seemed as if the switch had jammed, so the astronauts cut the cover of a flight plan to create a 'fly swatter', taped this to the end of the RMS, then dragged it across the switch to try to flip it; but the attempt failed, and they had to abandon the satellite. Once Hughes realised what had gone wrong, it devised a way to kick-start the satellite. After Discovery had deployed the next satellite, Leasat-4, it rendezvoused with Leasat-3 and two spacewalkers installed the bypass that brought the derelict back to life. Although Leasat-3 was successfully manoeuvred up to its assigned station, the network remained incomplete because Leasat-4 fell silent soon after reaching its operating station. As the terms of the lease did not oblige the Navy to pay for a satellite until it was on station and fully checked out, the time that Leasat-3 spent stranded in low orbit, and the loss of Leasat-4, were borne by Hughes. The spare had been booked on STS-24, but it was withdrawn for inspection. The plan was to reassign it as soon as it had been verified, but the shuttle manifest for 1986 was dominated by the missions to observe Halley's Comet, to launch the Galileo and Ulysses interplanetary probes, and to deploy the Hubble Space Telescope. The Challenger accident rendered this schedule irrelevant. It was not possible to offload Leasat-5 because it was far too wide for a rocket; it had to await the shuttle. However, the fact that it was not flown until almost 18 months after the shuttle had resumed operations showed its low priority. The Fltsatcom satellites had proven rather longer-lived than expected. The Fltsatcom and Leasats are now in the process of being superseded by the satellites of the UHF Follow-On (UFO) network.

The shuttle not only took Leasat in its stride; it showed its versatility by retrieving and activating a satellite which otherwise would have to have been written off, and it did so for free, because the repair was tacked onto the end of an already planned flight. The failure of Leasat-4 in geostationary orbit amply demonstrated the fact that the shuttle was limited to low orbit. If the OMV-tug had been available, it may have been able to retrieve Leasat-4, so that the shuttle could have returned it to Earth for repair and relaunch; and if these tasks had been integrated into established missions, this refurbishment could have been achieved on the fly.

EARLY WARNING

Although the first intercontinental-range ballistic missile (ICBM) was tested in secret by the Soviet Union, its power was dramatically demonstrated on 4 October 1957 when it was used to launch Sputnik, the world's first artificial satellite. From the Department of Defense's point of view, the threat it faced was transformed overnight.

Since the United States had lost its monopoly over the atomic bomb, it had developed a sophisticated air defence system integrating perimeter radars and missile-armed interceptors, and in the Strategic Air Command it had built up the world's most powerful strike force. It had been confident of its ability to detect Soviet bombers and shoot them down before they could do damage, and launch an overwhelming counterstrike. It had built a *fence* to protect itself. But the ICBM gave the Soviets a means to hop over this fence. Furthermore, it could not be intercepted. Even worse, its great speed cut the time between mounting an attack and having it delivered from half a day to half an hour, and this exposed the United States to a devastating surprise attack. The Air Force promptly made the ICBM bases prime targets for its strategic bombers, and accelerated development of its own missile, the Atlas. The ICBM threat also created a new strategic mission for the Department of Defense. It was crucial that it be able to provide the President with the greatest possible warning that an attack was underway. The first step was to build powerful radars to scan the approach routes, to detect missiles as they climbed over the horizon, but for the maximum warning-time it would be necessary to spot the missiles as they were launched, and this could be done only from space.

In early 1960, the Air Force started to send up MIssile Defense Alarm System (MIDAS) satellites to test sensors for detecting missiles as they rose out of the atmosphere. Detecting a plume of hot exhaust gases proved to be straightforward; what was difficult was the elimination of signals from forest fires and gas-burning from drilling platforms in oil fields, because signal-processing technology was rather crude. It was also necessary to experiment with operational strategies, so some satellites were put into low polar orbit, others into high polar orbit, and later, into geostationary orbit. It was 1970 before a mature system could be deployed operationally, and this was called the Integrated Missile Early Warning System (IMEWS). Five satellites had been sent up to geostationary orbit by 1975. This TRW-built satellite was a drum with a cruciform of fold-out solar panels augmenting the conformal cells, and it had a telescope projecting from the top of the drum. The telescope was offset from the spin axis, so that its narrow field of view could sweep a wide area, and there was a 2,000-element infrared detector array at its focus. The first Phase-II was launched in 1976, under the banner of the Defense Support Programme (DSP). It could detect a missile in the first few minutes of flight, pinpoint the site, and, as it tracked a missile, report its azimuth. And it was not as if this capability was difficult to test; in those days the Soviets were launching rockets with satellites every few days.

As the IMEWS satellites grew, the Titan 3C gave way to the more powerful Titan 34D. In the early 1980s the Air Force set out to take full advantage of the shuttle's cavernous

Fig. 5.1. The Defence Support Programme (DSP) 'early warning' satellite shortly before it was ejected from the payload bay.

payload bay for its Phase-III model, and it produced a 10-metre long giant that weighed in at 2,360 kg and required an IUS for the ride up to geostationary orbit. The problem with the IUS totally disrupted the Air Force's schedule so, unable to launch a Phase-III, it built a hybrid satellite in which the new 6,000-element detector and its associated signal processor were retrofitted into a left-over Phase-II, then launched it in December 1984 as a stopgap. As a result, when the IUS resumed flying in 1985, the Phase-III was a lower priority and, unfortunately, the shuttle was grounded by the loss of Challenger long before it was to be launched. Although the Air Force promptly began to offload satellites from the shuttle, the Phase-III was so heavy that it had to await the introduction of the Titan 4. The plan to send up another hybrid was frustrated by the grounding of the Titan 34D in 1986, so it did not appear until November 1987. The importance of orbiting the first Phase-III is evident from the fact that it was assigned to the maiden flight of the all-new and essentially untested Titan 4. Although delayed by six months of teething troubles, when it finally got off the ground in June 1989 the new rocket performed flawlessly. After the second DSP followed in 1990, it was announced that STS-44 would carry the third. Named Liberty by the crew, it was deployed without incident. In 1994 the Air Force considered launching another DSP on the shuttle, but this was not done; when this satellite appeared in early 1997, it rode the first of the upgraded Titan 4B rockets.

Superficially, therefore, the shuttle could be said to have played an insignificant part in the IMEWS/DSP programme, but this view fails to acknowledge the fact that this satellite, like Magnum and Lacrosse, had been designed with the shuttle in mind, and it was the *unavailability* of the shuttle that seriously affected the programme.

Having been quick to exploit the shuttle's payload bay capacity, the Air Force was then caught out when the shuttle was grounded. As none of its few remaining rockets could lift its new satellites, it had to wait-out the development of the Titan 4, which was intended to deliver the performance *originally* set for the shuttle. There can be little doubt that the Air Force would have preferred to develop the Titan 4 a decade earlier, and, in retrospect, this would have been wise, but providing rockets to *complement* the shuttle had been contrary to the entrenched policy at that time.

If the IUS had not suffered teething troubles, then the shuttle would have been able to carry many more of these Air Force satellites before it was itself grounded. With more of its high-priority reconnaissance, communications and early-warning satellites already in space, the crisis that faced the Department of Defense upon Challenger's loss would not have been nearly so serious.

NAVIGATION

When the Navy placed ballistic missiles in submarines, it had to devise a more accurate means of navigation. The great threat posed by a submarine was that it would remain out of sight, yet it had to be ready to strike at short notice. How could it navigate underwater with sufficient accuracy to be able to fire its missiles? Ironically, the Navy took advantage of an inertial navigation system that the Air Force had developed for one of its missiles. This was excellent by any standards, but such devices suffer from *drift*, and it had to be periodically updated. This reimposed the original problem, but with a much more demanding degree of accuracy. If the submarine needed the inertial system to find its approximate position, how was it to find its actual position with sufficient accuracy to cancel the inertial unit's drift!?

It was possible to navigate offshore by triangulating signals from several Loran radio beacons, but this was not practicable for a submarine submerged in mid-ocean. However, a system of satellites would allow radio triangulation. All the submarine had to do was to run up its radio mast. The Navy deployed the Transit system in the 1960s, and over successive generations this system served through to the late 1970s. Its utility was effectively limited to more or less stationary submarines, however, because it required bearings from several satellites over a period of about an hour to establish an accurate fix. Certainly it was no use to an aircraft in flight. But the Air Force wanted to find a way of navigating in real time with a very high degree of accuracy. Furthermore, whatever system it developed it would have to be usable by a compact and lightweight receiver because there was no room for an electronics cabinet on a fighter aircraft, a missile or a smart bomb. The design it selected was the NAVSTAR Global Positioning System (GPS). Several sets of satellites in 20,000-km orbits inclined at 55 degrees, each set in a different plane with respect to the equator, would provide an accurate fix, but the trick was to build the complexity into the satellites so that the receiver could be fairly simple. The first six satellites were sent up between 1978 and 1980 to test the feasibility of the system, and the results were excellent. Transit could provide a fix within 200 metres, and it was strictly two-dimensional. When in operation, the GPS network was to have sufficient satellites to fix the receiver's position to within 15 metres, do so on a continuously-updated basis, and also supply altitude. Even though it was intended for military use, it was clear from the start that such a system would have civilian applications, so a parallel triangulation algorithm was added to the system; one that did not operate in an encrypted mode, and had 100-metre accuracy. And, true to the requirement that the receiver be simple, it is the size of a pocket book and can be bought for a few hundred dollars. When airliners start using GPS to navigate and make all-weather landings, navigation will rival communications for its total dependence on satellites.

The operational GPS system was to employ six sets of four satellites. This was a major trucking contract for the shuttle. By ferrying satellites up four at a time, each flight would be able to fill in one plane of the network. Even though they were not destined for GSO, at 1,600 kg each these satellites would require the powerful PAM-D2 kick-motor. Deployment was to start in 1987, and the network was to be operational within two years. In the immediate aftermath of Challenger's loss, it was hoped to slip deployment by no more than a year, but, in January 1987, the Air Force offloaded the entire GPS constellation from the shuttle and placed an order for the new Delta 2, even though this would be able to install them only one at a time. The first went up in February 1989, and the last in March 1994. If the shuttle had not had to be grounded, and deployment had proceeded as planned, it is evident that the GPS system would have become available *five years* earlier than was actually the case.

STAR WARS

The first shuttle payload flown for the Department of Defense was the unimaginatively designated DoD82-1. Being classified, it was not mentioned in NASA's Press Kit. In fact, it carried two instruments: the Cryogenic InfraRed Radiance Instrument in Space (CIRRIS) and the Ultraviolet Horizon Scanner (UHS). These were mounted on a pallet in the payload bay, but so as not to reveal the nature of this payload, the video downlink from STS-4 was not allowed to show a view of its bay.

The UHS telescope measured the far-ultraviolet spectral characteristics of the horizon. It gave basic data for the development of a system to provide early warning of ballistic missiles by detecting them as they came into view across the horizon. This task was complicated by the fact that ballistic missiles from the Soviet Union would approach over the north pole, and so would be seen against emissions stimulated by the magnetic field. The Strategic Defense Initiative Organisation (SDIO), which was set up in 1983 specifically to develop a way to intercept ballistic missiles, would devote considerable effort to solving this problem.

The CIRRIS package was to look down at the Earth to record the *spectral signature* of the exhaust plumes from airbreathing engines. This data was required to calibrate another package, called Teal Ruby, which was then in an advanced stage of development and was itself to be tested on the shuttle before being deployed operationally as a satellite. Teal Ruby was to detect and track aircraft, and particularly cruise missiles, to provide early warning of an attack. The advantage of tracking aircraft *passively* from space was that, unlike tracking by radar, the target was not made aware of the fact that it had been detected. Furthermore, a constellation of such satellites in polar orbit could monitor the oceanic areas in which Soviet cruise-missile submarines were likely to operate. Furthermore, even if the missile's *launch* was not directly observed, its exhaust was expected to remain detectable for long enough to identify its point of origin, so that a hunter–killer submarine could be directed to track down and dispose of the launch platform. CIRRIS employed a telescope with an infrared detector cryogenically chilled by liquid helium. Sophisticated processing was required to distinguish an exhaust plume from the thermal background, but this was simplified by scanning open ocean. Unfortunately, the experiment was frustrated when the cover of the telescope failed to open.

The sensor technology needed for Teal Ruby involved a significant advancement of the state of the art. It used a 3-metre focal length telescope with a matrix of 250,000 individual detectors. Dividing the image into small cells provided high spatial resolution, and by using different spectral bands it was possible to identify the nature of any given source. From the intended 600-km operating altitude, the telescope would have had a 2-degree field of view and the matrix would have provided a surface resolution of about 100 metres. The use of a *staring array* detector was crucial because it continuously sensed the *entire* field of view, whereas a scanning detector, by its nature, would have sensed points within the field in sequence. It was not the time it took to sense a specific field of view that was important; it was the time the detector could devote to each cell within that area. In the time it took a scanner to sample the entire field of view, it would have sensed each cell for only a fraction of that time. With a staring array, each cell would have been sampled for that *entire* period. Since the sources were so faint, relative to the background, it was this ability to *integrate* that made detection possible. In 1977, in a climate of concern that its air defence system was vulnerable to the latest Soviet cruise missiles, the Air Force set out to test whether this space-based tracking system would work in practice.

With the shuttle expected to start flying in 1978, or 1979 at the latest, it was hoped to have Teal Ruby ready for a trial in 1980; it was assigned the payload designation DoD80-1. In the event, the shuttle's protracted development did not hold up Teal Ruby, because the cryogenic sensors proved more difficult to make than expected and it was not actually ready to fly until 1983. By this time, it had been decided to fly Teal Ruby from Vandenberg AFB in California, so that it could be realistically tested in a polar orbit. Unfortunately, work on Space Launch Complex 6 (SLC 6) had been postponed when it became evident that the shuttle would not be available for military service as early as originally expected. Although it was hoped to finish SLC 6 in 1985, this soon slipped to 1986. Nevertheless, as soon as everything was finalised, Teal Ruby was assigned to the first flight from SLC 6. As part of its deal with NASA, the Air Force was to have virtually exclusive use of Discovery. It was to make flight 62A in July 1986. The Teal Ruby test was regarded as so important that Air Force Under Secretary Pete Aldridge was to accompany it as an observer. It was to be deployed and left for a year, during which time its abilities would be exhaustively assessed, and then it would be retrieved.

In the aftermath of the loss of Challenger, NASA decided not to proceed with 109 per cent of rated thrust of the shuttle's main engines. It took more energy to attain a high-inclination orbit, and the Air Force had been relying on this to carry its heaviest payloads. Consequently, the Air Force lost interest in the shuttle, halted work on SLC 6 and, in a series of steps, mothballed and then decommissioned the facility.

When the shuttle resumed flying in 1988, Teal Ruby was manifested on STS-39, at that time slated for early 1990. As originally envisaged, it was to have served as a technology demonstrator in advance of being deployed operationally. However, the sensor that was so innovative when proposed in the mid-1970s, was, a decade later, essentially obsolete, so it was withdrawn. Six months were lost in 1990 when hydrogen leaks grounded two of the shuttles, so STS-39 slipped back to April 1991. The primary payload of this Department of Defense flight was CIRRIS, this time in support of the SDIO. It obtained spectral signature data on airglow and auroral emissions in the 2–25 μm wavelength range. In contrast to its first outing – when it was to have provided data to assist space-based tracking of cruise missiles – this time the goal was to identify wavelengths at which processes operating in the upper atmosphere would not interfere with tracking ballistic missiles during the coast phase of their trajectory, and, as luck had it, there was a major auroral display early in the flight.

The major new payload on STS-39 was the Infrared Background Signature Survey (IBSS). The SPAS free-flying pallet, which had flown twice on early missions, had been purchased from Germany and fitted with the detectors for this experiment. After it was deployed, the orbiter moved 10 km ahead of it. The IBSS carried a multispectral suite of imaging sensors and an imaging spectrograph sensitive in the infrared. It observed the orbiter perform a series of tight manoeuvres and measured the spectral characteristics of

Fig. 5.2. The deployment of the Infrared Background Signature Survey (IBSS) package. This had been mounted on the old SPAS free-flyer by the Strategic Defense Initiative Organisation (SDIO).

the thruster plumes. The orbiter fired its big OMS engines one at a time to fly lazy circles. Xenon, neon, carbon dioxide and nitrous oxide were sprayed in turn from canisters in the bay, while the IBSS observed their interaction with the Earth's magnetic field to establish whether exhaust gases generated a characteristic glow. The Chemical Release Observation (CRO) experiment involved ejecting canisters which released a selection of rocket propellants when they were 200 km away, so that the effect could be studied by IBSS, the orbiter, and airborne sensors. The aim of these experiments was to determine whether it would be feasible to spot a ballistic missile's bus as it manoeuvred to dispatch its warheads to their individual targets. This could be done either directly, by observing the bus' thruster plumes, or indirectly, by observing the glow from the exhaust products reacting with the Earth's magnetic field. Another objective was to observe the appearance of leakage of propellant from a damaged missile. It had been deemed unacceptable for an anti-missile system to rely on interception during the terminal phase of an attack. Without the technology to track ballistic missiles while they were above the atmosphere, interception in the last moments of the boost phase and in the post-boost cruise would be impracticable, so this testing was key to the SDIO's programme.

In February 1990 a Delta rocket launched the Laser Atmospheric Compensation Experiment (LACE) satellite into 550-km circular orbit for the SDIO. Its primary role, as its name suggested, was to serve as a target for ground-based laser beams. Once on station, it extended a cruciform of sensors. When a laser was aimed at the target, the sensors enabled

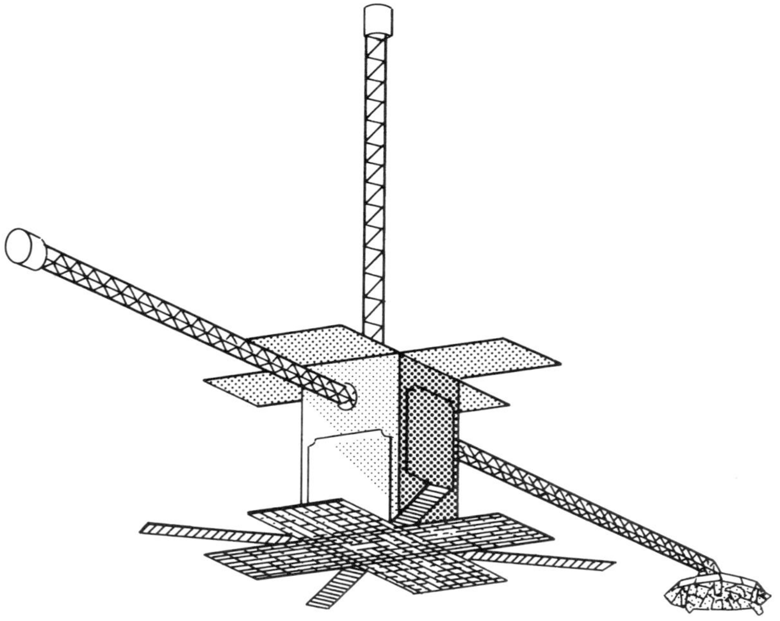

Fig. 5.3. A schematic of the Laser Atmospheric Compensation Experiment (LACE) satellite that the Strategic Defence Initiative Organisation (SDIO) sent into orbit to act as a target to refine a system for aiming lasers.

the extent to which the beam was disturbed by its passage out through the atmosphere to be measured, so that the efficiency of a corrective-optics system designed to stabilise the beam could be evaluated. The laser was fired by the Air Force's Maui Optical Station (AMOS) in Hawaii. The ability to direct a laser onto a small target in space was another objective of the Star Wars scheme. In June 1985, AMOS fired a laser at Discovery to test its ability to lock the beam on target. For this high-precision tracking experiment, the crew mounted a 20-cm wide corner-cube reflector in the porthole of the orbiter's hatch. AMOS aimed another laser at Discovery in December 1992. This time, the laser was aimed on the basis of the orbiter's predicted position, rather than on optical tracking. This Battlefield Laser Acquisition Sensor Test (BLAST) proved the practicality of the Army's plan to uplink information by laser, for secure communications in close proximity to an opposing force. In these tests, the orbiter's role was simply that of a convenient target of opportunity.

As a secondary payload, LACE carried the UltraViolet Plume Instrument (UVPI), a CCD imager designed to detect a rocket's exhaust plume. The effects of airglow and aurorae in the 2,300–3,450 Å range had been extensively studied, so it was hoped that UVPI would be able to see its targets against such sources of interference. Sounding rockets were fired to assess its performance. This followed up on earlier experiments. Back in February 1988, a Delta had launched a bus which ejected a series of packages and proceeded to track their exhaust plumes as they manoeuvred nearby. The following year, another satellite was launched. Sounding rockets were fired to assess the ability of its sensors to detect and track exhaust plumes. These early tests were technology trials. LACE's UVPI was a long-term engineering evaluation, and every opportunity was seized upon to test it with a well-defined target. It had attempted to spot thruster firings on STS-36 and STS-39, but had failed; and the plan to do so on STS-44 was curtailed when an inertial measurement unit failed and forced Atlantis to return to Earth early. It was rather more successful with subsequent missions. AMOS routinely observed shuttles in orbit. Its imaging system took spectral signature data from thruster firings and water dumps, and tried to track the vehicle by its interaction with atomic oxygen, in the so-called *shuttle glow* phenomenon.

Prior to the loss of Challenger, Atlantis had been scheduled for a May 1987 flight with a Spacelab devoted to SDIO experiments. This *Starlab* comprised a full-length pressurised module and a pallet. A laser in the main module was to have fired through an axial porthole in the aft bulkhead, then been redirected by a mirror on the pallet to assess its ability to hit a variety of test articles carrying laser detectors. After the laser had been calibrated by LACE, satellites known as *Starlet* were to have been released by the shuttle, and sounding rockets known as *Starbird* were to have been launched from the ground to serve as targets.

In the immediate aftermath of the accident, NASA had hoped to resume flying in early 1987, so Starlab was reassigned to June 1989. Once it became clear that the shuttle would not fly until 1988, Starlab slipped to March 1990. With the shuttle back in service, the hope of building up to monthly flights proved impracticable, so crucial payloads were brought forward in the schedule and others were either offloaded or permitted to slide. The Starlab was pushed back first to September 1991, then to January 1992, and was finally cancelled.

But by this time the strategic situation which had defined the SDIO's mission had undergone a remarkable transformation. The Soviet Union – the entity that Ronald Reagan had dubbed the 'Evil Empire' – had taken everyone by surprise, and collapsed. In 1993 the SDIO was renamed the Ballistic Missile Defense Organisation (BMDO) and refocused for Theatre rather than Strategic Defence. Terminal phase interception *is* viable, although only just, in such an attack, so the objective of developing an orbital battle-station was dropped. The space-based sensors, however, were retained to provide early warning, and the shuttle was to continue to play a key role in testing.

By February 1995, when the SPARTAN-204 free-flyer carried the Far-UltraViolet Imaging Spectrograph (FUVIS) for the Naval Research Laboratory (NRL), the state of international relations had been so transformed that it was a *Russian* cosmonaut, Vladimir Titov, who operated the RMS. He lifted the platform out of the bay and aimed the instrument towards the orbiter's tail so that it could monitor shuttle glow, the propellant that was leaking from one of the RCS thrusters, and the plumes from thruster firings; then it was set back in the bay. A few days later, however, he deployed it to carry out observations of the ultraviolet sky so as to characterise the background against which ballistic missile sensors will have to work.

SPACEFLIGHT ENGINEERS

In the mid-1960s the Air Force recruited pilots to fly the Manned Orbiting Laboratory. They were to use a Gemini spacecraft mated to a pressurised module, and ride the powerful Titan 3. Their primary task was to conduct military reconnaissance with the KH-10 camera. After a tour of duty lasting a month or so, they would pack up their film, vacate the module and return to Earth. The project was building up to its first test flight in 1969, when it was cancelled by the incoming Nixon administration because the automated form of the camera, the KH-9, promised to be able to achieve the same results at much reduced cost and greater operational flexibility. With this plan to run its own man-in-space programme on the rocks, the Air Force told its astronauts that they could either return to their old posts or transfer to NASA.

The youngest members of the team – Dick Truly, Karol Bobko, Bob Crippen, Gordon Fullerton, Hank Hartsfield, Bob Overmyer and Don Peterson – opted for NASA, even though Deke Slayton, who was responsible for crew assignments, made it clear that there was *no* chance of them flying Apollo or Skylab missions. They made themselves invaluable behind the scenes, on support crews, however, and when the Apollo astronauts retired in droves, these 'cuckoos' ruled the roost, and claimed the shuttle as their own. Unlike their predecessors who, since Mercury, had been derided by 'real' pilots as 'spam-in-the-can', they would fly a real spaceship, and land on a runway; the days of plucking a capsule out of the ocean were over. Only a few of the others stayed on, the most notable being Fred Haise, Vance Brand, Joe Engle, Jack Lousma and, of course, John Young, who went on to command the shuttle's first test flight.

In 1979 the Air Force recruited a new class of Spaceflight Engineer to ride the shuttle and supervise classified payloads. In effect, they were to be the military equivalent of what NASA dubbed Payload Specialists, who flew for institutional collaborators and commercial customers to oversee specific payloads. The Spaceflight Engineers would receive *complete* briefings on how to check out, deploy and, if necessary, spacewalk to fix secret payloads, so that it would *not* be necessary to supply such details to NASA. Each secret payload was to shepherded by at least one member of the Air Force's elite cadre.

The first was to be 'STS-10' in 1983, but the fault with the IUS led to its being delayed. In the interim, the Air Force assigned David Vidrine to the SolarMax repair mission so that he could observe how this ambitious spacewalk went. However, a month before this mission was due the Air Force decided that as there was to be live video of the spacewalk, *in situ* observation would have "no value", and Vidrine was withdrawn from the crew. It was January 1985 before Gary Payton became the first of the cadre to supervise the deployment of a classified payload. He was followed by William Pailes a few months later.

Once the shuttle was running on a more or less routine basis, NASA set out to broaden the public's sense of participation by instigating Teacher-In-Space and Journalist-In-Space, but the first two Civilian Observers (COs) were the chairmen of the Congressional committees which set NASA's budget, Senator Jake Garn and Congressman Bill Nelson. On the military side of the shuttle, Air Force Under Secretary Pete Aldridge ousted one of the assigned Spaceflight Engineers so that he could fly the first shuttle from Vandenberg and oversee the Teal Ruby, but he was grounded along with the shuttle on that fateful day in January 1986, the day that Christa McAuliffe, the Teacher-In-Space competition winner, perished with Challenger.

Upon his promotion to Air Force Secretary, General Aldridge initiated development of the Titan 4, ordered the Delta and the Atlas production lines restarted, scrapped SLC 6, and wound down the Air Force's commitment to the shuttle as fast as he could. The Spaceflight Engineers were disbanded in August 1988, just a month before the shuttle resumed flying, and the dream of 'blue suits' in space faded once again.

MILITARY-MAN-IN-SPACE

Even though the Air Force disbanded the Spaceflight Engineers, this did *not* mean that there was no place on the shuttle for a military officer. Astronauts had demonstrated that the best tool on a spacewalk was the human hand. Despite its reliance on imaging satellites, the Department of Defense put Tom Hennen on Atlantis in November 1991 to test the potential of the human eye as a reconnaissance sensor. Hennen was particularly well trained to make this assessment. As an Army photo-interpreter he had spent years scrutinising reconnaissance imagery. By going into orbit, he would not only be able to test his highly-trained vision; he would be able to establish whether a camera missed anything important. His flight was part of the Military-Man-In-Space programme.

Paul Scully-Power was a Navy oceanographer. When he flew on Challenger in 1984 to study oceanic phenomena, he was astonished by the subtle texture of the sea and by the fact that it had not been visible on the film he had studied. The human eye was far more adept at distinguishing subtle variation in hue than was film. Not all of his discoveries were in open ocean; he spotted a system of giant eddies in the Mediterranean that had not been noticed by seafarers. When STS-31 had climbed to 600 km to deploy the Hubble Space Telescope, its crew noted that the scope for visual study of oceanic phenomena was *improved* rather than degraded by the altitude, because larger-scale structures stood out better. From that altitude, to convey some idea of the scale of the view, Story Musgrave interrupted his repair activity to note that he had just turned around and discovered the *entire* Australian continent staring him in the face, with all the subtle shading of the outback on display.

Tom Hennen had the Direct-View Optical System (DVOS) for the Terra Scout project. In effect, this was a pair of binoculars on a stabilised mount that, once cued onto a target, compensated for the shuttle's motion. He had been assigned a variety of 'test sites' typical of sites studied by reconnaissance satellites, and was to follow a specific schedule of visual observations and note what he was able to see. His list included military bases and built-up areas, as well as oceanic areas. He was astonished by the detail he could resolve.

The rather blandly-designated M88-1 project was more ambitious, and it turned Atlantis into a sophisticated reconnaissance platform. Hennen used a telescopic CCD camera to take high-resolution imagery of ships at sea and in port, airfield operations, artillery firing, and manoeuvring armour. He had an image-processing system to analyse each site in real time. Mario Runco reported to on-site teams by a secure UHF radio called Night Mist. In return, the supervisor could ask Hennen to confirm specifics, but Hennen had to be wary because some sites were misleading. The pressure imposed by the *interactive* nature of this process was considered a vital part of the experiment. Way back in 1963, during his day in space on the final Mercury flight, Gordon Cooper had spotted trains in open country, but his observations had been dismissed as wishful thinking. On Cooper's Gemini flight two years later, Pete Conrad had been present to corroborate his claims. With his imaging system, Hennen had demonstrated not only that he could distinguish different types of transport aircraft, but also that he could produce militarily useful data in *real time*.

The Navy's main interest in the shuttle was in taking pictures of the ocean; but to be of any use, it was necessary to know latitude and longitude. How does an astronaut locate an object in open ocean? Clearly, an automated method of calculating the geographical location of the centre of an image was needed. The Latitude & Longitude Locator (L3) measured the angle between the horizon and the centre of the frame, and knowledge of the orbiter's path, to calculate the location of an image. Trials conducted on STS-28 and STS-32 showed this system to be accurate to within 12 km, but it was awkward to use, so the NRL developed a more sophisticated system. Hercules was a Nikon-based Electronic Still Camera (ESC) and a ring-laser gyroscope. On its first trial, on STS-53, it proved able to fix a site within 4 km. On STS-56 its digital imagery was relayed to Earth in real time. The initial form needed its inertial reference to be initialised by taking star sighting, but an upgraded version could use GPS to fix sites within 2 km. On a stabilised mount, Hercules could be set up to wait for a specific site to come into view, and then to snap a series of pictures. It was an improvement over L3, because it did not need the horizon to be in the frame, so it was suitable for nadir-viewing imagery.

For the Navy's Military Applications of Ship Tracks (MAST) the astronauts took high-resolution images on infrared-sensitive film of ships at sea. The objective was to determine how a ship affects its immediate environment. Most oceanic life lives in the topmost 200 metres. Some species emit light – bioluminescence due to a chemical reaction between two enzymes. At night, it was hoped to observe the bioluminescence from fauna feeding on the plankton stirred up by a ship's high-speed passage, and, during the day, the wispy cloud which forms in the wake of a ship due to increased evaporation from the sea surface of the dimethyl sulphide phytoplankton waste which seeds cloud formation. This work started on STS-65 in July 1994, and became routine on later missions. A long-term goal was to develop a way to track a *submerged* submarine from space. When a shallow sea of clear water is viewed at a low angle of lighting it becomes semi-transparent, and it is possible to see the shape of its floor; in such conditions it was possible to *see* a submerged submarine. In open ocean, a submarine running at 30 knots in shallow water might reveal itself by the bioluminescence in the wake.

The CLOUDS project made its debut on STS-12 in August 1984. It involved nothing more then using a Nikon 35-mm camera with an infrared filter to take high-resolution pictures of clouds, particularly those associated with severe weather and wispy cirrus, and recording each over as wide a range of viewing angles as possible, but its military nature was evident from the acronym (Cloud Logic to Optimise Use of Defence Systems) and the fact that it was subsequently pursued on STS-39, STS-45 and STS-53, all of which were Department of Defense missions. The procedure was for meteorologists to specify the cloud structures they wished recorded, and the crew would pick up the target the moment it came into view and track it until it disappeared, taking pictures at regular intervals using the motorised camera. The objective was to quantify *apparent* cloud cover as a function of the angle at which clouds of different types are viewed, to provide data to calibrate the sensors used by the Defense Meteorological Satellite Programme.

THE RETURN OF THE TITANS

The shuttle never really achieved the operational requirements set by the Air Force. The orbiter's bay was of the specified size, but its payload capacity was undermined by the fact that the stack was somewhat heavier than expected. NASA was progressively working on this, however. It was possible to test the orbiter's subsonic handling by carrying it on the back of the Boeing SCA and then releasing it at altitude, but there was no way to verify that it would be able to withstand the dynamic loads associated with hypersonic soaring until it was returning from orbit, so Columbia was heavily instrumented. Successive orbiters came off the production line successively lighter. Other measures addressed the rest of the stack: for example, the payload capacity was increased by 275 kg simply by *not* painting the ET.

Even though the shuttle had been declared operational following its fourth test flight, it was still undergoing development. At first it was criticised for not meeting its specifications, and was then criticised for undergoing continuous refinement. The progressive uprating of the SSME thrust was not taking 'unnecessary' risks; it was a process of *stretching the envelope* to work towards the optimum performance. This could be done inconspicuously by a research aircraft like the X-15 rocket plane, which was an end in itself; but the shuttle was a utility, and it had to be used as its performance was refined. Similarly, although the orbiter's crossrange was initially restricted, a series of trials progressively explored its true potential.

On the other hand, several of the Air Force's requirements turned out to be short-lived. The cancellation of the new reconnaissance camera in 1980 not only terminated the single-orbit mission, it also deleted the 30-day endurance requirement that was to have revived the Manned Orbiting Laboratory. As introduced, the orbiter was restricted to about 10 days in space, but this was just a consequence of the size of the cryogenic tanks carried. A wafer of stores was later installed to facilitate extended-duration flights for the Spacelab programme. One by one, therefore, NASA was addressing the requirements laid out by the Air Force so long ago. But it was not just NASA that was behind schedule. If the Air Force had *really* wanted to fly in polar orbit, it would have had its Vandenberg facility ready and waiting. It was not until 1985 that it signed an agreement with Chile to upgrade the airport on tiny Easter Island, 4,000 km off the coast, to take an orbiter making an emergency descent after it was beyond the point at which it could return to Vandenberg. As part of this deal, NASA was ordered to fly a Chilean national on the shuttle as an observer.

But how did the shuttle perform from the point of view of the Department of Defense? Its late introduction undoubtedly held up the novel payloads that had been built to exploit it, but after it was declared operational it was the IUS which imposed further delay. All of the satellites were successfully deployed. One that failed to start up was repaired, and one that failed in geostationary orbit was beyond the shuttle's reach, but neither of these failures was in any way attributable to the shuttle itself.

When Challenger was lost, and the fleet was grounded, the Air Force had only just begun to step up its use of the shuttle. Initially, NASA had expected to resume operations within a year. Pete Aldridge had secured a stockpile of a dozen Atlas and Titan 3 rockets as a "contingency", so the Air Force believed that it would be able to cover this gap. When the first of these Titans exploded in April, the US found itself in the unprecedented position of being unable to launch heavy satellites. Upon receiving the Rogers' Commission's report in June, the Reagan administration not only abandoned the shuttle-only policy and authorised the Air Force to order more rockets, it also initiated improvement programmes and, most significantly, resumed development of the powerful Titan 4 so that the National Security payloads would not be dependent on the shuttle. The newly-liberated Air Force effectively withdrew from the shuttle, first by mothballing and then writing off its Vandenberg facility, a decision which had the side-effect of denying NASA access to polar orbit. As a result of this divergence, the Air Force now has, in the shape of the Titan 4B, a rocket with a shroud even larger than the shuttle's cavernous payload bay; an awesome rocket that is all the more useful because it can attain polar orbit. In the end – as Robert Seamans told Congress – the shuttle was "not essential" for national security.

Part 2: Weightlessness

6

Materials processing

One much-touted use for the shuttle was the promotion of microgravity research and, through the perfection of methodologies, the facilitation of the commercial use of space. To prime this process, NASA, in partnership with industries and universities, sponsored a number of Centers for Commercial Development of Space (CCDS) in disciplines likely to lead to applications, and it offered to fly their experiments at heavily discounted rates.

The first commercial payload on the shuttle was a middeck unit called the Monodisperse Latex Reactor (MLR). Its function was very simple: it was to produce large numbers of tiny latex spheres, all exactly the same size. On Earth, gravity interferes with the manufacturing process, and it was hoped that a more uniform product (a monodisperse product) would result from making them in microgravity. Even on its first flight on STS-3, the reactor produced millions of perfect spheres 10 μm in diameter. Although all the output from each run was the *same* size, that size could be programmed anywhere between 1 and 100 μm. This polymerisation process was first employed by the Dow Chemical Company in the 1950s, but, for spheres greater than 1 μm in diamater, gravity caused distortion and promoted the formation of irregular lumps – a process called *creaming* – that ruined the product. The ability to make perfect spheres 100 times larger than was otherwise possible represented a significant step forward. Since the attainable size was a factor of time, some of the product from one flight was used to seed a subsequent run, in order to build up to the larger sizes without requiring overly-long flights. The apparatus took up three middeck lockers. It comprised four 30-cm tall reactor chambers and a large feeder tank, and was fully computerised; all an astronaut had to do was turn it on; it turned itself off.

Latex microspheres had a multitude of engineering and biotechnology applications, and the Bureau of Standards found a ready market for them. They were sold at the rate of $400 per 15 million spheres. One mundane but important use was to calibrate scientific instruments such as electron microscopes. On the medical front they could be used in cancer research to measure the pores in tumours so that spheres encapsulating anticancer drugs and primed for controlled release could be made to embed themselves on the surface of a malignant growth to deliver concentrated doses of drugs straight to the cells most in need, yet lower the total dose received by the body. Because there is a significant difference in size between the size of the pores in the lining of the stomach and the intestines, and the tumours that grow there, these cancers were particularly susceptible to this kind of directed dosage. This project was undertaken jointly by the Marshall SFC and Dr John Vanderhoff, who first formulated the process while at Dow. The reactor was successful, and its product was useful, but turning out little rubber spheres was hardly likely to excite the public.

The *holy grail* of the search for a microgravity application was the miraculous drug that would cure some deadly disease. If it could be made only in space, then so much the better. But this was no more than wishful thinking. Nevertheless, the pharmaceutical industry did offer the best hope of a commercially viable application. What was needed was an agent for which there was a proven market, for which a small-volume product could be sold at a premium rate, and for which the standard manufacturing process was disrupted by gravity. Electrophoresis fitted this criterion perfectly; indeed, it seemed to be an application which had been waiting for microgravity processing.

ELECTROPHORESIS

Electrophoresis is a technique for separating material suspended in a fluid by exploiting the net molecular electric charge. If an electric field is applied across the fluid, it causes the molecules to move with an acceleration proportional to charge, separating out by molecular weight. If a mixture of constituents in suspension flows along a tray across which an electric field is applied, the molecules will gradually separate, and will emerge from the other end at different points, where the purified material can be drawn off.

The problem with using electrophoresis on Earth is that gravity disturbs the flow; tiny convection currents cause mixing, which degrades the separation process. Convection does not take place in microgravity, so it was hoped to improve both the volume and the purity of a material by processing it in space. The pharmaceutical industry had been trying to use electrophoresis to purify biologically active agents, so Ortho Pharmaceuticals, a subsidiary of Johnson & Johnson, teamed up with McDonnell Douglas, one of the main developers of space hardware, to fabricate the Continuous-Flow Electrophoresis System (CFES). NASA was strongly supportive, and the Bioprocessing Laboratory at the Johnson SFC assisted in certifying the experiment for flight. Unlike many of the middeck experiments, it was not a simple box with an on–off switch; it was what NASA termed a *tended* package. It required significant attention by an astronaut. Its primary element was the 2-metre long rectangular-section electrophoresis flow tube, or *canal*. It was transparent so that the separation process could be photographed. The entire package of the canal, the pump, the refrigeration and the control electronics occupied a full cabinet and weighed 250 kg.

Starting with STS-3, NASA made most of the early shuttle flights available to test this experimental unit as rapidly as possible so that a production unit could be built. Initially, it was used to purify raw protein, the amino acid compounds that perform numerous critical roles in biochemical processes, but STS-8 marked a milestone by processing live cells. The cells were flown up in an incubator, purified, and then returned to the incubator for return to Earth. This was hand-carried onto the orbiter by a technician immediately prior to launch, and retrieved immediately upon landing. Three types of cell were processed: human kidney, rat pituitary and canine pancreas.

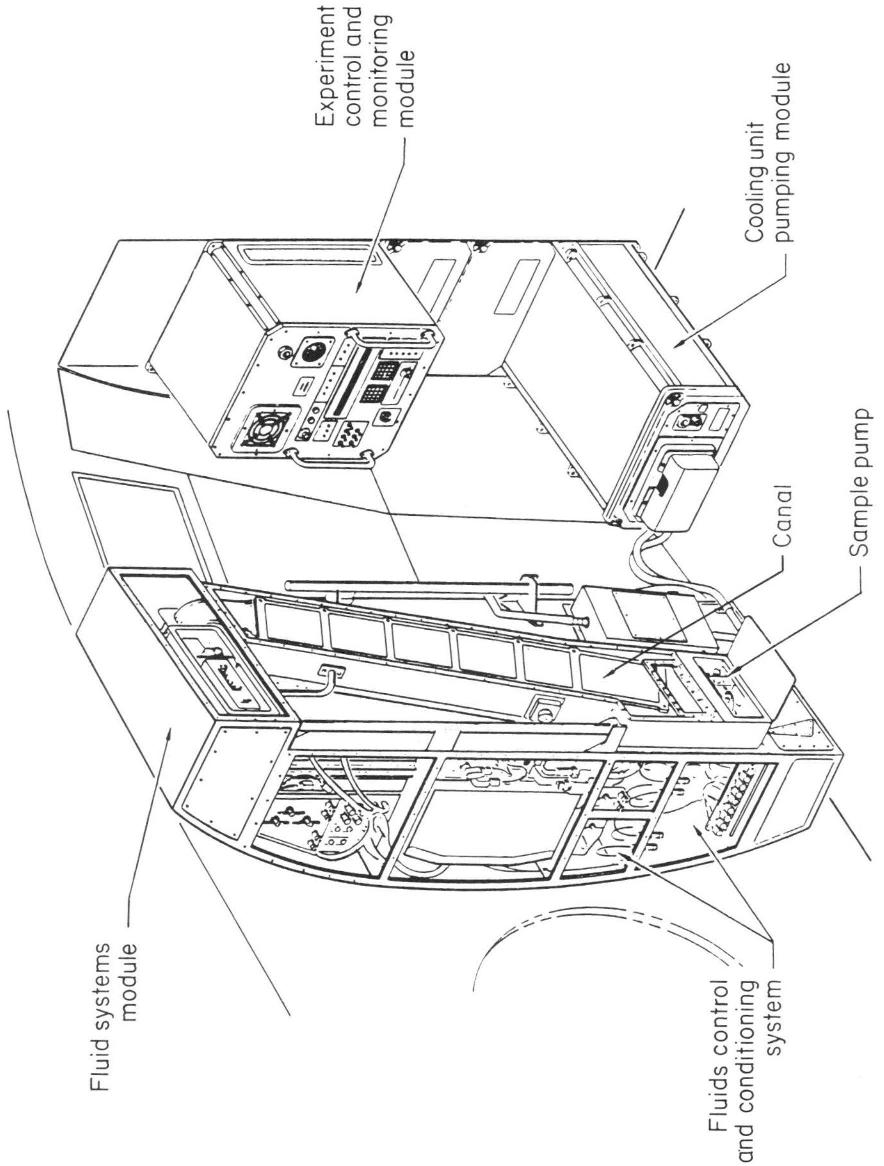

Fig. 6.1. A schematic of the Continuous Flow Electrophoresis System (CFES) which was installed on the orbiter's middeck.

In this first basic demonstration, kidney cells were separated from blood. An endocrine gland secretes a hormone directly into the blood stream in order to exert a specific effect on some other part of the body. A hormone is a chemical messenger – usually a protein – which turns an organic chemical reaction on or off. An enzyme – also a protein – drives the chemical processes of life in a catalytic manner, so a hormone had to trigger the production of only a tiny amount of enzyme to prompt a significant effect. The kidney produces erythropoietin, the hormone that regulates erythropoiesis, the process which makes erythrocytes – red blood cells. Back on Earth, the purified cells were cultured to extract the enzyme that dissolves blood clots. It was hoped to eventually derive a treatment for anæmia, the ailment in which the kidney fails to produce erythropoietin at the required rate. Red blood cells live for only about 120 days, so the body has to manufacture such cells at a rate matching the rate at which they die; in an anæmic patient this fails. The pituitary is an endocrine gland which secretes a polypeptide hormone that promotes growth. The beta cells in the pancreas secrete insulin, the hormone which regulates carbohydrate metabolism. Diabetes is the ailment whereby the body fails to make insulin. In principle, a *single* injection of a large quantity of beta cells would cure diabetes, but these cells were difficult to make. Transplanting cells had proven effective, but the procedure was not yet allowed as a treatment.

In the tests, the CFES processed *700 times* more sample in space than it had in ground trials, and produced a five-fold improvement in purity. This was very encouraging. Several dozen other agents, including hormones to stimulate bone growth, which could be used to treat osteoporosis, epidermal growth agents that could be used to treat burns, and interferon protein used by the immunological system to combat viral infection, held out the prospect of a commercial industry in space. On all these tests, career astronauts had dutifully loaded the solutions and drawn off the product, but it was clear that if the system was to be scaled up it would be necessary to fly a specialist to look after the equipment on a full-time basis.

So it was that Charles Walker became NASA's first Payload Specialist. The CFES had been conceived by McDonnell Douglas in 1976. It was pursued as a long-term research and development project so that if microgravity materials-processing was commercially viable, this giant of the aerospace industry would be ahead of the field. Upon joining the company in 1977, Walker was appointed as Chief Test Engineer for the CFES project. NASA charged the company a fee of only $40,000 to have one of its employees accompany its experiment on a shuttle mission, and Walker was the obvious candidate.

In the 12-month gap between the live-cell test on STS-8 and Walker's flight on STS-12 in August 1984, the CFES apparatus was upgraded so that it would be able to sustain 100 hours of continuous operation with about 27 litres of raw material being processed in an effort to make sufficient erythropoietin for Ortho to conduct all the clinical trials required to secure a licence from the Federal Drug Administration. As it turned out, Walker's knowledge of the system proved crucial, because he was able to spot problems early and intervene before the situation deteriorated. In particular, when a computerised degassing pump malfunctioned, he adapted the operating procedure in-flight. Unfortunately, his effort proved to be in vain. Although the hardware had been sterilised prior to launch, it had not been sufficient to kill a contaminant, and when Walker deactivated the system to attend to the faulty pump, the rise in temperature in the fluid let this bacterium grow, and the product was rendered useless for clinical trial. Nevertheless, from an engineering viewpoint the new hardware had operated successfully, and Walker had actually *seen* it operate in space. He repeated the run in April 1985, and this time returned with a clean sample. However, in September Ortho backed out of the project to pursue a revolutionary method of making erythropoietin which exploited a breakthrough in genetic engineering: a small Californian start-up called Amgen had figured out how to splice the appropriate genes into fast-growing bacteria. Undeterred, McDonnell Douglas sent up Walker again in November and started talks with the 3M Company's Riker Laboratory to use erythropoietin made in a scaled-up form of the CFES. Walker was now in the enviable position of having ridden the shuttle more often than most of NASA's career astronauts.

The *industrial scale* Electrophoresis Operations in Space (EOS) was to be carried on an MPESS in the bay. Its 4-metre long canal was to run for 100 hours to produce 24 times the output of the trials unit. Although it would eventually have to function autonomously, on its proving flight in July 1986, and on the next run in January 1987, the EOS was to be supervised by Robert Wood, a colleague of Walker's, who had overseen its development. The loss of Challenger in January 1986, and the shuttle's indefinite grounding, obliged McDonnell Douglas to place the project in abeyance. By the time flights resumed in 1988, Amgen's protein, called epogen, had become the biotechnology industry's most successful product. Microgravity electrophoresis had been rendered obsolete even before the EOS had shown commercial viability. Nevertheless, it represented a logical development of a known technology, and it would have worked. The gene-splicing breakthrough could not have been predicted, and, if it had not occurred, EOS would indeed have given McDonnell Douglas a pioneering lead.

A key factor in the development of a new pharmaceutical product was that it had to be subjected to comprehensive trials before the Federal Drug Administration could approve it. So, even if a miracle drug *was* discovered as a result of microgravity research, and even if it *could* be manufactured on Earth, it might take ten years for the product to be marketed.

Despite McDonnell Douglas' abandonment of its project, both NASDA and ESA picked up on the encouraging results and developed their own experimental apparatus. NASDA's Free-Flow Electrophoresis Unit (FFEU) was tested in 1992 on STS-47, the first Japanese-sponsored Spacelab. ESA's system was called RAMSES, in line with the French acronym for Applied Research on Separation Methods using Space Electrophoresis. It first flew in 1994 on the second International Microgravity Laboratory, the Spacelab mission on which the FFEU made its second flight. Unfortunately, repeated pump faults severely limited the RAMSES programme, and the FFEU developed gas bubbles, overheated and shut itself down. Ironically, the joint test programme included the purification of pituitary cells; exactly the kind of work that the Americans had turned away from.

It is worth noting that, following trials with a similar electrophoresis unit on Salyut 7 in 1984, the Soviet Union sent successively improved models to the Mir space station; then, once a satisfactory system had been perfected, an automated form was flown on the Foton satellites.

PHASE PARTITIONING

The Phase-Partitioning Experiment (PPE) was devised by an academic consortium led by the University of British Columbia in Vancouver, Canada, and managed by the Marshall SFC. Phase partitioning is a method used to separate different types of biological cell, and is used in the pharmaceutical industry. It is used to separate bone marrow cells for cancer treatment, and to purify transplant cells. Its main advantage is that it can be readily scaled up. Two immiscible liquids are used to separate the cells; in PPE, two polymers in saline solution were used. As they separate (or, in the vernacular of the field; demix), the cells in the mixture attach to one solution or the other, and demix with them. On Earth the liquid with the lighter molecular weight floats to the top, but in microgravity this does not happen; instead the two phases take on a structure which can be likened to an egg, with the yoke floating inside the white. On Earth, gravity-induced convection degrades the process, so the goal of the experiment was to observe the process when gravity was absent, in order to understand the role of gravity in the process.

PPE was a simple hand-held tray with a matrix of sample chambers, each of which was seeded with a two-phase system and then shaken to mix the phases. As the cells demixed, the chamber was repeatedly photographed. After the flight, a densitometer derived the rate and extent of the demixing. Each sample tested different physical parameters, such as viscosity. Its initial trial was on STS-16, in 1985. There was an enforced pause while the shuttle was grounded, but it then flew as one of the biotechnology experiments carried by STS-26, and later on the first International Microgravity Laboratory, but in this case with an electric field applied across the demixing samples.

PROTEIN CRYSTALS

A pharmaceutical is *designed* to induce or inhibit the chemical properties of a specific biological agent, so the first step is to determine its properties. The best way to discover the properties of an organic substance is to crystallise it, and then use X-ray crystallography to expose its three-dimensional structure. However, it is difficult to crystallise organic material on Earth, and a flawed crystal is not only difficult to analyse, it is not necessarily representative of the material's true structure. The crystallisation problem can be likened to a crowd filling a cinema: as soon as the door is opened the people rush in and, even though everyone may have a reserved seat, a few will inevitably sit in the wrong seats, causing confusion and disrupting the planned layout. Although humans might insist on sorting themselves out, this is not the case with protein molecules, and they often latch onto the incorrect point, creating a flawed crystal which can be misleading. In microgravity the whole process runs more slowly and there is no convection to disturb it, so a truly representative crystal can be more readily formed. In a protein crystal, it is not individual atoms that take their assigned slots, as is the case in an inorganic crystal such as salt; entire protein molecules join to form the matrix. The goal was to grow crystals the size of a salt grain. It was not necessary to produce them in bulk, as just a few would be sufficient to give an insight into the structure of one of the lesser understood proteins. The pioneering work in 1953 by James Watson and Francis Crick, which had revealed the molecular structure of DNA, had been from a limited amount of high-quality X-ray diffraction data, although it required a great deal of insight to read this data. Even with computer assistance, it took Max Perutz many years to fully study the structure of haemoglobin, the erythropoietic agent responsible for taking up and subsequently releasing oxygen. Progress in understanding proteins is dependent on the ability to crystallise them for analysis. On Earth, three methods of crystallising protein have been developed: vapour diffusion, liquid diffusion and dialysis. The crystallographers were eager to determine the response of each process to microgravity.

The first attempt, on the Spacelab 1 mission in November 1983, prompted considerable excitement. This 60-hour run rewarded the team at the University of Freiburg in Germany with a beta-galactosidase enzyme crystal fully 27 times larger than any previously available for study. The most amazing result of further experiments on Spacelab 3, and the German-sponsored D1 mission, was a lysozyme crystal *1,000 times* the size of any seen before. Unfortunately, the programme was then interrupted by the loss of Challenger.

The Center for Macromolecular Crystallography was set up on the Birmingham campus of the University of Alabama as one of NASA's Centers for the Commercial Development of Space (CCDS). It developed the Protein Crystal-Growth (PCG) experiment to assess the potential for crystallising hormones, enzymes and other proteins. The principal investigator was Charles Bugg, the director of the Center. The preliminary test hardware was flown by McDonnell Douglas. In addition to supervising the CFES on his STS-16 mission, Charles Walker tested this new experiment. In contrast to the massive electrophoresis unit, the PCG was a small hand-held package which comprised a dialysis unit and a pair of vapour diffusion trays. Dialysis is a process of selective diffusion across a membrane. The dialysis experiment was a small block of lexan in which there was a central cavity containing membrane 'buttons' and glass ampoules with precipitating agent. When the apparatus was shaken, the ampoules broke and the buttons were activated. As the protein emerged from the membrane it was crystallised by the precipitant. Unfortunately, the vibration during the shuttle's powered ascent triggered the experiment early, and ruined it. Each of the 35 × 8-cm rectangular vapour diffusion trays incorporated 24 chambers with porous liners saturated by precipitant to crystallise a drop of protein solution injected by an overlying tray of syringes. Some of these chambers had been seeded with a microscopic crystallised protein, to test whether they would resume growth. This apparatus survived the ride into orbit. Its most significant product was a crystal of lysozyme 1.5 mm across. This vapour diffusion package was flown again on STS-19, STS-23 and STS-24, after which the loss of Challenger pre-empted further tests. This simple hand-held apparatus had proven that large crystals could be grown, but, because the temperature was not regulated, thermodynamic fluid effects had degraded their quality, so it was decided to fabricate an isothermal unit.

While the shuttle was grounded, the initial tests stirred considerable interest as industry, university and government researchers studied the results. NASA's Office of Commercial Programmes sponsored a suite of materials-processing experiments on the first flight upon the shuttle's return to service. The more sophisticated apparatus – which occupied a middeck locker and provided a thermal regulator for a 20-chamber vapour diffusion tray – was flown on STS-26. The experiment processed 10 different proteins in 60 tests. Although not all of these samples produced usable crystals (either because they did not grow very much in the time available, or because they were damaged by the stresses of the return to Earth and subsequent handling), some were of exceptional size and quality, so the package was flown repeatedly to yield crystals for pharmaceutical, biotechnology and agrichemical studies.

In the 1980s the protein pharmaceutical industry greatly expanded its product range. It was widely expected that it would rapidly create cures for cancer and for the new scourge, the AIDS virus. And the proteins in insulin, interferon and human growth hormone were

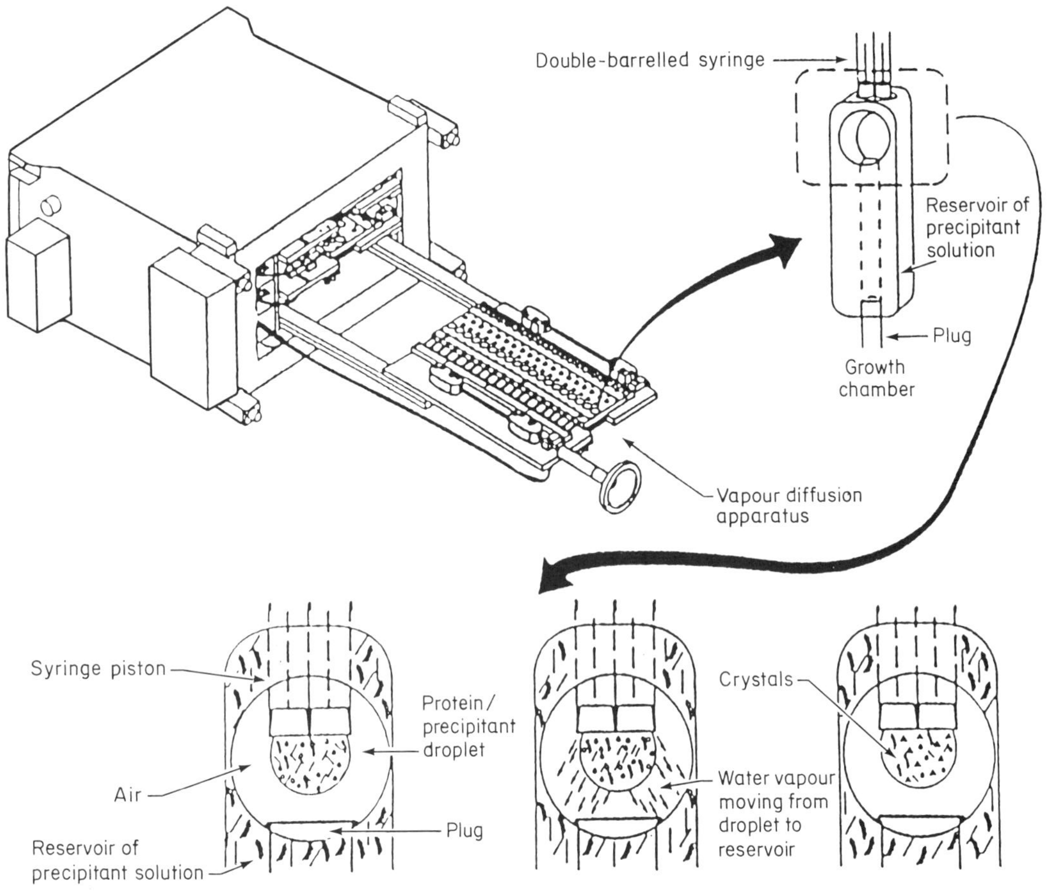

Fig 6.2. A schematic of the operation of the Protein Crystal Growth (PCG) experiment.

indeed developed into successful products. The commercialisation process was assisted by the fact that the Federal Drug Administration could approve drugs derived from pure proteins more readily than it could new artificial agents. It was in this mood of optimism that the crystallisation work was undertaken on the shuttle.

The vapour diffusion unit flew on STS-26, STS-29, STS-32 and STS-31. Then the Protein Crystallisation Facility (PCF) was introduced on STS-37. This used temperature to activate and control the growth of the crystals with the advantage that it eliminated thermally-induced convection. The refrigeration and incubation unit had been changed so that it could be programmed to follow a predefined temperature profile. It reduced the temperature from 40 °C to 22 °C (i.e. room temperature) over a 4-day period. This slow cooling increased the regularity of the crystal. With four large chambers, this could process much larger amounts of material, in batches. The results were encouraging. STS-48 carried the vapour diffusion unit, but the plugs on the syringes of raw material did not retract, and the droplets did not penetrate the growth chamber; nevertheless, small crystals did form on the plugs. STS-42 carried it as part of the first International Microgravity Laboratory. Then, for STS-49, the PCF was refined to make it more cost-effective, so that a smaller sample could produce a higher yield. The vapour diffusion unit had shown that useable crystals could be produced, and the PCF held out the prospect of increased yield. It was now time to *study* the process of crystallisation to determine the rates at which different proteins formed. It was suspected that fluid effects were degrading crystallisation, so the researchers were keen to observe the kinetics of the process. STS-50 carried the first US Microgravity Laboratory Spacelab, so Larry DeLucas, who was Charles Bugg's deputy, flew as a Payload Specialist to undertake a detailed study. To expose the crystallisation to inspection without jeopardising conditions for growth, the PCF was mounted in ESA's GloveBoX (GBX). This was an enclosed and pressurised workspace which had facilities for both photo-documentation and on-line video monitoring. Because this was the first Extended Duration Orbiter mission, there was time for DeLucas to vary the operational parameters as he went along, to take into account his observations. This was more akin to laboratory science. In addition, three of the standard vapour diffusion units were operated, one of them using the seeding technique. About 300 samples were processed in all. Of the 34 proteins samples, 60 per cent had been flown before. Some of the crystals produced were the largest ever seen. One was particularly welcome; it was an HIV-related protein, and it was hoped that understanding its structure would hasten the development of an anti-AIDS drug. "Crystals from space are of such high quality", noted DeLucas afterwards, "that you can see details, like hydrogen atoms, that you wouldn't see with Earth crystals".

By STS-66 in 1994, the vapour diffusion unit had been incorporated into the Crystal Observation System (COS), and the Space Acceleration Measurement System (SAMS) was being carried to monitor the microgravity environment. The COS resided in a double-locker Thermal Enclosure System (TES). In this form of the experiment, water evaporation was used to regulate the crystallisation rate by increasing the precipitant concentration. A video camera recorded it for later study, and allowed the astronauts to inspect its progress without removing the apparatus from its enclosure. Detailed study of the crystallisation would help refine the laboratory technique. In addition to proteins, the commercial sponsors' interest in the programme had expanded the test subjects to include other macromolecules such as viruses. The Protein Crystallisation Apparatus for Microgravity (PCAM) developed by the Marshall SFC in Huntsville, Alabama, made its debut on STS-63. It was undergoing trials for scaling up as a mass-production unit. Although it occupied only a single-locker TES, it comprised six cylinders, each of which contained nine trays, each with seven sample wells, so it could process 378 samples. When it accompanied the COS on STS-67, PCAM used the proven vapour diffusion technique. And in 1993, following up its earlier work, ESA returned to the field with its Advanced Protein Crystallisation Facility (APCF). This was a multifunction unit that could use both vapour and liquid–liquid diffusion as well as dialysis techniques. The entire process was videotaped for later analysis of the forces at work. In its various forms, the Alabama PCG experiment flew on more shuttle missions than any other commercial programme. Steady progress was therefore being made in both making the crystals aboard the shuttle and in analysing them on the ground.

By the early 1990s, a blood-related protein, an emphysema-related protein, the gamma interferon agent which was used to treat cancer and viral diseases, and the immunologically relevant Factor-D were all yielding their structures. The National Institutes of Health (NIH) worked closely with the pharmaceutical industry to exploit the protein discoveries. In 1994, the National Cancer Institute announced its intention to try to crystallise enzymes that work to prevent a cancer spreading from one part of the body to another via the blood stream – the process of metastasis.

Of the first 100 proteins tested by the Centre for Macromolecular Crystallography, 25 per cent yielded crystals which were substantially larger and more regular than had ever been grown on Earth. Although this might at first appear to be a rather low rate, it would be misleading to draw such a negative conclusion; the hit-rate on Earth was far worse. The objective was to make crystals for analysis, rather than to turn them out in bulk for commercial sale, and a few crystals were sufficient for substantive analysis. Some of the proteins grew too slowly to form large crystals in the limited time available on a shuttle mission. The introduction of the Extended Duration Orbiter (EDO) helped, but what was required was a run of several months, and this required a space station. With little prospect of NASA building its space station in the immediate future, US companies approached the Soviet Union to fly protein crystallisation experiments on Mir. However, the US Government's ban on exporting technology to the Eastern bloc meant that even proprietary apparatus could be 'exported' only if the package was completely sealed. Fortunately, it was possible to encase a simple protein package in a box which had only an on/off switch. Payload Systems Inc. served as agent for several pharmaceutical companies who opted to remain anonymous, but Boeing, which had never used the shuttle for such experiments, flew its unit openly. The Payload Systems package processed two enzymes in 112 samples using both the vapour diffusion and boundary layer diffusion techniques. Over periods of two or three months, these produced samples of the slower-growing proteins.

The International Space Station will transform the crystallisation programme. The research will no longer be limited to growth times lasting at most a fortnight, with only a few flight opportunities per year; it will become a continuous process in which crystals will be able to be grown for as long as necessary. The packages sent up to Mir had shown that *time* was a vital factor; some crystals grow extremely slowly. Unlike the *black boxes* flown on

Mir, however, the protein crystallisation facility on the ISS will be a real laboratory, and it will, at long last, facilitate traditional laboratory working practices.

GENERIC PROCESSORS

BioServe Space Technologies was established at the University of Colorado in 1987. It oversaw the development of a suite of experiment facilities for the shuttle. Instrumentation Technology Associates (ITA) of Exton, Pennsylvania, was one of BioServe's affiliates. It built the Materials Dispersion Apparatus (MDA) for biotechnology experiments employing fluid mixing. This processed a block of 100 materials samples at once, so it was meant for experiments in need of a great deal of data, as opposed to limited proof-of-concept studies. Although conceived as a high-capacity protein crystal growth unit, it was suitable for any process involving liquid–liquid diffusion, and the fact that it could be set up to mix as many as four fluids in each container made it a very flexible tool. It could be used for growth of alloy crystals, membrane casting and cell growth. BioServe was ITA's first customer, and the result was an integrated four-MDA middeck package called, obviously, BIMDA. It flew twice in 1991, and performed a wide range of experiments involving growing both organic and inorganic crystals, germinating seeds and fixing live cells. A few tests were aborted by mechanical failures on each occasion, but most were completed successfully.

The MDA was subsequently flown as a commercial unit, as the Commercial MDA-ITA Experiment (CMIX) for the Consortium for Materials Development in Space, a CCDS on the Huntsville campus of the University of Alabama. It made its debut in October 1992 on STS-52, and its 31 experiments covered the growth of protein crystal, thin-film membranes, and live cells. CMIX flew again in 1993, then twice in 1995. In each case, ITA traded 50 per cent of the MDA capacity to the Consortium in return for a flight opportunity, and then it leased the remaining capacity to the biotechnology industry. On its later flights, the CMIX unit was modified to accommodate customers requiring larger mixing chambers with 100 times the fluid volume for living cell cultures. These Bioprocessing Modules obviously carried fewer samples than the MDA block. An even larger chamber, the Liquid Mixing Apparatus, was introduced for experimenters requiring greater flexibility in the way that samples are processed, and this offered options such as laminar or turbulent flow when the fluids were injected.

It is worth summarising some of the projects undertaken with the MDA minilab in order to illustrate its tremendous scope. On a *single flight* in 1995, the Consortium for Materials Development in Space and its affiliates tackled the process of ageing, multi-drug effects on cells, neuro-muscular development, gravity-sensing and calcium metabolism, production of plant cell products and the usual protein crystal growth experiments; and ITA's commercial customers worked on microencapsulation of drugs and a treatment for breast cancer. In the case of ageing, it had been observed that microgravity affects individual cells, not just large organisms. Specifically, absence of gravity seemed to slow cell growth. To put this another way: many of the effects of ageing may ultimately derive from living in a gravity field. The experiment conducted in this case observed the growth of human lymphocytes, the primary type of cell flowing in the lymph system that distributes fluid around the body. The work on the effect of drugs on cells exploited this slowing down of cell metabolism. A drug has to cross a cell membrane to be effective, but after a while a resistance to drugs develops. In another case of membrane research, it had been noted that microgravity slows development of nerve and muscle cells. The nerve–muscle system is based on interactions between these types of cell, by the transmission of chemical messengers across their membranes. When the process fails, neuro-muscular disorders result. Studying the slowed metabolism of cells from frogs helped in the understanding of these vital functions. Calcium regulates cellular activities leading to growth, which is why milk, rich in calcium, is an important aspect of a child's diet. It was believed that gravity played an active part in this process. To understand the role of gravity better, an experiment was conducted in its absence. It is well established that plants are a rich source of pharmaceutical products. Soya bean produces an agent which could well fight colon cancer, but the yield of these cells is low. The aim of the experiment was to determine whether the yield of the secondary metabolites in these cells increased in microgravity, but xploitation would require space-based manufacturing. Urokinase is a protein, and although this was successfully crystallised on the first two CMIX missions, it turned out to be a slow-growing type. CMIX-3 flew on an EDO, so it was hoped the two-week flight would produce a urokinase crystal large enough to enable its three-dimensional structure to be derived. A cancer research centre was standing by to make a top priority effort to produce a drug which would inhibit urokinase, and so prevent breast cancer metastasis. Another commercial experiment tested a microencapsulated anti-tumour drug on mouse cells, preparatory to a programme of clinical trials. It is also worth noting that all of this research was done as a *secondary* objective, on a mission which was devoted to astronomical observation.

BioServe built the Commercial Generic Bioprocessing Apparatus (CGBA) for growing micro-organism cultures and plants in fluid-processing chambers. It could accommodate 500 individual samples for a wide selection of experiments, controlled its own temperature, and incorporated an isolation Glovebox for materials handling. It first flew in June 1992, as part of the first US Microgravity Laboratory's extremely varied experiment set. CGBA became another package that flew regularly, and conducted a variety of experiments for commercial customers. Osteoporosis is a disease in which the equilibrium between bone formation and bone loss tips in favour of bone loss. The reduced rate at which cells grow in microgravity applied to bone cells too, so astronauts lose bone mass in space. This is a calcium cycling process. Unlike most aspects of adaptation to weightlessness, decalcification appears not to bottom out; the longer an astronaut stays in space the greater the bone loss and the longer it takes to recover upon return to Earth. The loss rate is an order of magnitude faster in space than it is in a terrestrial osteoporosis sufferer, so the process could be more readily studied. Collagen is a fibrous protein found in connective tissue, such as skin and blood vessels. It is difficult to synthesise, so an experiment was conducted to test whether it could be made in microgravity. Other experiments concerned the development of biological materials for encapsulating drugs. The trick was to create the capsules. Protoplasm is the colloidal mix of protein composing the living material of a cell. Cytoplasm is the protoplasm within a cell, but external to its nucleus; this forms fatty droplets called liposomes. The objective of the experiment was to determine whether the process of assembling spherical capsules with liposomes was simplified in the absence of gravity. Other experiments provided insight into the growth of plants and brine shrimp with a view to developing a closed-cycle life-support system; what NASA had taken to calling an Ecological Life Support System.

POLYMERS

The 3M Company had a 10-year contract with NASA to operate a series of experiments to assess the viability of likely microgravity applications. In addition to biologically active agents it was interested in polymers. A polymer is an organic material composed of giant molecules. These long-chain molecules are formed by the union of smaller molecules, or monomers, in the process of *polymerisation*. The polymerisation of ethylene, for example, gives rise to the long-chain polyethylene molecule, more commonly known by its product name of polythene. The company was keen to find out how polymerisation was affected by microgravity.

The main middeck experiment on STS-14 was the Diffuse Mixing of Organic Solutions (DMOS) experiment which crystallised organic material. It had been built in-house by 3M. Each of its six stainless-steel reactors held three teflon-coated chemical chambers. Materials in the chambers were kept separate until electrically-controlled gates released to enable them to mix. These gates were opened slowly over a five-hour period so that their action would not to disturb the contents. On the proving flight, one reactor contained urea in solution, and this diffused into an incompatible solvent that crystallised the organic material. In two other reactors, the end chambers contained tetraethylammonium oxonol and cyanine tosylate in solution, and these were mixed in the central chamber. The other reactors had unannounced proprietary materials. On its second run, on STS-23, two of the reactors grew crystals to study molecular growth, two grew crystals to study how packing-density affected electro-optical properties, and two used dyes of different densities to investigate the physics of the mixing process. On Earth, the developing crystals settled on the base of the chamber, and were distorted. In space, DMOS yielded larger and more regular crystals which were better suited to X-ray crystallography.

The Physical Vapour Transport of Organic Solids (PVTOS) experiment was another of 3M's projects. Although it was packaged in the same way as DMOS for shuttle integration, it consisted of nine independent vacuum-insulated ampoules mounted within the container. Each ampoule contained an organic solid at one end, and a silicon wafer on which a special metal film had been deposited at the other, with a buffer gas between. The organic material was vaporised, transported by the gas to the other end of the ampoule and then condensed as a thin film on the temperature-controlled substrate. The thermal insulation was sufficient to allow internal temperatures as high as 400 °C without posing a problem for the middeck environmental system. The crystalline structure of each thin-film was studied for chemical, optical and electrical properties. This experiment flew on STS-20, but the follow-up had to wait until the shuttle resumed operations in the wake of the loss of Challenger. PVTOS was one of the set of materials-processing experiments on STS-26, however. On STS-34, the company introduced the Polymer Morphology experiment, which processed 20 samples including polyethylene, a form of nylon and a variety of polymer blends.

3M began a new investigation on the first International Microgravity Laboratory carried by STS-42 in 1992. This Gelation of Sols (GOS) study was part of the ongoing Applied Microgravity Research (AMR) programme, so the experiment was called GOSAMR.

A colloidal system is a mixture of two substances, one of which, the colloid, also called the dispersed phase, is uniformly distributed in a finely divided state through the dispersion medium. In a *sol* the dispersion medium may be gas, liquid or solid. In a *gel* this medium is semi-solid. A gel is therefore a thick colloidal suspension involving particles of less than 1,000 nanometres across. *Gelation* is the creation of a gel from a sol. The objective of the GOSAMR experiment was to investigate the influence of microgravity on the processing of gelled sols. Stoke's Law of fluid dynamics predicted that there would be less settling out of the larger particulates; in other words, less *sedimentation*. Specifically, 3M wanted to know whether composite ceramic precursors composed of large particulates and small colloidal sols could be produced in space with more structural uniformity, and whether this would in turn result in a finer matrix with correspondingly superior physical properties. The aim was not really to set up a manufacturing-in-space facility; rather, it was to determine the benefit to be derived from more regular structures. The data were to set a benchmark which efforts to improve terrestrial processing could seek to match in the knowledge that the investment would indeed be justified by enhanced applications. Advanced ceramic composites would improve the company's wide range of abrasive and metal-polishing products, so this use of the shuttle was directly related to the 3M Company's core business.

The GOSAMR locker held five modules, each with two banks of eight double-barrelled syringes containing the various sol and gelling agent combinations. The electrically-driven syringes were activated by switches on the control panel. Each module was programmed to shut down after a different predetermined time. The astronaut needed only to activate each module in turn. The apparatus was designed to create a precursor for an advanced ceramic material by chemical gelation, the process which forms a gel by disrupting a sol's stability. Although the precursor would have to be baked at a temperature of up to 1,650 °C to make a ceramic, this part of the process was not done on the shuttle; it was undertaken afterwards, and the properties of the result examined. Eighty samples were processed with a wide range of particulate sizes, sol sizes and gelation times. Colloidal silica sols with diamond particles, and colloidal alumina sols with zirconia particulates were processed, reflecting the products most likely to yield short-term commercial return (the diamond ceramic was used in metal polishing, and the zirconia-toughened alumina was a premium abrasive).

Other companies were also interested in microgravity polymers. The Battelle Laboratory in Columbus, Ohio, had set up the Center for Advanced Materials as a CCDS to investigate commercial opportunities for polymers, catalysts, electronic materials and superconductors. The grounding of the shuttle delayed this experimental programme, but STS-31, in 1990, marked the debut of its investigation into Polymer Membrane Processing (PMP).

Polymer membranes are widely used for separation applications such as desalination of water, atmospheric purification, medicinal purification and medical dialysis. The standard method of making a polymer membrane is to deposit a thin film of polymer solution across a cast, then let the solvent evaporate to leave the membrane in the required configuration. The membrane's porosity is determined by its structure, which is in turn influenced by the evaporation rate. Gravity-induced convection influences the evaporation process. The initial objective of the project was to investigate evaporation casting in microgravity, where such convection is absent, to better understand membrane morphology and, thereby, to establish whether it would be cost-effective to try to improve the standard manufacturing process.

The basic PMP unit comprised two containers, one far larger than the other, connected by a valve. The smaller container was filled with polymer solution, and the larger one was reduced to vacuum. Opening the valve in space caused the solvent to flash-evaporate due to the sudden drop in pressure. Two units were flown on a half-locker tray on the middeck. Unfortunately, the delicate membrane was degraded by the stresses of the landing and the subsequent handling, so the apparatus was modified to better protect its contents. Although the new unit was booked on the next shuttle, problems with hydrogen leaks grounded the fleet for months, and it did not fly until October. This time, the final step in the process was to inject water vapour into the container to quench the evaporation process. This eliminated further evolution of the membrane, as seemed to have happened on the first run, and served to insulate it from subsequent stress. The original polymers were reflown so that the results could be directly compared. The improved apparatus worked; nevertheless, it was another year before the project picked up, and then it made five flights within a year, during which experiments were made with the timing of the quenching following the flash evaporation. In one test the evaporation step was missed out altogether and the water injection was used to precipitate the membrane directly from the concentrated solution. There was then a further pause of a year before another series of experiments was conducted.

FLUIDS, MELTS AND INORGANIC CRYSTALS

The multidisciplinary Spacelab 1 mission provided the first significant opportunity to observe fluid dynamics on the shuttle. One of the packages in the materials-science double rack was Italy's Fluid Physics Module. In one experiment, for the University of Naples, a cylinder of silicon oil was floated between a pair of disks, and the response of the fluid was filmed. In another experiment, the sloshing of fluid was filmed, the better to understand the behaviour of propellant in a tank. The primary study investigated the marangoni effect. The advantage of processing materials in microgravity was that there was no gravity-induced sedimentation and no convection, but other forms of convection were possible and one was tested in this case. The marangoni effect was a manifestation of surface tension that induced fluid mixing. It could be studied on Earth, but only in microscopic samples in which other effects were minimised; but its *macro*scopic effects could be displayed in microgravity. It was studied in this case not just for scientific interest, but also to assess the extent to which it might disrupt microgravity materials-processing. A wide range of marangoni studies was pursued on the first German-sponsored Spacelab. The Japanese Spacelab had a high-speed camera to record the process for subsequent analysis and computer modelling. To facilitate even more detailed three-dimensional modelling of the fluid motion, the second German Spacelab used a holographic camera. A thorough

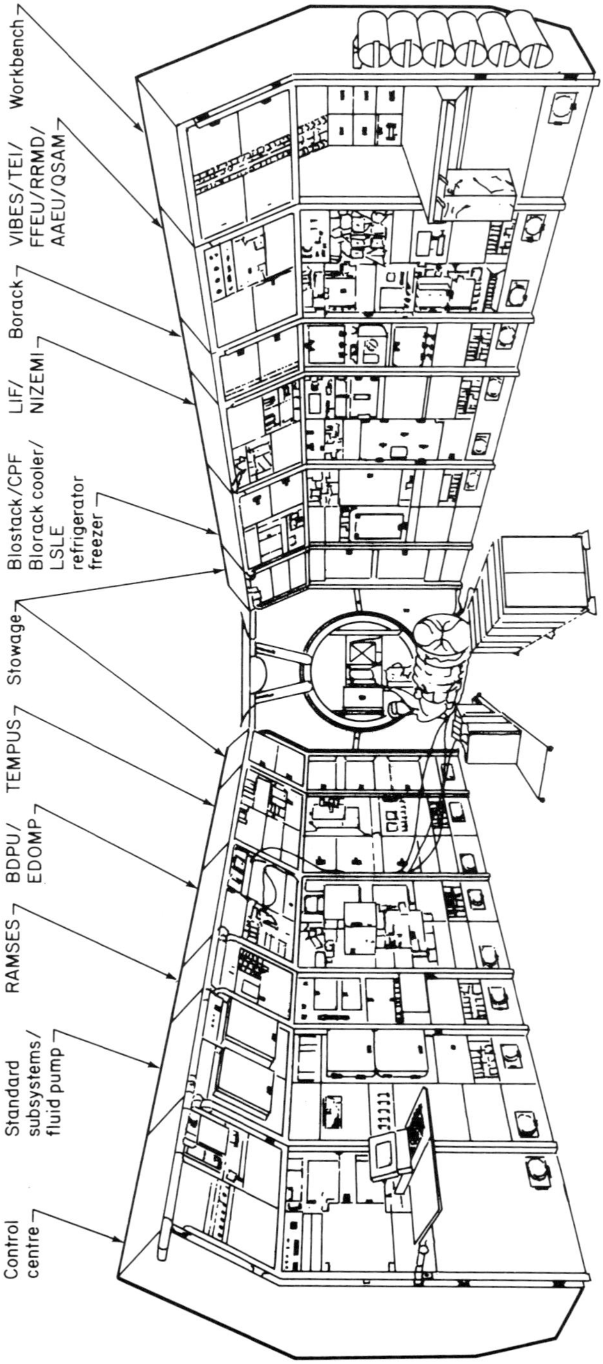

Fig. 6.3. A schematic of the layout of apparatus in the double-length Spacelab carrying the second International Microgravity Laboratory (IML).

understanding of this effect had long-term implications for the processing of materials in space, and not just because it was a source of interference; it could also be *exploited*. In 1994, on the second International Microgravity Laboratory, it was shown that different compounds of molten indium–gallium–antimony semiconductor could be mixed more rapidly and more uniformly by using marangoni convection than was otherwise possible. Marangoni mixing *within* a fluid was driven by a temperature gradient over the surface layer; controlling this gradient facilitated control over the convection. This was not possible on Earth because, as noted, the effect was almost completely masked by effects induced by gravity.

Another great advantage offered by microgravity is the ability to 'levitate' a sample in a crucible, so that it can be processed without coming into contact with its reaction chamber. On Earth, it is possible to use a gas jet to levitate a sample, but the gas flow also acts on the sample, distorting it. Only in microgravity can a sample remain stationary *without* a force acting on it. Several containerless crucibles were built to investigate manipulating materials in microgravity. Once the sample had been moved to a specific point by some active control system, it would stay in a state of *free drift*. Such a crucible would permit a material to be heated to a temperature above that which would melt a conventional crucible. It would also facilitate highly-corrosive reactions without suffering contamination from the crucible, and without threatening to breach its containment.

The Drop Physics Module (DPM), developed at JPL, used ultrasonics to manipulate its contents. Taylor Wang, who had designed the DPM, accompanied it on its Spacelab 3 trial. In the event, his presence proved fortunate, because no sooner was it activated than it broke down. As soon as Wang opened the cabinet, he realised that one of the power systems had failed. Given his familiarity with the apparatus, it would be a simple matter to strip it down, and rewire it to run off the reserve power system. Surely this was an excellent opportunity to demonstrate the merit in flying investigators as Payload Specialists? However, it took two days for the support team to establish that what Wang had proposed would not only repair the DPM, but would also not jeopardise the other Spacelab apparatus. While he waited, Wang assisted with other experiments as best he could. Finally given the go-ahead, he by-passed the failed power system without difficulty, and the DPM operated perfectly. Wang worked overtime for the rest of the flight to make up for the lost time. When the DPM flew again on the first US Microgravity Laboratory, it was operated by Eugene Trinh, another member of the team. The scientific objective of the DPM project was to test and verify the theories that described the behaviour of a vibrating liquid drop stimulated by sound pressure; to measure the physical properties of the surface of a drop; to study how a drop changes its shape as it is made to spin by a directed acoustic beam, and how it ultimately splits apart; and, conversely, how small droplets merge to create larger drops. All this was basic scientific work that could not be performed on Earth. The 3-Axis Acoustic Levitator (3AAL), which was also devised by JPL, flew on STS-24 as part of the second Materials Science Laboratory. On STS-47, the Acoustic Levitation Furnace (ALF), developed by Japan, suffered fluctuations which let its contents come into contact with the wall of the container.

Germany's Tempus furnace, tested on STS-55, used a magnetic field to manipulate its contents. It was used to study solidification of metallic melts free of container effects, with a consequently ultra-high purity. It used an electric current flowing through coils of copper tubes to form a spot of minimum magnetic field strength in which a sample could be held.

By varying the field, the sample could be moved, so that it could be precisely located to be processed in a thermal gradient. This field configuration had the advantage that the sample was in the weakest part of the field, so the solidification process would be least affected by the field's use. The ElectroMagnetic Levitator (EML) flown on STS-24 for MIT, also used an electromagnetic field. The objective of this particular experiment was to study the effects of flow during solidification of a material which was melted by induction heating as a side-effect of being held in the cusp-shaped field. The advantage of an electromagnetic container over an acoustic container was that the former allowed processing in a vacuum, whereas the latter clearly relied upon the presence of a gaseous medium to effect its control.

On the first International Microgravity Laboratory in 1992, ESA introduced its Critical Point Facility (CPF). This was a generic facility designed for the optical study of the behaviour of fluids at and near their *critical points.* Any pure fluid possesses a liquid-vapour critical point, which is uniquely defined by a temperature, pressure and density state in thermodynamics. For states with either temperature, pressure or density greater than the critical values, liquid and vapour are indistinguishable. At the critical point, a fluid fluctuates back and forth in small volumes from the liquid to the gaseous phase. The fluid is also highly compressible and, on Earth, observations are hampered because it is difficult to avoid compression of the critical region which literally collapses under the weight of the overlying fluid. As soon as the vapour begins to liquify and form droplets, gravity pulls the drops down. The closer to the critical temperature, the more compressible the fluid, and the thinner is the critical zone. At some temperature, the zone is too small to use any known experimental probe to measure the thermodynamic properties. In a microgravity environment, the weight of the overlying fluid on itself is reduced, and the critical zone is widened for a given temperature. It also allows a closer approach to the critical temperature before experimental probe limits are reached.

Scientists are interested in what happens to materials at their critical points because critical point phenomena are universally common to many different materials. Physically different systems act very similarly near their critical points. Several experiments were performed using sulphur hexafluoride, a gas which can be produced in extremely pure form. As the fluid moved through its critical point, the density and the heat-and-mass-transport processes were measured by using interferometric and light-scattering techniques. The Critical Fluid Light Scattering Experiment, also known as 'Zeno', which was performed on STS-62, was particularly successful in this respect. It proved possible to make accurate measurements to within a temperature 100 times closer to xenon's critical point than previously possible, so that the *onset* of the state of equilibrium between the phases could be studied in detail. Robert Gammon, the principal investigator, who operated the experiment by telescience, was astounded by the data, which represented "a dream" that he had "spent his career waiting to see".

Another basic physics investigation concerned how helium transformed from a fluid to a *superfluid* state. This change of phase occurs at 2.2 K. This temperature is known as the *lambda* point, so the experiment performed on STS-52 was the Lambda Point Experiment. The transition is easier to observe in microgravity because the process is very sensitive to pressure, and gravity causes the pressure on the bottom of a sample to be slightly higher than at the top, thereby imposing a temperature gradient across the sample which causes the process to creep. Before launch, the sample was chilled below the lambda point and kept in a cryostat. In orbit, the temperature was controlled to within a billionth of a degree as it was slowly raised back through the transition so that the fluid's heat capacity could be measured as it changed state. Helium is the only element to have this superfluidity property. Because superfluid helium flows so readily, it is extremely efficient at conducting heat. This is why it is used to chill infrared telescope detectors. However, its zero viscosity also means that it can pass freely through the tiniest of pores, so it will leak from *any* container. This is why the service life of such telescopes is so short. The SHOOT experiment that was performed on STS-57 tested a way of replenishing a telescope's coolant. The final Great Observatory – the Space InfraRed Telescope Facility (SIRTF) – will, however, be far beyond the reach of the shuttle.

The Canadian Queen's University Experiment in Liquid Diffusion (QUELD) was to measure the diffusion coefficients of a variety of metals in their liquid phase. The samples were heated in a furnace until molten, allowed to diffuse, and then rapidly solidified so that their state of diffusion could be determined after the flight. The experiment was performed in space because on Earth gravity-induced convection causes mixing that masks the degree of diffusion. The objective was to provide high-quality data with which to calibrate a model for the diffusion process, so that coefficients could be predicted for a wide range of metals, knowledge of which would assist industrial manufacturing processes. After being proven on STS-52, the QUELD apparatus was based on Mir for a more comprehensive study. The Optizon LIquid-Phase Sintering Experiment (OLIPSE) was another NASA experiment set up on Mir. In this case though, Mir's Optizon furnace was used. This non-crucible furnace heated its contents by focusing radiant energy from an electric lamp. Although the sintering process caused a powdered metal to fuse by heating, it did *not* melt it; instead, it relied on high pressure to create a coherent bonded mass. It is used industrially to make steel from a mixture of iron powder and graphite. The Equipment for Controlled LIquid Phase Sintering Experiment (ECLIPSE) that the University of Alabama had built for use on sounding rockets had been adapted for carriage on the shuttle; it flew on STS-60, to assess the properties of a range of materials believed to be likely candidates for strong, lightweight and yet extremely durable metals.

Cosmonauts on the Mir space station (and, indeed, on its Salyut precursors) had shown that semiconductor crystallised in space was of very high quality. Knowing this, NASA set out to study the process, in order to find ways to improve the technique. The Crystal Vapour Transport Experiment (CVTE) it sponsored flew on STS-52 in 1992. NASA saw this experiment as a precursor of the type of industrial research that would be undertaken on its space station: assisting terrestrial manufacturing by studying fundamental processes in the absence of gravity. In effect a furnace, it heated an ampoule of cadmium telluride to 850 °C, to evaporate and dissociate the compound into gaseous cadmium and gaseous tellurium. A temperature gradient caused the material to vaporise at one end of the tube, be transported to the other end, then recrystallise. Because the objective was to study the process and not just exploit it, the apparatus had a window so that an astronaut could observe the crystallisation process and adjust the set-up to test different operating conditions. Bill Shepherd and Mike Baker took turns looking after it. They could reinforce their visual observations with video and still-pictures. The astronauts had been shown the process on Earth, so that they would be able to recognise altered behaviour in microgravity.

This task turned out to be something of an ordeal. One of the two furnaces shut down early in the flight, and then the window on the other fogged up sufficiently to prevent the video camera from viewing the crystallisation; so the astronauts resorted to sketching the process and then displayed these to the camera to enable the researchers on the ground to follow the progress of the crystal's growth.

The Automated Directional Solidification Furnace (ADSF) that was tested for Grumman Aerospace on STS-18 was subsequently used on a number of missions. Its purpose was to investigate the possibility of producing lightweight magnetic composite materials in space. Alloy samples were first melted and then solidified for examination following the flight. It operated by progressively working its way along the length of a sample; hence *directional* processing. The ADSF had been adapted from an apparatus previously flown on sounding rockets, and whereas it had previously been restricted to five minutes of operation, it could now process materials more slowly over a period of hours. Mephisto, carried on STS-52, was a French furnace to study directional solidification of metals. This study involved *in situ* observation of the process and focused on the so-called solid–liquid interface: the point in the solidification process at which the solid and liquid forms were in contact. Directional solidification was exploited on the first US Microgravity Laboratory mission to evaluate the extent to which, on Earth, gravitationally-induced fluid-flow during crystallisation disrupts the microstructure of mercury–zinc telluride, an alloy that is highly prized for its sensitivity to infrared radiation. The intention was not to develop a precursor to manufacturing-in-space, but to establish the extent to which the terrestrial manufacturing process is flawed, and thus set the benchmark for improvement. The test was performed in the Crystal Growth Furnace (CGF), which was the first furnace flown on the shuttle able to operate at temperatures as high as 1,350 °C. This reconfigurable apparatus could be used for a variety of experiments. It was also used to produce mercury–cadmium telluride, which is another infrared-sensitive semiconductor, but this time it employed the vapour transport technique. Unlike CVTE, no attempt was made to observe the crystallisation process. The crystals were studied after the flight to determine their properties, again to assess the degree to which convective flow degrades the terrestrial process.

The Center for Commercial Crystal Growth in Space – a CCDS established at Clarkston University at Potsdam, New York – teamed up with Battelle's Advanced Materials Center to develop the Zeolite Crystal Facility (ZCF), which was tested on STS-50. The project's principal investigator, Albert Sacco, of the Worcester Polytechnic Institute, accompanied this furnace when it flew on STS-73. A mix of silica and alumina, zeolite has the useful property that it has an open crystalline structure, which makes it selectively porous. Since it can function as a molecular sieve, it is used in catalysts, filters, absorbents and ion-exchange systems. It is used by the petrochemical industry in *cracking*, for example, to increase the gasoline yield by filtering out large hydrocarbon molecules. Unfortunately, zeolite crystals made on Earth are small and irregular, both of which factors complicate its use in absorption, separation and ion-exchange processes. It was hoped that microgravity would yield larger and more regular crystals, and this proved to be the case. The first stage in exploitation of this fact was to study these new crystals to determine their properties, and assess whether it would be practicable to improve the terrestrial manufacturing process, but the wider objective was to assess the possibility of microgravity production. The ZCF experiments therefore set out to establish the best procedure for zeolite manufacturing-in-space. The commercial viability of making zeolite in space derived from the fact that there was a ready market for the product, that natural crystal is rare, and that synthesis in gravity is difficult. Microgravity zeolite would have another advantage over its artificial terrestrial counterpart in that it would not be necessary to use chemical additives during the manufacturing process, so clean membranes could create a lucrative new market. Zeolite may well become one of the materials produced in bulk on a free-flyer alongside the International Space Station.

COMBUSTION

With STS-41 in 1990, NASA began to study the process of microgravity combustion. The outbreak of fire is a perennial danger facing a spacecraft. However, microgravity and the reduced air pressure in the cabin conspire to reduce this threat; a fire would tend to burn itself out without spreading. The Solid Surface Combustion Experiment (SSCE) developed by Robert Altenkirch, of Mississippi State University, was the first of what was to become a comprehensive series of combustion experiments. On Earth, flame is strongly influenced by gravity; heated gas loses density, and it rises above a flame by *buoyant* convection. On the one hand, the cold gas which replaces it serves to cool the flame, but this circulating air flow also provides fresh oxygen, and this not only sustains the flame but also keeps it hot. Combustion, therefore, represents a balance between these heating and cooling factors. The outcome is determined by the speed of the airflow and the scale of the fire. A match, for example, will be snuffed out by a gust, but a camp fire will be strengthened.

The specific objective of the SSCE was to study the spreading of flame over solid fuels in the absence of buoyant convection. Since air motion would be eliminated, oxygen would not be renewed. Combustion on a localised site would soon burn itself out through lack of oxygen. The only way that combustion could be sustained would be for the flame to spread *away* from the hot efflux, into fresh air. With airflow absent, the balance between heating and cooling effects was calibrated by adjusting the oxygen content of the air in the container in which the experiment was done. In the shuttle's cabin, as on Earth, the oxygen content was 21 per cent, but for the combustion tests it was raised, and varied between 35 and 50 per cent. The process was recorded by video cameras, and the temperatures of the air and of the fuel (an ashless filter paper) were measured for detailed analysis. There was a brilliant glow when the paper first lit, but this diminished as the spherical flame slowly travelled along the length of the paper. In later tests, small samples of plexiglass were burned. The initial bright orange flash gave way to the steady blue flame. These materials were used because a comprehensive database for their combustion already existed. The SSCE flew on eight shuttles and was then sent to Mir for further studies on a wider range of materials.

The effects of buoyant convection can be calibrated using controlled gusts, or *forced* convection. However, because the forced flow can only augment the ambient flow, it is not possible in gravity to extend this calibration to speeds *lower* than that of the buoyant flow. In space, however, in the absence of gravity-induced convection, it would be practicable to measure the effects of such minimal forced flow. This was a remarkable opportunity to increase our understanding of this fundamental process. Achieving mastery of fire stands as one of the most significant achievements of our early ancestors. Combustion engineering is the key to our modern way of living; that it should be possible to discover *anything* new about the process after it had been applied for so many years served to demonstrate that our view of the world is conditioned by the many and varied effects of gravity. A refinement of the SSCE was therefore to assess how forced flow affected solid-fuel combustion; this was the Forced-Flow Flame Test (FFFT). Once the apparatus had been tested, it was sent up to Mir for extensive trials.

The Candle Flame Experiment (CFE), supplied by NASA's Lewis Research Center, set out to test whether a candle flame could be sustained in the absence of buoyant convection. It would have to rely on diffusive mixing of the essentially stationary air surrounding the flame. Characterising this effect was a precursor to moving on to test a flame using forced flows at speeds *lower* than that of the buoyant flow such a candle would have produced on Earth, to extend the calibration into this previously unattainable zone. With this done, flame propagation was tested by mounting candles at different spacings, then lighting one. Surprisingly, even when the candles were in close proximity the flame did not jump across the gap. Unlike the SSCE, which was conducted in a sealed apparatus, the CFE experiment was performed in the ambient atmosphere, using the Glovebox for physical isolation. It too was later sent up to Mir.

The University of California's Smoldering Combustion Experiment (SCE) assessed the smouldering characteristics of polyurethane foam, both with and without a forced flow, and with different geometries. The smouldering-to-flame and smouldering-to-extinction processes were recorded to identify the conditions which led to each transition. Having studied direct flame propagation, NASA turned to *radiative* effects, in the form of the Radiative Ignition and Transition to Spread Investigation (RITSI).

The International Symposium on Combustion is the principal forum for research in the field; in 1996 some 10 per cent of the papers presented reported microgravity work. The pace was clearly picking up.

The Materials Sciences Laboratory, flown in 1997, concentrated on combustion. The crew included Donald Thomas, an astronaut whose speciality was materials sciences; Roger Crouch, the chief of NASA's Microgravity Science & Applications Division; and Gregory Linteris, a combustion engineer at the US National Institute of Standards & Technology. A variety of materials were burned. Individual droplets of heptane and ethanol were ignited in the Glovebox. The Droplet Combustion Apparatus was used for a University of California study of the ignition properties of assorted hydrocarbon-based fuels. The Lewis Research Center's Combustion Module had a gas chromatograph. A Californian experiment used this to study the process by which a flame, instead of moving over a linear front, broke off into a flame-sphere; this phenomenon was very difficult to study on Earth. The mechanisms that affect flame stability and extinction in mixed gases were poorly understood, despite the fact that they were basic to applications involving internal combustion engines, so this research held out the prospect of a significant commercial return. A University of Michigan experiment used a laser-based instrument to study soot concentrations. Radiation from soot determines the durability of combustion. The Combustion Module included temperature, pressure and radiation sensors to measure these dynamics as propane and ethylene were burned; the soot was retrieved for post-flight analysis. Lest it be thought that little can be learnt from soot, it need only be noted that it was very recently that Fullerenes, the carbon-60 form of carbon, were discovered in soot. Burning fluids in space required special care, so additional fire-fighting facilities were carried in the Spacelab module as a precaution.

On Earth, overloaded electrical wiring is a common source of fire. To varying degrees, different types of wire resist passing an electrical current, and transform this energy to heat. The construction industry uses codes of practice based on extensive trials relating current to temperature in different wire configurations. Lewis focused on the development of systems for spacecraft. Its Wire Insulation Flammability Experiment (WIFE) was to determine the outgassing, flammability and flame-spreading characteristics of wires, to isolate any effects present in microgravity, as a step towards formulating a 'building code' for the laboratories on the International Space Station. These facilities are to be generic so that the experimental suite can be revised, so, to reduce the risk of fire breaking out on the station it was vital to know whether the electrical properties of power supplies were different in microgravity.

LIMITATIONS OF MICROGRAVITY

What is *micro*gravity? Although, in general parlance, it is used synonymously with the terms zero-gravity and weightlessness, strictly speaking the 'micro' prefix actually refers to a gravitational force of 10^{-6} of Earth-standard.

The quality of the ideal microgravity environment offered by flying in space is degraded by the orbiter's manoeuvres, by the crew's activities and by vibration from other apparatus. Ironically, firing the RCS thrusters to hold the orbiter stable can disturb sensitive processes such as crystallisation. This can be overcome either by leaving the vehicle in free-drift or by placing it in an inherently stable attitude. As free-drift is inherently *un*stable, it is not often used. However, if a nonsymmetrical vehicle is orientated with its major axis vertical, it can exploit differential gravity, which tugs very slightly more on the end nearer the Earth, to maintain a stable attitude without firing thrusters. So for the most sensitive experiments, the orbiter is placed in this gravity-gradient attitude with its tail aimed towards the ground. The *best* location for a sensitive experiment is at the orbiter's centre of mass, so a Spacelab far aft in the bay is an inherently stable platform. However, these flights tend to carry a lot of apparatus that operates simultaneously, and vibrations transmitted through the structure can pollute the state of the microgravity environment. In addition, sensitive experiments can be upset by the activities of the crew.

To help researchers interpret the results of their experiments, NASA developed devices to measure the state of the microgravity environment. The first was the Space Acceleration Measurement System (SAMS), a very sensitive three-axis accelerometer which could be run continuously to monitor vibrations during the operation of a nearby experiment. It was introduced on STS-40, on the first Life Sciences Spacelab. On the second International Microgravity Laboratory – the German version – the Quasi-Steady Acceleration Measurement System (QSAMS) was tested. Having noted the vibrations most disruptive to experiments, the next task was to try to remedy the situation. Isolation mounts were devised to damp out threatening vibrations, so that extremely sensitive experiments could be undertaken aboard a 'noisy' *micro*gravity environment, with the long-term goal of utilising acceleration states approaching *nano*gravity, or 10^{-9} g. NASA is slowly but surely reducing the influence of the crew. On flights needing the best microgravity conditions, the stationary bicycle and the treadmill float on springs to damp out the vibration as the crew exercises. A free-flyer would not have this problem, but it needs to be fully automated, and many (although by no means all) of the experiments flown on the shuttle rely upon some element of astronaut supervision. However, at the rate at which telescience techniques are being developed, it may be practicable to remotely operate equipment on a free-flyer from a nearby shuttle, from the space station or from the ground. At best, the orbiter can provide about 10^{-5} g. A free-flyer can yield an order of magnitude improvement. A *multipurpose* free-flyer, running a variety of apparatus simultaneously, would, however, still be susceptible to interference.

As NASA built up its research facility on Mir, it augmented the experimental apparatus with a Glovebox for materials handling and a SAMS to monitor the local environment. One new piece of apparatus was the Microgravity Isolation Mount (MIM). This was a Canadian device that used a magnetic field to isolate an experiment from microvibrations in the range 0.01–100 Hz. It was placed in the Glovebox so that residual vibrations transmitted onto the surface of a liquid in a test chamber could be recorded. SAMS simultaneously measured the *ambient* vibrations so that the MIM's ability to filter them out could be determined. Within the Spacehab module carried by STS-79 to resupply Mir, the Active-Rack Isolation System (ARIS) was tested. This had been designed by Boeing for use on the International Space Station. It used a purely mechanical suspension system to damp out micro-accelerations. On the second International Microgravity Laboratory, Japan tried out the Vibration Isolation Box Experiment System (VIBES) for possible use in the laboratory that it is to build for the space station.

More subtle effects pose more difficult long-term problems. Calibration tests performed on Mir demonstrated that the gravity gradient can produce a measurable effect on a sensitive experiment; it is a low amplitude force, but it acts *continuously*. These tests also found that the atmospheric drag that causes a satellite's orbit to decay produces a detectable effect. These forces represent a quite different form of pollution of the microgravity environment, and they will be very difficult to eliminate.

Those opposed to a human presence in space argue that having humans on board a space station leads to 'pollution' of the microgravity environment. Clearly this is an over-simplified argument since humans are not the sole source of such pollution. Therefore, developing an ideal 'weightless' platform is a far more complex task than simply banishing astronauts from the facility.

7

Facilities for commercial research

As soon as the shuttle began commercial operations it became apparent that there was scope for service-providers to exploit the shuttle's ability to fly small microgravity payloads on a commercial basis.

NASA's simplest experiment package was the Small Self-Contained Payload. This was more commonly known as the Get Away Special (GAS). The 0.5-metre diameter canisters were carried in the payload bay, mounted either individually on the sidewall, or a dozen at a time on the GAS Bridge Assembly (GBA), an MPESS running across the bay.

The cost of the GAS payloads was deliberately held down. Different categories of user were offered different rates. An educational institution could book a canister for a few thousand dollars. When the shuttle first flew, there was already a long waiting list, which was serviced on a 'first-come-first-served' basis. However, when Challenger was lost, only 53 GAS 'cans' had been 'serviced'. The 100th did not fly until 1994, on STS-60. GAS even offered possibilities for a commercial operator willing to configure an experiment to fit into a can and then prepare and certify it fit-for-flight, because NASA had rather strict regulations for such matters.

The German company Messerschmitt–Bolkow–Blohm eagerly seized the opportunity to service the European market by developing a bridge structure that could be lifted out of the bay by the RMS and then released as a short-term free-flyer, in order to provide its experiments with a higher quality of microgravity than was practicable within the bay of an orbiter involved in other operations. MBB named its platform the Shuttle Pallet Applications Satellite (SPAS). This bridge was attached to a triangular support truss of carbon-fibre tubing. It had a three-axis control system and nitrogen gas-jets for attitude control, and 60 per cent of its 1,500-kg mass was available for payload. When it first flew on STS-7 in June 1983 it became the first payload to be released and then retrieved by the shuttle. NASA charged a minimal fee for flying the SPAS in return for the opportunity to mount a camera on it, to provide the first views of an orbiter in space. To mark the occasion, Sally Ride fixed the RMS in the shape of a '7'. The SPAS had seven GAS cans for ESA and Germany's Ministry of Research and Technology. This varied payload included two materials-processing experiments, a solar-transducer test, a heat pipe, a pneumatic conveyor, a remote-sensing camera, and a spectrometer to note the gaseous environment as the orbiter manoeuvred close alongside. The SPAS flew again on STS-10, but a problem with the RMS meant that it could not be deployed, and the tests had to be conducted with it mounted in the bay.

The SPAS was not flown again prior to the loss of Challenger, and in 1988 it was sold to the US Department of Defense. The newly-established German Space Research Agency (DARA) had a new free-flyer under development, in the form of the European Retrievable Carrier (EURECA). This was essentially a double-SPAS bridge without the triangular support truss. Unlike the SPAS, which had to be retrieved within a few hours because it relied on a battery for power, EURECA had solar panels for extended operations. It was to be released by one shuttle and retrieved by another, months later. This new free-flyer was far more than a scaled-up SPAS, and it was aimed at a different class of microgravity payload: those which would take a long time to run their course.

EURECA was released by STS-46 in July 1992, then retrieved by STS-57 almost a year later. ESA astronaut Claude Nicollier used the RMS to lift it off its cradle, and then held it high over the bay while it unfolded its long solar panels. After it had been released, to reduce orbital drag EURECA fired a hydrazine thruster to boost its orbit to 500 km. Over the next fortnight its 15 experiments were checked out and started up. Most had completed by the end of the year, but because its retrieval had been pushed down the shuttle manifest, those able to benefit from additional time were kept running. When retrieval was imminent, the payload was shut down, and the platform lowered its orbit back to 300 km for pick-up. David Low grabbed it with the arm, then held it while it retracted its appendages. Although the solar panels readily folded away, the two long whip antennas remained extended. As it turned out, there was an unexpected time constraint. An electrical connector had been fitted into the end-effector of the RMS so that EURECA could draw power for its thermal regulation system whilst it was on the arm. This feed had been incorrectly wired, however, so it was imperative that the spacecraft be placed onto its cradle, from which it would be able to draw power. The flight rules prohibited placing the spacecraft in the bay while its antennas were extended because they could easily strike the payload mounted on the bridge immediately in front. Yet, unless this was done, it was possible that the propellant in EURECA's tanks would freeze, split the tank, and contaminate the payload bay. The regulation was waived and the satellite was gingerly lowered into the bay. The following day, two astronauts went out and manually stowed the antennas. EURECA's 1,000 kg of payload included a pair of furnaces, a protein crystallisation facility, a radiation monitor, an instrument to measure the total energy output of the Sun, another to measure the energy spectrum of the Sun, a telescope for high-energy astrophysical observations, and a variety of technology demonstrations of apparatus intended for use on future spacecraft. In most cases the apparatus functioned satisfactorily. One experiment to assess a new type of thruster was crippled by a short circuit, but this did not occur until useful data had been acquired.

The operational concept for EURECA envisaged five flights over a 10-year period, each of about six months duration. NASA allocated a second flight opportunity on the manifest for 1995, but ESA let this lapse due to lack of funding. Another flight opportunity in 1997 was also turned down.

The EURECA project serves to highlight an unfortunate trend. Like the original SPAS and the LDEF, EURECA was designed to be *reusable*, but was then used only once. A more constructive outlook, however, is that if ever there is an urgent need for a small free-flying platform, EURECA will be available.

SPACEHAB

The middeck of the orbiter's cabin had 42 locker spaces, but 80 per cent of this capacity was routinely reserved for crew equipment. Experiments that would not fit into a GAS can and required some element of astronaut oversight could be flown in the free lockers, but the list of such experiments for educational and commercial users grew far faster than it could be serviced. Spacehab Incorporated was established in 1983 to service precisely this queue of payloads with a *middeck augmentation* module.

With NASA's enthusiastic support, Spacehab raised the venture capital needed to place a contract with McDonnell Douglas to develop a pressurised module for carriage at the front of the bay. It was to be connected to the cabin by a short tunnel. It was to sit at the front of the bay in order to leave the rear of the bay free for other payloads. Ironically, this position constrained its own payload. Set far forward, it would upset the orbiter's centre of mass if its overall mass was excessive, and this would jeopardise the orbiter's aerodynamics in the final approach to a landing. Consequently, the 5-tonne module was limited to 1,360 kg of cargo. Although it would occupy less than 25 per cent of the bay, this 4-metre wide, 3-metre long flat-topped compartment nevertheless *doubled* the orbiter's habitable volume. More to the point, by carrying an additional 60 lockers it increased the shuttle's experiment capacity so significantly that it prompted the addition of a Mission Specialist to the crew specifically to supervise its operations.

It had always been assumed that the first customers would be government agencies, but in the longer term it was the commercial market that Spacehab was designed to service. The belief was that providing *ready access* to space would promote experimentation and, later, exploitation of microgravity. The loss of Challenger did not directly affect development, but it did little to encourage industry to place bookings. The initial plan had called for entry into service in 1991, then a rapid build-up to five flights a year. This had been when NASA had hoped to fly on at least a fortnightly basis, however. The economics of Spacehab were the same as for the shuttle, to the extent that profitability required a high flight-rate. By the time the shuttle resumed operations in 1988, it was evident that the flight rate would never exceed a dozen flights per annum, and, with so many high-priority payloads on this limited manifest, it was also clear that Spacehab would be lucky to secure one flight per year. With Spacehab's prospects looking grim, NASA guaranteed the first flight as planned, so that the company could firm up its existing commercial options, and it offered a lifeline in the shape of a $184-million booking for 200 lockers distributed over the first six flights. NASA intended to make these lockers available to the CCDS in order to remove the potential bottleneck in testing hardware for these projects. In fact, the module was not finished until early 1992. It flew in June 1993 on STS-57, the mission which retrieved the EURECA free-flyer. Two dozen of the experiments aboard were supplied by NASA. They covered materials and life sciences, and an engineering test of a waste-water recycling unit under development for the space station. The commercial experiments that the company had managed to secure filled most of the rest of the lockers, but it was already clear that this favourable payload mix would be difficult to repeat.

Despite having built Spacehab in order to provide ready access to microgravity, the fact that it typically took four years to run a middeck payload through NASA's certification procedure was a considerable disincentive. A *service-provider* such as Spacehab Incorporated, familiar with NASA's payload integration procedures, and already approved by

NASA as a supplier, was able to reduce this to two years, but even this was far from 'ready access'. And, since 'time is money', this delay prohibitively inflated the cost of what would otherwise be cost-effective low-budget experiments. No matter what it carried, Spacehab Incorporated faced the problem that its income was extremely erratic: it did not receive its final payment until after the payload flew, and it was therefore a 'hostage' to any slippage in the shuttle schedule. From a commercial viewpoint, therefore, it was a rather risky business. In effect, NASA was carrying the module for free, as indeed it was the CCDS experiments it carried, because the fee which it charged the company for each flight was exactly one-sixth the value of the contract it had placed for lockers on six flights. The idea had been to establish a flight schedule so that the company would be able to generate revenue from leasing out the remaining lockers on a commercial basis.

By the time Spacehab had its second outing, in February 1994, NASA had renegotiated its contract. Since the company was not attracting sufficient commercial interest to use the lockers *not* used by CCDS, NASA agreed to reassign all of its remaining utilisation to the next two flights. Thereafter, however, the company would be on its own.

As it happened, Spacehab Incorporated was saved by the transformation in the political situation following the collapse of the Soviet Union. When NASA decided to fly shuttles to the Mir space station, and to ferry supplies to astronauts working aboard on projects which would lead up to the new International Space Station, it looked for a logistics carrier. There were only two real options: Spacelab and Spacehab. In winning this contract to ferry cargo up to Mir, Spacehab secured its short-term future. The convenience of the locker system had been one factor in Spacehab's favour (Spacelab used a laboratory-style instrument rack which was less flexible), but the vital factor had been the ability to turn the module around within the planned four-monthly flight cycle. The contract obliged the company to affix the pressure hull of its engineering test article to the rear of one of its flight modules, to create a double-length module for increased capacity. There were two operational modules, so this left one for the already agreed missions. Fittingly, Spacehab's third outing was in February 1995 – the mission that rendezvoused with Mir to pave the way for the first docking set for later in the year. The module had been fitted with a roof window and a camera to offer the shuttle pilot a viewpoint equivalent to that which would be provided by the special docking unit, in order to rehearse close-proximity manoeuvres alongside Mir.

Spacehab 3 was notable for Charlotte, a spider-like tele-operator system. It was strung on wires stretching across the module, and it could locate itself in front of any given locker. It had an arm with a gripper, and it could flip switches and load samples and data cassettes. A video system allowed its supervisor, on the ground, to observe its activity. It was under development by McDonnell Douglas for possible use on the space station. By its fourth flight, Spacehab was no longer news in itself, but was just part of the shuttle infrastructure. Nevertheless, it kept the astronauts busy with a broad range of experiments supplied by the CCDS. These included the Commercial Float Zone Furnace, the Space Experiment Facility – a high-temperature furnace for semiconductors – and the Advanced Separation Processor for organic materials. But it was the SPARTAN carrier which deployed an inflatable antenna that captured the headlines.

The first Spacehab flight as a cargo carrier was on STS-76, in March 1996. The double module was not yet ready, so this was the standard format. However, because the docking

Table 7.1. Spacehab missions

#	Mission	Date	Lockers (NASA)
1	STS-57	Jun 1993	22
2	STS-60	Feb 1994	44
3	STS-63	Feb 1995	49
4	STS-77	May 1996	50

system had to be at the front of the bay, the module was placed well-aft to restore the centre of mass. A tunnel ran from the middeck airlock to the base of the docking system, and another tunnel ran from the other side of the docking system to the module. All the hatches were on the same axis, so this was a straightforward configuration. Placing the module further back had the advantage of increasing its allowed payload to 2,000 kg. Some 880 kg of materials was delivered for Mir, and 740 kg for NASA's science activities on Mir. An ESA Biorack of microgravity experiments was also carried, together with a freezer in which to return the biomedical samples that had accumulated on Mir. A total of 500 kg of scientific results and assorted apparatus was off-loaded from Mir, for return to Earth. The to-Mir cargo was very welcome, but the delivery could easily have been accomplished by a single Progress ferry, several of which routinely docked with Mir each year to deliver up to 1.5 tonnes of dry cargo as well as 1 tonne of propellant. The to-Earth cargo, however, was far more significant, because it was practicable to return only 120 kg of compact items in a Soyuz descent module, and there were only two flights a year (Mir crews served 6-month tours), so the shuttle provided Mir with a return capacity which had previously been sorely missed. It was ironic that a module developed to facilitate microgravity research had found its niche as a bulk cargo hauler. But it was *business*, and in 1995 Spacehab Incorporated turned in a profit for the first time, because it was genuinely selling its services to NASA, and that was what really mattered.

The double Spacehab was effectively a hybrid; only the front half – the flight article – had power, and the extension was simply a storage rack; but then, that was all that was required. It was still possible to carry experiments in the powered section, so even logistics missions carried secondary payloads with long-term significance.

It is clear that if it had not been for the Shuttle–Mir programme, Spacehab would have been a commercial failure. The basic concept was valid, but the raw commercial market did not develop as expected. Nevertheless, Spacehab was successful in clearing the backlog of middeck payloads, and it proved invaluable in supporting the semi-industrial research of the CCDS, but its long-term prospects seem rather bleak. After the completion of its Mir resupply role, and while the shuttle is fully occupied in assembling the International Space Station, there will be few flights with sufficient spare capacity to carry Spacehab, and once the station is operational there will be little call for a microgravity module in the bay. The concept of middeck augmentation evidently had built-in obsolescence, so if the company is to survive it will have to compete with Italy's Alenia Spazio for contracts to ferry cargo to and from the International Space Station. NASA has already committed to use the Alenia pressurised logistics module to deliver racks of equipment for outfitting its laboratory, but there is a great deal of apparatus to be mounted outside the station, and Spacehab is proposing an unpressurised Integrated Cargo Carrier as an adjunct to its pressurised module. This would be lighter than the Spacelab pallet that NASA had intended to use. Meanwhile, the company spotted another potentially commercially-viable niche. NASA still has to clear its backlog of GAS cans, and Spacehab offered to build a *raft* to span the bay in the otherwise unusable volume around the tunnel leading to either a Spacehab or a Spacelab. This would enable the cans to be flown in batches of up to 60 at a time. Considered together, all these facilities provided the company with a very flexible strategy for servicing the evolving requirements of the programme. Despite the evaporation of the early optimism in commercialising access to space, Spacehab Incorporated has proved itself to be a survivor.

INDUSTRIAL SPACE FACILITY

Spacehab was explicitly planned as a middeck augmentation module for experimental packages already designed to fit the shuttle, but, early on, a need was also perceived for a semi-permanent *free-flying* facility on which longer-term experiments could be performed, or, better yet, applications run. The rapid progress being made by McDonnell Douglas in 1985 with Electrophoresis Operations in Space seemed to be a clear signal for the future, and sooner rather than later the company would need an orbital factory. Would it, perhaps, lease space on a multipurpose platform? It seemed that by the early 1990s there would be a demand for space on orbital platforms. In this lucrative service-provider market, it would clearly be an advantage to be first in the field, and it would also help to be seen to have close links with NASA.

Space Industries Inc. had been set up in Houston in 1982 by Max Faget, the designer of the Mercury capsule. When Joe Allen resigned as an astronaut, he had joined the company. SII was therefore intimately familiar with NASA's ethos; nevertheless, it set out to transform the way that microgravity payloads were developed. A decade before Dan Goldin made it the norm, SII advocated a 'faster–cheaper–better' approach. It had some success with small packages for carriage on the COMmercial Experiment Transporter (COMET) developed by the Center for Space Transportation and Applied Research, a CCDS at the University of Tennessee. In the mid-1980s it noted the future for a large orbital facility for long-running experiments, and so proposed the Industrial Space Facility (ISF) as a shuttle-deployed, and shuttle-tended, automated materials-processing factory. The concept called for a pressurised hull configured to exploit the shuttle's payload bay. It would have a pair of solar panels of the type tested on STS-12; the 12 kW generated would run both services and applications. A docking system would be carried in the forward part of the bay, and would be linked to the cabin's airlock by a short tunnel. The ISF would have a hatch so that the shuttle could mate with it. Unlike the elaborate space station that NASA was planning to develop, the ISF was to be self-contained. However, a logistics module was to be carried during follow-up visits to deliver raw material and to retrieve the finished product and, if necessary, replace broken apparatus. It would even be practicable to replace one customer's apparatus with that of another. The design was sufficiently flexible to allow several utility modules to be joined together, and if necessary the entire facility could be returned to Earth for refurbishment. In the long term, the design held out the prospect of attachment of a crew module. It might even be integrated into the space station. Anything was possible.

The ISF proposal was well received, and Westinghouse, Boeing and Lockheed backed engineering studies. In August 1985, NASA announced an agreement which guaranteed the ISF two flight opportunities, so that SII would be able to assure prospective customers that it had NASA's support. To obviate the need for the company to raise capital for launch costs, NASA introduced a 'fly-now-and-pay-later' deal whereby SII would reimburse the cost of its flights from the revenues earned from renting time on the ISF. The company's close association with NASA had really paid off. In effect, however, this revolutionary deal had its origins in the Reagan administration's July 1984 announcement promoting the commercialisation of space operations, and this was NASA's way of helping private sector start-ups make commercial headway during their most crucial formative years.

The 1986 grounding of the shuttle did little to dampen interest in the ISF. Indeed, some in Congress argued that it made more sense to deploy a number of shuttle-tended ISFs than it did to construct a space station; the unoccupied factories would have a better microgravity environment, and they would produce a significant return for a far smaller investment. This was not what NASA wanted to hear.

The ISF generated considerable interest in the aerospace industry that would construct it, but, as the Spacehab experience confirmed, the commercial sector did not have a need for a generic manufacturing-in-space platform, so the project was eventually abandoned. There was nothing wrong with the basic concept; it was simply several decades ahead of its time. This was the direct result of it having been proposed to satisfy the great demand for microgravity suggested by the hype which was prevalent in the early years of shuttle operations. SII moved on to develop a range of packages for specific microgravity applications, one of which looks as if it could lead to the first successful space industry – making semiconductors.

WAKE SHIELD FACILITY

The Space Vacuum Epitaxy Center was set up at the University of Houston in 1985 as a CCDS to focus on commercial manufacturing of crystals in microgravity using epitaxy, the process by which the crystal is grown on a substrate in such a way that its structure is parallel to the substrate's lattice. This crystal was to be built up layer by layer, using a spray (atom-by-atom) of raw material. It was hoped that by exploiting the space vacuum, the crystallisation would be more controlled than was feasible on Earth, to produce a crystal that was both larger and more regular. All of the previous experiments growing inorganic crystals had exploited the microgravity of space, but this epitaxy process would also rely upon the *vacuum* of space. It was evident from the very beginning, therefore, that the Center would have to develop an automated package for external operation, and that this would be a major undertaking.

The task was complicated by the fact that the apparatus could not be built in the form of a pallet that would operate within the payload bay. This was because the orbiter polluted its immediate surroundings with outgassing, water, and thruster efflux, which would interfere with the epitaxy. The SVEC therefore devised the Wake Shield Facility (WSF). This was a remarkably simple idea. As a spacecraft moves through the rarefied atmosphere at orbital altitude, it leaves a semi-vacuum in its wake. The orbiter did this, but it also trailed a cloud of its own debris. It had been hoped that if the RMS held the WSF well clear of the bay the environment would be sufficiently clean, but tests demonstrated that this was not the case. The WSF would therefore have to be a free-flyer that would operate away from the orbiter. This made the package rather more complex. It would either have to operate autonomously, or it would have to carry a sophisticated remote-control system. It was decided to make it fully automated.

There is no well-defined boundary to the atmosphere; its density merely decreases into what is generally regarded as the vacuum of space. Air pressure is traditionally measured in terms of millimetres of mercury in a capillary tube: a unit which is also known as the *torr*. On this scale, the pressure at sea level is 760 torr. At an altitude of 350 km, the pressure is a mere 10^{-7} torr. Although this is low, it is insufficient for vacuum epitaxy. The best manufacturing facilities on Earth can attain 10^{-11} torr, and the hope was that the vacuum in the wake of the WSF would be 10^{-14} torr. With the benefit of microgravity, this would enable a crystal to be built up one layer at a time in a perfect matrix.

The WSF was an attractive project from NASA's point of view, because once it had been shown that it could make a premium-price product it could be commercialised. Even better, because the process relied upon the space environment, the product would not be able to be matched by terrestrial facilities, so it would be a genuine manufacturing-in-space application.

Although it was hoped to be able to manufacture a variety of thin-film products, it was decided to manufacture semiconductors first because there was a proven market for higher-than-usual-purity chips that could be sold at a considerable premium. Gallium arsenide was chosen for the trial run because it is difficult to make large impurity-free gallium arsenide wafers on Earth. Other candidate materials were indium phosphate, zinc selenide and indium–gallium–antimonide. Each of these offered an order of magnitude improvement in performance over traditional silicon. Crystal purity was the key. The exotic new semiconductor technologies held out the prospect of integrating digital and analogue logic and incorporating microwave and photonic interfaces with a single material. This was *in addition* to making conventional computers faster.

The WSF hardware was developed by SII, which was one of SVEC's affiliates. It was a 3.7-metre stainless steel disk carrying the attitude-control system on its leading face and the epitaxy apparatus on its rear, and was carried flat on a trestle spanning the payload bay. It would be hoisted out by the RMS and appropriately orientated for release, and a nitrogen thruster would then fire to start the separation manoeuvre. When it was well away from the orbiter, ovens would evaporate cells of gallium and arsenic, and a beam of atoms would be sprayed onto the substrate of the wafer. Even though the orbiter would be 75 km away during this processing, it was not to fire its thrusters in case air-drag swept the efflux down its wake to the free-flyer trailing behind. Instead, the orbiter would sit in the gravity gradient, and since this was ideal for microgravity research, testing the WSF was to be combined with suitable secondary payloads.

The WSF test was assigned to STS-60 in early 1994. As it happened, this mission was notable for the fact that Sergei Krikalev, a Russian cosmonaut, was a member of the crew.

One of the early investigators on the project, Ron Sega, was also an astronaut; his inclusion in the crew proved fortunate. The plan was for the WSF to be deployed on Day 2, and then spend two days growing half a dozen wafers of gallium arsenide up to 6-μm thick. The Air Force's Geophysics Department had provided instrumentation to measure conditions in the wake during this trial. When the orbiter returned to retrieve the WSF, it would perform a series of proximity manoeuvres so that the sensors could measure the degree to which its thrusters impinged on the now-deactivated epitaxy package.

The experiment started well. Jan Davis used the RMS to lift the 1,800-kg WSF from its cradle, turned it around, and held it over the bay so that it could be checked. It was heavily computerised so that it could operate autonomously, but it generated telemetry and could be set up from the ground. Unfortunately, a firm radio-link could not be maintained. Evidently there was interference from other apparatus in the bay. This was not a common fault, but it was not the first time that interference had hit. For two days, Sega worked in the confident expectation that the experiment would be able to proceed; then a horizon sensor failed and deployment became impracticable. The best that could be done was to hold the WSF high above the bay in order to verify the functionality of the epitaxy apparatus. As it turned out, the wafers made in this less-than-ideal test were of comparable quality to the best that could be produced using the same technique in vacuum chambers on Earth. All in all, this was an encouraging start.

The next test flight was scheduled for early 1995, so SII set about modifying the WSF with some urgency. Extensive testing revealed that the sensor problem was due to electrical interference from a nearby power cable, so the wiring was rearranged. It turned out that the reflection was due to interference between the transmitter on the WSF and the relay system mounted on the cradle. The WSF was soon ready for its reflight, but NASA was having its own problems. Knock-on effects due to slippage in the schedule, the hiatus in the summer of 1995 caused by uncertainty over when to send STS-71 for its historic link-up with Mir, problems with the SRBs, and a problem with woodpeckers trying to nest on STS-70, pushed back STS-69 by six months.

Despite the rewiring of the communications relay, it took two hours of trouble-shooting before Jim Newman was allowed to release the WSF. It made quite a picture; and seen edge-on it was reminiscent of the fabled 'flying saucer'. When it fired its thruster, the WSF became the first payload to perform its own separation manoeuvre; previously, it had been the orbiter that had moved away. The telemetry link fell silent a few minutes later, so, reluctantly, the orbiter was ordered to fire its thrusters to rotate to give the relay unit a better line of sight, and the link was restored. As before, the task was to make gallium arsenide wafers. For the first 16 hours of processing, everything ran smoothly, but then the cooling system malfunctioned and the attitude controller began to overheat. A magnetorque system was used to maintain orientation, as thrusters would have polluted the wake. When the disk pitched over about 12 degrees this degraded the wake, so the wafer-building underway at that time was abandoned. Houston allowed the apparatus to cool for twelve hours before resuming operations. To give the WSF time to complete its programme, the retrieval was put back a day. It was not long before the platform pitched over again, but this time it was left to make a full 360-degree flip. When the orbiter drew up alongside it, the WSF was flying edge-on. It was stable, however, so was easy to recover. Although the WSF had turned out to be difficult to operate, it had successfully manufactured four 75-cm diameter wafers of semiconductor. The wake had never bettered 10^{-13} torr, but this was 100 times better than that attainable in a terrestrial vacuum chamber, and the product was the purest gallium arsenide ever manufactured.

After this flight the communications system was completely revamped, and in an effort to overcome the heat build-up which had prompted the attitude-control failures the thermal control system was upgraded. These modifications increased the WSF's mass to 2,150 kg for its next outing. This test in November 1996 began with an alarming near-collision. As the end-effector of the RMS disengaged the pin on the front face of the WSF, it imparted a slow roll. The separation thruster could not be fired until this rate had been cancelled by the attitude-control system, but this took time and, during that time, differential gravity caused the relative geometry of the two vehicles to change, with the result that when the thruster finally fired, the big disk skimmed over the roof of the orbiter's cabin as it departed. If the orbiter had manoeuvred clear, the efflux from its thrusters would have polluted the surface of the experiment. This incident prompted a review of the dangers of having self-separating payloads. This time the WSF performed flawlessly and completed its full programme, and on this occasion wafers of aluminium–gallium arsenide and indium–gallium were produced in addition to gallium arsenide.

NASA had allowed the WSF three flight opportunities to demonstrate its capabilities. In effect, it had taken this entire allocation to achieve the objective set for the first flight. It had always been recognised that further flights would be contingent upon securing commercial backing. The troubled trials had hindered the timescale for commercial development, but it was also clear that even if the WSF was successful and the semiconductor industry did buy into the project, there would be limited opportunity to fly the package because the shuttle's manifest would be dominated by flights devoted to the assembly of the International Space Station. This was the problem with a manufacturing-in-space application: the shuttle simply could not support any application that placed a heavy load on the manifest. Carried flat, the WSF took up too much of the bay to be carried as a secondary payload-of-opportunity. In a commercial form it would carry a carousel so that it could produce up to 200 wafers, and it would have solar panels to support extended operations, so it would be even bigger. But the WSF is not a shuttle payload at all. In its operational form it will be a platform to be flown alongside a permanent facility, from which it can be deployed, and to which it can return for servicing, and only then will it be able to turn out product at its maximum rate. As the semiconductor industry is one market sector which *can* impose a premium for a product of exceptional quality, it has been estimated that if that premium were to be 10 per cent, then a yield of only 200 kg of wafers per annum would be required to make the venture profitable.

If the WSF does secure commercial backing, it will not fly until the International Space Station has been built, but it should then be able to make wafers on an industrial scale, and thus become a true manufacturing-in-space application.

ALL DUE CREDIT

Some have labelled microgravity research as ivory tower science with little commercial value, but much of the work has been undertaken by academics in concert with industry to ensure that the results stand the best chance of being applied. It is true that the potential for commercialisation has not lived up to the initial hype, but hype is exactly what the word implies. Much of the blame for this early over-optimistic expectation must be borne by NASA, for the way in which it promoted first the shuttle and then, in the early 1980s, the role of a space station as a laboratory in space. Even if NASA did not originate some of the wilder concepts, it did little to correct such false promises.

Whilst it is the case that McDonnell Douglas pulled out, the 3M Company methodically pursued its 10-year programme even after the popular mood turned against commercialised microgravity. Other companies, in a wide variety of industries, signed up as affiliates of the CCDS, so their involvement has a fairly low profile. To appreciate *why* these companies are investing in microgravity research, it is essential to understand that the result does not *have* to be space-based manufacturing in order for the investment to be worthwhile. It is true that the goal of the CFES project was a space factory, but the protein crystal research's goal is to increase our understanding of biological systems so that drugs can be designed and manufactured on Earth, and the polymerisation and thin-film membrane work set out to seek insights to improve terrestrial manufacturing. If there was no commercial benefit to be derived from growing proteins in microgravity, then the biotechnology companies would long ago have given up sending up crystallisation packages on the shuttle, and they would not have seized upon the opportunity to use the Mir space station.

Unfortunately for NASA, there is no public awareness of whether any given drug owes its design to knowledge derived from analysis of a crystal grown on the shuttle. In a sense, however, this is a direct result of the shuttle's success in serving industry.

When the explorers of earlier centuries sailed on voyages of discovery, they came back with looted baubles and carcinogenic compounds. The shuttle returns knowledge, and the space station will do the same, but to a very much greater extent.

8

Biology

In the long-term, the objective of NASA's life sciences programme is nothing less than to understand the fundamental nature and origin of life. In the shorter term, however, it has two imperatives: firstly to understand the way in which life is adapted to the Earth – in particular the relationship between life and gravity – and secondly to develop the medical and biological systems required to enable humans to live and work in space. But we exist in the context of a larger biological system, not just for food but also for environmental and waste management, and if we are to establish orbital bases and to colonise the inhospitable lunar surface we will need to take crucial elements of this environment with us, so it will first be necessary to develop the technology to sustain such an environment in space.

FLORA

It is not really practicable to *grow* plants on the shuttle because its orbital endurance is so limited. However, it can be used to *test* apparatus in anticipation of the establishment of a semi-permanent facility. In the event, this opportunity for research and development proved to be fortunate because, as work on the Mir space station had shown, cultivators optimised for microgravity can fail for a multiplicity of reasons.

In preparation for its experiment on the first Spacelab, the University of Pennsylvania flew the Hyflex Bioengineering Test (HBT). It was a sealed unit of 72 small chambers with dwarf sunflower seeds (*helianthus annus*) to test how different degrees of moisture affected growth. Unfortunately, the return of STS-2 after only two days, rather than the planned five, curtailed the test, so it was reassigned to STS-3; this 8-day flight was sufficient to determine the optimum set-up of the apparatus.

The Biorack was developed by ESA as a multipurpose research facility for Spacelab. It was introduced on STS-22, which was the first German-sponsored shuttle flight. This full-height cabinet could be filled according to experimental requirements. On its first outing it supported a variety of experiments with two incubators at different operating temperatures, a freezer, a hermetically sealed Glovebox and a microscope with which to examine samples. It later flew on the first International Microgravity Laboratory, which also carried Germany's Biostack. This was a sandwich of a variety of life forms within a stack of detector plates to record the passage of radiation. It was to assess radiation effects, and was tested with bacteria, fungal spores, cress seeds, and shrimp eggs.

The Plant Growth Unit (PGU) flew on STS-3, with a type of pine, mung-bean and oat, to investigate the formation of lignin, the agent that stiffens the stem of a plant so that it can withstand the pull of gravity. The objective was to test whether a wood-plant develops with less of this agent in microgravity. It had long-term implications for building self-sustaining habitats in space. The human gut cannot digest a lignin-cellulose cell so, even though such a plant is rich in carbohydrate and protein, it cannot serve as a food. If a lignin-free variety could be grown in space, it would broaden the range of foods available. This experiment (fully called Gravity-Induced Lignification In Higher Plants) flew again on STS-19. After the shuttle resumed flying, following the loss of Challenger, the PGU was used for the Chromex experiment, which was to determine whether the roots of a plant in microgravity developed in the same way as did those on Earth. Root-free shoots were grown to measure the rate, frequency and patterning of cell-division in the root tips and the genetic make-up to determine whether this was upset by exposure to the microgravity and radiation in the space environment. The plants grew well; indeed, the rate of root-tip growth was faster than that back on Earth, but the roots were disorientated and they grew in all directions. Of far greater significance, for this experiment, was the discovery of a substantially lower level of cellular division in root tips and of chromosomal abnormalities such as breakage and fusion. Data from radiation monitors flown alongside suggested that the radiation dose was insufficient to cause the up-to-30 per cent aberration rate seen in dividing cell chromosomes. The only way to separate the effects of microgravity and radiation was to fly a centrifuge to reintroduce the effect of gravity in the radiation environment, but this has yet to be done (although the Russians built a cultivator of this type, called the Biogravistat). The PGU flew regularly, to build up the Chromex database.

The real challenge was to plant a seed, grow it, and have the plant flower and deliver a new seed which would then be planted to restart the cycle, because it is only by running through successive generations that genetic mutation is made manifest. Quite apart from the study of space genetics, the successful growth of plants will be crucial to the establishment of a fully self-sufficient orbital habitat, not just for food but also for atmospheric processing, because plants consume carbon dioxide and liberate oxygen. Unfortunately, even with its extended-duration facility, a shuttle cannot remain in space long enough for a plant to complete its life cycle; this requires a semi-permanent orbital habitat.

After many disappointments the Russians finally successfully ran arabidopsis, a weed, through its life cycle. This had not been done with a staple such as wheat, so when NASA was invited to have a succession of astronauts serve tours aboard Mir, it assigned them this task. It had taken the Russians several years to perfect a successful cultivator. This process of trial and error had been undertaken on the Salyut stations which had preceded Mir. The first flowering had been achieved in 1980, but it was another two years before seed pods fully developed; an event which greatly pleased the botanists. When NASA set out to grow wheat on Mir, it decided to use the Russian cultivator rather than try to use one of its own. The Svet cultivator in the Kristall module had been supplied by Bulgaria. It had a high-intensity lamp and grew the seed in a substrate infused with nutrient. It had successfully grown radish and lettuce. NASA installed instrumentation to monitor physical parameters.

John Blaha was to have started this Greenhouse Experiment, but Shannon Lucid set it up and planted the first wheat seeds while she waited for the delayed shuttle to retrieve her and drop off Blaha. He tended the crop and harvested it just before he left. Jerry Linenger, who replaced him, replanted some of the seeds and tended them throughout their cycle, and Michael Foale continued the process. Successfully growing a staple food was a real milestone in space biology. Later, Foale managed to grow a type of dwarf turnip and, after 48 days, it too produced seed.

The Wisconsin Center for Space Automation and Robotics, one of the CCDS, built the Astroculture cultivator specifically for long-run plant experiments. It was computerised, so its initial flights in 1992 and 1993 were engineering tests of the water and nutrient delivery, the lighting and the humidity control systems. It was a self-contained unit within a middeck locker, and comprised two growth chambers. An inert material served as the root matrix. One porous steel tube was embedded in the matrix to supply water and another to recover water. The water flow in the supply tube was pumped, but the nutrient solution emerged from the tube by capillary action. The trick was to propagate the nutrients through the matrix without having it become saturated. Growing plants in a fluid-infused matrix is called hydroponics. The need to control fluids made this the most technically demanding approach. In this initial form, the apparatus was heavily instrumented to measure flow rates, and it was extensively tested to define its operating parameters. Astroculture was developed specifically for use on the International Space Station, and plant growth will form a major part of that programme.

FAUNA

The first animal experiment, flown on STS-3, was the winning proposal for the Student Involvement Programme which NASA instigated to foster a sense of popular involvement in the shuttle. It was a self-contained unit with bees and moths, which were filmed for later analysis of their motion in microgravity. These species were selected because of their very different ratios of body-mass to wing-area.

Several thousand 12-day old bees flew on STS-11, in order to test their ability to make a honeycomb. On Earth, bees make a hexagonal cross-section, but this is not simply the optimum shape for holding honey; it is structurally stronger than a triangular or square section matrix. The container mimicked the diurnal light cycle, and it maintained a hot environment. The bees adjusted rapidly to microgravity and made a recognisable comb. This experiment was reminiscent of that on Skylab with the spider, Arabella; it had initially been confused, but had then woven commendable webs. A colony of ants flew on STS-7, to see how they adapted. Some 10,000 students had submitted proposals, a dozen or which were selected and industrial sponsorship arranged, but by the time all these experiment were flown, their originators had long-since left college. However, these investigations were performed as secondary tasks, as life sciences research was not to become the *focus* of a mission until the Spacelab 3 mission. Physiological changes run their course faster in animals than in humans, so more can be learnt from a short flight by studying animals.

NASA's Ames Research Center took the lead in developing facilities for conducting animal studies on the shuttle. Its Animal Enclosure Module (AEM) was a middeck drawer unit that provided its occupants with food, water, air and light. It was tested on STS-8 using six rats, which were flown to observe their reaction to microgravity. When it flew again on STS-10, it was to determine whether microgravity relieved the symptoms of arthritis. The Animal Holding Facility (AHF) introduced on STS-17 was a Spacelab 3 cabinet. On this occasion the rat enclosure was combined with a monkey enclosure.

The monkeys had been designated 'A' and 'B' in the hope that they would not become anthropomorphised by the public. Nevertheless, this mission was condemned by Animal Rights activists.

Although monkeys had flown in space before – notably in the early days of testing the Mercury capsule – this was the first time that humans and monkeys had flown together. A young squirrel monkey called Miss Baker made a sub-orbital flight in 1959, then went on to live to the grand old age of 27, so it had clearly not found the experience too traumatic. On this occasion, however, the monkeys were rather withdrawn, and they declined to eat from their food dispenser. The animal experiments were supervised on alternate shifts by the two medical specialists. As Norman Thagard's speciality was human medicine, Bill Thornton worked the shift when the monkeys were awake. In fact, he had worked with NASA Ames on the AHF project since its inception. He took pity on the monkeys and opened their enclosure to feed them by hand; this contact enlivened them considerably, and thereafter they were more active.

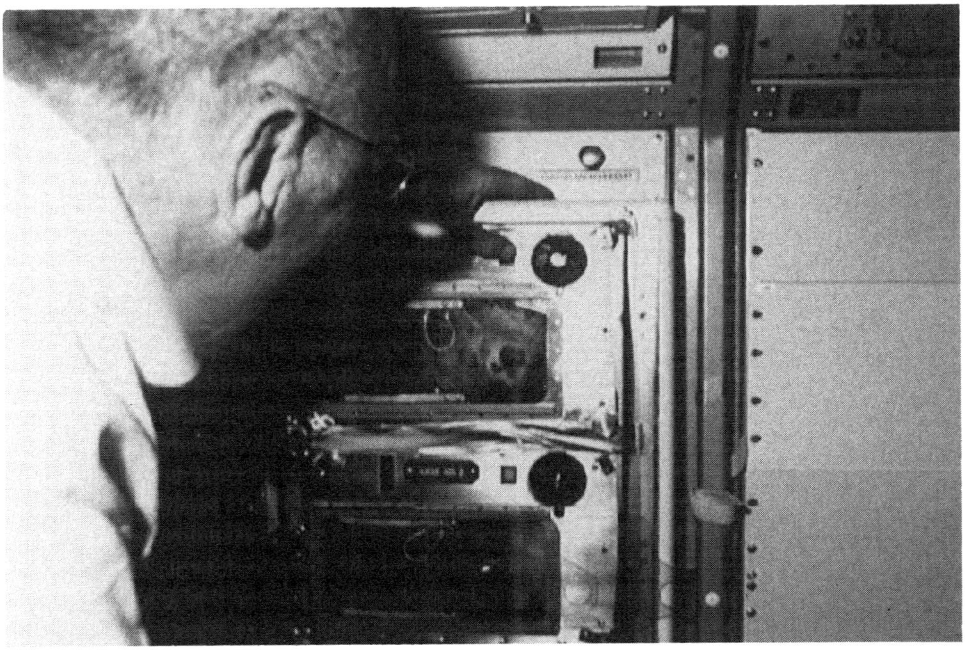

Fig. 8.1. One of the squirrel monkeys which flew on STS-51B. Although monkeys had flown in space before, this was the first time that they had flown alongside humans.

Thornton's intervention was contrary to the experimental procedure which called for the animals to be fed according to a schedule, using only the food trays. Opening the enclosure had been explicitly forbidden in order to preclude loose material from escaping. However, the AHF proved deficient in that when a food tray was removed for refilling, loose material floated out from the enclosure. This debris was not simply uneaten food; it included pieces of excrement and droplets of urine. The environmental system had been configured so that the air-flow was from the orbiter's cabin into the Spacelab, to retain particulate debris in the laboratory, but this failed to prevent debris from reaching the cabin. One memorable scene on the video downlink was the look of disgust on shuttle

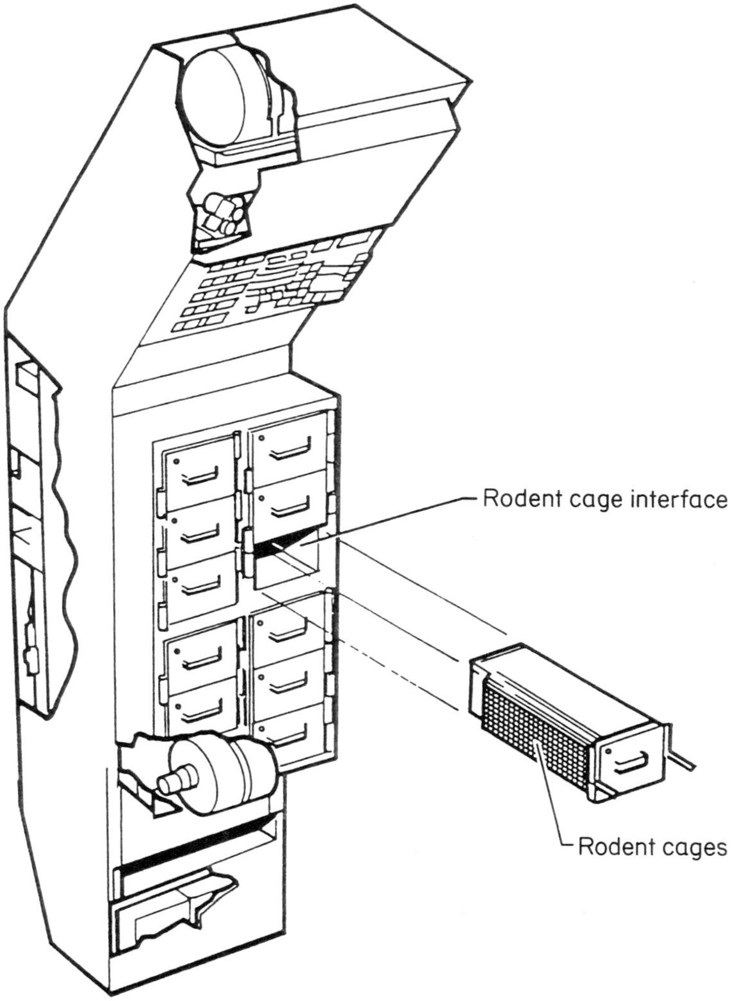

Fig. 8.2. A schematic of the Research Animal Holding Facility (RAHF) designed to enable animals to be carried in middeck lockers.

commander Robert Overmyer's face when a monkey dropping drifted in front of his face as he sat at his seat on the flight deck. This particulate represented a health hazard, because it could be swallowed, inhaled, or trapped in the eye or ear. Worst of all for this pioneering crew, though, was the pungent aroma that emanated from the animal enclosure. But the animals were not the only source of unpleasant emissions. With an all-male crew, it had been a simple exercise to modify the toilet to collect urine for a human biology experiment, but it persistently malfunctioned, and sprayed the fluid into the cabin. To overcome the problem of leakage from the AHF, it was modified to employ a *ten-day* feeder tray, which precluded the need for the cabinet to be opened.

Two dozen rats flown on Spacelab 3 had been surgically implanted with sensors so that biotelemetry could record physiological data during the flight. Immediately after landing, before the process of readaptation to gravity could set in, they were dissected to extract tissue to reveal its state of adaptation to microgravity.

To appreciate the range of biomedically pertinent data which can be derived from studying rats, the following should be considered. Skeletal muscles can be classified in terms of their rates of reaction to stimulus by nerve impulses. Those that react most rapidly are related to body motion, and those that react slowly are involved in maintaining posture in the presence of gravity. Support and posture are not so essential in microgravity. The response timing is governed by a protein, and by dissecting a rat, this state of adaptation can be related to the production of this protein. After seven days in microgravity, a rat will have lost 40 per cent of the mass in the muscles of the hind legs. These muscles would normally have been active in opposing gravity, but in space there is nothing for them to do. There is a marked decrease in the diameters of the muscle fibres, and there is an almost total absence of muscle tone. These were all physical measurements. But there are also biochemical changes: the process which generates energy in muscle cells is almost totally absent too. In space, the slow-acting load-bearing and postural muscles atrophy. On the other hand, the fast-acting motion-control muscles can be significantly enhanced. The musculo-skeletal system rapidly adapts to its new environment, both upon going into space and upon returning to Earth. This confirmed observations by astronauts who had flown on Skylab for three months, and by the cosmonauts who undertook even longer flights aboard the Mir space station. This long-term process is difficult to study in detail in the case of the human crew because they exercise specifically to overcome this deterioration (and because they cannot be dissected upon their return), so the rats enabled the biochemical processes to be identified.

Stamina is decreased, too. It was suspected that in space the reduced muscle activity not only inhibits the protein that controls response time, but also enzymes which are associated with metabolism. Metabolism is the process by which food is broken down and converted into energy. In the absence of energy released by metabolism, the muscles rapidly exhaust the limited energy available and then begin to consume the glycogen that is stored in non-load-bearing muscle tissue, primarily the liver. Glycogen is a polysaccharide stored as a reserve carbohydrate. It is a *poly*saccharide in that it is a mixture of *mono*saccharides, the most common of which is glucose (otherwise known simply as 'sugar') and it plays a significant role in metabolism. The first flight of the Physiological and Anatomical Rodent Experiment (PARE) focused on glycogen's role. Specifically, it sought to find out how microgravity affects insulin control of glucose transport in the soleus, a calf muscle which

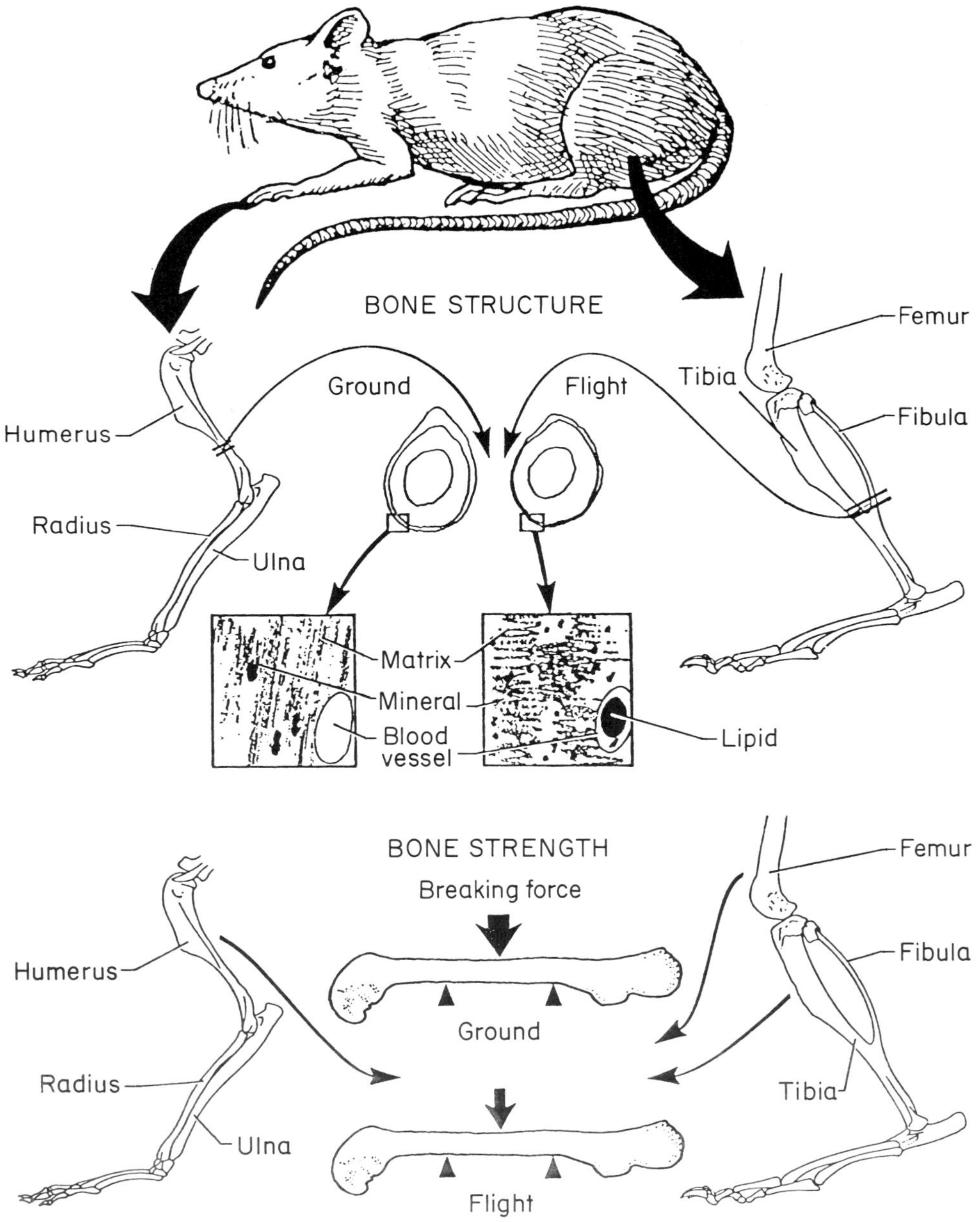

Fig. 8.3. A schematic of the way in which rodents are able to contribute to studies of the effect of weightlessness on bone.

is involved in overcoming the pull of gravity *as well as* in movement. It was found that non-load-bearing tissue stored *additional* amounts of glycogen as a result of the altered regulation of glucose metabolism. Because this test was done on STS-48, the mission which deployed the Upper Atmosphere Research Satellite, the AEM was fitted as a mid-deck locker so it could fly only eight rats and, because it was a five-day mission, very young rats were flown because they would be undergoing development, and so should be more susceptible to this change in the short time available. The PARE was carried frequently thereafter, each time focusing on a different aspect of adaptation. These biochemical studies showed that the body adapts in a remarkably short timescale to a significant change in its gravitational environment.

It had long been known that bones atrophy in weightlessness. Data from rats indicated that this loss is worst in the case of load-bearing bones, that it results from a slowing in the rate of bone formation, and that the inhibition of the process of calcification is linked to a reduction of a protein secreted by bone-forming cells. But how did the cells detect the absence of gravity? Two PARE experiments on STS-56 set out to study bone development in detail. It was clear that gravity is essential for 'normal' skeletal development, because the role of bone is to bear the load imposed by the rest of the body, and posture reflects this. One test set out to study the configuration that the body adopts in the absence of gravity. The bone mass, mineralisation rates and bone strength were all measured. Earth-based experiments in *unloading* the skeleton of rats by hanging them by their tail throughout the formative stage in their growth had shown that, although bone continued to be produced, the cells formed a structure with little structural strength. In space, however, this process is inhibited. The test was to find out whether this resulted indirectly, from reduced muscular stimulation, or was directly attributable to the bone. Both the sites and mechanisms involved were examined. Regardless of the cause, it was hoped to identify those parts of the bone most affected. The second experiment studied the osteoblasts – the cells that make bone. The production of pre-osteoblast cells is mechanically sensitive. The goal was to figure out *how* such cells sense gravity, how the process is switched off when gravity is removed, and how it is restarted immediately upon returning to Earth. Osteoblast proliferation was tracked by a marker for DNA synthesis. The test used load-bearing shin bones as well as non-load-bearing jaws, to determine whether the process was selective. A student experiment on STS-29 had flown rats in which non-load-bearing bones had been fractured and then reset, so that the effect of exposure to microgravity on the healing process could be studied. It transpired that the tiny amount of bone grown in space had little structural strength, and it began to look as if the long-term prospects for humans living in space would be limited by bone demineralisation. This work also has implications for osteoporosis sufferers.

In addition to rats, the first Space Life Sciences Spacelab had an aquarium with 2,500 baby jellyfish (*aurelia ephyra*), one of the simplest organisms to possess a nervous system. In the sea, they maintain orientation by using a statolith sensor analogous to the mammalian vestibular apparatus. They swam about in circles, with spasmodic bursts of activity; maybe they too had low stamina? A variety of other aquatic species had already been tested. STS-5 had carried sponges, an organless animal which is characterised by the presence of a canal system and a series of chambers, through which water is first drawn in and then expelled. Frogs flew on the first German Spacelab, and the Japanese Spacelab hatched tadpoles in 1992. Those fertilised prior to launch were disorientated and corkscrewed, but those hatched from eggs *fertilised* in space were content in microgravity. How did they know? Two years earlier, a Japanese visitor to Mir had flown small tree frogs, to see if the suckers on their feet enabled them to orientate themselves in space, which they did. Another tank on the Japanese Spacelab studied the adaptation of two carp, one of which had had its otolithic organs removed. All of these experiments provided valuable information for application to humans because the sensory apparatus is common; namely, tiny hairs in the inner ear with a calcified node at the far end, whose reaction to stimulus provides a sense of direction.

The follow-up Space Life Sciences mission employed the Extended Duration Orbiter to sustain a 14-day programme. This time the AHF housed 48 rats. They had been dosed with tracers so that the process of adaptation to microgravity could be followed in greater detail. This time six of the rats were to be dissected *in space* so that their tissue and bone could be sampled *in situ*. Some of the others were to be killed immediately after landing, so that the effects of descent could be isolated. The rest were to be killed at intervals thereafter in order to snapshot the process of readaptation. Martin Fettman, a veterinarian, was included in the crew to undertake this delicate task. He used a guillotine to kill the rats by decapitation. The blood oozed from the carcass and floated free, so the dirty job was done in the Glovebox. If it ever proves necessary to carry out an invasive medical procedure on an injured astronaut, *that* will be the day when space medicine comes of age.

The Army's Institute of Research based at the Walter Reed Hospital had developed the Space Tissue Loss (STL) experiment to study the effects of spaceflight on the body. Rather than study animals, it had built a middeck apparatus to grow cultures of live cells. The great advantage of this approach was that it was possible to study the cells directly, rather than as part of an organism. In addition, because the cells were in a culture it was feasible to inject drugs. This highly automated apparatus, tested as a secondary payload on STS-45, grew both bone and muscle cells, and determined parameters such as those of cell shape and membrane integrity. The effects of enzymes were recorded *in situ*, then the cells were fixed to preserve them for return to Earth. On STS-53, in addition to the ongoing studies of muscle disintegration and bone demineralisation, the STL apparatus was used to investigate the response of white blood cells (associated with the immune system) to antigens produced by infectious agents and tumours. It was found that muscle growth in microgravity is impaired by disruption of the ability of precursor cells to fuse to make fibres. Just as the bone studies could ultimately lead to a treatment for osteoporosis, it was possible that an understanding of the process of muscle wastage would help devise a treatment for muscular dystrophy.

The National Institutes of Health (NIH) introduced a comprehensive suite of rat tests on STS-66 as NIH-R ('R' for rat). In fact, this was an international effort, and one of the experiments was conducted by Moscow's Institute of Biomedical Problems. Each of the two AEM carried on the middeck housed five pregnant rats, so this mission pushed the study of the body's development back to the fœtal stage. Immediately following the ten-day flight, all the rats were taken to a laboratory. The test exploited the fact that a rat has *two* wombs; one fœtus was surgically removed, and the other was allowed to run its full term. The mothers were then dissected. The objectives included studies of placental development as a result of this prenatal exposure to microgravity, the role of gravity in the formation of the optic nerve, the vestibular otolithic apparatus, the immunological system, the regulation of body fluids, the tendon attachment process, and the formation of the muscle spindles which serve as sensory receptors. A second suite of experiments in the STL – the NIH-C – grew cultures of cells from chicken embryos. Cartilage is a connective fibrous tissue, and if it ossifies it transforms into bone. This study of cartilage mineralisation focused on the process of calcification. Previously, on STS-29 as a student experiment, and then again on the Spacelab J1 mission, chicken eggs fertilised at specific times prior to launch had shown that the resulting chicks did indeed develop weak bones. Quail were hatched aboard the Mir space station, but the chicks were badly disorientated.

The Center for Cell Research was established at Pennsylvania State University as one of the CCDS to focus on commercial projects involving physiological testing. It developed the Physiological Systems Experiment (PSE) which flew first on STS-41. This used rats in AEMs to determine whether biological effects induced by microgravity mimicked terrestrial medical conditions sufficiently closely to facilitate testing of new pharmaceutical products. Tumour cells had already been flown to evaluate anti-cancer drugs, but it was now decided to address a wider range of ailments. PSE offered a way to test drugs for bone- and muscle-wasting diseases, organ tissue regeneration and immunological diseases. It became one of those middeck payloads that flew as a low-profile secondary payloads several times a year. Its first flight was sponsored by Genentech, the Californian company which specialised in using recombinant DNA-based products to replicate natural proteins. The second flight was for Merck & Company to test a potential treatment for osteoporosis induced by oestrogen depletion in menopausal women. This could be extended to patients immobilised for months in bed, during which time, just as in space, the load-bearing bones atrophy. The NASA interest, of course, was in any possible application to inhibit the bone-loss that will afflict astronauts on tours of duty on the International Space Station.

There have been several notable cases in which a system is not fully appreciated until it becomes possible to view it from outside. It was no coincidence that the vulnerability of the Earth was not recognised until it was observed from orbit, and the fact that an astronaut on the Moon could mask out the Earth with a thumb held at arm's length served only to reinforce this realisation. The shallowness of the atmosphere that harbours life is strikingly shown in the pictures of the Earth's limb taken from orbit. Without cameras in geosynchronous orbit, the vast scale of the weather system would be difficult to see. The early history of the Earth remained a mystery until spacecraft revealed the cratered surfaces of the other bodies and so demonstrated that the Earth must have undergone similar bombardment. It should come as no surprise, therefore, that an understanding of life's dependence on gravity, as a result of its evolution on the surface of a planet, is being revealed only now that we can observe life in the absence of gravity. Microgravity is highlighting effects that take place on Earth when the body *fails* to accommodate gravity, so this space research is not simply intended to enable astronauts to remain in space for ever-longer periods and then return to Earth without long-term debilitation; it is about understanding ailments and developing new treatments, and so it must be judged in this wider context.

HUMAN

When the newly-established NASA set out to send a man into space, medical specialists suggested numerous ways in which the astronaut might become disabled by the absence of gravity. Although Yuri Gagarin's single orbit robbed NASA of the glory of being first, the cosmonaut's survival was welcome news, because it silenced the prophets of doom and gloom. Clearly, the human body could function perfectly adequately in space, and Alan Shepard's all-too-brief sub-orbital hop a few months later revealed nothing to contradict this view. The fact that Gherman Titov, who spent an entire day in space, reported symptoms of nausea was written off by NASA as an aberration, especially when Gordon Cooper made a similar flight without ill effects on the final mission of the Mercury programme in May 1963. One of the primary objectives of the Gemini programme was to demonstrate that the human body could endure long enough to mount a lunar mission, and when Frank Borman and Jim Lovell spent almost two weeks in space on Gemini 7 in December 1965, NASA was sure that the medical risks of spaceflight had been exaggerated. It came as a shock, therefore, when the first Apollo crew suffered symptoms similar to those reported by Titov. Worse, 'marathon man' Frank Borman vomited on his way to the Moon in Apollo 8. The Mercury and Gemini spacecraft had been too small for their occupants to move around, but Apollo had sufficient volume for its crew to stow their couches. As they performed gymnastics to explore their environment, the unfamiliar motions and viewpoints disturbed their vestibular organs, with the result that they experienced motion sickness. It transpired that, unlike Gagarin, Titov had unstrapped, and tried such antics in the confines of his 2.2-metre wide spherical Vostok capsule. Evidently the human body was more sensitive to weightlessness than NASA had believed. The Soviet Union had sent up a doctor, Boris Yegerov, in 1964, but he was squeezed into the Voskhod 1 capsule alongside two colleagues, so opportunities for studies of motion sickness were necessarily rather limited.

After Apollo, on Skylab, NASA repeatedly set records for exposure to weightlessness. Fittingly, the first crew included Joe Kerwin, a physician. The most serious incident was at the start of the third, and final, expedition, when Commander Gerald Carr suffered a particularly severe bout of sickness within hours of taking up residence; nevertheless, he went on to complete an 84-day mission. Skylab showed that although the human body was indeed initially disturbed, it *adapted* to the space environment. The simple instruments available provided tantalising clues into how the heart, muscles, bones and blood responded, but the fragmented data were insufficient to draw conclusions about the interrelationships.

The major trends were evident, however: the nausea derived from the brain receiving discordant visual and inner-ear otolithic signals. Only about 30 per cent of astronauts suffered any ill effects, but in the worst cases it prompted a brief bout of vomiting. It struck during the first few hours of initiating 'free-floating'. Susceptibility could generally be reduced by restricting sudden head motion, avoiding orientations that offered views likely to induce vertigo, and by refraining from weightless antics. This first phase of adaptation is complicated by the pooling of blood in the upper torso, as gravity is no longer drawing it to the legs, resulting in a sensation of lightheadedness and a puffing up of the face, especially around the eyes. It takes a few days for the vestibular disorientation to abate, evidently because the brain filters out the otolithic signals, and about a week to overcome the pooling of blood in the upper torso, during which time the body increases urination to shed what it takes to be excessive body fluid. It is during this time that the heart comes to terms with the fact that it is no longer pumping blood against the tug of gravity. Astronauts have to ingest extra water to preclude dehydration. As the body adjusts, the capacity of the cardiovascular system *decreases*, and the heart migrates up into the chest cavity. After a month, the body has settled into a state which is compatible with its new environment. It seemed that strict exercise ameliorated the atrophying of the load-bearing muscles, but the most alarming trend was that the slight loss of bone mass which had been detected following short flights evidently did not level out, with the result that after three months in space the astronauts displayed an effect comparable with a bone-wasting disease. Unlike an osteoporosis sufferer, however, the astronauts' bones slowly recovered following their return to gravity.

As the shuttle neared the point at which it would start flying, NASA picked up on what had become known as 'stomach awareness', relabelled it Space Adaptive Syndrome (SAS) so as to encompass other factors influencing the body's adaptation to weightlessness, and worked up a plan to conduct a comprehensive study of the body as a *system*.

By this point, the endurance record was 185 days, set aboard Salyut 6 in 1980. With the launch of Salyut 7 two years later, the record was extended to 211 days. Then it was announced that a cosmonaut doctor would accompany the crew which would attempt to break this record, but it was to be 1984 before Oleg Atkov played his part in pushing the limit to 237 days on Soyuz T10. Atkov had a multifunction biomedical test kit, and by regularly repeating a set of tests he was to track amenable parameters in unprecedented detail. Salyut crews on long missions hosted a succession of visitors for periods of a week. Tests were devised to note – both subjectively and objectively – every aspect of the early phase of adaptation. This programme recorded vestibular, hormonal, chromosomal and immunological changes; the capacity of the cardiovascular system; the heart's rhythm, structure and migration inside the chest; the composition and distribution of body fluid; the capacity of the respiratory system; bone loss and muscle degradation. Other topics included posture; skin sensitivity; sources of physical irritation; changes to the senses of hearing, taste and visual acuity; aspects of brain activity and cognitive function; psychological questionnaires tracked self-assessment of working efficiency, relationships amongst the crew, and with the control centre; and the video downlink was studied by behavioural psychologists to independently assess the state of mind of each member of a crew.

Meanwhile, NASA was gearing up the shuttle flight rate and demonstrating its ability to deploy commercial satellites. Back then, the maximum time that the orbiter could spend in orbit was about 10 days, so the SAS research was limited to the initial phase of adaptation to weightlessness. Upon reaching orbit on STS-3, Jack Lousma vomited, as he had on his Skylab mission. This was only a minor inconvenience, but when Bill Lenoir was sick on STS-5 and the first attempt to spacewalk had to be put off, NASA assigned physicians Norman Thagard and Bill Thornton to STS-7 and STS-8 respectively, to conduct their own tests. It subsequently assigned them both to the Spacelab 3 mission on which the rats and monkeys were flown. Although the shuttle's endurance was severely limited in comparison to the Soviet Union's space stations, with *tonnes* of apparatus in a Spacelab it transformed into a biomedical laboratory.

The SAS programme therefore developed parallel strands. The early subjective study of human reaction to space was supplemented with a detailed study of how the body regulated its physiological processes. As the tests became more sophisticated, testing started prior to launch, and continued after landing in order to study *re*adaptation to gravity. And this in turn was supplemented by a detailed analysis of animal tissue taken during and after a flight. The objective was to monitor the temporal variation of physiological parameters to track *in detail* the body's response to the absence of gravity.

The blood cells are carried along in the blood stream by the plasma, the rich solution of proteins, nutrients, electrolytes, hormones and assorted metabolic wastes. In fact, there are two types of blood cell: the red blood cells – the erythrocytes – that contain the hæmoglobin that selectively absorbs and releases oxygen, and the white blood cells – the leukocytes – that provide the basis of the immunological system. The body has about six litres of blood, and the plasma constitutes more than half of it. Sampling of blood is a very effective method for studying a whole range of factors affecting body chemistry. In the

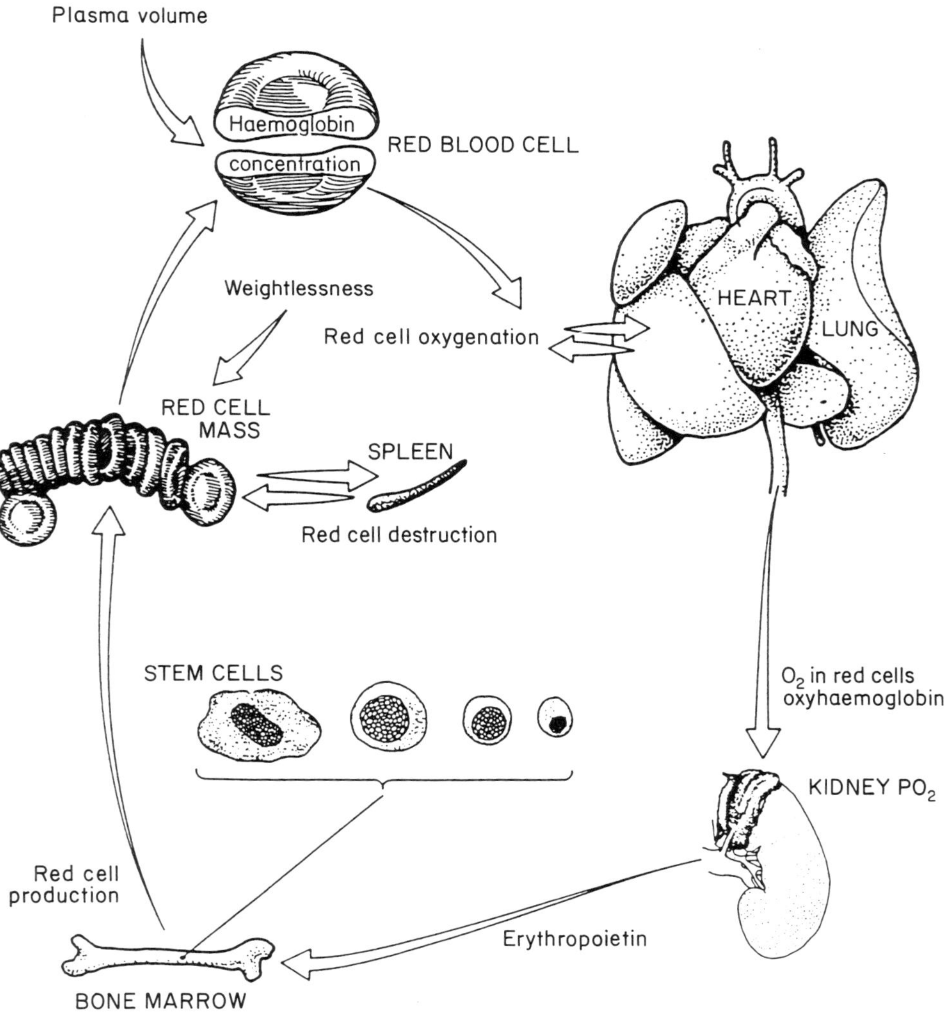

Fig. 8.4. A schematic of the regulation of red blood cell production (erythropoiesis).

cardiovascular system, veins carry oxygen-deficient blood to the heart's right ventricle, and it is then pumped to the lungs, where it exchanges carbon dioxide for oxygen. Pulmonary veins take the now oxygen-rich blood to the heart's left ventricle, which pumps it out through the aorta – the primary artery – so that it can refresh the body's tissue.

The migration of fluid to the upper torso upon entry to weightlessness causes the heart to expand to handle the increased blood flow. The lungs are also sensitive to gravity, which causes ventilation, blood flow, gas exchange and pressure to vary in different parts of these organs. In space, lung capacity decreases, possibly in response to the pooling of blood in the upper torso. As pressure in the arteries rises, nerve cells clustered in the heart, the aorta and the carotid artery, which is in the neck, signal the brain to adjust the heart rate to restore the blood pressure, but because the blood is no longer evenly distributed the only way to reduce the pressure is to reduce the capacity. The body's water balance is controlled by the kidneys, which regulate blood volume and electrolytes content, and remove the waste products discharged into the blood stream by other organs. This kidneys' action is known as the renal system. The hormone-secreting glands work with the kidneys to regulate the body's processes, and selectively secrete hormones into the blood stream in order to control the rate of chemical reactions in other organs. This control action is known as the endocrine system. In weightlessness the renal and endocrine systems increase urination to reduce the total body fluid and electrolytes and, thereby, the total blood volume. This adapted state is characterised by a reduced heart volume, an overall reduction in blood volume but a greater proportion of that which is present within the upper torso rather than in the legs, a slight increase in the rate of the heart when at rest, and a reduction in the heart's ability to respond to strenuous exercise, with a consequent loss of stamina. However, none of these changes impairs cardiac function.

By measuring cardiovascular and cardiopulmonary parameters, and by sampling blood and urine to follow the endocrinic (hormone), erythropoietic (red blood cell) and immunological (white blood cell) evolution, this process can be tracked in detail. Analysis confirmed the earlier observations of an overall redistribution of body fluid and a reduction in plasma, the red cell count and the white cell count. Although there are several types of white blood cell, they all work in the same way in that they maintain a low concentration until they detect a foreign substance in the blood; then they rapidly proliferate to create sufficient antibodies to attack the invader. However, in space, not only is the cell count reduced, lymphocytes (one type of such cell) proved barely able to proliferate when a sample was suitably stimulated, and produced only 3 per cent as much product as the control experiment conducted on the Earth. It is uncertain whether this really means that the body is more susceptible to infection in space. The only direct evidence is that none of the resident crews of the Mir space station has been infected by any of the visiting crews.

In addition to the cardiopulmonary investigation, attention has focused on the effect of weightlessness on the musculo-skeletal system. The load-bearing muscles atrophy because they are not used, and there is an overall decrease in muscle strength. The loss of muscle mass is particularly pronounced in the calves. Cosmonauts have dubbed the weakened state of their upper legs as 'chicken leg' syndrome. However, many cosmonauts return with greatly *enhanced* forearm muscles, and studies with rats have shown an *increase* in the fibres of muscles used for rapid action. Skylab data had shown that the the most significant muscle-loss took place in the first month or so, and that although rigorous exercise

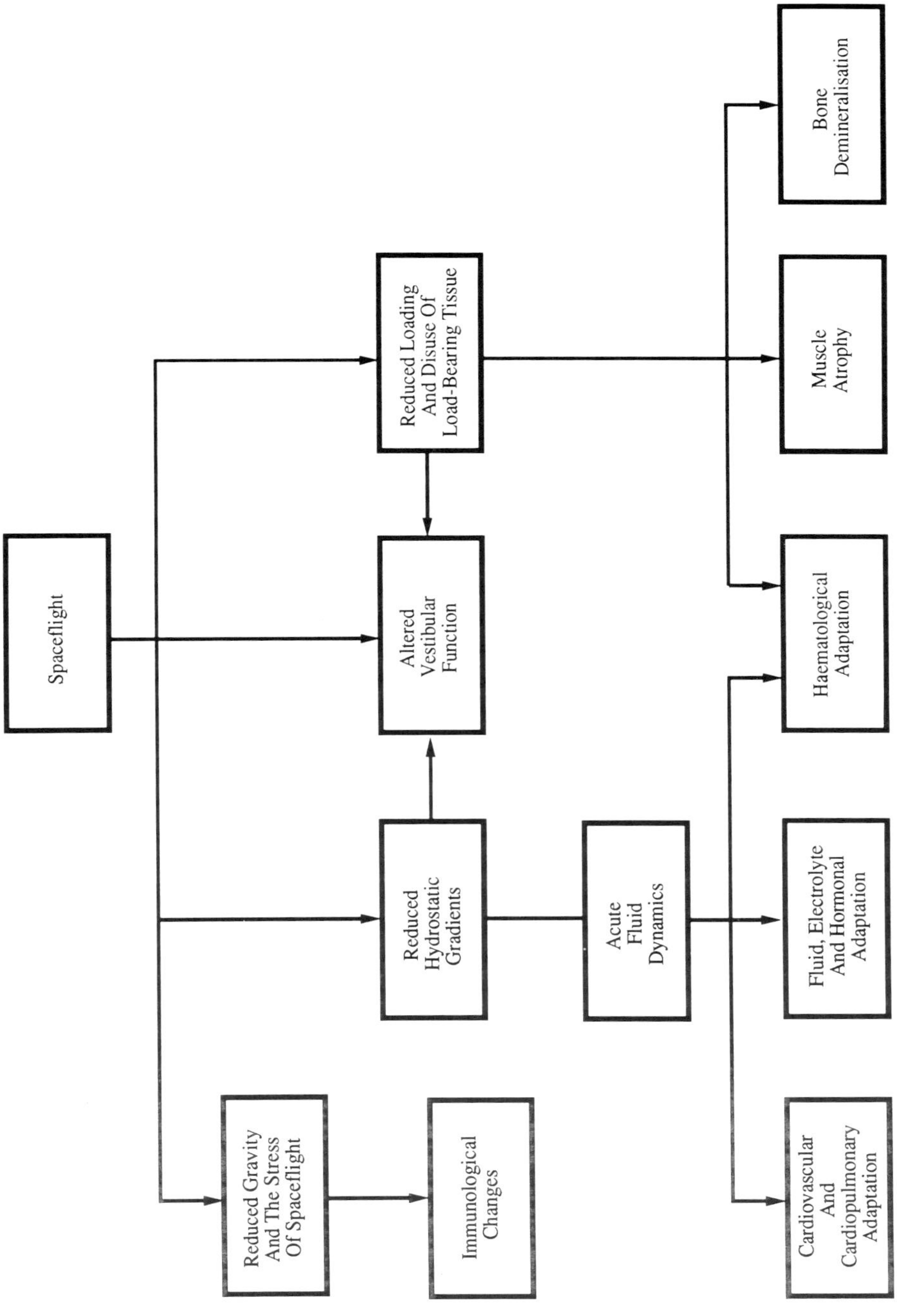

Fig. 8.5. A schematic of the various ways in which weightlessness affects the body.

greatly slowed the atrophying, it did not completely halt it. Load-bearing bones also atrophy in space. This takes the form of a gradual demineralisation in which calcium and phosphorus leach into the bloodstream, and the most significant loss of bone mass occurs in the leg and the spine. As with the blood, bone is in a continual state of regeneration, and this departure from the terrestrial norm is actually the result of a change in the *balance* between the respective rates of processes of production and destruction. Rodent studies had linked this to reduced levels of osteocalcin, the protein which is secreted by bone-forming cells, and this suggested that bone *production* is inhibited. In the longer term, this demineralisation might pose the risk of kidney stones. The bone-loss is like osteoporosis, but much more pronounced. On Earth a sufferer typically loses bone mass at 2 per cent per annum; it is *an order of magnitude* greater in space. But bone loss is not a straightforward correlation with duration. One cosmonaut lost 8 per cent after six months, which was the same as was lost by the Skylab crew in three months; and whereas one cosmonaut lost fully 20 per cent in five months, another lost only 15 per cent in seven months, so it is not an easy effect to study. Once this decalcification mechanism is understood, it may be feasible to make a drug either to inhibit the loss rate or to enhance the regeneration rate.

Each of the various functions undergoes a phase of *acute* adaptation, and then slowly settles down to its adapted state, but each takes a different time to peak. With so many functions simultaneously reacting to the onset of weightlessness, the body's adaptation is particularly acute during the first day. By the end of the first fortnight, most of the acute evolution has run its course. It is another few weeks, however, before the functions which are most sensitive to weightlessness achieve their fully adapted state. After that, the slower-acting functions gradually depart from their terrestrial norm and become the dominant issue of concern.

Thus, even though its endurance was severely restricted, the shuttle was able to make a significant contribution to understanding the body's adaptation by tracing the acute phase

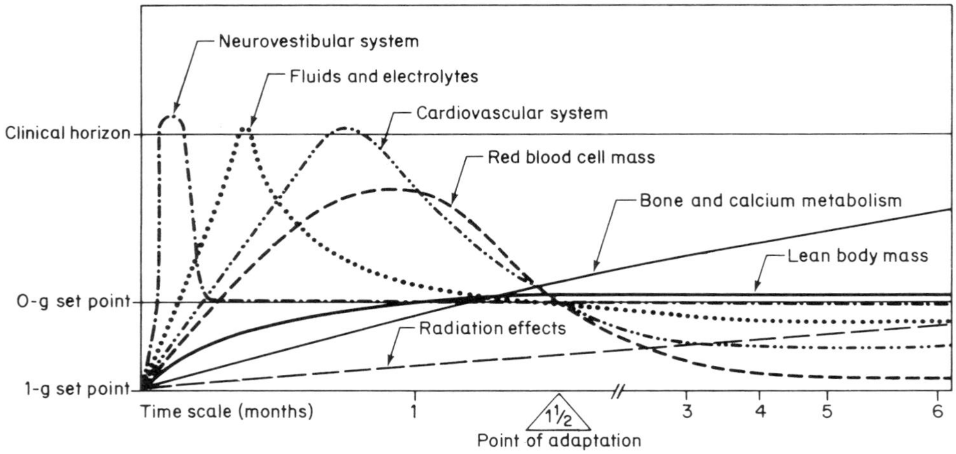

8.6. A schematic of the timescales of the various ways in which weightlessness affects the mammalian body.

in unprecedented detail. With its space station evidently stalled, NASA had to indefinitely postpone follow-up studies of the fully adapted state. But the decision to cooperate with the Russians, which led to the Shuttle–Mir programme, provided an unexpected opportunity to catch up. When Atlantis made its first visit to Mir, it carried a life sciences Spacelab so that Norman Thagard and his two Russian partners Vladimir Dezhurov and Gennadi Strekalov, who had spent *four months* in space, could be subjected to a battery of tests to record their adapted state. Previous work had demonstrated that even the g-forces endured during re-entry of the atmosphere significantly upset the body's adapted state, so this *in situ* snapshot was extremely valuable.

Many questions remain, however. For example, when 'excess' plasma is shed by the kidneys in the initial phase of adaptation, this increases the proportion of cellular material in the residue, so the body gradually reduces the red and white cell count to compensate. How is this done? The seemingly static red cell count actually reflects the result of an equilibrium between two processes, one producing and the other destroying erythrocytes. Erythrocytes are continuously produced by erythropoiesis, a process that takes place within bone marrow. It is regulated by the erythropoietin hormone. Is the destruction rate increased, or is the production rate reduced? If the production rate is reduced, is this due to a signal sent via erythropoietin, or as a side effect of changes affecting the bone? In space, biochemical reactions run at a reduced pace, so how is the regulatory system itself affected? There is evidence that hormone secretion is different, but the resultant effects on the kidneys, blood vessels, and heart have yet to be isolated. It is possible that a deeper understanding of the regulatory system could shed light on related terrestrial diseases, such as excessive blood pressure and heart failure.

In the final weeks of a six-month tour of duty on Mir, cosmonauts drink saline solution laced with electrolytes to build up body fluids, spend an hour a day in their Lower-Body Negative Pressure (LBNP) chamber to draw blood into their legs so as to encourage the heart to increase the cardiovascular capacity, take stamina-enhancing drugs, and undertake rigorous exercise, both to place an increased load on the heart and to develop their muscles.

The first problem encountered upon returning to Earth is that gravity drains blood into the legs. Even though the heart rate increases, the blood *pressure* falls due to the reduction in volume in adaptation to weightlessness. At the very least, this orthostatic intolerance can prompt a blackout. Dizziness resulting from lack of blood flow to the brain is compounded by the transition to an otolithically-driven vestibular system, and by the generally weakened state of the load-bearing muscles. Upon their return, cosmonauts display a distinctive gait. The heart and the sense of balance recover within days, but the muscles take rather longer to recover. To assist the heart to adjust, they wear inflatable leggings designed to enable pilots of high-performance jets to endure high-g turns. They also swim because neutral buoyancy relieves the physical loading. Nevertheless, cosmonauts are usually walking normally within a week. Furthermore, as cosmonaut physician Valeri Poliakov pointed out at a press conference shortly after his return from 438 days (more than 14 months) aboard Mir, his very presence indicated that the time required to readapt to gravity had been *decoupled* from the length of the flight.

As yet, no irreversible effects have been identified, and none of the observed effects are life-threatening, either during the course of a mission or upon return to Earth.

Perhaps the most serious issue facing long-stay crews, or those venturing far from the Earth, will be the radiation environment. Apart from the threat of accumulating a lethal dose of radiation from solar flares, the most serious threat will be from cosmic rays, the heavily-ionised atoms that travel at relativistic speeds. Astronauts have reported 'seeing' flashes of light as these high-energy particles fly through their heads. If such a particle strikes a cell's nucleus, it can damage the DNA within. The ultimate threat to space travellers could, therefore, be unpredictable genetic diseases, and the effects of weightlessness may turn out to have been but a minor inconvenience.

Part 3: Exploration

9

The Hubble Space Telescope

In April 1966, an Atlas rocket sent the first Orbiting Astronomical Observatory (OAO-1) into low Earth orbit. This satellite was instrumented to make precision astronomical observations at ultraviolet, X-ray and gamma ray wavelengths, and it offered an unprecedented vantage point above the atmosphere. Ultraviolet light, for example, is almost totally absorbed by the atmosphere, so to observe at these wavelengths astronomers had been restricted to flying instruments on stratospheric balloons and sounding rockets. It was hoped that OAO-1's instruments would open up new 'windows on the sky'. Unfortunately, just two days into its systems testing, the spacecraft was crippled by a power supply system failure and high voltage arcing in the star trackers, and the experiments were never activated.

The loss of OAO-1 was a terrible blow to the astronomical community, and to NASA. How magnificent it would have been if a Gemini spacecraft could have rendezvoused with the observatory, and a spacewalking astronaut serviced its damaged electronics. In fact, the second spacewalk of the Gemini programme was scheduled for June 1966. A great deal was expected of Eugene Cernan: he was to exit Gemini 9, make his way to a bay at the rear of the spacecraft, retrieve a rocket-powered backpack, don it, and make a series of manoeuvres to simulate crossing over to a satellite. In the event, he floundered about outside, and was unable to don the backpack. It wasn't his fault; the difficulty of simply maintaining a given position outside, let alone doing useful work, had not been anticipated. Repairing a satellite such as OAO-1 was clearly beyond the state of the art at that time; in any case, it had not been seriously considered. The satellite was simply written off.

Astronomers finally obtained the necessary instrumentation to make observations in the ultraviolet from orbit in December 1968, with the launch of OAO-2, which carried 11 UV telescopes, with photometers and scanning spectrometers. It completed the first survey of the ultraviolet sky, and observed specific objects, discovering ultraviolet emission from Uranus, and a huge cloud of hydrogen around comet Tago–Sato–Kosaka. OAO-2 remained in use until February 1973. By that time the third, and final, satellite in the series, OAO-3, had been added, which carried an 80-cm aperture UV telescope. When OAO-3 was named Copernicus, it established a precedent for naming space observatories after pioneering scientists.

THE GREAT OBSERVATORY

The OAO results were so impressive that the astronomers lobbied for a larger telescope – one that would be better than the best ground-based telescopes. To provide the light grasp required to detect extremely faint sources, and the power to resolve their fine structure, this new telescope would require a much larger mirror than that of OAO-3. A 3-metre telescope, it was suggested, would be unrivalled.

Despite the certainty of budget cuts in the post-Apollo era, NASA established a panel to define the requirements for this Large Space Telescope. The report in 1975 was favourable, but it recommended reducing the mirror's diameter to 2.4 metres in order to make the instrument – by then just the Space Telescope (ST) – a more manageable payload. In 1977, Congress authorised the funding, Lockheed was assigned prime contractorship, and Perkin Elmer beat Kodak in the competition for the contract to supply the mirror. It was to be the first in a series of Great Observatories which would, between them, span the electromagnetic spectrum from the far-infrared to the highest-energy gamma rays.

In October 1983 the ST was officially christened the Hubble Space Telescope (HST), in honour of the American astronomer Edwin Hubble. In the early 1920s, Hubble had used the then new 100-inch telescope on Mt Wilson, near Los Angeles, to prove that the so-called 'spiral nebulae' were remote galaxies, as had long been suspected. He then went on to reveal that in general the galaxies were moving away from us at a speed that is proportional to their distance and that this was most easily explained if the Universe was expanding, a possibility which had been analysed mathematically but never considered seriously as a model for reality. Hubble's astounding discovery meant that the Universe had a distinct beginning, an event which the English astrophysicist Fred Hoyle, who hated the idea, was later to derisively dub the Big Bang. Because the speed of light is finite, light from the most distant objects has taken so long to reach us that we see the objects as they were at a much earlier stage in the Universe's development. So, by seeing far (in astronomical jargon, 'deep') into space, we see back in time. For the last 50 years astronomers have been assiduously trying to measure precisely how fast the Universe is expanding, and, thereby, establish the Universe's age.

It was indeed fitting that the ST be named for Hubble because, even though it does not have a larger mirror than many ground-based telescopes, its location above the atmosphere offers sufficient light grasp for it to see much fainter sources than is practicable from the ground, and this, following Hubble's discovery that the Universe is expanding, enables it to see much further back in time. Indeed, by being able to see down to 30th magnitude, this gain in the logarithmic brightness scale corresponds to a *fifty-fold* increase in luminosity, or ability to see faint objects; and this, by the inverse-square law, enables it to see *seven* times deeper into space than is practicable using the best ground-based telescopes.

The HST's primary objective, therefore, was to be nothing less than to see all the way to the edge of the Universe in order to establish its size, its rate of expansion, and its age. So much was expected of it that NASA's Public Affairs Office said that the HST was likely to be "the single most important science payload" the shuttle would ever carry.

YEARS OF FRUSTRATION

The initial schedule for the development of the shuttle called for a test flight in 1978 or 1979, but it was soon realised that this would not be feasible. As the test date progressively slipped, a number of science satellites were either deferred or cancelled outright in order to ameliorate the cost overruns on the shuttle. As the National Science Foundation's favourite project, the HST was safe. Nevertheless, its development was slowed to match the delay in completing the shuttle, and in the long term this extended phasing had the effect of driving up its cost. By 1982, when the National Space Transportation System became operational, an element of reality had set in, but it was still optimistically believed that the shuttle would soon be flying monthly missions.

Organising the shuttle manifest was a tricky task. The HST could not be launched until the TDRS geostationary satellites were operational, because it would rely upon their relays. The first two of these satellites were to have been launched in 1983, but a fault in the IUS stage carrying the first prompted the postponement of the second, and, consequently, of the HST. It took longer than expected to recertify the IUS. The first opportunity to dispatch the second TDRS relay was on flight 51L, in January 1986. That summer's manifest was reserved for observations of Halley's Comet and for the launch of the Galileo and Ulysses spacecraft, so the HST was assigned the first available slot thereafter, but the loss of Challenger, which had been carrying that TDRS satellite, rendered the carefully coordinated manifest obsolete and grounded the entire fleet for 34 months.

DEPLOYMENT

Despite lobbying from the astronomical community for the HST to be launched as soon as the TDRS network was completed and the Magellan and Galileo spacecraft had been sent on their way, NASA was obliged to accommodate the priority payloads for the Department of Defense.

Fig. 9.1. An artist's impression of the Hubble Space Telescope (HST) in orbit, the first of NASA's Great Observatories.

The HST was initially assigned to December 1989, but then it was realised that the orbit of the Long-Duration Exposure Facility (LDEF) was decaying more rapidly than expected, due to expansion of the Earth's upper atmosphere as a consequence of recent enhanced solar activity; and unless its retrieval was brought forward, the satellite and the samples of materials it carried would be lost, so the manifest was revised, with the result that the launch of the HST slipped back to April 1990.

The 1986 HST deployment flight was to have been commanded by John Young, but he had retired soon after the Challenger accident, so Loren Shriver inherited the mission and its crew. The flight of STS-31 was exceptionally smooth, and the 600-km orbit was the highest that any shuttle had yet achieved. Discovery orientated itself upside down and tail first so that when Steven Hawley, operating the RMS, hoisted the 12-tonne spacecraft out of the bay it would not be exposed to the Sun until its thermal regulation system was running. It was not to be deployed until all of its appendages had been unstowed and the engineers at the Goddard SFC had verified its telemetry. Only a jammed solar panel caused any concern, but once this had been coaxed out, the HST was released. Before operational control could be yielded to the Space Telescope Science Institute (STScI), which had been established in Baltimore to manage the HST as an observatory, three months was to be spent first verifying the spacecraft and then calibrating its scientific instruments.

INSTRUMENTS

The telescope had five main instruments. One was set off-axis just behind the mirror, and the other four were in a cluster further back around the central axis. The suite comprised a pair of CCD-technology cameras, two spectrographs and a photometer. The output was initially stored on a high-capacity magnetic tape deck, and subsequently downloaded via the TDRS relay network.

The Wide Field/Planetary Camera (WF/PC), built by JPL, was sensitive across the entire spectrum from the near-infrared to the far-ultraviolet. As its name suggested, it was a dual-purpose camera. Its f/13 focus provided a 2.7 × 2.7 arcmin field of view which, for an astronomical telescope, is fairly wide. The f/30 focus gave a narrower field of view of 1.2 × 1.2 arcmin, with a correspondingly higher magnification, mainly for planetary work. A comprehensive range of filters, prisms, gratings and polarisers could be inserted into the light path to highlight specific aspects of each image. Although the WF/PC could make exposures lasting many hours, in most cases these would have to be built up by summing many shorter exposures, because the field of view might be blocked periodically by the Earth, as the telescope orbited every 95 minutes.

The European Space Agency's Faint Object Camera (FOC) was designed to take maximum advantage of the HST's light grasp and resolving power. It offered focal ratios of f/48, f/96 and f/288, with fields of view of 22 × 22, 11 × 11 and 4 × 4 arcsec. It was optimised for the blue and ultraviolet spectral regions, so was ideally suited to following up the pioneering work done by the OAOs.

The two light-splitters were the Faint Object Spectrograph (FOS) supplied by the University of California and the Goddard High-Resolution Spectrometer (GHRS) built by the Goddard SFC. The FOS was sensitive across the entire visual range, but the GHRS was restricted to the ultraviolet. On the other hand, the GHRS offered much higher spectral resolution within this narrow range. The High-Speed Photometer (HSP), built by the University of Wisconsin, was a photon counter capable of measuring the brightness of an object with a time resolution of 16 μs – ideal for monitoring rapid fluctuations. Taken as a whole, this was an awesome array of instruments.

ABERRATION

The schedule started with step-by-step calibration tests. In the process, instrument data were returned, and no output was more eagerly awaited than the 'first light' images from the cameras. The first picture from the WF/PC was received on 20 May. It was of the open star cluster NGC 3532 in Carina, and astronomers were delighted to see that it had resolved many previously unsuspected double stars. The FOC was brought into action on 6 June, and sent back a picture of the galaxy NGC 7457. It revealed that it had *100 times* more stars in it than had been believed. Although very impressed, the astronomers noted that the output from both cameras was slightly fuzzy, and they took a series of pictures to refine the telescope's focus as part of the ongoing commissioning process. Shockingly, the telescope could *not* be focused. Analysis revealed that the telescope was suffering from spherical aberration: the light from the centre of its main mirror reached a different focus than the light from around its periphery, with the result that there was not a *single* focus. The inescapable conclusion was that the mirror was the wrong figure. Although this was only a tiny difference, it was crucial because it smeared out the light of a star and, with closely spaced stars, resulted in a pattern of interference fringes which corrupted the image.

The HST's optics had been designed to place 70 per cent of the light in a star's image within a circle of diameter 0.2 arcsec. At best, the flawed mirror placed only about 12 per cent of the light on target; the rest was smeared over a 1-arcsec diameter circle. This not only reduced the telescope's ability to resolve fine detail, but by not concentrating the light as intended it would be unable to detect objects as faint as planned. In effect, the flawed mirror had degraded the otherwise excellent 'seeing' derived from its vantage point in space.

A commission of inquiry set up by NASA under the chairmanship of Lew Allen, the JPL director, reported in November that the laser interferometer used to test the primary mirror had been incorrectly installed. The error had been committed in 1981. The mirror had been extremely precisely polished, but to the wrong figure. To save money, no integrated testing had been undertaken, so there had been no opportunity to discover the spherical aberration. This was the same integration problem that had caused so many headaches during the development of the shuttle itself. Eliminating preflight testing was a risky business. Whatever the cost of an integrated test of the telescope's optics, it would be nothing compared with the cost of overcoming the flaw now that the telescope was in space.

Unfortunately, the two cameras, which were to be expected to provide the most readily publicly appreciated results, were the instruments whose performance was most degraded by the mirror's spherical aberration. Nevertheless, for suitable objects computer processing could be used to bring out much finer image detail than could be achieved with the best ground-based telescopes, which made the flawed mirror all the more frustrating. Unfortunately, computer processing could not be used for very faint objects since too much light was scattered by the aberration to permit reliable image reconstruction. Although the spectrometers operated effectively, their output was too esoteric to be widely appreciated. The photometer's performance was unaffected by the mirror, but its light curves were unlikely to impress the American tax-payer. The STScI hastily reorganised its observing programme to make the best use of the telescope's capabilities. Originally, it had been intended to give 40 per cent of the observing time to the WF/PC, but in the revised programme it received less than 10 per cent. Without its promised outstanding pictures, the HST was a public-relations disaster. Instead of being a priceless national asset, it was seen as an enormously expensive folly.

Although NASA could not know it, the Galileo spacecraft which it had dispatched on a mission to explore the Jovian system would not be able to return the planned 50,000 images, because its yet-to-be-unfurled high-gain antenna was stuck. An astronaut could readily pull it out, but Galileo was beyond help. However, there *was* a real chance that the flaw in the HST's mirror could be corrected.

CORRECTIVE OPTICS

The original operational concept was that the HST would be placed in orbit by a shuttle, and periodically returned to Earth for refurbishment. But, upon reflection, it was concluded that this would be impracticable. Two shuttle flights would be needed for each servicing on the ground, and the astronomical community was likely to be denied its telescope for up to a year. To enable the instruments to be upgraded, to exploit the inevitable advancements in the state of the art of detector technology during its projected 15-year operating life, the spacecraft was designed to be serviceable in orbit. This strategy represented a considerable risk, because it required spacewalking astronauts to exchange the bulky and very delicate instruments without damaging the telescope. But NASA had high hopes for astronauts working in space, and, in the event that an in-orbit service proved impossible, it retained the option of returning the telescope to Earth.

Routine servicing in orbit was one thing, but overcoming the flawed mirror promised to be an unprecedented task for a spacewalk. The light path to each of the primary instruments would need to be corrected. The WF/PC was a broad wedge-shaped package just behind the primary mirror, and the other instruments were further back, in telephone-booth sized packages clustered about the axis. JPL had planned to replace its WF/PC on the first service call, so it installed corrective optics directly into its new camera. It was deemed impracticable to replace *all* the instruments, so it was decided to sacrifice the High Speed Photometer in order to be able to install a new corrective optics package that would restore the vision of the remaining three instruments without modification.

The job facing the astronauts was awe-inspiring. The COSTAR and WF/PC-2 packages were to be delivered on a pair of Spacelab pallets. These 300-kg units were to be moved by hand. The 2.2-metre tall axial replacement unit needed to be positioned to an accuracy of a millimetre if it was to mate with its guide rails. After extensive training in the WETF, the astronauts exuded confidence, but others were not so sure that the spacewalkers would not wreck the telescope. In early 1993, Congress considered prohibiting NASA from trying to repair the telescope. The astronauts were eager to try, and NASA relished the prospect of success, but it feared that failure would provide sufficient ammunition for its Congressional opponents to kill the space station that was to rely on astronauts for much of its assembly. To ensure that it had covered all of its bases, NASA asked Joe Shea at MIT to undertake an independent review of the plan. This panel reported in May, and recommended that NASA be allowed to proceed.

In November 1992, NASA had lost the option of just retaining the HST in orbit with its degraded optics. The HST had six gyroscopes for attitude control, of which one failed in

December 1990, and a second 'died' in June 1991. Now a third had ceased operating, and if another failed, the telescope would have to be shut down and locked in a stable attitude. NASA hoped that the HST would survive until the shuttle arrived in December 1993. Even as the MIT panel was considering the triple-EVA plan, one of the three Fine-Guidance Sensors started to misbehave, as did one of the electronic drive controllers for the solar arrays, both of which limited the utility of the telescope. So, with each passing month the scale of the repair mission expanded. The number of spacewalks required increased to five on the 'optimistic' plan, and seven on the 'pessimistic' plan. To cover itself, NASA tentatively assigned a follow-up mission in late 1994 to finish the repair and to rectify any faults caused on the first attempt.

To minimise the risk of being caught out, the tools designed to operate on the telescope were tested on two shuttle missions in 1993. The astronauts assigned the repair spacewalks underwent extremely thorough training in the WETF, to rehearse absolutely every aspect of the task. As a result, when, on STS-61, Endeavour's RMS eased the HST onto the FSS at the rear of its bay and then *every* assigned task was completed, NASA was jubilant.

On alternate days during five lengthy spacewalks, Story Musgrave and Jeff Hoffman, and Kathryn Thornton and Tom Akers, progressively repaired the HST. On the first excursion, a new set of gyroscopes was installed. The following day, the solar panels were removed. One was so badly distorted that it would not roll up, so Thornton cast it adrift. The other was returned for examination. Improved panels of the same configuration were mounted on the telescope, but not unrolled. On the third excursion, the old WF/PC was removed and the new one installed. Then the photometer was removed, and COSTAR put in its place. Finally, a failed power supply on the GHRS was replaced, the faulty solar array drive controller was replaced, and the new solar arrays were unrolled. The fault with the fine-guidance sensor had been overcome, so the plan to replace this had been cancelled. Almost everything had gone to plan: the tools had worked, the bulky instruments had come out on their rails exactly as designed to, and the astronauts had been able to align the new ones for insertion. The stuck solar panel was not a serious problem, and only a problem door on the axial-instrument compartment, that proved tricky to shut, had caused any concern. The HST, whose configuration was dominated by the requirement that it be serviceable by astronauts, had proven itself to be exceptionally well designed.

The corrective optics fully restored the performance of the telescope by focusing 75 per cent of the incident light into the central core of each point image, which immediately added three magnitudes to the cameras' capability. The improvement in resolution was appreciated all the more because it had been denied for so long. Barbara Mikulski, Chair of the Senate committee that set NASA's budget, gave the HST the official seal of approval. "The trouble with Hubble is over", she pronounced. The contingency mission was dropped from the manifest very quietly, as if it had never been there. For the next three years, the HST made a series of startling discoveries, and the STScI was heavily oversubscribed with requests for observing time. Meanwhile, Ball Aerospace in Boulder, Colorado, which had built COSTAR, was fabricating two new instruments.

In February 1997, STS-82 replaced failed gyroscopes, fine-guidance sensors and tape recorders, removed the two spectrometers and installed the new instruments. Once again, the spacewalks went well. When it was discovered that the reflective foil insulation on the side of the telescope which faces the Sun had cracked, an extra spacewalk was made to tape improvised covers over the worn insulation to preserve it until the next visit (scheduled for late 1999 or early 2000). The second service was performed by Mark Lee and Steven Smith, and Greg Harbaugh and Joe Tanner. The man who had originally deployed the HST, Steven Hawley, was on the RMS. The new instruments were the Near-Infrared Camera and Multi-Object Spectrometer (NICMOS) and the Space Telescope Imaging Spectrometer (STIS). The NICMOS, supplied by the University of Arizona, was for extremely-high-resolution imagery, but it also had a spectroscopic capacity. Unfortunately, it suffered problems with one of its three cameras and consumed coolant at such an alarming rate that it was unlikely to run through until the next shuttle visit. Goddard's STIS multi-role camera immediately proved perfect for recording rotation curves of galaxies, giving crucial data about the black holes at their cores and 'dark matter' in their outer regions. Work has already started on an even more advanced multi-role camera with an unprecedentedly dense pixel array to fully exploit the telescope's resolving power. Such instrument substitutions ensure that the HST remains a state-of-the-art facility. Unless something catastrophic disables it, therefore, the HST is assured of a dominant role in astronomy for some time.

OBSERVATIONS

As the "most important" payload the shuttle ever placed into orbit, it is worth examining the HST's output in some detail. On 22 June 1996, it yielded its 100,000th image. In all, it had examined some 10,000 individual objects, ranging from those lying relatively nearby to some of the most distant known objects in the Universe. The object that the astronomers were most eager to take a look at though, no longer actually existed.

TOP OF THE LIST!

The supernova officially designated SN1987A, detected on 23 February 1987, sent the astronomical community into a frenzy of activity. Although it was in the Large Magellanic Cloud (LMC), a small satellite galaxy of our own, it was the closest such stellar explosion to be observed since the invention of the telescope. If it had not been for the loss of Challenger only a few months before it was to have been launched, the HST would have been available to make timely observations of SN1987A. By the time the engineers at Goddard SFC finally released the HST to the STScI, the blast wave from the explosion had been expanding for well over three years. The astronomers were eager to see if the telescope could resolve any structure. Early imagery showed a shell of luminescent gas around the site of the explosion, thereby confirming the spectroscopic data from the International Ultraviolet Explorer satellite (IUE). Despite being degraded, this imagery contained more detail than could be seen by terrestrial telescopes, so the HST was by no means the joke that it was being portrayed as in the popular Press.

The real surprise of SN1987A was that examination of earlier photographs revealed that the progenitor had been listed in a catalogue as Sanduleak –69° 202; a *blue* star with a mass estimated at about twenty times that of the Sun. The more massive a star, the hotter its core, and so the more rapidly it consumes its nuclear fuel. The vast majority of stars are no more massive than the Sun, and they burn for *billions* of years; the Sun has been around for five billion years, and is only half way through its life-cycle. But the rate at which a star uses its nuclear fuel is proportional to the cube of its initial mass. At the rate that this particular star would have consumed its energy, it could not have been more than several million years old. This much was accepted. The real issue was that astronomers had not suspected that such a star could 'go supernova'; or, more precisely, not at such an apparently early phase in its development.

In fact, two types of supernova had been detected over the years, and they were called, rather blandly, Type I and Type II. It had been believed that a large star had to evolve into a red giant before it consumed its last nuclear fuel, thereby triggering the core collapse which led to the explosion. But this had not been a red giant. The other case, the Type I, occurs in a close binary system in which the larger star evolves, and then, as such stars do, ejects a large part of its mass in the form of a stellar wind, before finally shrinking into a white dwarf. Much later, when the originally less massive companion has evolved, and begun to shed mass, sufficient of this is accreted onto the white dwarf to cause its mass to slip over the limit that its structure could support, and *it* detonates as a supernova. But this situation did not apply either; there could hardly be a massive blue star – a relatively young star – in company with an evolved white dwarf.

As luck had it, therefore, just as astronomers believed they would have the opportunity to confirm their understanding of stellar evolution, this strange new type of supernova burst onto the scene. Computer modelling eventually revealed that it *was* possible for a massive blue star to explode because, if the circumstances were right, it could actually be a lot older than it appeared. All of this was determined long before the HST entered service, however. Its introduction enabled the astronomers to see the remnant of the detonation, and its effects on its immediate environs. The site of the explosion was concealed by an expanding nebula, and when this clears it is likely that the collapsed core of the star, in the form of a rapidly-rotating neutron star, will be observed. Before the HSP was removed to accommodate the COSTAR package, it searched for evidence of the regular flashing that would confirm the presence of a pulsar, and although it sampled down to 24th magnitude, at periods as short as 0.0002 seconds, no flashes were found.

Ground-based telescopes had hinted at the presence of a ring around the site of the detonation, but it was difficult to be certain. In August 1990 the HST returned a high-resolution image that clearly showed this ring to be about 1.4 light years in diameter. Unfortunately, the aberration smeared the light from another star in the field across the ring, so it was difficult to resolve any fine structure. Only after the optics had been rectified did it become clear just how badly the combination of the mirror's flaw, and the computer processing developed to correct it, had degraded the image of the supernova with a web of ray-like artefacts. It was not until early 1994, therefore, that the first really clear view was secured. It appeared that there were two other, wider, fainter rings forming an '8', overlapping the inner ring. They cannot be the result of any physical interaction with the supernova ejecta, because the HST was unable to detect *this* until 1996, when it resolved a bipolar dumbbell-shaped nebula expanding at 3,000 km/s. It will be many years before this rams into the surrounding gas, but when it does the shock-heating effect should be spectacular.

But the supernova was merely a timely target-of-opportunity. The HST had a variety of programmed objects to investigate, and these ranged from those which were comparatively nearby right out to those at the furthest reaches of the Universe.

IN OUR OWN BACKYARD

Within the Solar System, the HST was able to follow up on the results of spacecraft that had already visited the planets, and to carry out advance surveys for those still to arrive. It therefore played a key role in NASA's ongoing mission of exploration. However, it was never turned anywhere near the Sun, for fear that it would burn out its cameras in the event that its attitude-control system momentarily lost control, so although it has observed Venus, it has not studied the planet Mercury.

In 1997, Mars made a particularly favourable opposition pass of the Earth at a distance of only 100 million km. The HST was able to resolve the planet with a 22-km resolution (a capability exceeded only by spacecraft that had orbited the planet), and thereby monitor the weather over an extended period as the planet rotated every 24.6 hours. It was spring in the northern hemisphere, so the sublimating carbon dioxide polar cap was caught in the act of retreating to expose the smaller water-ice cap. Later observations followed a dust storm that encroached on the ice. This was the first time that a high-latitude dust storm had been seen. Measurements indicated that on average, the Martian atmosphere was rather colder and with less suspended dust than it had been when the Viking spacecraft had landed more than 20 years earlier. These observations were made to establish the planet's climatic trends in advance of the arrival of Mars Global Surveyor, which was to start monitoring from orbit later in the year.

When Jupiter was suitably placed, the HST was able to resolve detail with a resolution of 250 km. This was adequate to monitor the giant planet's cloud structures. These weather forecasts were used to refine knowledge of the likely conditions at the impact point of the atmospheric probe that Galileo was to release prior to entering orbit. The HST produced the best images of the atmospheric disturbances that resulted from the impacts of the fragments of the disrupted comet Shoemaker–Levy 9. It was able to make ultraviolet studies of the extensive aurorae on Jupiter, and to see the four Galilean moons with a resolution equivalent to the naked-eye view of the Moon from the Earth. It could see sufficient detail on Io to note changes in its volcanically active surface since the Voyagers imaged it in 1979, and was even able to detect several active vents. As for the other moons, it found evidence of ozone on the surface of Ganymede, the largest moon in the Solar System, and a tenuous haze of atomic oxygen on Europa (this latter discovery was significant because it made Europa only the fourth satellite to have any trace of an atmosphere, the others being Io, Titan at Saturn, and Triton at Neptune). Apparently, the oxygen is released as a result of the dissociation of water molecules sublimated from the ice sheet that covers the entire surface of the moon. However, Callisto, on which the HST detected what appeared to be fresh ice, showed no sign of a gaseous envelope. The main item of interest at Saturn was the large moon, Titan, on which, for the first time, detail was observed. Although the resolution was only 360 km, this was sufficient to produce an albedo map in the near-infrared, a wavelength at which the thick nitrogen and methane atmosphere is semi-transparent, and this suggested that the surface is of variable composition; so, if there really are hydrocarbon oceans, it may be feasible to aim Huygens, the lander to be released by Cassini after it achieves orbit around Saturn, onto a shoreline to sample a varied environment. The HST offered the best views of Uranus since the Voyager 2 flyby in January 1986. It could see the clouds on Neptune, and revealed that the Great Dark Spot which Voyager 2 had found had since disappeared. It had been thought that, as with the

Jupiter's Great Red Spot, this massive storm would be a semi-permanent feature. A series of HST images over the years showed that spots came and went. Contrary to expectation, the weather system on Neptune is characterised by a powerful jetstream, and despite being so far from the Sun – much further than bland Uranus – it is very dynamic, probably because it has an internal heat source and Uranus does not.

Beyond was Pluto. A decade earlier, Pluto had been found to have a moon so large that it could more correctly be regarded as a double planet. This moon had been named Charon. On the discovery pictures, Charon had been visible only as a distortion of the outline of the planet's disk. Better ground-based images had managed to separate the two bodies, but the HST was able to resolve them in sufficient detail to produce basic albedo maps. As *no* spacecraft are likely to visit this system until well into the next century, the HST is the best instrument for studying these bodies.

In 1995 the HST observed Comet Borrelly. It revealed the nucleus to be highly elongated (3 × 8 km), noted the rotational period to be just over 24 hours, and suggested that no more than 10 per cent of the surface was active, spewing out gas and dust. Later that year, the HST went on to provide the first direct evidence that a population of comets orbit in a broad disk beyond Pluto, the so-called Kuiper Belt. Although more than 50 large bodies have so far been discovered in this frozen outer fringe of the Solar System, the first in 1992, no dormant comets, which are far smaller, had been detected. The HST took a succession of 10-minute exposures of a single area of sky in this direction. They were processed by computer to delete the fixed objects, so that the remaining objects, of which there were many, could be analysed for movement consistent with orbits in the outer Solar System. The preliminary results were very promising. The Kuiper Belt may be several times wider than the planetary system it surrounds. It is believed to be the source of short-period comets having periods of less than about 200 years. The comets within it are believed to have been ejected there by planetary perturbations after falling out of the Oort Cloud, the roughly spherical shell of primordial material much further out that is believed to be the original source of comets. However, not even the HST will be able to detect such tiny bodies way out there.

Although the HST made a significant contribution to Solar System studies, the real challenges lay beyond, and it found a great variety of unusual objects for theoreticians to ponder.

STARS, CLUSTERS, GALAXIES AND QUASARS

Eta Carinae is a highly luminous blue variable that coalesced as a 'supermassive' (about 100 solar masses) star, ran through its hydrogen 'burning' phase on the main sequence in only a few million years, became unstable, then shed mass in a series of sporadic pulsations which left its exposed core embedded in a dense nebula. The intense stellar wind it radiates excites this surrounding gas, and the HST showed the fine structure of this interaction in awesome detail. Images taken 18 months apart demonstrated that the polar plumes are expanding at 700 km/s. At a distance of 7,500 light years, this corresponded with the outburst of the star which was noted in the 1840s; evidently the plumes were ejected only about 150 years ago.

R136, a dense star cluster at the core of NGC 2070, the Tarantula nebula in the LMC, was studied in ultraviolet by the HST after it had been repaired, and, for the first time, was able to resolve *hundreds* of luminous blue stars packed into a region only four light years across. This corresponds to the distance between the Sun and its nearest neighbour; an observer on a planet around one of these stars would never know a dark sky. In the centre of the globular cluster M15 in Pegasus, the HST found an unsuspected group of 15 extremely hot stars in the central one light year of the cluster. This was a surprise, because a globular cluster is an old cluster of red giant stars. It detected these hot blue giants by observing in the ultraviolet, a wavelength at which all the other stars were extremely faint. The favoured model for their origin exploited the fact that they were bound together in a knot. Were they the exposed cores of red giants which had lost their atmospheres as a result of close encounters? The HST was able to see the hot but faint white dwarfs at the heart of the globular cluster, 47 Tucanae, but only after it had been repaired; it had not been able to detect them when their light had been smeared out by the aberration. The white dwarfs had been expected, and their observation was welcomed. Once added to the Hertzsprung–Russell plot of luminosity against temperature, they refined the age of the cluster. White dwarfs are the burnt-out remnants of stars like the Sun, whose evolution is believed to be thoroughly understood. They must have been created in the first wave of star formation, and we know this because it takes a long time for such a star to evolve to that point. Setting the age of white dwarfs in globular clusters can therefore establish a lower limit for the age of the Universe. These observations show that the Universe is at least 11 billion years old.

In another globular cluster, NGC 6624 in Sagittarius, the HST found the optical counterpart of 4U1820-30, an object that had been unmatched since being first noted by the Uhuru X-ray satellite in the early 1970s. The field was unremarkable in visible light, but in the ultraviolet the HST found a star blazing brilliantly at exactly the right position. This turned out to be a white dwarf with a neutron star in an 11-minute orbit. The continuous ultraviolet emission is due to the infall of material onto the accretion disk surrounding the neutron star. From time to time, however, material from the disk falls onto the star, and the flare as this material ignites produces an intense burst of X-rays.

The Cartwheel galaxy in Sculptor was known to be a disrupted spiral, but by observing it in the ultraviolet the HST was able to observe dense clumps of star formation along the ring surrounding the core, as well as in the spokes that link this ring to the nucleus. Particularly intense star formation at the base of the spokes suggested that the spokes were falling back, but the ring had escaped. The HST image revealed the structure to be the result of a collision between a spiral (target) galaxy and a smaller (projectile) galaxy which had scored a direct hit leaving behind the ring of gas which shows extensive star formation. Star formation in the nucleus, in contrast, was notable by its absence.

A single image of NGC 4261, an elliptical galaxy in the Virgo Cluster, was sufficient to confirm what had been suspected for 30 years. There really *was* a dark doughnut-shaped disk of gas and dust encircling a bright core at the very heart of the galaxy. This had been predicted by the standard model for active galaxies, in which an accretion disk surrounds a supermassive black hole. The associated polar jets of extremely hot gas excite the chilly intergalactic gas, and in fact, it was the radio emission from these jets that first revealed the active nature of these otherwise seemingly normal star systems. Upon imaging the centre of the companion of M87, the giant elliptical galaxy in the Virgo Cluster, the HST saw a *double* nucleus and, intriguingly, found that neither component was at the dynamical centre of the system; the two knots of mass were some 40 light years apart, and in orbit around a common centre. This double nucleus was suggestive of collisions early in the history of that galaxy, and the dynamical behaviour indicated the presence of a black hole at its heart. Incredibly, it seemed that this object must contain a mass equivalent to 2.5 billion solar masses within a volume only slightly larger than the Solar System. Another surprising discovery was made in the heart of M31 in Andromeda, the main spiral galaxy in the Local Group; it, too, had a double nucleus. The clump previously identified as its nucleus is really some five light years from the dynamical centre, and there is a fainter knot on the far side. It seemed that a satellite galaxy had long ago fallen into the heart of M31, and was now circulating in tandem with the original core around a central black hole. Unlike the black holes at the centre of the active galaxies, that in M31 is dormant. A pleasant surprise was the discovery of clouds of hydrogen surrounding normal galaxies, extending out to 20 times the radius of the stellar structure. Rotation curves for spiral galaxies had shown that their outer regions were moving as if they were within a larger structure rather than on the periphery, so this was a welcome detection. The HST confirmed that the abundance of deuterium and the ionisation state of primordial helium were exactly as predicted by the standard model of the Big Bang.

Almost everywhere the HST looked, it made an amazing discovery. And there was no shortage of long-standing enigmas for it to resolve, especially in the furthest reaches of space.

A LONG TIME AGO, IN A GALAXY FAR, FAR AWAY...

With the HST's optics fixed, key projects needing extremely high-resolution images of remote galaxies could finally be undertaken, but even with its restored light grasp, the HST would need to make very long exposures to record such faint objects.

One project involved surveying the morphology of early galaxies. Three clusters, at distances which would span the early history of the Universe, were selected. The nearest of these clusters, at a distance of five billion light years, had a far higher than expected percentage of spirals, but many were clearly interacting, and fragments seemed to have been ripped off by collisions. Evidently, long ago, when galaxies were more densely packed, in a much smaller Universe, interactions were commonplace. This could explain why spirals are now relatively scarce compared with ellipticals, as well as why so many galaxies have small companions. Although these fragments were small they could easily be seen because collisions caused shockwaves in the hydrogen concentrated in the spiral arms, prompting intense bursts of star formation which made them very luminous. The cluster at a distance of nine billion light years away was a shock, because the only recognisable galaxies were ellipticals. But there were many fragments, which were dubbed 'blue dwarfs' because they radiated intensely in the blue despite being heavily redshifted. The third cluster, at 12 billion light years, had only a few ellipticals, and a quasar. Evidently, therefore, ellipticals are ancient structures; spirals did form early on, but until the Universe expanded sufficiently for the interaction rate to fall they were ripped apart, and relatively few survived.

In 1995 a tiny patch of sky in Ursa Major was observed for 120 hours by combining 342 exposures taken over a 10-day period. The area selected had three virtues: it was out of the Galactic plane, so was clear of the obscuration that limits our view near the plane of the Milky Way; it was uncluttered by stars; and most crucially, it did not overlap any known cluster. The goal was to peer as far into space as possible to see, at last, what the faintest, most remote, and hence, the earliest galaxies, looked like. The result, which is known as the Hubble Deep Field, revealed an astonishingly rich field of 2,000 objects, many of which were irregular 'clumps' of stars rather than galaxies. Analysis revealed that these were gravitationally bound into clusters that were destined to coalesce into galaxies. This was significant, because it indicated that the clumps collapsed out of the plasma left by the Big Bang and that they formed stars even before they merged to create galaxies. Further study revealed that some of the recognisable galaxies were actually further away than the most distant known quasar. This was very significant too, because it was the first direct evidence that galaxy formation predated the onset of the quasar phenomenon. This also correlated with the observation showing quasars in the cores of galaxies. It had long been a matter of debate whether quasars sat at the heart of 'host' galaxies or whether they were a distinct class of object. The name is actually a contraction of 'quasi-stellar radio sources'. When these radio sources were first noted, astronomers looked to see what they corresponded to at optical wavelengths, but they saw only star-like points. In the early 1960s, it was realised that they had redshifts implying that they were extragalactic. As more were found, it was noted that the luminous point-sources sat in a faint nebulosity, but no ground-based telescope had been able to resolve stars in the faint glow. It was therefore a major step forward when the repaired HST revealed that in four cases a quasar was located deep within a faint elliptical galaxy; and it did not escape the notice of the delighted astronomers that in each case there was a companion close by. Perhaps an interaction between the two had triggered the onset of the quasar phenomena. A subsequent study of a much larger sample revealed that not only were the quasars resident within galaxies, but that the range of galaxy types was remarkably varied, and most were indeed in the process of interacting with small satellite galaxies.

The early evolution of the Universe was slowly being revealed, and the HST was playing a key role.

THE BIG ISSUE

Measuring the rate of expansion of the Universe involved the determination of the value of what is called the Hubble Constant. Because this is a measure of acceleration, it requires finding both the distance and speed of recession of a galaxy to obtain a specific measurement, and then sampling over as wide a range of distances as possible. The usual units for H_0 are kilometres per second per megaparsec (km/s/Mpc), where a parsec is 3.26 light years (by definition, it is the distance at which the orbit of the Earth around the Sun would produce a parallax of, or appear to subtend an angle in the sky of, one arcsecond). A problem in determining the value of H_0 is the uncertainty in the measurements; the typical 65 per cent error-bar offered considerable scope for debate.

Astronomers have argued over the exact value of H_0 ever since Hubble discovered that the Universe is expanding, with estimates ranging from 50 to 100 km/s/Mpc. Unfortunately, it had not developed into a trend towards a compromise of 75 km/s/Mpc. Two groups had formed and settled on values towards the extremes of this range, with mutually-exclusive error-bars. Each group used different methods and was extremely critical of the other, and each hoped the HST would support its chain of reasoning. A lot was riding on the derivation of H_0, because a value of 100 km/s/Mpc implied an age for the Universe less than that computed for globular clusters (around 12 billion years); and if this was the case then either the theories of stellar evolution were wrong or the expansion of the Universe had not been as simple as believed.

If the Big Bang is accepted, then three possible outcomes are possible, distinguishable by the overall mass of the Universe. Gravitation serves to slow the expansion. Put simply, if there is enough mass, the expansion will be brought to a halt, and then reversed. If there is insufficient mass, the expansion rate will diminish, but the Universe will expand for ever. The interesting case though, is when there is *just* enough mass to halt the expansion, but it will take forever to do so. For a number of reasons, it is believed that the Universe is finely balanced. For a given initial rate of expansion, the mass required to halt it can be computed. The ratio between the actual mass of the Universe and this special value is known as Ω. The finely balanced case is when $\Omega = 1$, also dubbed the 'flat' universe. $\Omega < 1$ is 'open', and $\Omega > 1$ is 'closed'. But even though it is believed that $\Omega = 1$, there is insufficient detectable mass by at least an order of magnitude, and many astronomers are involved in the quest for what has come to be known as the 'hidden' or 'missing' mass. If there really is *no* mass missing, and if the universe is indeed open, a high value of H_0 will not impose a youthful Universe, but if $\Omega = 1$ then the Universe makes sense only if the value of H_0 is low. Maybe the most amazing fact of all, however, is that $\Omega = 1$ corresponds to an average density throughout the universe of just *one* hydrogen atom per cubic metre.

Discovering the fate of the Universe was no small objective, but no worthwhile imagery could be secured until the HST had been repaired. Thereafter, this work was given priority for observing time.

Crucial to the process of anchoring the distance scale was the period–luminosity law of the pulsating variable stars called Cepheids after the first recognised case, δ Cephei. If such stars could be identified in a galaxy, its distance could be derived. The problem was that individual stars were difficult to resolve at intergalactic distances. It had been Hubble's observations of such stars in the Andromeda galaxy, the nearest spiral to our own, which had first demonstrated the great size of the Universe. Using ground-based telescopes, Cepheids could be used to measure the distances of only nearby galaxies, and in some cases only one or two Cepheids could be identified. For more remote galaxies a series of successively more luminous *standard candles* was used, but the near end of this cosmic distance scale was calibrated by the Cepheids. Firming-up this scale by extending the distance out to which Cepheids could be used was one of the HST's primary objectives. Wendy Freedman of the Carnegie Observatories led one of the competing teams, and Allan Sandage of Hale Observatories led the other. It was more than just the fate of the Universe that was at stake.

Freedman was first with a measurement. This was based on a sample of 20 Cepheids in M100, a spiral galaxy in the Virgo Cluster. This was an amazingly rich sample, considering earlier studies. Varying from 25th to 27th magnitude, the stars had been within the HST's light grasp throughout their cycles, and a series of images had established periods of up to 65 days. Knowing the period gave the absolute luminosity, and this gave the distance. Such a large sample enabled a firm estimate to be made, as M100 lies at a distance of only 56 million light years (±10%). This gave a value of H_0 of 80 km/s/Mpc, and this in turn produced a figure of 8–12 billion years for the age of the Universe. Of course, this was just one estimate. Distance to other galaxies in the cluster would have to be made, to smooth out any peculiar motions of M100 within the cluster, but even then other clusters would have to be measured to eliminate systematic motions, because, on the largest scale, the Universe is a 'seething mass'.

This comprehensive study clearly required years of work. In 1996, Freedman offered a 'mid-term' report: based on a sample of a dozen key galaxies, she had settled on a value of H_0 of 73 km/s/Mpc, which equated to an age of only 8–9 billion years. Sandage, however, using a different set of galaxies (in order to calibrate his Type Ia supernova standard candle) gave a value of 57 km/s/Mpc, which increased the figure for the age of the Universe to 12 billion years. The two teams were slowly converging, but with error bars of 5 per cent they were still mutually exclusive. A team of astronomers at Harvard measured H_0 at 64 km/s/Mpc on the basis of a study of Type Ia supernovae, which, because of the way in which they are triggered, all seem to reach the same peak luminosity. A Lick Observatory team derived an identical value by a thorough survey of globular clusters in the giant elliptical galaxy NGC 5846. Astonishingly, an H_0 of 65 km/s/Mpc did not fall within the error-bars of either Freedman's or Sandage's results, but it did exactly average them.

While the HST was making its observations of the most distant reaches of the Universe, another satellite – one which had merited little public attention – had been precisely measuring the distance to every star in our immediate neighbourhood. The preliminary results of ESA's Hipparcos astrometric satellite, published in early 1997, revealed a flaw in the foundation stone of the distance scale – the luminosity of the Cepheids. Trigonometric derivation of the distances of the local Cepheids showed that they were 10 per cent more distant than had been believed. Fortunately, correcting for the fact that the Cepheids are more luminous works in favour of making sense of the Universe. If the Universe is larger than had been believed, then for any given H_0 it is correspondingly older. Furthermore, more luminous Cepheids meant that the age of the oldest stars in our own Galaxy could be cut to 11 billion years. Everything was beginning to make sense: the Universe was evidently not, after all, a crazy place.

OPERATIONAL STRATEGY

Firstly, the HST is a magnificent success. Even with flawed optics, it made significant discoveries and, once the spherical aberration had been corrected, its results were outstanding. Secondly, it should not be regarded as an expensive satellite that was delivered late and massively over budget, but rather as an unique, and hence priceless, long-term asset.

Although it was not strictly necessary to have the HST launched on the shuttle, it is also evident that, irrespective of the flawed mirror, the rate at which the gyroscopes failed meant it would have been rendered unusable within five years. It was the capability to *service* the HST in orbit that made it viable in the long term. Consequently, it cannot be considered in isolation from the shuttle, and the shuttle cannot be properly assessed without understanding its vital contribution to making the Hubble Space Telescope such a great success.

10

The Gamma Ray Observatory

Prior to the Space Age, celestial gamma rays could be observed only by flying detectors on stratospheric balloons, because the atmosphere absorbs high-frequency electromagnetic radiation, but in April 1961, just after Yuri Gagarin made his pioneering orbit of the Earth, NASA launched Explorer 11. The gamma ray detector that it carried saw no evidence of the continuous matter–antimatter creation that was needed by Fred Hoyle's Steady State theory, thereby undermining it, and this left the Big Bang as the clear favourite for the origin of the Universe.

When the Orbiting Astronomical Observatory was built to fly an ultraviolet telescope in space, astronomers put detectors on the satellite to monitor the level of gamma ray activity on a longer-term basis. The results were sufficiently interesting for a series of specialised satellites to be designed for gamma ray astronomy. The first such satellite was Explorer 48, also known as the second Small Astronomy Satellite (SAS-2). Placed into orbit in 1972, it carried a large spark chamber. In this, gamma rays are detected as a result of the electron–positron production which is induced by their passing near the nucleus of an atom. An electric field was applied across the chamber, so the charged particles which caused a momentary short-circuit could be measured. The spark chamber did not actually produce an image of the sky, but its data could be processed to locate sources sufficiently accurately to enable an all-sky survey to be produced. It identified both the Crab and the Vela supernova remnants as strong sources. The COS-B satellite that was launched by ESA a few years later carried a similar instrument, and this was able to refine the positions of several dozen discrete sources.

Although the fact that intense bursts of gamma rays wash across the Solar System was first noted in 1973, it was discovered in the process of analysing Vela output accumulated since 1967. These Department of Defense satellites had been placed in extremely high orbit to police the various treaties prohibiting the detonation of nuclear devices in the atmosphere and in space. It had been boasted that if anybody went to Mars to test a bomb, the sensitive Vela satellites would detect it. When the satellites reported intense gamma ray flashes from seemingly random points in the sky, it came as quite a shock. In 1979, NASA launched its third High-Energy Astrophysics Observatory (HEAO-3), but despite the best efforts of its gamma ray spectrometer, the nature of the bursts remained a mystery. The Granat satellite launched by the Soviet Union in 1989 carried an international payload for high-energy astronomy, and had several instruments to determine the spectral and temporal characteristics of the gamma ray bursts. The sky appears very different in various parts of the spectrum, and only the most energetic objects emit gamma rays. The Sun, for example, dominates the sky at lower energies but is barely detectable in gamma rays. The problem in identifying the bursters, on the other hand, was that there did not seem to be any corresponding lower energy outburst, so their nature was still a matter of speculation when NASA launched its Great Observatory for high-energy astrophysics, the Gamma Ray Observatory (GRO).

The GRO was conceived in 1978, in the heady days which followed the funding of the HST. Like the HST, it was to be launched and serviced by the shuttle. The original concept was to release it into a 300-km orbit and have it climb to its operating altitude, so an orbital manoeuvring engine was installed. Since it would have to lower itself for servicing, it was decided that the servicing activity would have to include replenishing the propellant supply. At a later date, however, it was decided to compromise, have the shuttle deploy the GRO at 450 km, operate it there, and simply have the satellite periodically overcome the decay in its orbit. It was the most sophisticated satellite ever built for this region of the electromagnetic spectrum. Upon being commissioned, a year after the HST, it was named in honour of Arthur Holly Compton, the American Nobel Prize winning physicist who carried out pioneering work in high-energy physics in general, and cosmic rays in particular.

INSTRUMENTS

The challenge of the GRO's designers was to sense the entire gamma ray spectrum with an order of magnitude improvement in detector sensitivity and spectral, temporal and spatial resolution. The energy imparted to an electron when it is subjected to an electric potential of one volt is known as an electron-volt. All electromagnetic radiation can be described by its characteristic wavelength and, by using the speed of light, in terms of its frequency. But by using Planck's constant it can also be described in terms of its energy.

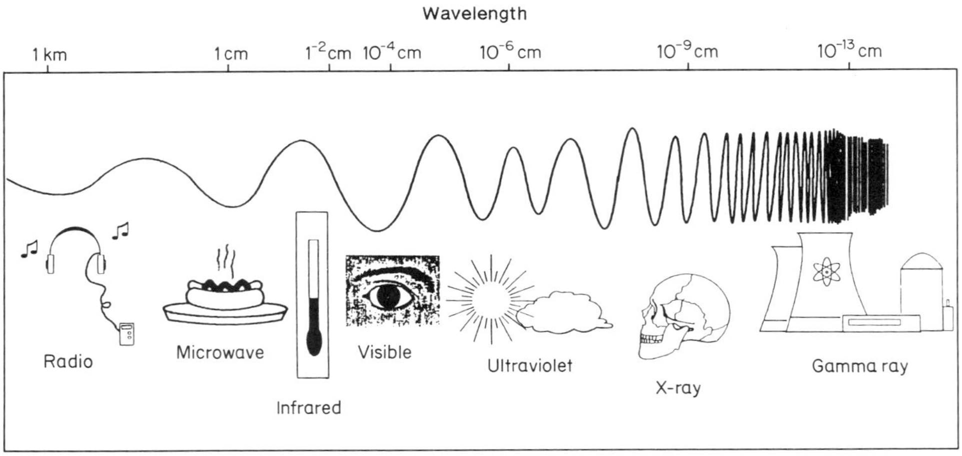

Fig. 10.1. A schematic of the electromagnetic spectrum.

For visible light this energy is a few eV, but radiation of ever-shorter wavelengths corresponds to a thousand (keV), a million (MeV), a billion (BeV), or even a million million (GeV) times greater. It is more convenient, therefore, to describe X-rays and gamma rays in terms of their energy. In the case of the GRO, the objective was to span the range 50 keV to 1 BeV. Since no single detector could cover the whole of this range, a suite of four instruments was developed.

Although the distinction between X-rays and gamma rays is arbitrary, it is significant in terms of detector technology. The wavelength of gamma rays is shorter than the diameter of an atom, so they cannot be focused using a mirror; it simply absorbs them. In striving for an imaging instrument, the best that can be achieved is to construct a directionally-sensitive detector, track individual photons, and rely on computer processing to construct a picture.

The Oriented Scintillation Spectrometer Experiment (OSSE) comprised four identical phoswich detectors which operated in pairs. It was built by Ball Aerospace for the Naval Research Laboratory (NRL). Being a scintillation device, it measured gamma rays indirectly by the optical flash which is generated within a crystal upon absorbing a gamma ray. Since it was sensitive in the range 0.1–10 MeV, it could see gamma rays released by atoms undergoing radioactive decay, and, because this process creates emission lines, OSSE could determine chemical composition. Unfortunately, by its nature such a detector does not offer very high spatial resolution, so its field of view of 3.8×11.4 degrees set the limit to which it could pinpoint sources. In addition to being used to study energetic processes in discrete objects such as pulsars and quasars, it was a powerful tool for investigating the enrichment of the interstellar medium by supernovae ejecting nucleosynthesised elements.

The Compton Telescope (COMPTEL), which was the GRO's primary imaging system, was built by the Max Planck Institut, Germany. By using two planar arrays of scintillation crystals mounted 1.5 metres apart it was able to achieve an angular resolution of two degrees, which was exceptionally good for a gamma ray telescope. It was sensitive in the 1–30 MeV range, and could distinguish gamma rays from cosmic rays by the manner in which they affected the detector. Because it had a wide field of view, its primary function was to produce a sky survey of unprecedented clarity and detail.

It was hoped that the Burst and Transient Source Experiment (BATSE), supplied by the Marshall SFC, would finally reveal the nature of the enigmatic gamma ray bursters. It used eight wide-field scintillation detectors, one on each corner of the box-like satellite. These were able to monitor the whole sky, yet, by coordinating their output, immediately pinpoint a specific source to within a few degrees. BATSE was sensitive in the 20 keV to 2 MeV range, and had sufficient temporal resolution to trace a light curve for each transient. Some bursts evidently lasted for only a few milliseconds, most lasted for a few seconds, and a few had been observed to last for several minutes. The fastest formed a single pulse, but the lengthy ones tended to degrade into a gradually diminishing flickering after the initial flash. BATSE was to measure these light curves in unprecedented detail.

The Goddard SFC's Energetic Gamma Ray Experiment Telescope (EGRET) was built to detect gamma rays from 20 MeV all the way to the observable limit (30,000 MeV). Any process capable of radiating at extreme energies is likely to occur so rarely that most of the time there will be nothing to see but black sky, but if anything was active EGRET would detect it. Its spark chamber was an order of magnitude larger and more sensitive than any gamma ray detector previously flown. Its primary function was to complement COMPTEL, and map the sky, but it was also to extend BATSE's coverage of the bursters to the most extreme energies.

SPACECRAFT

The GRO was a 3.2 × 6.3-metre aluminium tray some 0.6 metres deep, with the instruments around the edge and on top, propulsion and communications on the base, and a solar panel on either side. The BATSE detectors were mounted on the corners, COMPTEL was centrally situated, EGRET was towards one end, and OSSE was towards the other. With its various appendages stowed, it was almost 8 metres long and 4 metres wide. It was also extremely heavy – some 16 tonnes at release. It was launched on STS-37. When it was being checked out on the end of the RMS arm it deployed its solar panels, but the boom carrying its TDRS antenna remained stuck fast. The antenna was vital because, like the HST, the GRO was to relay its data via the TDRS network. Once two astronauts had manually released this boom, the satellite was deployed.

Fig. 10.2. The deployment of the Gamma Ray Observatory (GRO).

The GRO ran into serious problems in early 1992, when its data storage system failed. Its established method of operation was to collect data for two orbits, store data on tape, and then dump them via TDRS high-rate S-Band. The primary tape unit had failed early on, so it had just lost its backup; there was no tertiary unit. Without the ability to store data, the GRO would be usable only when it was able to send its data to the ground in real time. Unfortunately, the TDRS network provided a relay for only 65 per cent of the

GRO's orbit; over the Far East it was out of contact. NASA repositioned the 10-year old TDRS-A satellite at 85° E to plug the gap, and set up a ground station at Tidbinbilla, in Australia, to restore the GRO's capacity.

If the GRO had been manoeuvred into a high orbit as originally intended, it would have been safe from atmospheric drag, but by May 1993 its compromise orbit had been cut by 100 km. Unfortunately, when it tried to raise its orbit its manoeuvring engine failed, and if nothing was done, the spiralling path would result in this uniquely capable observatory plummeting into the atmosphere. A test was made to determine whether the attitude-control engines could be fired in concert to achieve the same effect, but during the test one of the thrusters misfired and the spacecraft began to tumble. Fortunately, the attitude-control system managed to recover the situation. Time was of the essence. If the orbit decayed too far, the low-thrust engines would not be able to compensate.

NASA balanced the risks of resuming the attempt to raise the orbit, and possibly losing the observatory, against leaving it to decay and gathering data in the interim. It was decided to try again. The apogee was successfully raised in October, over a period of a fortnight, and circularised by a second series of engine firings a month later, which set the GRO up for another three years of operations. As the Sun entered its period of minimum activity the Earth's atmosphere contracted, significantly decreasing the drag on the spacecraft, so it did not need a reboost until mid-1997, at which time its altitude was increased to 500 km, higher than it had ever been.

RESULTS

The GRO's baseline mission was to produce its sky survey while watching out for the transient bursters. This was completed in November 1992. Although the general structure of the galaxy was evident, the survey showed no indication of a central supermassive black hole, which was rather a surprise. Most of the discrete sources turned out to be pulsars, and the rest, randomly distributed across the sky, were quasars. The overall picture was consistent with the COS-B results of the late 1970s. The *real* value of the GRO's survey was in the fine detail.

COSMIC RAYS

In 1911 came the discovery of an energetic type of radiation which appeared to be composed of individual atoms, in various states of ionisation, travelling at relativistic speeds. Since they seemed to come from space, they were called cosmic rays. Their origin was still a mystery when the GRO was launched.

EGRET's first six months of data were used to test the hypothesis that cosmic rays are produced in the intergalactic medium. The data did not support this; it indicated that the flux in excess of 100 MeV was created 'locally', in our own Galaxy. However, a long-running study by the so-called 'fish eye' detector in Utah, which could detect only the rare cosmic rays able to penetrate the atmosphere – those with billions of times greater energy than those sampled by the GRO – had found no such bias towards the Galactic plane. The energy that a charged particle picks up by accelerating in a magnetic field is dependent on the scale of the field. It seemed that the highest energy cosmic rays were accelerated in the vast expanses between the galaxies, while those with lower energies were accelerated in the more compact fields within the Galaxy. The material in the plane of our Galaxy radiates gamma rays due to cosmic rays exciting the atoms. The glow in the LMC showed that the flux of cosmic rays is the same there as it is here. A finer-resolution tool for measuring cosmic ray flux at specific locations within the Galaxy was demonstrated when OSSE found gamma ray emission lines from cosmic rays exciting carbon and oxygen atoms in the Orion nebula. A comprehensive survey would reveal 'hot spots', and possibly even reveal where the cosmic rays are being accelerated.

SUPERNOVA REMNANTS

Early on, the GRO looked for pulsed gamma rays from the supernova in the LMC, but found nothing; if the explosion produced a pulsar, its emissions were being masked by the nebula. OSSE did see emission lines characteristic of the radioactive decay of a very short half-life cobalt isotope in the nebula. It also found a titanium isotope in the Cassiopeia A remnant, formed by a supernova 350 years ago. In the much older Vela supernova remnant this was absent, but there were lines of a much longer-lived aluminium isotope. As well as enabling ancient supernova remnants to be 'dated', the study of radioactive decay products can check up on models of nucleosynthesis.

BLACK HOLES

In June 1992, BATSE saw a series of low-energy pulses from an object in Aquila that had been noted by an earlier satellite. Because its outburst did not have the harsh spectrum of a burster, it was dubbed a 'soft gamma ray repeater'. It was only the third such object to be discovered. Classification was easy, but figuring out what they were was far more difficult.

In 1993, however, Shrinivas Kulkarni, of the California Institute of Technology, and Dale Frail, of the National Radio Astronomy Observatory, used the Very Large Array (VLA) in New Mexico to locate the radio counterpart of SGR1806-20, one of the other repeaters. It seemed that the gamma rays originated from a young pulsar in the G10.0-0.3 supernova remnant in Sagittarius. Kulkarni used the Palomar Observatory to identify it in the infrared. It was a long way off, somewhere in the heart of our Galaxy. The other object, SGR0526-66, appeared to be associated with a much more distant supernova remnant in the LMC. There the matter rested until, over the course of March–April 1994, the VLA witnessed a pair of relativistic gas plumes emerge from the Aquila object. Evidently, hot gas was being driven out along the polar axes of the accretion disk system by the compact object within. A few months later, the GRO detected a source coincident with a newly-discovered X-ray nova in Scorpius. Eclipses in the light curve confirmed this to be a close binary system. Follow-up observations by the VLA detected relativistic plumes.

The soft gamma ray repeaters had finally begun to make sense. In the favoured model, the polar plumes were a by-product of the outburst which resulted from an accretion disk dumping material onto a neutron star. Close binary star systems possessing accretion disks are common. It was now clear that the energy of the outburst activity which they undergo is dependent on the nature of the star within the disk. In the classical novae it is a white dwarf; in the more energetic soft gamma ray repeaters, it is a neutron star. Any system in which there is a stellar-mass black hole could be expected to be even more violent at high energies.

Ever on the alert for the onset of a transient burster, on 5 August 1992, BATSE detected a comparatively slowly brightening object in Perseus. Over the next few days it gradually became the brightest object in the sky in the 20–300 keV range. It was soon spotted visually and at infrared at a point in the sky at which there had been no recorded object. It peaked a week later, and slowly faded over the following three months. Although it was clearly a nova, the fact that it blazed at high energies indicated that a compact star, rather than a white dwarf, resided within the accretion disk. Because they radiated in X-rays when in outburst, they had been dubbed soft X-ray transients so, by analogy, this was a 'soft gamma ray transient'. It was catalogued as GROJ0422+32, and was put on the list of binaries suspected of containing stellar-mass black holes (this was a very short list, comprising just Cygnus X-1, V404 Cygni, LMC X-1, LMC X-3, A0620-00 in Monoceros, and Nova Muscae 1991). The determination of the orbital period was the key to finding out whether a compact object is a neutron star or a black hole, because the period of the orbit and an estimate of its size yields the masses of the stars. It had already been established that most of the other candidates were indeed black holes.

THE BURSTER ENIGMA

All the time that the GRO was conducting its sky survey (and, indeed, ever since) it was collecting data on the strange gamma ray transients. BATSE was detecting them at an average rate of one per day, so by early 1994 it had data on 1,000 – just sufficient for a sensible statistical analysis.

A close examination of the light curves, energy spectrum and distribution in the sky of the gamma ray transients, prompted Robert Nemiroff and Jay Norris of the Goddard SFC to conclude that they were extragalactic. Establishing that they were randomly distributed in the sky mitigated against them being within the Galactic disk. A random distribution did not prove that they were extragalactic; they could be in an extended halo around the Galaxy, or even be sitting in the Oort Cloud around the Sun. The case for their being extremely remote was based on the observation that the shape of the light curves of the fainter bursters was basically the same shape as the brighter ones, but with the time base stretched by a factor of two. Nemiroff and Norris suggested that this was a time-dilation effect, attributable to the recessional velocities which would apply if the bursters were at cosmological distances. If they were *that* far away, their transiency implied that their source must be physically small, because an extended object cannot react quickly. Bursts were clearly caused by something that switched on almost instantaneously, prompted an enormous release of energy in a very short time, and rapidly faded. The immediately favoured model involved a pair of neutron stars in a binary system. Such binary pulsars have been detected. Their orbits decay as they radiate energy. It was suggested that a burst occurs when the two stars meet, and coalesce. A neutron star cannot have a mass greater than three solar masses, so coalescence would instantaneously create a black hole. Such an event should produce substantial gravitational radiation, so we might be able to test for it.

The fact that the number of faint bursters declined more rapidly than the inverse-square law was consistent with the time-dilation effect, because it meant that the spread of distances to the bursters was quite narrow. This implied that if the bursters were cosmological, we were unlikely to see any in nearby galaxies. In fact, the very absence of any clustering of bursts in nearby galaxies had been cited by some as clear evidence for their being local to our own Galaxy. Nemiroff and Norris took this to indicate that the bursters were in the far reaches of the Universe.

The astounding implication of the bursts being cosmological was that the redshift effect meant that their radiation, already the highest frequency ever seen, must have been radiated at a higher energy, and so have been even harsher. The crucial fact for the theoreticians was that the amount of energy liberated in a few seconds in such an event, was equivalent to the accumulated output of a star like the Sun during its entire 10 billion-year life cycle. While some astronomers tried to ascertain what could trigger such an outburst, others maintained that it was a crazy dilemma.

In particular, Don Lamb of the University of Chicago questioned the assertion that they were randomly distributed in the sky; the spatial resolution of the GRO's detectors was not sufficient to make such a definitive statement. His analysis of the BATSE data suggested a statistically significant temporal relationship. Although the 38 brightest flashes were indeed randomly distributed, 30 per cent of these had been 'followed' by a fainter flash within a few months. However, Bohdan Paczynski of Princeton argued that this was spurious. BATSE was able to localise to within the advertised 2 degrees only the brightest bursters. Fainter transients could not be located to better than 12 degrees. This was such a large uncertainty that all manner of spurious alignment coincidences were likely to occur for faint sources.

Undeterred, Lamb argued for the bursts being flares of material dumped on the surface of a neutron star – a much less violent event than a neutron star coalescence. This had the advantage that it was repeatable. It was noted that the neutron star did not have to be in a binary system, and that the flare could be even from a comet spiralling in. Others suggested a variety of 'starquake' phenomena in a neutron star, all of which were non-catastrophic, and hence repeatable.

There *was* tentative evidence that bursters might recur. COMPTEL had seen a burst on 1 March 1994 that was on the 'same' line of sight as a burst noted the previous year. Most mysteriously of all, in October 1996, the GRO saw *four* bursts over about two days, all of which appeared to be from the 'same' direction. The Ulysses spacecraft also saw them. The spectrum was not that of a soft repeater. Was this a class of hard repeater, distinct from the catastrophic bursters, or were the hard bursters inherently recurrent?

An astonishing burst was observed by BATSE on 17 February 1994; instead of lasting for a few seconds, it flared for an hour and a half. Only a fraction of the bursts recorded by BATSE are detected by EGRET, but just before this sustained flare finally faded, EGRET recorded a few extremely high-energy gamma rays, one at an unprecedented 18 GeV. This 'superburst' was 1,000 times more powerful than the usual event. Ulysses spotted it, too, but what was it? Unfortunately, the HST was still undergoing systems trials after its repair, so could not be turned to try to find an optical counterpart.

It was not until 28 February 1997 that an optical counterpart was finally spotted. The Italian–Dutch BeppoSAX satellite, which had been launched a few months earlier, was able to record a gamma ray burster's fading X-rays. The higher resolution of the X-ray telescope enabled the direction to be determined sufficiently accurately for an optical search to be made, and within hours the William Herschel Telescope in the Canary Islands found a faint star-like point of light, an observation soon verified by the nearby Isaac Newton Telescope. Analysis of previous survey images of that part of the sky revealed no progenitor, and over the next few days, it faded from view. A week after the event, the New Technology Telescope at La Silla in Chile managed to resolve a very faint extended source, and this was verified by the Keck Telescope in Hawaii. On 26 March, the HST, reintroduced to service after its recent servicing, finally took a look and saw a point-like object embedded in a diffuse background glow. Now we knew what a burster looked like after it had 'popped', but what was it?

It was tempting to draw the line, and say that bursters are indeed exceedingly far away, and thus enormously violent events, but the evidence was still not *conclusive*. The final proof came two months later. On 8 May, BeppoSAX pinpointed another gamma ray burster at X-ray wavelengths, on the 10th the optical counterpart was spotted from the Kitt Peak National Observatory, and the day after the Keck's 10-metre light grasp secured a spectrum of it which indicated a cosmological redshift. Gamma ray bursters *are* enormously violent events in the farthest reaches of the early Universe.

MEANWHILE, BACK HOME

In May 1994, Gerald Fishman, of the Marshall SFC, reported that BATSE had observed gamma ray flashes from the Earth. Correlation with meteorological satellites showed that in each case the flash had occurred high above a thunderhead.

As it happened, NASA had on the shuttle an on-going programme to monitor lightning effects – the Mesoscale Lightning Experiment (MLE). In 1993, to follow-up airline pilots' reports of optical flashes above storms, it flew its DC8 Airborne Laboratory over the most severe thunderstorms it could find, and low-light cameras spotted two types of event, dubbed 'jets' and 'sprites'. The bright blue jets shot up from the storm at a speed of about 100 km/s, but faded away by the time they reached an altitude of 40 km. The sprites were more complex; they lingered at an altitude of 100 km, and resembled jellyfish with a red body 10 km wide and 40 km deep, and with blue tendrils linking to the storm beneath. This phenomenon had probably not been spotted earlier because pilots tend to detour to avoid storms, although emissions above thunderstorms had been detected by sensors on the Salyut space stations. It was the GRO's data which was most revealing, because such electromagnetic discharges to the ionosphere were an order of magnitude more energetic than conventional lightning. The mechanism creating the discharges to the ionosphere was unknown. Clearly, there was a lot still to be learnt about the atmosphere of our own planet.

A JOY TO WORK WITH

Paul Pashby, the GRO's project manager, proclaimed the spacecraft to have been "a joy to work with", despite having lost its data storage capacity. Although the instruments are still functional, the spacecraft is running low on propellant. It has a replenishment system, but no shuttle service is planned.

11

Spacelabs and free-flyers

In addition to designing satellites to be deployed by the shuttle, astronomers devised an observatory that could be kept in the shuttle's payload bay and be used directly by the crew for as long as the shuttle remained in orbit. This ASTRO project was undertaken in the context of the Spacelab programme.

In the beginning, NASA had hoped to be able to operate shuttles in orbit for periods of up to a month, but, given the difficulties it encountered in making the shuttle ready to fly at all, the development of the extended-duration capability was assigned a low priority. It also became apparent that the shuttle would never be able to fly as often as had been hoped, and that the ASTRO work would have to be packed into a small number of extremely intensive missions. With few flight opportunities, it was hoped that repeatedly using a single payload would justify the amortised development cost.

Despite the success of the Orbiting Astronomical Observatory (OAO) satellites, and the International Ultraviolet Explorer (IUE), ultraviolet astronomy was still in a state of relative infancy compared with optical astronomy, so there was every chance of making important discoveries, even on a short flight.

INSTRUMENTS

A battery of optical telescopes was developed for ASTRO. These were instrumented for ultraviolet astronomy, and were aimed together by the Instrument Pointing System (IPS). When this was introduced in 1985, on STS-19, it had carried instruments to study the Sun. In effect, this Spacelab was equivalent to mounting an OAO in the payload bay. The IPS package, with three telescopes, sat on two Spacelab pallets, so there was room at the rear of the bay for another complementary payload.

The Wisconsin Ultraviolet Photo-Polarimetry Experiment was for taking spectra of active galaxies and quasars in the far- and extreme-ultraviolet. It had a 90-cm mirror with a fast f/2 focus, and its spectrograph was optimised for a spectral resolution of 3Å in the 900–1200 Å range.

The Wisconsin University Polarimetric Photometer (WUPPE) had a 50-cm f/10 mirror with a 4-arcmin field, and its spectropolarimeter provided a spectral resolution of 6 Å in the 1400–3200 Å range. Its main targets were hot stars in general, and white dwarfs in particular.

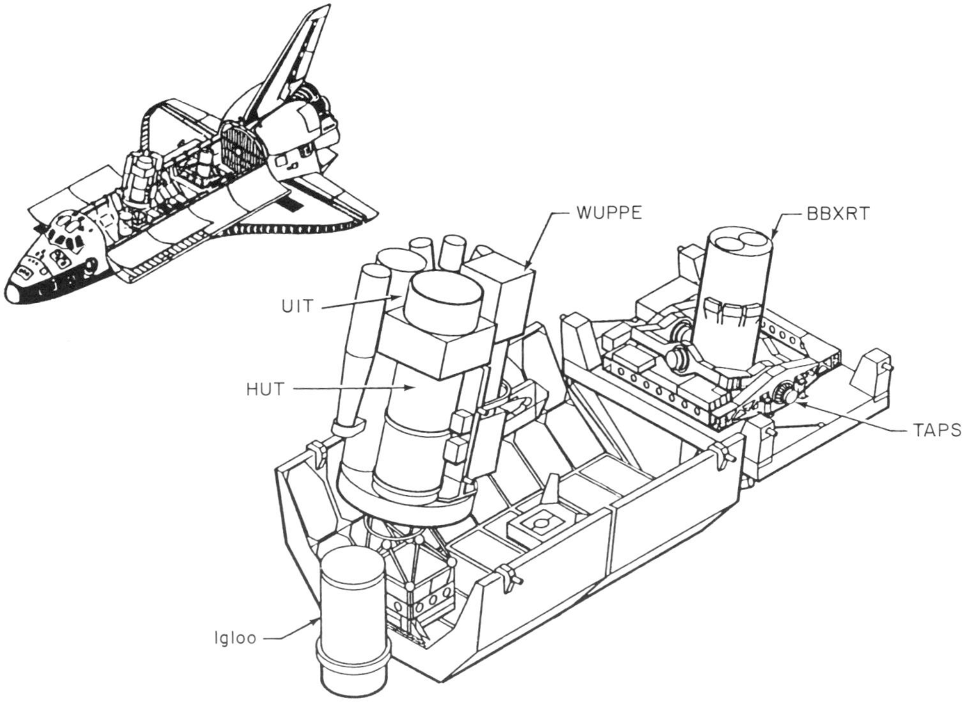

Fig. 11.1. A schematic of the telescopes (ASTRO and BBXRT) as they were installed in the payload bay for the STS-35 ASTRO-1 mission.

The Ultraviolet Imaging Telescope (UIT), supplied by the Goddard SFC, was sensitive across the 1200–2300 Å range. The output of its image intensifier was recorded on 70-mm film, which had a capacity of 1,000 exposures. This 38-cm f/9 camera had a relatively wide 40-arcmin field of view and a 2-arcsec angular resolution. (The IPS was designed for a pointing accuracy of 1 arcsec, but to overcome any vibration from the orbiter which could not be damped out by the IPS, the UIT had its own high-resolution stabilisation system.)

Although small, the UIT had the light-grasp to reach 25th magnitude using a 30-minute exposure. This was sufficient to record anything the other instruments were likely to detect. Its wide field was to be utilised both to document the observing sequence and to search for new interesting objects. Although the Hubble Space Telescope could observe in the same ultraviolet band as the UIT, its extremely narrow field of view (only 2-arcmin at its widest) meant that it could image extended objects only by overlapping a series of exposures, whereas the UIT, with a field wider than the disk of the Moon, could capture most such objects in a single exposure, so it was perfect for surveying the sky.

MISSIONS

In 1984 two astronomers, Sam Durrance and Ron Parise, were named as the Payload Specialists on ASTRO-1, which was to fly in March 1986 to combine its ultraviolet studies with observations of Halley's Comet by the Large Format Camera in the bay aft of the IPS. In October, ASTRO-2 would have deployed the SPARTAN free-flyer, and the next mission, in 1987, was to have released the CRRES magnetospheric research satellite. The loss of Challenger rendered all of these plans obsolete, and when the shuttle resumed operations the ASTRO mission was manifested for mid-1990. The Large Format Camera was no longer needed, so its slot at the rear of the bay was given to the Goddard SFC's Broad-Band X-Ray Telescope (BBXRT), in order to observe the supernova in the Large Magellanic Cloud (LMC).

In the event, the first flight, as STS-35, slipped six months due to persistent problems with hydrogen leaks, and once it reached orbit it was not a smooth mission. No sooner had the observations started than the primary instrument controller failed. Fortunately there was a backup, but this also failed soon after. *Time* was the most critical commodity, so the crew worked hastily to improvise an unrehearsed manual procedure to enable observations to resume. Although, with practice, they proved adept at slewing the IPS onto a target and stepping through the telescope procedure, they were not as fast as the automatic system. It had been hoped to examine 230 individual objects, by picking off a few on each orbit, but the list had to be pared to the top priority items. It is worth noting, though, that if ASTRO had been deployed as an independent satellite, it would have been a write-off. The crew split into two shifts to keep the telescopes running 24 hours a day, and when the results were examined, it was realised that 70 per cent of the objectives had been achieved.

Fig. 11.2. A view of the Hopkins Ultraviolet Telescope (HUT), one of the telescopes of the ASTRO package, set against the constellation of Orion.

In early 1990, before ASTRO even made its first flight, its reflight, set for late 1992, was cancelled. It was reinstated, however, once the results of the first flight were analysed, but with a low priority that placed it far down the manifest. On the other hand, because the Extended Duration Orbiter had finally been introduced, this second flight was to be twice as long as its predecessor. The EDO wafer sat in the rear of the bay, so no secondary payload could be carried. ASTRO-2 went extremely smoothly, and the only problem was a fault in the UIT that ruined several of its images. Overall, this mission was deemed to have been a 98 per cent success.

The ASTRO project was curtailed after two outings, in part due to cost, but primarily because NASA had to sacrifice half a dozen Spacelab flights in order to free shuttles for the Shuttle–Mir missions that emerged from the new spirit of cooperation with the Russians.

VARIED PROGRAMME

The HUT's primary targets were extragalactic: it was to yield high-resolution spectra of active galaxies and quasars. Top of the list was 3C 273, the brightest quasar in the sky, and also a priority was the giant elliptical galaxy M87, in which the HST had recently shown there to be a supermassive central black hole. Although Markarian 66 was just a dwarf galaxy, it was so riddled with star formation that it had been dubbed a 'starburst' galaxy.

By chance, on the day before ASTRO-2 lifted off, astronomers at NASA's Infra-Red Telescope Facility on Mauna Kea in Hawaii were observing Io, Jupiter's volcanic moon, and they detected the onset of a volcanic eruption. The astronauts were alerted immediately, and pointed the HUT to it. The spectrograph was able to record the ultraviolet emission lines from the eruption. Io had been observed by the first flight, so baseline data on the sulphur and oxygen, whose lines are prominent in the ultraviolet, was on file for comparison. One objective of this serendipitous study was to establish whether Io's tenuous envelope was maintained as a direct result of intermittent volcanic activity, or by ongoing sublimation of ices from its sulphur dioxide encrusted surface. The infusion of material into the sulphur torus trailing the moon around its orbit could also be assessed. Jupiter was in any case on the target list to observe its aurora, which was strong in the ultraviolet, for information on the magnetic fields which would be encountered by the Galileo spacecraft when it skimmed the planet to brake into orbit later in the year.

On its second outing, the HUT made a fundamental observation which confirmed a key prediction of the Big Bang theory. In taking the spectrum of an extremely remote quasar it was able to measure an ultraviolet absorption feature attributable to helium. The redshift indicated that the absorption being measured was from intergalactic helium rather than from local helium, and the profile of the absorption feature revealed the density and temperature of the material. It was therefore possible to be sure that this was primordial helium between galaxies at the outer fringe of the Universe. According to theory, the Big Bang should have created a Universe containing, by mass, 75 per cent hydrogen, 23 per cent helium and 2 per cent other elements. Stars begin by 'burning' hydrogen to form helium, and then, if they are hot enough, manufacture all the other atomic elements by the process of nucleosynthesis. This test had been planned for the first flight, but because of the problems there had not been time for it. In the interim, the HST had observed intergalactic helium using the same technique, but its spectral resolution had been insufficient to perform the observation accurately. An entire generation of astronomers had waited patiently for this prediction to be proven.

WUPPE's targets were different. Polarisation reveals the physical conditions both at an emission source and also in the interstellar medium in the line of sight. Some nebulae shine because they emit radiation; others because they reflect starlight. Emission nebulae radiate only in spectral lines characteristic of the elements they contain, but reflection nebulae offer a broad-band stellar spectrum. The nebulosity surrounding the stars of the Pleiades is a well-known example of a reflection nebula. Polarisation of starlight in the ultraviolet indicates the presence of unevenly distributed reflective material, and by measuring both the energy spectrum and the polarisation in reflection nebulae, WUPPE was able to distinguish the different materials in the nebulae by the way that they scattered the starlight. The results showed that there were two types of 'dust': one was composed of grains of graphite, the other of iron-impregnated silicates, all of which had been shed by evolved stars during the later stages of their lives. WUPPE also studied dust in the LMC; studied the active chromosphere of Betelgeuse, in Orion, which, although a relatively cool red giant, undergoes intense flare activity which shows in the ultraviolet, the light of which is strongly polarised by the magnetic fields of the flares; and found evidence of iron-rich disks of dust encircling, and shed by, rapidly rotating blue emission-line stars.

Like the HUT, WUPPE was presented a target of opportunity in the form of Nova Aquilae 1995, which erupted a few days before its second flight. Novae are binary stars in which an otherwise normal star is extremely distended by its close proximity to a white dwarf, and 'dumps' hydrogen onto it. The outburst is due to the infalling hydrogen igniting on the surface of the white dwarf. The eruption is not particularly violent so, unlike the case of a supernova, the star is not disrupted. The infall is repeated intermittently. Ultraviolet observations of novae were welcome because they were so few.

In addition to photographing everything that the other instruments saw, the UIT had its own programme, the most important part of which was to support the 'deep field' work of the HST. Although the very furthest galaxies that the HST was imaging were detected at visual wavelengths, it actually *saw* ultraviolet light which had been redshifted to longer wavelengths as a result of the expansion of the Universe 'stretching' the light, so the structure seen by the HST in such galaxies was not that which *radiated* in the visual, but rather that which shone in the ultraviolet. The UIT's role was to map the detailed structure of nearby face-on spirals, to compile an 'atlas' to be used to help interpret the HST imagery. The features that shone in the ultraviolet were regions of star formation. The HST could not do this itself, because it had such a tight field of view that it would need a mosaic of 250 frames to build up a field which the UIT could capture on a single exposure, and that would not be making effective use of the HST's time. Other key targets for the UIT were globular clusters in our own Galaxy, to find the white dwarfs needed to set a lower limit to the age of the Universe. The UIT, therefore, was serving in a vital support role for the HST in achieving its most important objective.

Secondary targets for the UIT included mapping the thin filaments of the Cygnus Loop, which is the remnant of an old supernova; catching comet Levy as it emerged from behind the Sun; taking the first high-resolution far-ultraviolet images of the Moon to provide data on its surface chemistry; and demonstrating whether such remote-sensing was a viable way of determining the composition of the asteroids. Many objects were being studied in detail at ultraviolet wavelengths for the first time.

The BBXRT was an innovative X-ray telescope. The difficulty in designing a telescope for X-ray astronomy is that the photons do not reflect back from a conventional mirror, but are absorbed. X-rays bounce off mirrors only at shallow angles, so a telescope using such a grazing-incidence mirror tends to have a very long focus, which makes it bulky. Also, the higher the energy, the shallower the angle, so the larger and more polished the mirror. The BBXRT, however, was to spot photons at the softer end (0.3–12 keV) of the X-ray range, so it could use a smaller mirror – and, rather than use a heavy, highly polished mirror, it used thin layers of gold-coated aluminium foil, which made it fairly compact with a 3.8-metre focal length and a 40-cm aperture. In the event, its solid-state detector proved rather more sensitive than expected, so its results were excellent. It was able to focus on the physical processes operating in energetic objects, and provide high-resolution spectra. Its role was to observe the most interesting objects in the survey that had been made by the Einstein Observatory in the late 1970s. Most of its targets were active galaxies and quasars, but it also looked for evidence of hot gas between remote galaxies, to determine whether this accounted for the so-called 'missing mass' of the Universe; it found no evidence of this gas. BBXRT's target of opportunity was SN1987A in the LMC, but it also observed older supernova remnants, including the Crab Nebula. By measuring spectral lines in the X-ray region, it was able to confirm the extent to which nucleosynthesised elements enriched the interstellar medium.

BBXRT flew only the problem-plagued first ASTRO mission. Although it had pointing problems, the fact that it was controlled directly from Goddard meant that it did not impose on the astronauts, who were fully occupied with the ailing IPS package. It was rebuilt, and relaunched in 1993 on the Japanese Asuka satellite to conduct a longer-term survey.

In terms of their scientific yield, the ASTRO Spacelab missions were highly successful, but in the first case this was only because the astronauts, ably assisted by the support teams at the Johnson and Marshall SFCs, were able to overcome severe equipment malfunctions. The inescapable conclusion was that human beings were definitely a key part of the system.

CLEARING UP AN OUTSTANDING MYSTERY

When Endeavour flew as STS-54 in January 1993, its primary mission was to deploy a TDRS satellite. Once this had been achieved, the crew turned its attention to the secondary payloads. Most were on the middeck, but one, the Diffuse X-ray Spectrometer (DXS), was in the bay.

Astronomers had been aware for some time that the sky glows at *soft*, or relatively low energy, X-ray wavelengths, but they did not know why. The problem was compounded by the fact that the source could not be far away – at least in terms of the scale of the Galaxy – because this radiation should be readily absorbed by the hydrogen gas which permeates the spiral arm through which the Sun is passing. DXS was to measure the energy spectrum in the 0.15–0.28 keV range to identify the nature of this diffuse all-sky glow. It was supplied by the University of Wisconsin, which had built the WUPPE detector flown alongside the ASTRO-1 package. It had originally been assigned to fly with the BBXRT as a part of the Shuttle High-Energy Astrophysics Laboratory (SHEAL), but this payload was split up after the loss of Challenger. DXS employed a pair of detectors, one on either side of the bay, and almost immediately after being switched on they yielded anomalous readings. Rather than abandon the experiment, gas was pumped in and heated to try to stabilise the detectors; this worked, and thereafter 24 hours of excellent data were collected.

The results confirmed general suspicions, but also contained a surprise. The glow is due to faint emission from the interstellar medium, heated by expanding supernova remnants. However, two very different signatures were identified: one was seemingly due to a vast, slowly cooling shell of gas ejected by a supernova millions of years ago; the other, hotter shell was a closer and much more recent event. Furthermore, there was evidence to link it to a mysterious object that had first been detected in a gamma ray survey by a Small Astronomy Satellite two decades earlier. Even though it was the second brightest discrete source in the sky, there was no obvious optical counterpart, so the Italian astronomers who had discovered it named it Geminga – for 'missing object'. Suddenly, the pieces of the puzzle were falling into place.

The ROentgen SATellite (ROSAT) was an advanced multinational X-ray astronomy facility launched by NASA in 1990. It saw Geminga immediately, but it was not until 1992 that it was noted to be flashing with a 0.237-second period. The Gamma Ray Observatory confirmed this pulsation rate, at higher energies. Such flashing was a sure sign of a rotating neutron star. That Geminga eventually turned out to be a pulsar was no great surprise, but what was unexpected was that it was spinning so *slowly*. Because pulsars radiate so intensely at high energies, they slow down with age. At three times slower than the Vela pulsar, and seven times slower than the Crab pulsar, this pulse rate indicated that the Geminga pulsar was relatively old. The supernova which made it must have exploded some 300,000 years ago. The gaseous shell that it ejected would be widely dispersed, but it should still be visible, so where was it? Its scale against the sky was dependent on its distance as well as its age and expansion rate. In the optical field, attention was drawn to a blue star near the point of the X-ray object, but there was nothing to confirm that it was the pulsar. However, historical records revealed that the star had a high angular motion across the sky, which implied that it had to be nearby; possibly only 500 light years distant. With this vital clue it was realised that the expanding shell of gas from the supernova had *washed right over* the Solar System.

The local source detected by DXS turns out to be the still-propagating shock-front from the Geminga supernova remnant. It has created a spherical void in the interstellar medium, and, because the Sun is *inside* it, this has been dubbed the 'Local Bubble'. The soft X-ray flux is able to cross such long distances because the gas density is reduced in this cavity in the interstellar medium. This transparency also means that stars inside the cavity can be observed at extreme-ultraviolet wavelengths without the interstellar medium attenuating the radiation, and this is a real boon for astronomers.

The passage through the Solar System of the shock-front leading the supernova remnant would have had interesting terrestrial effects. While at its peak, the ionising radiation would have significantly depleted the Earth's ozone layer. It has been speculated that supernovae might prompt a variety of effects, ranging from triggering ice ages to extinguishing most of the life on the planet, so we could well be in trouble were one to explode nearby. However, only massive stars can produce supernovae, and at the moment there are none in our immediate vicinity. Although the supernova which created the Crab Nebula occurred barely 1,000 years ago, and that in Vela 10,000 years ago, they are so far away that their ejecta will never engulf us. Nevertheless, it has been computed that were a supernova to occur 33 light years away, the Earth's atmosphere would be totally stripped of ozone, land life would be fried, global fires would break out, and all manner of chemical and climatic effects would ensue.

Following up on this DXS data, Jeff Hester of Arizona State University closely studied HST imagery of the filamentary structure of the Cygnus Loop. This is a supernova remnant which spans a 3-degree field because (at 2,500 light years) it is relatively nearby. He was able to confirm that the brightest filaments *are* due to a supersonic shockwave expanding through the cool gas of the interstellar medium, with hot gas following close behind the shock-front. In fact, the shock-front is so thin that it is detectable only when viewed edge on. Also known as the Veil Nebula, this shell has been expanding for 25,000 years. It was imaged by the ASTRO ultraviolet telescopes, and the gas was found to be at a temperature of 100,000 °C. It was the unprecedentedly high resolving power of the HST which made possible direct observation of the thermal structure of the mechanism in sufficient detail to prove that it precisely matched what had been predicted by theory. It provided a working model for the process at work in the Local Bubble. The DXS results demonstrated that it was not just the high-profile payloads carried by the shuttle which made important discoveries.

FREE-FLYERS

In addition to the large telescopes of the Spacelab, which relied on human operators for their flexibility, a number of astronomical instruments were built into small satellites which could be released and retrieved during a shuttle mission; but this type of instrument had, of necessity, to operate essentially autonomously and follow a predefined programme.

SPARTAN

The Shuttle-Pointed Autonomous Research Tool for Astronomy (SPARTAN) programme was conceived in the mid-1970s to develop a platform that would be carried by the shuttle as a *secondary* payload, released to conduct observations and then retrieved and brought back to Earth. It would carry instruments which would normally have been developed to be flown on sounding rockets. To a community accustomed to observations lasting only a few minutes, a two-day run represented a considerable improvement. The project was managed jointly by the Naval Research Laboratory (NRL) and the Goddard SFC. SPARTAN contained its own battery for power, thermal control and data storage, and upon being oriented by the RMS and released, its attitude-control system supervised a series of reorientations with a 3-arcmin pointing accuracy to undertake a pre-programmed observational programme.

SPARTAN-101 flew in June 1985. NRL's X-ray telescope successfully mapped emission in the direction of the centre of our own Galaxy, and from hot gas between the individual members of the vast supercluster of galaxies in Perseus.

Although its bus was designed to be reusable, SPARTAN-101 was sent to the Smithsonian Air & Space Museum in Washington because Goddard had a new bus ready. The 100-class bus could accommodate only 140 kg of payload; its successor could carry 500 kg. The first 200-class bus, which was designated SPARTAN-203, had two ultraviolet spectrographs built by the University of Colorado to study Halley's Comet. It was loaded aboard Challenger to be launched in January 1986 as flight 51L, but was destroyed. Following Return-To-Flight, SPARTAN-201 had a series of outings with sophisticated instruments to observe the Sun, and others carried an assortment of other payloads, none of which related to astronomy.

ORFEUS

Germany had sold the original SPAS free-flyer to the Department of Defense in order to fund the development of a new heavyweight version. Of its 3.6-tonne mass, fully 50 per cent was payload. In its expanded form, the SPAS was able to sustain up to a week of independent activities. A 50-Gigabit tape unit was fitted to store observations during this extended time, and an S-Band system relayed telemetry via the shuttle. Two of these new

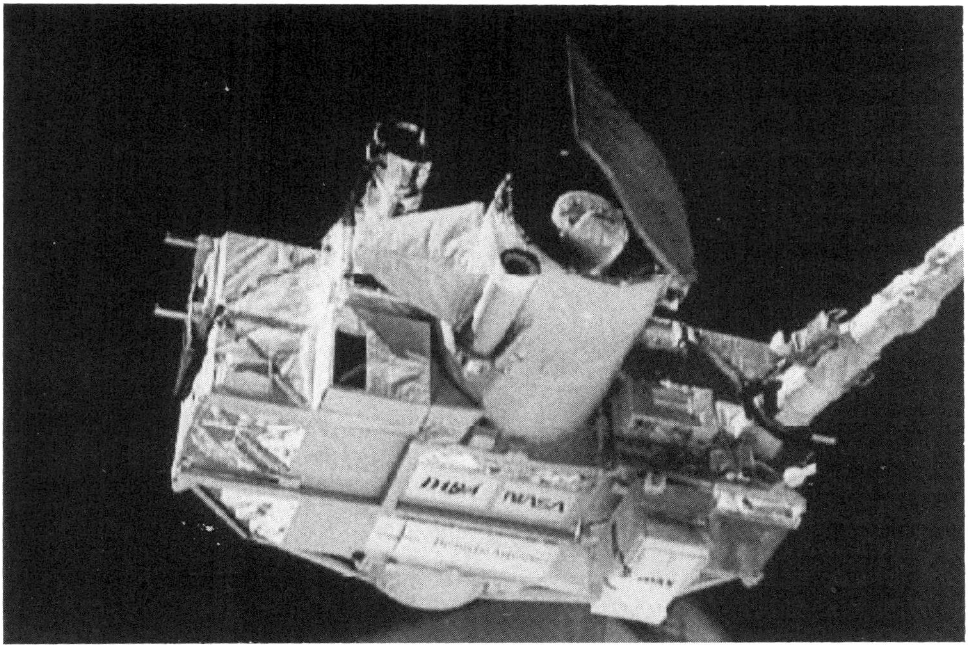

Fig. 11.3. Deploying the Orbiting Retrievable Far and Extreme-Ultraviolet Spectrometer (ORFEUS), one of the two configurations to employ the new heavyweight SPAS free-flyer.

platforms were developed: one was to monitor the Earth's atmosphere, and the Orbiting, Retrievable Far and Extreme-Ultraviolet Spectrometer (ORFEUS) was to observe the ultraviolet characteristics of a wide variety of star types in order to refine models of stellar evolution.

The primary instrument was a 1-metre diameter f/2.5 telescope. A collimator alternated its light between the Far-UltraViolet Echelle Spectrograph (FUVES) built by the University of Tubingen and the University of California's Extreme-UltraViolet Spectrograph (EUVS). Between them, the spectrometers covered the 400–1200 Å range. The rotary mirror of the collimator refused to start, but it worked when the entire system was reset. Diagnosis was complicated by radio interference from other apparatus in the bay, and it was later discovered that the FUVES results were marred by stray light within the telescope.

The spectrometers were to observe the energetic processes on and near young hot stars. The target list focused on those stars which are currently shedding mass by way of a strong stellar wind. The chemical composition, state of excitation, temperature, density and radial velocity of the hot gas can be deduced from the intensity, shape and exact wavelength of its spectral lines. The target list did not consist of only young hot stars, however; cool stars with flares and solar-type stars with rarefied but exceedingly hot coronae that radiate in the ultraviolet were also studied. One target of opportunity was the outburst of a dwarf nova in Sagittarius, a member of a binary system in which one star undergoes a sudden temporary brightening when hydrogen falling onto its surface from its companion is ignited.

As a secondary payload, the SPAS carried Princeton's Interstellar Medium Absorption Profile Spectrograph (IMAPS), sensitive in the 900–1,150 Å range. The advantage of studying the ultraviolet was that this part of the spectrum contained the most interesting hydrogen lines; in particular, the Lyman-α line at 912 Å. IMAPS was able to measure a radial velocity to within 1 km/s. This was sufficient resolution to reveal the fine structure of the hydrogen occupying the spaces between the stars, by its absorption of light from more remote objects.

By combining the stability of an independent satellite with the low development cost of tightly focused objectives and a limited mission duration, the free-flyers proved invaluable, and the International Space Station will hopefully enable free-flying operations to be greatly expanded.

12

Forthcoming attractions

To complement the shuttle, which would test astronomical instruments on brief flights, NASA proposed what it called the *Great Observatories*, a series of high-cost, big-science telescopes to revolutionise astronomy. The Hubble Space Telescope and the Gamma Ray Observatory have proven to be extremely capable; the third satellite is about to be launched, and the fourth is in the pipeline.

ADVANCED X-RAY ASTROPHYSICS FACILITY (AXAF)

In the early days, the only way to study X-rays arriving from space was to fly detectors for brief periods on sounding rockets, or to send them up on stratospheric balloons. At best though, these detectors provided only a tantalising glimpse of the sky. The only viable way to make long-term observations was to put the detectors in space. Piggy-back payloads for X-ray astronomy were therefore bolted onto many of the early research satellites. The first Orbiting Astronomical Observatory, OAO-1, in addition to its ultraviolet and gamma ray instruments, carried an X-ray detector.

The first satellite specifically devoted to X-ray astronomy was Explorer 42, the first of the Small Astronomy Satellite (SAS) series of lightweight satellites. Also known as Uhuru, SAS-1 was launched in 1970. It was a magnificent success. In surveying the entire sky, it identified several hundred discrete sources.

A flotilla of satellites followed up on the Uhuru survey. These included SAS-2 in 1972, the Dutch ANS and the British Ariel 5 in 1974, SAS-3 in 1975 and Ariel 6 in 1979. These revealed that many of the discrete objects were binary systems which flared up as unusually energetic novae containing neutron stars. A few binary systems seemed likely candidates to contain stellar-mass black holes. NASA's High Energy Astrophysics Observatory (HEAO) was so big that it required an Atlas rocket to launch it. Its sophisticated detectors enabled it to make great advances, however.

All this work was carried out using wide-angle detectors that relied on analysis of intersecting scans to localise sources, a technique that was simply not capable of providing high spatial resolution. What was needed was an imaging system – a real X-ray telescope. The problem was that a conventional mirror does not reflect X-rays; it absorbs them. An alternative way had to be devised for focusing X-rays, using an unconventional type of mirror. X-rays *are* reflected by a finely-polished mirror if they strike it at a grazing angle of incidence, and, by arranging nested mirrors concentrically, it proved feasible to create an image. Its focus was behind the mirror, not in front of it, so the mirror sat at the aperture. The first instrument had a 0.6-metre diameter mirror with a 3-metre focus. It was launched in 1978 as HEAO-2, and proved so successful that it was named the Einstein Observatory.

Most notable amongst the recent satellites was ROSAT, a multinational facility named for the discoverer of X-rays, Wilhelm Roentgen. It used a sensitive wide-field detector to make an improved sky-survey, and a telescope to make high-resolution observations of specific objects. ROSAT was to have been deployed by the shuttle, but the loss of the Challenger so delayed it that it was offloaded onto a Delta rocket and finally launched in 1990.

The Advanced X-Ray Astrophysics Facility (AXAF) will be the third of NASA's Great Observatories. It was to have operated as ROSAT's contemporary, but a safety-aware NASA cancelled the progressive upgrading of the shuttle to increase its capacity, which meant that when the shuttle finally resumed operations it could no longer lift the satellite as it had been conceived. It was cut back to a high-resolution spectrograph and an imaging system, using CCD technology, and slated for launch in 1996. However, when the advanced solid rocket motor was cancelled in 1993, NASA was forced to recast the AXAF as a pair of satellites, each with a single instrument. Congress promptly withdrew funding for the satellite for the Goddard X-Ray Spectrograph, so NASA redesigned it for carriage on a yet-to-be-launched Japanese satellite. With a tight field of view and 1-arcsec spatial resolution, the camera will be the best imaging X-ray telescope ever built. Long after it made the HST's optics, Perkin Elmer sold its mirror-making factory to Danbury Optical Systems, a subsidiary of Hughes, and this won the contract for the AXAF's optics. To be sure that it did not launch a lemon, NASA ordered that the optics be verified with an *integrated* test because, once it has been dropped off in low orbit by the shuttle, the AXAF is to be boosted by an IUS stage into a high orbit that will minimise the area of sky that is masked by the Earth, but which will also take the telescope beyond the shuttle's reach. With each design change, the AXAF suffered additional delay. Its final configuration is a 14-metre long, 3-metre wide tube, with all the spacecraft systems in a ring around the aperture of the telescope and the camera at the focus at the far end of the tube. AXAF is currently scheduled to be launched in late 1998. It should be able to resolve the energetic X-ray sky with a clarity comparable to the optical view from the HST, which should finally relieve the handicap that has been the bane of X-ray astronomy.

SPACE INFRARED TELESCOPE FACILITY (SIRTF)

Electromagnetic radiation at wavelengths beyond the blue end of the visible spectrum is completely absorbed by the atmosphere so, to carry out high-energy astronomy, detectors have to be placed in space. Radiation at wavelengths beyond red in the visible spectrum is transmitted by the atmosphere, but only selectively so, because the various chemical constituents of the atmosphere absorb in it characteristic bands. Not only that, the atmosphere re-radiates in the infrared much of the energy it absorbs in the visible, for neither is it completely transparent in that part of the spectrum.

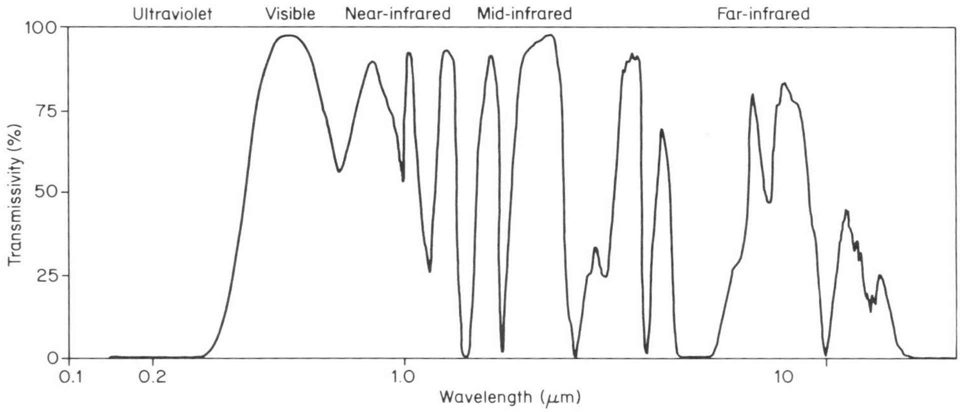

Fig. 12.1. A schematic of the transmissivity of the Earth's atmosphere in the infrared region of the electromagnetic spectrum.

Consequently, infrared astronomy *can* be performed by ground-based telescopes, but only in those wavelength bands in which the atmosphere is transparent. Although a conventional mirror can focus infrared, a telescope still needs advanced technology. To detect a faint celestial source through the atmosphere, when the atmosphere is aglow and the telescope is hot, is comparable to attempting to see a firefly against the glare of an arc lamp. To see through the atmosphere, astronomers cool an infrared detector to as low a temperature as possible and rock the telescope back and forth across the target in order, quite literally, to *subtract* the sky. Furthermore, because water vapour is a strong absorber across almost the entire range, infrared telescopes are placed only on the highest mountain observatories, above as much of the troposphere as practicable.

Astronomers were therefore understandably eager to put an infrared telescope in space. This offered two advantages; firstly, the detector could be chilled to a *much* lower temperature, which would make it more sensitive, and secondly, there would no longer be the equivalent of an arc lamp blinding the telescope. The first satellite, the InfraRed Astronomical Satellite (IRAS), was an international effort. It was launched on a Delta rocket in January 1983. Although its supply of cryogenic coolant lasted only a year, the rich detail in its sky survey amazed the astronomers, and there was an immediate clamour for a follow-up.

A small InfraRed Telescope (IRT) was carried on Spacelab 2 in 1985. It was cooled to 10 K by liquid helium, which was sufficient to make its detector 50 times more sensitive than that on IRAS. Whereas IRAS had the conventional narrow field of view, the IRT has such a wide field that even on a brief flight it was able to make a substantial survey of the sky. The IRT was to study *extended* sources spanning a large fraction of the sky. The dust clouds which litter the spiral arms of the Galaxy run in a band across the sky, coincident with the Milky Way, and it was readily able to survey these with a few exposures. A narrower field of view would have required so many exposures to build an equivalent mosaic that it would not have been possible within the time constraints. It could be operated either by the astronauts, or remotely by the ground. On this test, its output was marred by a small plastic sunshield that apparently broke loose due to the thermal stress and partially blocked the field of view. An important result from this test was the demonstration that the shuttle's bay was not suitable for a cryogenic instrument. NASA was developing an IRAS-like telescope with a better detector for the ASTRO Spacelab programme. This was to be mounted on the universal joint of the Instrument Pointing System. It was now clear that if this was operated within the bay, it would suffer thermal interference, and the revised plan was to deploy it on a free-flyer which would operate independently until it exhausted its coolant, at which time it would be returned to Earth for replenishment. However, the basic telescope soon attracted supplementary instruments, and the larger it grew the more attractive it became to leave it in space as a permanent satellite, and service it *in situ*. This would primarily involve pumping superfluid helium into its coolant dewar, and experiments have since been conducted to prove that this is feasible. This Space InfraRed Telescope Facility (SIRTF) was to have an 85-cm diameter beryllium mirror chilled down to a mere 1.5 K, and be fitted with a wide selection of instruments.

Construction of the SIRTF had yet to start when Challenger was lost, so it was put in a state of limbo until the shuttle resumed operations. During that time, NASA decided to offload it to the military's new Titan 4 rocket. Next, the mirror was reduced to 70 cm so that the telescope, around which the dewar would be wrapped, would fit on a cheaper Atlas. In 1993 it was realised that if the SIRTF was placed in a 100,000-km high elliptical orbit the reduction in the heat it absorbed from the Earth would reduce the rate at which the coolant was consumed; and because the dewar could be smaller, the telescope could be restored to its full size. Even though it had been touted as one of the Great Observatories, only at this point were funds awarded to begin hardware development. In 1994, the plan was revised again. To further reduce thermal contamination the telescope is to be placed in *solar* orbit, trailing the Earth by a few million kilometres. The new slim dewar will enable the SIRTF to fit on a cheaper, Delta 2 rocket. The SIRTF is to be eased away from the Earth, to spend a few weeks passively cooling until it is at about 50 K, then the cryogenics will chill the detector to its operating temperature. It will have sufficient coolant to run its instruments for three years.

SIRTF will have two instruments. The Multi-band Imaging Photometer System (MIPS) supplied by the University of Arizona will be sensitive across a wide range, but particularly so in the far-infrared (30–180 μm), a region of the spectrum not available with any sensitivity using traditional silicon detectors because they do not work at the temperatures needed to sense such wavelengths. MIPS will use a new germanium technology that can be frozen, and should be 100 times more sensitive in this range than was previously practicable. Cornell University's InfraRed Spectrometer (IRS) will return high-resolution spectra in the 4–40 μm range. The crucial dewar will be built by Ball Aerospace, which made the unit used on IRAS. The Smithsonian InfraRed Array Camera (SIRAC), which was deleted during the 1994 review, was to have complemented MIPS in the 2–30 μm range, but a similar camera was flown on the Infrared Space Observatory (ISO) launched by ESA in 1995 into a highly elliptical orbit like that once considered for the SIRTF. However, ISO will have run out of coolant long before the SIRTF is launched in 2002.

The primary objective of the SIRTF is to conduct a sky survey with at least 1,000 times greater sensitivity than that produced by IRAS, and it should see down to the effective observable limit. With 0.5-arcsec resolution, its imaging capability should mimic that of the HST. It should be able to see through the dust that masks so much of our Galaxy, resolve previously unobservable structure in galaxies all the way out to the fringes of the Universe, and follow up on the Cosmic Background Explorer (COBE) by measuring anisotropy in the highly-redshifted 2.7 K background radiation left over from the Big Bang.

VLA-PLUS

It will be difficult for the shuttle to make a significant contribution to the other region of the electromagnetic spectrum in which astronomers are traditionally active – the radio region – because the assembly of the large antennas would be a very complex construction task. The radio astronomers already have extremely powerful facilities – notably the Very Large Array in New Mexico – which combine antennas electronically to operate as an interferometer with fine spatial resolution. By combining with a radio telescope on the far side of the planet, the VLA can exploit the longer baseline to increase its resolution. The greatest contribution that a space-based antenna could make, therefore, would be to work in concert with a facility on the ground. The longest baseline would be provided by setting up an antenna on the Moon, but the lunar cycle would complicate coordination with the ground. A compromise would be to place the antenna in geostationary orbit, so that it would remain 38,000 km above the VLA. However, both of these locations are well beyond the reach of the shuttle.

WHITHER THE SHUTTLE

The scientific results from the HST would have remained degraded by the flaw in the mirror if it had not been possible for the shuttle to ferry up spare parts and, most crucially, for the astronauts to carry out the necessary repairs. Even if the mirror had been perfect, the HST's gyroscopes were failing at such a rate that it would not have been able to fulfil its decades-long mission. The gyroscopes apart, advances in detector technology would have progressively reduced the telescope's effectiveness. By facilitating upgrades, the shuttle is therefore a vital part of the HST's overall life-cycle.

Whilst it is possible that the GRO's stuck antenna boom would eventually have shaken itself loose and deployed, it might not have done, and that would have been crippling. Yet it took only a minute for an astronaut to free it. Since being deployed, the GRO has lost its data storage facility and has suffered propulsion problems, but its instruments are working well. Although this spacecraft was also designed to be serviced, and it is low on propellant, NASA is unlikely to pay it a visit.

Having been conceived at a time when NASA was obliged to transfer all of its payloads off expendable rockets onto its shuttle in order to assure sufficient work for the weekly-flying shuttle, the SIRTF was offloaded once it became evident that the shuttle would be too busy to launch and service it. What had started as a telescope for operating in the shuttle's bay had progressively been transformed into a free-flyer, a low-orbit satellite, a high-orbit satellite and, finally, a deep-space facility.

Whilst it is certainly the case that the shuttle has proven itself highly capable when it comes to operating captive packages of sensory instruments for studying the Earth, the

Sun and the sky in the form of Spacelab, it is also undoubtedly true that the shuttle has served only to complicate the development of facilities such as the AXAF and the SIRTF, first by failing to live up to its early hype, and then as a result of the turnabout in operating policy following the loss of Challenger.

The Galileo spacecraft was considerably handicapped by the failure of its large high-gain antenna to unfurl. It was a trivial fault that an astronaut could have fixed in seconds, but Galileo was beyond the reach of the shuttle when the fault developed. All the appendages of the HST and the GRO were extended prior to deployment in order to ensure that, as Great Observatories, these satellites were fit for service. If either the AXAF or the SIRTF develops a problem, even one that an astronaut could overcome, its position beyond the shuttle's reach places it beyond repair. A depressingly large number of satellites have been lost recently in their first few hours of checkout. Let us hope that neither of the final Great Observatories is added to this list of orbital debris.

13

Galileo's ordeal

When NASA set out to develop the shuttle, it had several interplanetary spacecraft in the pipeline. Those that were to be launched before the shuttle was expected to enter service were to be launched on rockets, but those that were to follow were to be *designed* to go up as shuttle payloads.

Pioneer 10 passed Jupiter in December 1973. This Pioneer really lived up to its name. It was the first spacecraft to pass through the asteroid belt *en route* to Jupiter and the first to probe the intense radiation belts within the giant planet's enormous magnetosphere, and its simple scanner sent back the first close-up pictures of the planet, revealing an unexpectedly intricate structure of vast storms. After its brief encounter with Jupiter, it headed out of the Solar System. Its successor, Pioneer 11, following a year later, used Jupiter's gravity to set it on a curved trajectory high above the ecliptic and across the Solar System to Saturn, the next planet out from the Sun, which it passed in September 1979.

These deep-space probes had been designed by NASA's Ames Research Center. It had built drum-shaped spacecraft because it was primarily interested in the particles and fields in space, for which a rotating sensor was most effective. All of the earlier probes in this series had been confined to the inner Solar System, so the imaging system on these later craft was a departure for Ames. Planetary imaging was the speciality of the Jet Propulsion Laboratory (JPL). It had developed the Mariner spacecraft as three-axis stabilised platforms with state-of-the-art cameras, and these had mapped Mars. In the early 1970s, JPL promoted a plan to dispatch a series of Mark 2 Mariners on what it dubbed the Grand Tour of the outer Solar System. It obtained funding for a single mission, but following the standard operating procedure of that time, dispatched two spacecraft in case ill-fortune befell one. These were launched in 1977. In recognition of the travel times – decades instead of years – and the tremendous distances involved, this new programme was renamed Voyager.

Ames had a plan to place one of its spacecraft into orbit around Jupiter and drop a probe into the atmosphere; it called this Pioneer Jupiter Orbiter and Probe (PJOP). In 1975, Ames was awarded funds to start development work. At that time, it was expected that the shuttle would start test flying in 1978, and enter service within 12 months. NASA placed PJOP on the manifest for 1982, but a few months later decided to assign responsibility for planetary spacecraft to JPL, so PJOP was transferred. Although, at that time, the Voyagers were still being constructed, JPL had already begun to plan a follow-up in which a similar spacecraft would go into orbit around Jupiter and thoroughly explore its family of moons. The orbiter proposed by Ames was to have been a spinning craft with limited imaging capability. In order to carry a sophisticated camera, JPL required a stabilised platform; however, it compromised with Ames, and devised a vehicle combining both spinning and non-spinning sections. This complicated the system, but despun antenna mounts had proved effective on communications satellites, so there was little technological risk. The compromise produced a far larger vehicle than either Center would have constructed on its own. To facilitate manoeuvring in Jovian orbit, JPL added a tank of propellant. This transformed it into a heavyweight, but the shuttle was being touted as the space equivalent of a large truck, so this did not look like being a problem.

In 1976 the joint project became Voyager Jupiter Orbiter and Probe (VJOP), which Congress approved in July 1977. JPL decided to exploit the 1982 dispatch to make a close pass by Mars *en route*. This would make the cruise a little longer, pushing arrival back from 1984 to 1985, but the energy picked up from the encounter allowed the spacecraft to carry a bigger propellant tank, which would benefit the primary mission. However, this decision also took the overall mass of the spacecraft and the rocket stage needed to boost it to escape velocity to the very limit of the payload capacity predicted for the shuttle. In 1978, VJOP was renamed Galileo to honour the Paduan scientist who, upon pointing his first telescope at the planet in 1610, was surprised to observe that it was accompanied by four large moons – subsequently named the Galilean satellites.

Even though by this stage it was obvious that the shuttle would not fly in 1978, NASA was confident that it would in 1979, and it assured JPL that it would reorder the manifest to launch Galileo on time. Unfortunately, the shuttle's engines and thermal protection suffered development problems and NASA had to acknowledge that it would be lucky to start flying by 1980. In July 1979 it informed JPL that Galileo would *not* make the 1982 window. This meant that when the spacecraft was eventually launched it would not be able to exploit a Mars flyby. Galileo was to be sent on its way by a three-stage 'planetary' form of the Inertial Upper Stage (IUS). Without the gravitational slingshot, the IUS did not have the energy to send the enlarged spacecraft all the way out to Jupiter. This presented JPL with a dilemma. It argued for a transfer to a Titan-Centaur, the most powerful conventional launcher, so that it could use the favourable 1982 window, but a cost-benefit analysis of the shuttle relied on a busy schedule to force down overheads, so to offload such a prestigious payload would set a bad precedent. As a result, JPL was told to strip Galileo to suit the IUS's capabilities. By this point, however, it was also clear that the shuttle would not meet its nominal capacity in its initial configuration. This meant that JPL's task was not simply a matter of refitting the smaller tank; instruments would also have to be deleted. The programme was in a state of flux, however, and a few months later NASA said that if JPL could split its vehicle into an orbiter and probe, each of which could be expanded to exploit the shuttle's capacity, it could have *two* launches in 1984. However, the three-stage IUS was cancelled in December 1980. This meant that JPL would have to rebuild its spacecraft to use the two-stage variant, which had been intended only to place heavy satellites in geostationary orbit.

By the end of 1979, pressure had begun to develop for a more powerful planetary stage for the shuttle. The Department of Defense, which had ordered the IUS, had specified solid propellant so that it would be simple to operate. But a solid rocket is not as effective as an engine burning high-energy liquids. As the shuttle's prime function was to ferry mass into low orbit, it made sense to make every kilogram count. The Centaur had proved itself as an upper stage for Atlas and Titan rockets, so surely it should be possible to make it compatible with the shuttle's bay? It owed its high energy to burning hydrogen and oxygen, however, so it would be a more difficult payload to service. In January 1981 Congress told NASA to cancel the IUS, and to use the Centaur instead. JPL was immediately told to reintegrate the orbiter and probe, because the Centaur would be able to send this straight out to Jupiter. The penalty was a slip in launch to 1985, because it would take that long to adapt the Centaur.

Although JPL made good progress, its work was being undermined by NASA politics. The Marshall SFC supervised NASA's interest in the IUS. The Lewis Research Center had built the original Centaur with General Dynamics. Switching to the Centaur was a boon to Lewis, but it hurt Marshall. Lobbyists set to work. In November, Congress reversed its decision; the IUS was back, and the Centaur was out. Galileo was in trouble again. There was hope, however, as General Dynamics also had lobbyists. The Californian economy was dominated by Aerospace, and certain members of Congress were facing re-election in 1982. The Centaur was reinstated in July, but this time the IUS was not cancelled; both would be built so as to offer flexibility. Galileo was safe! However, the time wasted by political manoeuvring had delayed the Centaur, so Galileo's launch was slipped to the May 1986 window.

Galileo was paying a stiff price for its brief ride into orbit aboard the shuttle, but it had fared better than several other projects which had been cancelled outright, and their funding soaked up by the shuttle's overruns. The policy of forcing all payloads onto the shuttle was playing havoc with the planetary science programme. Instead of the shuttle picking up the load in a vibrant programme when it belatedly entered service, by causing delay and redesign it was imposing cost overruns on this programme, and effectively 'killing off' its customers. In this phase of its development, the shuttle was a predatory beast.

As soon as the commitment to Centaur looked secure, JPL allowed Galileo to expand to take full advantage of the Centaur's capacity. Unconstrained by a Mars encounter, it sought opportunities in the asteroid belt. A flyby of Amphitrite on 6 December 1986 was possible; no asteroid had yet been inspected close-up, so this *first* represented a welcome bonus for JPL's hardworking staff.

Over the next few years, the shuttle ran through its test flights and became operational. During this time of stability, Galileo's development proceeded apace. By 28 January 1986, the spacecraft was at the Kennedy Space Center, about to be mounted on its Centaur as the final phase of its preparation for launch on Atlantis on 21 May.

WHICH WAY?

In the immediate aftermath of Challenger's loss, JPL was told to prepare Galileo for the next launch window of June 1987, on the assumption that by then the shuttle would be back in service.

The first sign of trouble came when NASA decided not to run the shuttle's engines at 109 per cent of their rated thrust. The SSMEs could be throttled over a wide range of thrust. They ran at 100 per cent for take-off, throttled back to 65 per cent while the shuttle passed through Mach 1, because that was when the aerodynamic loads on the structure were most intense, and then throttled up again afterwards. On the early flights they had run back up to 100 per cent, but over a series of tests they had reached first 104 per cent and then 106 per cent. The higher the thrust, the larger the payload capacity. A Centaur with Galileo on its nose would require 109 per cent. The SSMEs had played no part in the Challenger accident, but safety had assumed a higher profile, and having lost a shuttle NASA was in no position to stretch the operational envelope.

In the light of this, JPL decided to fly a Centaur with a partial propellant load and make up the energy deficit by a gravitational flyby. Since Mars was no longer favourably placed, the only viable target was the Earth. The Centaur would boost Galileo out on a long ellipse that would produce an encounter with the Earth which would toss the spacecraft all the way out to Jupiter. This roundabout route, dubbed the Earth Gravity-Assist (EGA), would get Galileo-as-built to Jupiter, but the initial loop out to the asteroid belt would add *three years* to its flight time.

In June, following a thorough review of ascent-phase abort modes, NASA decided that having a Centaur in the bay of an orbiter attempting a RTLS emergency landing posed too great a risk. Even if the orbiter managed to make the runway, the stress of a rough landing would likely fracture the Centaur, and if the liquid oxygen spilled it would inevitably cause a devastating explosion. The Centaur was cancelled again, this time with no possibility of being reprieved. At first, it seemed that Galileo would have to be cancelled too. The three-stage IUS had long ago left the scene, and Galileo had been put on too much mass for a two-stage IUS to boost it very far out into the Solar System. When NASA realised that it would need much longer than a year to get the shuttle back into service, it told JPL to prepare Galileo for the September 1989 Jovian window.

With the straightforward EGA now beyond reach, the wizards at JPL began to consider multiple-encounter routes. Even a second Earth flyby did not add sufficient energy to make up for the difference in capability between the IUS and the *partially*-fuelled Centaur. In a moment of desperation, the orbital mechanics team looked at what would happen if Galileo set off in the *other* direction, towards the Sun, and it came up with the VEEGA (Venus-Earth-Earth Gravity Assist) trajectory. It would allow an IUS to send Galileo-as-built out to Jupiter, but at a terrible penalty in flight time. It would now be launched towards Venus in October 1989, and make the flyby in February 1990, which would accelerate it and send it to encounter the Earth in December 1990, which in turn would put it into an ellipse that would take it out to the asteroid belt. It would then return for a second Earth encounter in December 1992, at which time it would pick up sufficient speed to loop out for a rendezvous with Jupiter in December 1995. If this could be achieved, it would be a truly inspired solution. It offered not only encounters with Venus and the Earth, but also new opportunities in the asteroid belt, so it had its compensations. Encounters with Gaspra in October 1991 and with Ida in August 1993 were possible, the first roughly midway between the two Earth encounters, and the second while *en route* to Jupiter.

Having adopted the VEEGA solution, JPL set out to modify Galileo to survive the extra heat, because insolation was twice as intense at Venus as it was at the Earth's distance from the Sun. Galileo had been built to survive way out at Jupiter, where insolation was just 4 per cent of that at the Earth. A change in procedure was also required. Under normal circumstances, as soon as the spacecraft was safely on its way to the outer Solar System it would unfurl the umbrella-like High-Gain Antenna (HGA) and orientate itself to maintain this facing towards the Earth. The antenna was too delicate to be heated by intense sunlight, so it was decided that it would be kept furled up around its axial support until Galileo made its second pass of the Earth and finally set off for Jupiter. The simplest way to protect the antenna was to mount a small disk-shaped sunshield on top of its support column, and then orientate the spacecraft so that it maintained its axis towards the Sun. This meant that while it was in the inner part of the Solar System, Galileo would have to rely upon its omni antenna, but since this would be required only to carry telemetry its low data rate was not seen as being a serious problem. A wide disk was mounted across the main body so that the rest of the instruments could be protected in its shade.

As before, as soon as the commitment had been made, a sense of stability returned, and preparations proceeded steadily. The launch window for Venus opened on 12 October and lasted until 21 November. The first countdown was scrubbed when a fault showed up in one of the SSMEs, long before the astronauts were due to board Atlantis. Bad weather intervened at the next launch attempt, but Galileo finally lifted off on 18 October 1989. Over the next few hours, Shannon Lucid verified that the IUS and its payload were both healthy, and then Ellen Baker and Franklin Chang-Diaz suited up and entered the airlock to deal with any issue that might interfere with the deployment. But everything went as planned, and Galileo was dispatched at the appointed time. The shuttle's contribution to Galileo's mission had taken only *six* hours. Had this delivery service been worth the wait? NASA *could* have launched Galileo in 1982 on a Titan-Centaur. The windows for the direct route which Pioneer 10 had flown recurred every 13 months. Even at this late stage the Titan 3, although in short supply, was still in use. If Galileo had been flown in its original configuration, it could have reached its destination within two years. As it was, Galileo would not even *start* its exploration of the Jovian system until late 1995.

The needless stretching out by a full decade of Galileo's budget transformed it into the most expensive planetary probe ever, and it prompted Dan Goldin, appointed as NASA's Administrator in 1992, to castigate it as being the perfect example of the wrong way to do planetary science. What was needed, he said, was a type of spacecraft that could be developed and flown on a short timescale. It is ironic that JPL had hoped to exploit its experience with Voyager to mount an early follow-up mission. It is worth following Galileo's encounters in some detail because they are the direct result of its close coupling with the shuttle.

THE LONG HAUL

When released by Atlantis, the IUS/Galileo stack was 19 tonnes. Galileo itself was just 2,560 kg, and of this, 2,220 kg was the orbiter. Fully half of this was the engine which would execute the Jupiter-Orbit Insertion (JOI) burn. The orbiter's scientific payload was a mere 120 kg.

The key instrument, the Solid State Imaging System (SSIS), was an 800 × 800 pixel CCD camera with a 1,500-mm f/8.5 lens. It had a broader spectral range and faster action than the vidicon on Voyager. The Near-Infrared Mapping Spectrometer (NIMS) measured reflected sunlight and thermal emission. It was for infrared studies of the Jovian atmospheric composition and temperature, and surface compositions of the asteroids and Jovian moons that the spacecraft would encounter. The UltraViolet Spectrometer (UVS) and Extreme-UltraViolet Spectrometer (EUVS) were to measure gas and aerosols in Jupiter's atmosphere, monitor Jovian aurorae, and measure the plasma in the Io torus. The Photo-Polarimeter-Radiometer (PPR) would determine the distribution and character of atmospheric particles and compare the flux of thermal radiation to incoming solar levels. These four remote-sensing instruments were carried on the non-spinning section of the

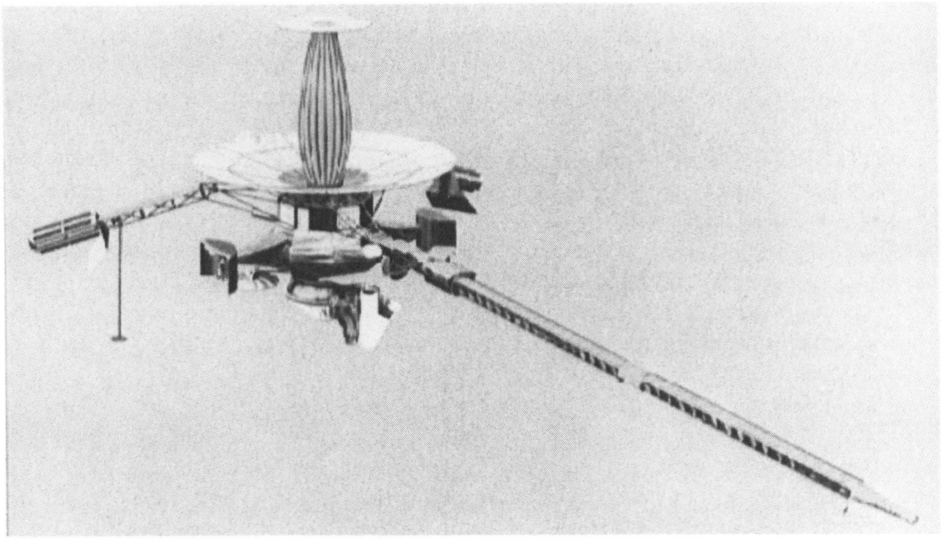

(a)

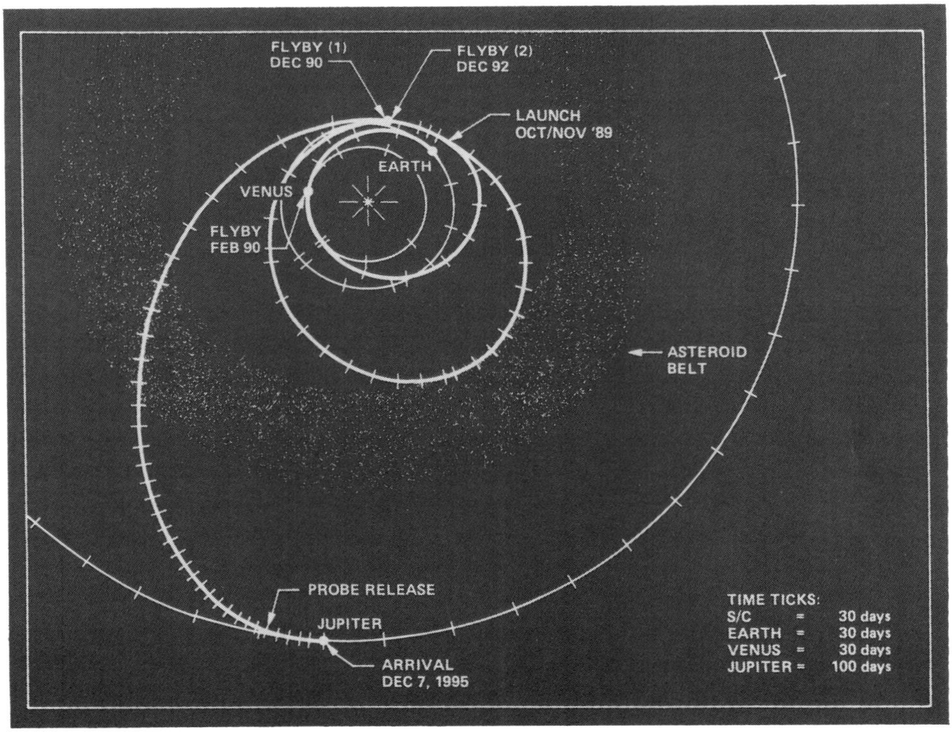

(b)

Fig. 13.1. Galileo: (a) an artist's impression of the spacecraft with its boom extended, its large dish antenna furled, and (b) a schematic of the long interplanetary cruise to Jupiter.

orbiter. Two Radioisotope Thermal Generators (RTGs) on 5-metre booms converted the heat released by 10.9 kg of plutonium dioxide into electrical power. However, although Galileo was fully self-contained, it had to make do with an average of just 500 watts from each of the RTGs. The fields and particle experiments comprising a magnetometer (MAG), heavy ion counter (HIC), energetic particle detector (EPD), plasma detector (PLS), plasma wave instrument (PWS), and dust detector (DDS) were mounted on or near a single 11-metre boom, on the spinning section of the spacecraft. In addition to studying the magnetic fields, energetic particles, plasma and dust in Jupiter's magnetosphere and near the satellites, these instruments operated during the interplanetary cruise, where they measured electromagnetic fields, dust, cosmic rays and the physical and chemical properties of the solar wind. Data were sent back once a week via the omni transmitter. The lengthy detour route facilitated study of the interplanetary medium at a wide variety of locations over a long timescale.

On 26 October the probe which was to penetrate Jupiter's atmosphere was subjected to a six-hour diagnostic test, and then powered down; it would be given such a check annually. On 9 November and again on 22 December, Galileo pulsed its low-thrust motors to make a small Mid-Course Correction (MCC), to refine its trajectory to Venus. It was discovered that the thrusters were actually slightly more efficient than expected, and so it would be possible to save propellant for the primary mission in Jovian space.

Galileo passed 16,000 km from Venus on 10 February 1990. It was within 5 km of the centre of the corridor required for the slingshot back to Earth; the encounter added 8 km/s to the spacecraft's speed. In all, 81 images were recorded over an 8-day period. In particular, NIMS provided original data on the atmosphere, and its imagery of the cloud system was of the highest resolution yet achieved. This enabled small-scale structures over mid-latitudes to be studied for the first time. Although the planet rotates exceptionally slowly, (243 days) the atmosphere is dominated by a rapid jetstream with a 4-day cycle. This new data gave a new insight into this circulation. The atmosphere's primary constituent is carbon dioxide. It was found that the atmosphere is semi-transparent in the micrometer bands characteristic of this gas, and the visible detail could be correlated with known surface features. It was also possible to use the plasma-wave instrument to confirm the presence of lightning within the dense atmosphere.

For the next few weeks, Galileo continued within Venus' orbit. It reached its 0.7 AU perihelion on 25 February 1990, and then began to swing back out. It kept its most delicate systems hidden behind the sunshades. It made MCCs on 10 April and 11 May, and then made observations of comet Austin. Another MCC on 17 July refined the trajectory for the Earth encounter. So little propellant had been used thus far that the Gaspra flyby was authorised. Aphelion at 1.28 AU was on 23 August 1990. While making observations of Comet Levy, the EUVS detected hydrogen in the comet's coma and tails.

When Galileo was both far from the Earth and close to the Sun in the sky, the 70-metre antennas of the Deep Space Network were barely able to distinguish the weak omni signal from the radio noise from the Sun, so the data-rate had to be restricted to just 10 bps. As a result, the data taken during the Venus encounter were stored on tape. An image was sent to verify that the camera had worked, but this took six hours to relay. When Galileo made its first approach to the Earth, the data-rate could be increased. Over a period of a week in late November, the entire tape was replayed, and the planetary scientists who had had to wait almost a year for this data yearned for the day that Galileo would be able to unfurl its HGA.

Galileo's first return to the Earth occurred on 8 December 1990. As it passed 1,600 km above Africa, it recorded useful data about the state of the Earth's magnetosphere. It was an ideal opportunity to test the instruments on the scan-platform. As it withdrew Galileo took a picture of the Earth every minute for a day to make a movie of the Earth turning on its axis. Up to this time, only satellites in geostationary orbit had been able to provide hemispherical views, but because such satellites remain stationary it was not possible to record the Earth's rotation, and this movie was therefore unprecedented. Galileo was also able to make important mineralogical observations of the Moon's far side.

The late Carl Sagan, then at Cornell University, used Galileo's data and employed 'first principles' to try to establish unambiguous evidence for life on the Earth. The imagery did not show any artificial structures such as cities, highways or fields, nor any lights on the dark side of the planet. The 21 per cent oxygen content of the atmosphere and the presence of water in abundance were consistent with carbon-based life, but did not imply it. The only unambiguous signal of *intelligent* life was the noise emanating from the planet in the radio spectrum.

The spacecraft had performed flawlessly throughout its interplanetary cruise and at both of its encounters. Thanks to the IUS's very accurate initial trajectory, and the 102 per cent efficiency of its own thrusters, Galileo had sufficient propellant margin to accommodate its secondary objectives without jeopardising its main mission in Jovian space. An MCC was performed on 19 December to refine the aim for Gaspra, and on 11 April 1991 Galileo was ordered to deploy its main antenna.

The HGA was based on the structure developed for the TDRS relay satellites. It was an umbrella configuration comprised of 18 graphite-epoxy ribs supporting a thin molybdenum mesh. It had been drawn in tight against its central column since launch to hide it behind its sunshield. When unfurled, it was to create a 4.8-metre wide parabolic dish which would be able to transmit in the X-Band at a rate of 134 kilobits per second (kbps). Unfortunately, when the motor cycled the latches to release the pins that held the tips of the ribs against the column, several of the pins stuck fast. The distorted umbrella was useless as an antenna. Overnight. It looked as if Galileo would be unable to report its findings when it finally reached its destination!

Reluctant to give up the primary mission, JPL studied the design of the mechanism for opening the antenna. Perhaps it had been insufficiently lubricated, and some of the latches had cold-bonded to their pins during the long time spent in the shade. Could this bond be broken? In May, Galileo turned to bake the pins in the full glare of the Sun, in the hope that thermal expansion would crack the bond. It didn't. In July, the pins were cold-soaked for 32 hours in the hope that thermal contraction would crack the bond. It didn't. The pins were frozen for 50 hours in August, but to no effect. Maybe the shock of an engine firing would shake them loose. Perhaps the next MCC would do the trick. In September, Galileo took a series of long-range images of Gaspra in order to refine the asteroid's orbit. Tracking from the Earth could localise Gaspra only to within a 200-km sphere; Galileo pinned it down to a 50-km sphere. This was essential, because the scan-platform did not actually recognise and track its target; it ran a pre-programmed sequence. These new data ensured that the camera would be able to be aimed properly.

The Gaspra encounter occurred on 29 October 1991. When 16,000 km out and with 30 minutes to go to the moment of closest approach, Galileo took a mosaic of nine colour frames, each in three colours. Although the resolution was only 160 metres, it was virtually certain that the camera would be able to capture the asteroid in the frame. Over the next 15 minutes, as the asteroid loomed larger, but ironically as the chances of a misalignment increased, a series of 49 monochrome frames were taken. The last picture, a few minutes before the moment of closest approach, gave a partial view with 50-metre resolution. The scan-platform's best rate of 1 degree per second was insufficient to track the asteroid as Galileo sped by at 8 km/s, so no pictures were taken at the 1,600 km closest point. The colour mosaic was transmitted at 40 bps in November. When Galileo achieved aphelion at 2.3 AU, in December 1991, the pins on the HGA were cold soaked again, but to no avail. The Gaspra data could have been replayed in minutes if the antenna had finally deployed, but it hadn't and so JPL had to wait until Galileo was on its way back to the Earth before it was able to transmit over the omni, and it took a month in May–June 1992 to replay the full data.

Gaspra turned out to be an 18 × 10 × 9-km irregularly-shaped object, probably created by collisional fragmentation of a parent body. It had many small craters, but relatively few big ones. The distribution of crater sizes can be used to determine the age of an exposed surface, and it appears that the surface of Gaspra is relatively young, on the order of 200 million years. Linear groove-like depressions extending across a substantial fraction of its surface are likely evidence of nearly catastrophic impacts on the parent body from which it was formed. It had been classified as an S-type (stony) asteroid. NIMS showed that its surface was not exposed bedrock, but a regolith of pulverised material, and the weathering of the crater rims indicated that this was fairly deep. There are many similarities between Gaspra and Mars' primary moon, Phobos, which is believed to be a captured asteroid.

The surprise was that Gaspra had a magnetic field. A minute before closest approach, Galileo's magnetometer noted a mild shock in the solar wind, and the magnetic field rotated to aim at the asteroid. Three minutes later, when Galileo emerged from the asteroid's wake, the field realigned. This was an unexpected discovery, because only a heat-processed body can develop a coherent magnetic field. Gravitational compression stresses rock, which heats it. If a body is massive enough, its core will melt and its structure will differentiate. If a nickel–iron core forms, it will create a magnetic field that will be 'fossilised' as the mantle rock crystallises. If such a body is subsequently fragmented by an impact, any rock which is melted from the shock will lose this ancient field. On a body that is insufficiently massive for gravitational heating to take place, melting can be induced only by the shock of a high-energy impact. In such an event, the recrystallised melt will acquire a magnetic field only if a field is extant. Iron meteorites are magnetised, so clearly a fraction of the asteroid population is composed of fragments of differentiated cores of larger bodies. As a stony asteroid, Gaspra seemed to be a fragment of mantle from a differentiated progenitor which had been sufficiently rich in metals to create its own magnetic field. NIMS indicated that Gaspra's surface was rich in metallic silicates, which was consistent with its being a chunk of mantle chipped off a larger body.

The scientific yield from the Gaspra flyby had been so impressive that on 25 June 1992 JPL authorised Galileo to make a close pass of Ida. First, however, it had to pick up more energy from the Earth. Manoeuvres were made on 9 October and on 13 November to refine the trajectory, but the HGA remained stuck fast. Early on 8 December 1992, Galileo flew over the Moon's north pole. This manoeuvre reset the plane of the spacecraft's orbit around the Sun so that it could reach Ida. Twelve hours later it skimmed just 305 km above the South Atlantic, only narrowly missing dipping into the atmosphere, and then sped away in a newly extended elliptical orbit which would take it to its rendezvous with Jupiter. On 16 December, Galileo took a family portrait of the Earth and its Moon, and this is the best such picture ever taken. It far surpassed the pictures taken by the Voyagers, when they set off for Jupiter. Over the next few months the pins on the HGA were heat-soaked and their motor cycled 13,000 times to try to release them, but again to no avail. On 11 March 1993, almost as a last resort, the spacecraft was spun up to 10 rpm and the mechanism was recycled, but the centrifugal force was insufficient to pull out the pins. JPL finally came to terms with the fact that the primary mission would not be able to proceed as planned. It had been intended to return 50,000 images during its two years of exploration of the Jovian system, but the low data rate of the omni antenna made this impossible. So long as the spacecraft was healthy, a drastically reduced and extremely selective imaging programme was still viable. Also, JPL had a few ideas for making the omni more effective. Galileo was reprogrammed to filter its images prior to transmission to avoid returning images of blank sky, and to use data compression in formatting the rest. In addition, the DSN was upgraded so that the rate on the omni could be increased without degrading the signal-to-noise ratio. The combined result was to make Galileo 100 times more effective. To put the magnitude of JPL's dilemma into context: the imaging system on Mariner 4, which had snapped two dozen pictures during its Mars flyby in 1965, had taken 8 hours to transmit its 200 × 200 pixel matrix at 8 bps. Each SSIS image had *sixteen times* as many elements. How could Galileo possibly explore a miniature Solar System employing only its omni antenna!?

The new system was to be tested during the Ida encounter. The trajectory was refined on 13 and 26 August 1993 and then Ida loomed large on 28 August. With about four hours to go, Galileo inexplicably stowed its camera and resumed its cruise mode. JPL commanded it to set up again, but the recovery process was exacerbated by the fact that it took an hour for each command to reach the spacecraft and its confirmation to be reported. Fortunately, Galileo was fully restored when, at a distance of 2,410 km, it sped by Ida at 12 km/s. As before, a colour mosaic was taken early on to gain a complete view of one side of the asteroid with a resolution of 35 metres. The final monochrome frame was a partial view with 24-metre resolution. At the time of the encounter, Ida was close to the Sun as viewed from Earth, so there was only just time to transmit the mosaic before the signal from the omni degraded. Even at 40 bps, it took *30 hours* to send each of the five frames. JPL had to wait until the Earth had moved round to the far side of the Sun to replay the rest, a process that ran from mid-February through to late June 1994, at which time the spacecraft was called upon to watch out for momentous events on Jupiter.

At 56 × 24 × 21 km, Ida was slightly larger than previously believed, and it was rather heavily cratered. Most craters were simple concavities, but some had flattened floors and a few had central mounds. The worn rims and the presence of a few ray-craters indicated a fairly thick regolith, and several linear features could be fractures. There were even a few chains of craters. A full analysis of all the images revealed that one side of Ida was a rectangular block, while the other side was irregular and was dominated by a single depression some 28 km wide. Spectroscopic data revealed surface chemistry which supported the suggestion that it was a composite of several distinct objects. The surface was more heavily cratered than expected. This implied an age of at least one billion years. Like Gaspra, Ida had its own magnetic field.

The truly amazing discovery was that Ida has a satellite. It was spotted on 17 February, as the first day's downloaded imagery was examined. It orbited approximately 100 km from Ida's centre of mass, was less than 2 km across, and was surprisingly spheroidal for such a small body. It was later named Dactyl, for the mythical bird which is reputed to have been seen by Zeus on Mount Ida.

Some suggested that Dactyl had formed independently, and had become gravitationally bound to Ida as a result of a low-energy encounter. However, the overall cratering density suggests that it has shared a common history with Ida. Although tiny, it has a few craters 300 metres across. NIMS confirmed that Ida and Dactyl are chemically similar, which argues for a common origin. Either Dactyl is a piece chipped off Ida or both are fragments of a shattered precursor. If they became associated so long ago, it is astonishing that this rather weak gravitational bond proved so stable.

In early October 1993, Galileo made a series of manoeuvres to set up its trajectory for Jupiter, and then in February 1994 it refined this into a collision course. It was not the only object heading for the giant planet, however. Split into a string of fragments following a recent close encounter with Jupiter, comet Shoemaker–Levy 9 was in an orbit that would lead to its fragments impacting the planet in July 1994. Observers on Earth had to wait 10 minutes for the planet's rotation to bring the impact sites into view, but Galileo had a clear view. At a distance of 238 million km, however, its resolution was low. The first image was sent in mid-August, but the sequence was not completed until early 1995, by which time the spacecraft was preparing for its own encounter.

The first task was to line up and release the probe that was to dive into the atmosphere. The Hubble Space Telescope had been monitoring Jupiter ever since Shoemaker–Levy 9, to track the rate at which the impact scars healed. This also provided a long-term study of the atmosphere to help in selecting the best location for the probe to sample. Full knowledge of the prevailing weather system would also assist in interpreting the probe's results. The final refinement of the collisional trajectory was on 23 June 1995. The probe was checked out on 11 July, and then its umbilical was withdrawn. Galileo increased its spin from 3 rpm to 10 rpm so that the probe would remain stable as it plunged into Jupiter's atmosphere. It was finally released on 13 July. Apart from a timer, the probe was to remain passive during the five months that it would take to fall the final 80 million km. The faint hope that the shock of the separation would deploy the HGA proved to be wishful thinking. The successful release of the probe was a relief because it had covered the main engine's nozzle. The firing of this liquid rocket on 27 July, to diverge from the probe's collision course, enabled its thrust to be calibrated so that the exact duration required for the JOI-burn could be calculated. Even the shock of this impulse failed to shake loose the stuck HGA, but there was still a slight hope that the main burn might do the trick. How wonderful it would be if Galileo re-emerged from behind the planet with the HGA fully deployed!

As it drew within 40 million km of Jupiter in early September, Galileo was battered by dust. The typical interplanetary rate had been one impact every few days, but over a period of weeks the rate soared to 20,000 hits a day. The origin of the dust was unclear, but it was speculated that it may have been ejected by Shoemaker–Levy 9. The tiny grains did not really threaten the spacecraft, however.

Galileo's prospects took a serious turn for the worse on 11 October when it was found that the tape deck had failed to halt at the end of its tape, and had seized. Without the tape, even the probe's data would be lost. After extensive trials on an identical unit at JPL, it was decided that it would probably be safe if it was operated at slow speed and avoided the bald segment of tape. The tape deck was to have stored up to 200 images. Galileo's capacity had just been halved.

THE BIG DAY

Jupiter first made its presence felt on 16 November. Galileo's magnetometer detected a fluctuation in the field when the planet's magnetosphere suddenly ballooned in response to a lull in the solar wind, and over the next few days it recorded the turbulence as the solar wind piled up in front of the magnetopause. On 26 November, still 9 million km from the planet, it crossed the bow shock. It was now officially in Jovian space.

The Big Day came on 7 December 1995, and a very busy schedule had been planned for its final run-in. Galileo's inbound trajectory took it to within 880 km of Io, the innermost of the four Galilean moons. This would be the most favourable opportunity for imagery. Until the tape deck had developed a problem, Galileo was to have recorded the passage, but JPL was reluctant to risk losing the tape deck just before it would be required to store the data uplink from the atmospheric probe, so the Io imagery was sacrificed. However, the EPD recorded the plasma torus which trails the moon around its orbit, at an altitude of 350,000 km above Jupiter's cloud tops, and it reported that the sulphurous cloud had been enriched by a recent volcanic outburst.

The probe 'woke up' six hours before it reached Jupiter. Spacecraft on flyby missions had steered clear of the worst of the radiation trapped in the vast magnetic field, but the probe *had* to pass through the radiation to reach the planet, so it measured the flux. Its trajectory intersected the planet just above the Jovian equator, near latitude 7° north. It needed to enter the atmosphere at an angle of between 7 and 10 degrees to the horizon. It would bounce off if it was any shallower, and it would be incinerated if it was any steeper. It struck the atmosphere at 47 km/s. This was almost *five times faster* than any spacecraft had ever entered the Earth's atmosphere. Accelerometers detected the entry interface and triggered a timer. At 250 g, the deceleration was harsher than expected, indicating that the gas density was higher than expected. The heatshield accounted for fully 220 kg of the probe's 340-kg mass. The thick carbon phenolic outer layer of the shield was seared by the peak temperature of 12,000 °C, but Ames had designed it conservatively, and it survived. As soon as the probe had been slowed, it jettisoned its shield and released a 2.5-metre wide dacron parachute. As it slowly sank into the ever-thickening atmosphere, the probe's 26 kg of instruments measured the temperature, pressure, density and composition of the gas, measured ambient thermal and solar energy, determined the cloud structure and sensed lightning. The data were transmitted in real time at 28 kbps by an L-Band radio-link to Galileo, which was passing 100,000 km overhead. After almost an hour, by which time it had reached a depth of about 650 km, the probe was crushed by the pressure, which was 20 times that at the surface of the Earth.

As soon as the probe's transmission ceased, Galileo turned away to prepare for the vital JOI burn by aiming its main engine forward to act as a brake, and then spinning up to 10 rpm for stability. The burn came an hour later. It was made while around the far side, and lasted 40 minutes. Everything had gone exactly as intended, but the HGA had not been deployed by the shock. The initial orbit was highly elliptical with a high point 20 million km out and a seven-month duration. From Galileo's viewpoint, the Earth was about to go around the Sun, so productive work could not begin until early in the new year. As a contingency against the tape failing in the interim, a summary of the probe's data was transmitted on 9 December; the full data set was not played back until March 1997.

The probe saw only one cloud layer rather than the triple-deck structure expected. This was surprisingly thin, and was mainly ammonium hydrosulphide. At 650 km/s, the winds were much stronger than expected, and they were uniform, rather than gusty, and persisted below the cloud layer, which suggested that they were driven by internal heat rather than by insolation. A substantial amount of water had been expected at depth, but it proved to be absent. It turned out that the probe had penetrated the junction between two of the primary atmospheric bands, and right in the centre of the 'upwelling plume' which had appeared in October and was still a strong 'hot spot' in images taken with NASA's InfraRed Telescope Facility in Hawaii. It was later realised that these regions are particularly arid. That the atmosphere below the cloud layer was free of haze was also a surprise, but this too is consistent with the effect of an upwelling plume of hot, dry gas, as indeed is the lack of lightning. Lightning is important, because it would be the primary means of stimulating the formation of organic molecules. The mass spectrometer found no evidence of prebiotic material – a blow to those who believed Jupiter was a benign environment for the development of the basic structures of life.

MISSION OF EXPLORATION

Galileo's exploration of the Jovian system involved monitoring the planet's atmosphere and performing a series of flybys of the four Galilean moons, in some cases passing within 200 km of their surfaces, to produce a *100-fold* improvement in resolution over the best Voyager imagery.

On 14 March 1996, at the top of its initial orbit, Galileo fired its main engine for the last time to lift its low point above the worst radiation. The first few encounters were with the largest moon, Ganymede. The first flyby, in June, was used to draw the spacecraft into the orbital plane of the moons, and the next few passes dropped the high point of the orbit and cut its period to about 35 days.

Each orbit was broken down into a seven-day encounter phase, a long cruise to transmit the imagery, and an update on the state of the planet's weather system at the lowest point in the orbit. Not every orbit yielded a really close encounter, but imaging with a fidelity at least as good as that of the Voyagers was possible each time around. As long as its attitude-control propellant held out, Galileo would be able to continue to fill the gaps in its mapping, and so eventually fulfil its original mission.

LOST OPPORTUNITIES

The Galileo mission has been closely linked with the shuttle throughout. The spacecraft would never have taken such a long detour if it had been able to exploit the Centaur. It was the decision to use the IUS which imposed the VEEGA trajectory on Galileo, and this must be borne in mind when assessing the mission.

Whilst it is true that the spacecraft achieved a great deal during its long cruise, and it is possible to regard its secondary discoveries as benefits directly resulting from the shuttle's operating policy, the problem with the HGA is one of the penalties paid by Galileo for that association. If it had not been for the need to keep the antenna furled during the time that the spacecraft was close to the Sun, it is extremely unlikely that the pins would have bonded to the latches. Under normal circumstances, the antenna would have been unfurled soon after the escape stage was jettisoned.

To sum up, therefore, the close coupling between the spacecraft and the shuttle delayed the mission and invoked a circuitous route which expanded the scientific yield with secondary objectives but also jeopardised the primary mission. Despite being restricted to 10 per cent of the hoped-for 50,000 images, JPL is nevertheless confident that it will be able to achieve 70 per cent of its mapping objectives, 60 per cent of the magnetospheric studies, and 80 per cent of the atmospheric studies.

Finally, consider this. If Galileo had been assigned to a Titan-Centaur in 1982, it would have begun its exploration in 1985, but in 1985 it was in storage awaiting a Centaur. If it were not for the loss of Challenger it would have ridden a Centaur in 1986, taken the direct route to Jupiter, arrived in 1988, acquired its 50,000 images by 1990, possibly have had its mission extended, and then been deactivated. But would it have been possible to revive it to record the impact of comet Shoemaker–Levy 9?

14

Dante's inferno

The Spacelab 2 mission in July 1985 was the first to use an all-pallet configuration, and its Instrument Pointing System (IPS) carried three instruments to study the Sun.

The Coronal Helium Abundance Spacelab Experiment (CHASE) was a telescope with a spectrometer. Built by the Rutherford Appleton and Mullard Space Science Laboratories in England, it worked perfectly. Its primary objective was to measure the hydrogen-to-helium abundance ratio in the solar corona, the extremely high-temperature, but low-density, outer atmosphere that is the source of the solar wind. This data enabled ultraviolet images to be formed which showed not just the solar disk but also far out into the corona. This part of the spectrum also demonstrated the presence of emission lines from a number of exotic highly-ionised forms of silicon, sulphur and oxygen.

The High-Resolution Telescope and Spectrograph (HRTS), a telescopic spectrograph supplied by the Naval Research Laboratory, studied the outer layers of the Sun in the ultraviolet. The Sun spasmodically ejects plumes of matter into space, and by studying violent flows spectroscopically, the physical properties of the environment and the forces at work could be characterised.

The Solar Optical Universal Polarimeter (SOUP) was built by Lockheed. It had no filter to select a specific wavelength; it was a white-light telescope fitted with a polarimeter and a video camera. It had its own stabilisation system to cancel shuttle vibrations not damped by the IPS. When such an instrument is employed on the ground, haze and turbulence in the atmosphere degrade its view. Able to monitor its output, the astronauts described its view of the solar disk as "stunning". Its imagery had sufficient detail to permit measurements of the Sun's magnetic field. Unfortunately, a power fault disabled it within hours, but when it mysteriously restarted on the final day of the mission it was hastily put to productive use.

Initially, the IPS encountered problems because its star trackers proved to be rather too sensitive, and they could not identify assigned guide stars, but it was reprogrammable and, once reset, the IPS yielded the required 1-arcsec pointing accuracy. These telescopes were complementary instruments which provided *simultaneous* data on magnetic, structural and compositional phenomena.

To attain a completely new perspective on the Sun, however, it would be necessary to send a spacecraft out of the plane of ecliptic in order to inspect the Sun's magnetic poles.

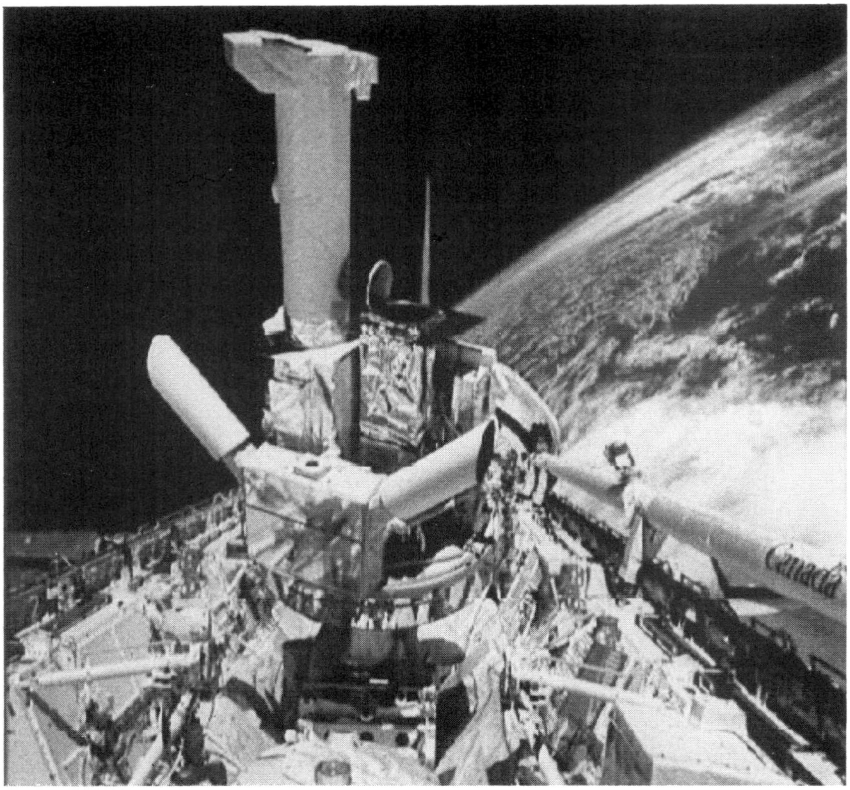

Fig. 14.1. The battery of solar telescopes (CHASE, HRTS, and SOUP) which were mounted on the high-precision universally-jointed Instrument Pointing System (IPS).

ULYSSES

In 1974, NASA and ESA conducted a project definition study of an International Solar Polar Mission (ISPM) in which they would each develop a spacecraft (to a different design) to make simultaneous close passes over opposite poles of the Sun to measure the Sun's radiation, particle flow, magnetic field and surface activity in hitherto unobserved locations, so as to provide insight into the solar–terrestrial relationship. Development began as soon as the agreement had been signed in March 1978. The ISPM was to be launched on the shuttle, with both spacecraft mounted on a single three-stage planetary variant of the IUS. To achieve the desired high-inclination solar orbits, they were to make a detour via Jupiter. The two spacecraft were to fly in formation until they were through the asteroid belt, then they were to slowly diverge in order to make *opposing* polar passes of the giant planet so that the resulting gravitational slingshots would hurl one above, and the other below, the ecliptic plane to facilitate simultaneous study of both of the Sun's poles. This ambitious new mission was assigned to the 1983 manifest.

The cancellation of the three-stage IUS, in December 1980, meant that the ISPM spacecraft would have to be reduced to fit the capabilities of the two-stage IUS, but the policy switch to the Centaur in January 1981 suddenly reversed this. The plan now was to dispatch Galileo and Ulysses on Centaurs, and to do this in the *same* 1985 window. The sting in the tail came a few months later, when Congress axed the funds for NASA's spacecraft. ESA decided to continue, and named its spacecraft Ulysses after the hero of Homer's *Odyssey* who, as told in Dante's *Inferno*, explored "an uninhabited world behind the Sun". Simultaneous studies of both solar poles would be impossible with a single spacecraft, so Ulysses would have to loop around to study them a year or so apart.

As events transpired, even if Ulysses had stayed with the IUS in order to meet its 1983 window, it would have had to have been postponed because the first shuttle-IUS, dispatched in April 1983, malfunctioned and stranded its TDRS payload in the wrong orbit, and it did not fly again until 1985. Ulysses did not fare better on the Centaur, though; work on the Centaur was delayed by political infighting, so it did not become available in time for the 1985 window for Jupiter, and Ulysses and Galileo were reassigned to May 1986. If Challenger had not been lost, its next payload would have been Ulysses.

In the immediate aftermath of the loss of Challenger, both Galileo and Ulysses were reassigned to June 1987, but it soon became clear that the shuttle would not return to service in time to meet that window. Furthermore, it was decided that trying to get two shuttles off the ground in the space of a fortnight would be too heavy an operational load. NASA and ESA reached an agreement in April 1987: Galileo would take the 1989 window, and Ulysses would follow a year later.

Ulysses was not as badly affected as Galileo by the cancellation of the Centaur and the plan to run the SSMEs at less than peak thrust. At 370 kg, it was hardly 15 per cent of Galileo's mass. However, even though it had been designed for the capabilities of the planetary-IUS, it was too heavy for the two-stage variant to dispatch on the necessary high-speed trajectory to Jupiter. It was so lightweight though, that it was possible to install a 2,200-kg PAM to help to accelerate the spacecraft away from the Earth, once the IUS had achieved escape velocity.

The initial plan had been for each spacecraft to carry a multi-national payload, and in practice this was still the case, even though there was now only one spacecraft. It had 11 instruments to measure the speed, temperature, density and composition of the solar wind, the strength of the magnetic field carried by the wind, the physical properties of the dust in the interplanetary medium, and gamma ray emission from short-lived but extremely violent events on the Sun (this also detected the enigmatic celestial transients known as bursters). It had been intended for NASA's spacecraft to carry a camera to record the solar disk and the corona, but this could not be transferred, because it would have made the remaining spacecraft too large for even the enhanced-IUS, so Ulysses had no imaging capability. Another of the instruments developed for the NASA spacecraft was repackaged and included on the Solar, Anomalous and Magnetospheric Particle Explorer (SAMPEX) and, in high Earth orbit, this instrument revealed the true nature of the radiation belt in the vicinity of the South Atlantic Anomaly.

SLINGSHOT

The Jupiter window in 1990 opened on 5 October. It had been a long, hot summer for NASA, which had been forced to ground Atlantis and Columbia due to persistent hydro-

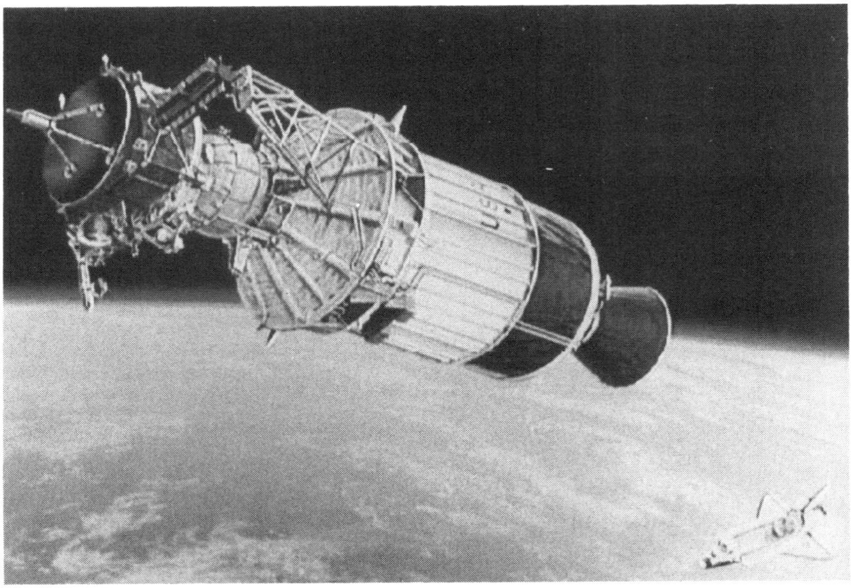

(a)

(b)

Fig. 14.2. Ulysses: (a) an artist's conception of the spacecraft mounted on its hybrid booster which combined the Inertial Upper Stage (IUS) with the Payload Assist Module (PAM), and (b) a depiction of its trajectory out to Jupiter to make a close polar pass aligned to slingshot it out of the ecliptic plane and back into the inner Solar System, to look down on the poles of the Sun.

gen leaks. Luckily, Ulysses had been booked on Discovery, and its launch on 6 October brought to a conclusion five months of institutional frustration. As with Galileo, Ulysses was checked out and dispatched within the first six hours of the shuttle achieving orbit. As Ulysses did not have a motor of its own with which to make substantial mid-course corrections, it relied on the IUS to place it on the proper trajectory. Furthermore, in order for Ulysses to fly its gravitational slingshot, the IUS had to deliver an unprecedented degree of accuracy. The IUS initiated the slow roll needed to even out irregularities in the burning of its solid rocket motors. The first stage motor fired for 148 seconds, shut down, and was jettisoned. After a 125-second coast, the second stage motor ignited, and 108 seconds later it released its payload. Although Ulysses was now travelling at 11 km/s, this was only *just* sufficient to escape the Earth. Four solid charges distributed around the waist of the PAM motor fired to spin-up to 70 rpm for enhanced stability during the motor's 88-second burn. The PAM then released a pair of small masses on 12-metre wires, conservation of angular momentum reduced the spin rate to 7 rpm, and a few seconds later the motor released the spacecraft. Now travelling at 15.4 km/s, Ulysses raced away from the Earth at a much higher speed than had any previous spacecraft dispatched into deep space.

Ulysses did not have to unfurl its HGA, because this was a built-in dish, but it did have to deploy several wire antennas and a boom projecting to the side. The 6-metre radial boom was deployed the next day, its extension cutting the spin to 5 rpm. This spin was to be retained in order to stabilise the spacecraft during its long cruise and to even out thermal loads. The small thrusters were fired to refine the trajectory on 15 October, and over the following week the instruments were started up one by one and checked out. On 4 November, the 8-metre boom was deployed. When fully unfolded, it projected along the axis on the opposite side of the spacecraft to the main antenna. Almost immediately, contact was lost with Ulysses; or rather, it became unreliable. Evidently nothing catastrophic had befallen the spacecraft, but something had disturbed it and it was unable to keep its antenna pointing towards the Earth. Analysis of the intermittent periods of contact indicated that the spin had developed a nutation. It was clearly caused by the axial boom. The nutation rapidly built up to 6.5 degrees, but then slowly decayed back to 3 degrees. This wobble was considerably greater than the 0.75-degree fine adjustment capability of the antenna. Fortunately, Ulysses had an active closed-loop control system, and when this was activated it *just* managed to overcome the nutation. It was finally concluded that the axial boom flexed in response to thermal stresses. As hoped, this effect diminished as the spacecraft pulled away from the Sun. The big antennas of NASA's DSN communicated with the spacecraft, and the European Space Research and Technology Centre (ESTEC) in Holland teamed up with JPL to operate it. On its high-speed run out to Jupiter, Ulysses monitored the interplanetary medium and the solar wind. Its data complemented that from other spacecraft orbiting within the plane of the ecliptic, some of which, notably Pioneers 6, 7 and 8, had been operating continuously for decades.

As Ulysses moved out beyond 3 AU, it detected increased dust. Most of the dust in the inner Solar System orbits in prograde fashion, which suggests that it is left over from short-period comets; but the hyperbolic retrograde orbits of the dust particles found by Ulysses implied that it had an interstellar origin. Galileo carried a similar instrument, but that spacecraft was still in the inner Solar System. Cassini, launched in October 1997, *en route* to an encounter with Saturn in 2004, also carries such a detector.

On 8 July 1991 another small correction was made to refine the trajectory. Ulysses had to approach Jupiter within a 160-km wide corridor for the over-the-pole slingshot to divert it onto the required exit-trajectory. If the IUS-PAM had been used to eject Ulysses directly out of the ecliptic plane, it would not have been able to exceed an inclination of 23 degrees to the ecliptic. The gravitational slingshot at Jupiter, however, could easily flip the plane of Ulysses' orbit perpendicular to the ecliptic, and transform the outward trajectory into an enormous ellipse. It is encounters such as these which turn itinerant comets into short-period comets.

On 2 February 1992, Ulysses penetrated the bow-shock in the solar wind, and passed into Jupiter's magnetosphere. The crucial slingshot took place on 8 February. Its trajectory took it 375,000 km above the cloud tops, which was so deep within the radiation field that it flew through Io's plasma torus. Although this was a fleeting event, most of the instruments were switched off in order to protect their sensitive electronics. Only the magnetometer and the radiation monitors were left on. Ulysses found that Jupiter's radiation field is more oblate than the Earth's, as a result of the planet's rapid rotation. The encounter was perfect, and Ulysses emerged in a five-year highly-inclined orbit around the Sun. The Pioneer and Voyager spacecraft which had visited Jupiter were now out of the ecliptic plane, but because they were in the outer reaches of the Solar System they could not climb to high solar latitude. Although these widely distributed spacecraft were able to monitor the large-scale and long-term effects of solar variability on the interplanetary medium, only Ulysses, by virtue of having doubled back from Jupiter, would be able to view the Sun's poles.

PRIMARY MISSION

With each passing month, Ulysses would penetrate further into unexplored territory. In June 1992 it reached the transition zone marking the boundary of the turbulent solar wind, which is confined to a disk in the equatorial zone, and passed into a simpler flow present at mid-latitudes. At 800 km/s, this wind was twice as fast, less dense and much smoother than in the equatorial disk. The transition zone was ragged, however, due to the inclined axis of the Sun, and over a 13-month period Ulysses flew through whorls and eddies in the solar wind; only in July 1993 did it finally leave the last trace of the equatorial zone behind. Its real objective, the polar zone, was still a year away. It passed 70° S on 26 June 1994. This was the *official* start of its polar passage. All the time, it was closing on the Sun. By 13 September, when it reached 80° S, it was 2.3 AU distant, which, although this was the highest latitude, was not the perihelion. It left the south polar zone on 6 November, and followed its tight ellipse in towards the ecliptic. For a month on either side of the ecliptic plane crossing on 5 March, it was back in the equatorial disk. On 12 March, at 1.34 AU, it made its perihelion passage. On 19 June it crossed 70° N and began its second polar pass. On 31 July it reached 80° N, and its departure from the northern zone on 30 September marked the official completion of its primary mission. A SPARTAN free-flyer was operated during STS-64 in September 1994 and STS-69 in September 1995 to collect data for correlation with Ulysses' two polar passes. It was a coordinated study of the Sun and its effects on the Earth.

Whatever Ulysses achieved now would be a bonus. The fact that its instruments were all working well held out the prospect of a second solar encounter. Its orbit carried it back out to Jupiter's orbit, but when the spacecraft reaches aphelion in April 1998, the planet will be on the far side of the Sun. Unperturbed, Ulysses will return to the Sun at the beginning of the twenty-first century. This encounter will be especially valuable, because whereas the first encounter coincided with solar minimum, the second will occur at the maximum level of activity, and Ulysses will be able to study the Sun in the state that it would have done if this mission had been flown as originally scheduled.

The plasma streaming away from the Sun carries with it an embedded magnetic field. This extended solar atmosphere is the corona. If the density of the plasma falls, so does the field strength. The low-density regions are called coronal holes, and at the equatorial latitudes dominated by spot activity, these are transient features. However, each polar zone is effectively a vast, persistent coronal hole. Ulysses was the first spacecraft to observe the polar holes directly, and it found that the solar wind emanates from the polar regions essentially undisturbed and radiates straight out into space. If the Sun did not spin, its magnetic field would resemble a simple bipolar field like that of a bar magnet. At low latitudes, however, angular momentum forces the plasma into a disk in the plane of the equator, and since the plasma carries the field with it, the field is wound up into a spiral. This spiralling decreases with increasing latitude, and is totally absent in the polar zones.

The solar wind leaves the Sun at 800 km/s, but in the highly turbulent equatorial disk this is slowed to 400 km/s. Ulysses showed that outside the disk this wind retains its high speed and, although it is less dense, it blows more smoothly. In fact, at the poles the field is 'open', and its field lines effectively radiate all the way to the heliopause – the outer edge of the Solar System, believed to be at a distance of at least 150 AU.

Violent events on the surface of the Sun eject dense masses of plasma into the corona at 1,000 km/s. Since the speed of the solar wind is reduced in the equatorial disk, such coronal mass ejections plough into the already turbulent plasma, and since such ejections often coincide with solar flares, when the shock-wave interacts with the Earth's magnetosphere we see all manner of magnetic disturbances. In the Sun's polar zones, however, where the solar wind radiates straight out at 800 km/s, a coronal mass ejection is much less disruptive; it produces a shock-wave, but this is less violent because the wind is tenuous, and the material is easily able to expand in a flow called 'unevolved turbulence'.

Towards the end of a solar cycle – as sunspot activity draws to a minimum, and the latitudes at which spots tend to develop moves towards the equator – the polar coronal holes encroach upon the mid-latitudes. When the new cycle begins, its spots appear in mid-latitudes, and this causes the polar holes to retreat. If Ulysses returns at or near solar maximum, it should find that things are rather different.

MEANWHILE

On STS-56, Discovery released a SPARTAN free-flyer. It was the first flight for this type of satellite since the resumption of shuttle operations following the loss of the Challenger, a mission on which, as it happened, a SPARTAN had also been lost.

This new SPARTAN-201 free-flyer was for solar science. The White-Light Coronagraph (WLC) was developed by the National Center for Atmospheric Research in Colorado, and the UltraViolet Coronal Spectrograph (UVCS) was developed by the Harvard Smithsonian Astrophysical Observatory. Both instruments had flown on sounding rockets, but 48 hours of *continuous* data would be a major advance. The goal was to pinpoint coronal holes and streamers and to measure the forces imparted on the solar wind, in order to study how this plasma emerged from the corona. The only time that the corona can be observed from the surface of the Earth is during a total solar eclipse, but this is a brief event, and the corona is visible only at wavelengths passed by the Earth's atmosphere. A coronagraph in space can arrange its own eclipse by occulting the solar disk; it can sustain this indefinitely, and can observe the unfiltered spectrum. The region of most interest was the extended corona – a 2 million-km thick layer of space above the photosphere – because this is where the solar wind is accelerated. It can be studied *only* from space, and the SPARTAN was a cost-effective means of conducting such research. Ellen Ochoa lifted it out of the bay on the RMS. There were no tests to perform, so she turned the SPARTAN to the appropriate attitude and released it, thereby activating its sequencer. The autonomous platform stored its data, so it produced no telemetry, and the only indication that it was active was the slight twitch as its attitude-control system refined its orientation. Its instruments performed flawlessly.

Fig. 14.3. Deploying the SPARTAN-201 free-flyer. The Shuttle-Pointed Autonomous Research Tool for Astronomy (SPARTAN) programme was intended to carry a class of instruments which previously would have been sent up on sounding rockets.

The WLC observed the entire solar disk to locate coronal holes and streamers where the already-accelerated solar wind emerged from the extended corona; then it keyed the UVCS, which noted the ultraviolet energy spectrum so that the composition, temperature, density and velocity of the ions in the plasma could be calculated for each site. The WLC itself was able to measure electron and proton densities, these particles being the debris of dissociated hydrogen, the most common constituent of the solar wind. The WLC took 1,000 coronal images, and the UVCS took 200 energy spectra. These data made a significant contribution to the theories that sought to explain why the tenuous corona is hundreds of times hotter than the photosphere, and its observations of *coronal mass ejections* helped to explain how the solar wind emerges from the corona. These relatively dense streamers interact with the turbulent zone in the disk of material caught up in the spiralling magnetic field, and create shockwaves as they are slowed. It is not really solar *flares* which lead to bursts in the solar wind; it is the mass ejections associated with coronal holes.

SPARTAN-201 was to be reflown as the Ulysses spacecraft made its two passes across the Sun's poles. The optimum times for these missions would have been May 1994 and March 1995, but events conspired to delay each flight by several months. Its spectroscopic studies of the corona, as viewed from near the solar equator, complemented the measurements by Ulysses of the physical properties of the high-speed wind streaming off first the southern, and then the northern, polar coronal holes.

MISSION ACCOMPLISHED

From a scientific viewpoint, Ulysses has clearly been a remarkable success, and is well-placed to deliver even more. From the point of view of the original ISPM plan, the shuttle was clearly predatory; it was the shuttle's cost overrun that led to the cancellation of NASA's spacecraft. The best that can be said of Ulysses' later coupling to the shuttle is that at least it did not suffer as badly as did Galileo from the IUS-versus-Centaur fiasco, nor from the changes imposed after the loss of Challenger. From the viewpoint of operating the shuttle, of course, sending Ulysses on its way was simply another mission successfully accomplished.

15

Magellan's triumph

Having successfully despatched two Pioneers into the outer reaches of the Solar System, Ames turned its attention to Venus. In 1978 it launched the first of a pair of Pioneer spacecraft to Venus. Although they were launched three months apart, they arrived within days of one another, in December. The first went into orbit, and its companion released several small probes which plunged into the dense cloud-laden atmosphere. After relaying the data from the probes, the orbiter settled down to long-term monitoring of Venus' atmosphere while its radio-altimeter built up a topographic map. Because its surface is forever masked by the ubiquitous cloud, the entire planet was *terra incognita*. Crude though Pioneer's map was, it nevertheless marked a significant advance. However, to really understand the geomorphology, it would be necessary to map with greater resolution.

In 1978 JPL adapted the side-looking Synthetic-Aperture Radar (SAR) used for airborne reconnaissance for use on a satellite. The resulting SeaSat mission was a joint venture with the US Navy. Although a power failure crippled the satellite after only a few months, JPL was so impressed by the quality of the data that it decided to modify the radar for carriage on a spacecraft that would map Venus. With its penchant for names that expressed intent, JPL called this the Venus Orbiting Imaging Radar (VOIR).

PROJECT DEFINITION

The policy was that all NASA spacecraft would ride the shuttle; but by 1978 it was evident that the shuttle was taking longer to develop than expected, so JPL allowed what it believed would be an adequate margin, assumed an ideal window (these least-energy windows recur at 19-month intervals) and scheduled VOIR for 1983.

Although it did not know it, JPL had competition. The Soviets had landed a number of small Venera probes on the surface of Venus, and these had sent back panoramic pictures which revealed a basaltic plain. Given this evidence of volcanism, the basic issue to be resolved concerned how the planet shed its internal heat. Specifically, was plate tectonics the dominant agent, as it was on the Earth? The two planets were almost identical, so surely the same mechanism was at work. In 1983, long before VOIR could be made ready, a pair of Venera spacecraft achieved Venus orbit, but although these mapped the northern hemisphere, and their data were highly suggestive, the maps were too low-resolution to resolve the plate tectonics issue. With the arrival of the Veneras, VOIR had seemed superfluous, but the map from the Venera probes raised more questions than it answered. JPL argued that with a resolution of 100 metres, VOIR should provide answers.

But the problem was in getting VOIR a firm shuttle booking; the hoped-for 1983 date had not lasted long. Continual shuttle delays, and the planned full manifest, both imposed delay after delay. In 1979, VOIR slipped to 1984, in 1980 it was rescheduled for 1986, and in 1981, when the shuttle finally made its first flight, it was pushed back to 1988. With every year, VOIR became less imminent! The latter review was more than a schedule slip; it was a project redefinition. The overrun on the shuttle was so significant that science projects – in particular the planetary spacecraft – had to be scaled back to ease NASA's cash flow, and VOIR was redesigned to make it smaller and cheaper by using as much off-the-shelf hardware as possible. Whereas the original plan had been to develop a large spacecraft with half a dozen instruments to perform atmospheric and magnetospheric studies in addition to mapping the surface, the revised design was lightweight, with a *single* instrument: the SAR. This was appropriately retitled the Venus Radar Mapper (VRM). As soon as it was approved it was named Magellan in honour of the sixteenth century Portuguese explorer, Ferdinand Magellan.

Whereas VOIR would have settled into a low circular polar orbit, continuously scanned the surface as the planet rotated beneath it, and relayed its data in real time, Magellan would have just one antenna, which it would have to use both for mapping and communications. Clearly, it could not do both at the same time, so the operating strategy had to be redefined. It was decided to put Magellan into an elliptical orbit. It would point its antenna at the surface during its passage through periapsis (closest approach to the planet), in order to scan a 24-km wide strip down that side of the planet, store the data on tape, and then, as it withdrew from the far side, would turn to aim the antenna at the Earth so that it could transmit its data. The period of this orbit was defined by the time needed for this replay. The resolution of an imaging radar is dependent on the size of its antenna as well as on the frequency used. Magellan had only a small dish, and it relied on the SAR technique, which exploited the fact that the spacecraft was moving to *synthesise* a larger antenna. To construct an image, the real antenna had to record the phase of the radar reflection at each point as well as its intensity. Consequently, each radar scan generated an enormous data stream. Although this meant adding a 900-megabit tape unit, it was possible to cut costs by using a spare HGA from Voyager as the dual-role antenna, and a computer left over from the Galileo spacecraft. On the way out to apoapsis (greatest distance from the planet), therefore, Magellan would transmit half of its stored data, make a star sighting at apoapsis to verify its inertial platform, then transmit the rest of its data as it fell back for the next radar pass. The disadvantage of the new proposal was that, because Magellan would have to allow Venus to make a full turn of its axis in order to scan the entire planet, its basic mission would take twice as long. This was significant, because for some unknown reason the axial rotation period of Venus is 243 Earth-days. To further cut costs, it was decided not to fit this stripped-down spacecraft with its own propulsion. Instead, the crucial Venus Orbit Insertion (VOI) burn was to be performed by a PAM solid-rocket motor. Designed to deliver a perigee-kick to send a small satellite up to geostationary orbit, this motor had always been fired within hours of launch; but in the role of Magellan's braking motor it would need to endure months of exposure to space, and relying on it was a calculated risk.

VOIR had been intended to employ the three-stage planetary IUS, but ironically, as soon as the scaled-down mission was approved it was reassigned to the more powerful Centaur! It remained in this configuration until Challenger was lost. Although there seemed initially to be every chance that it would still meet its projected 1988 window, the cancellation of the Centaur in the summer of 1986 forced JPL to adapt Magellan to match the limitations of the two-stage IUS. This was fairly straightforward, because cash-strapped Magellan had not been allowed to grow to exploit the Centaur. As it happened, the IUS would only *just* be able to accelerate the 1,150-kg spacecraft and its 2,200-kg braking motor to escape velocity. Also, the 20-tonne stack represented a full load for a shuttle with its SSMEs restricted to 104 per cent of rated thrust. The slimmed-down VOIR was, therefore, only just within the shuttle's capability.

AWAY AT LAST

STS-30's launch on 28 April 1989 – the day the window opened – was scrubbed at T-31 seconds; as soon as the onboard sequencer was assigned control, it had found a fault in one of the main engines. The next attempt on 4 May was completed without incident, however, and Atlantis became the first shuttle to carry a planetary spacecraft into orbit. But this was not the optimum window for Venus. Because the only way the IUS could send Galileo to Jupiter was via a circuitous route which began with a Venus flyby, it had been decided to assign the minimum-energy October window to Galileo. Ironically, although Magellan was being dispatched early, the slow 15-month trajectory that it was to take meant that it

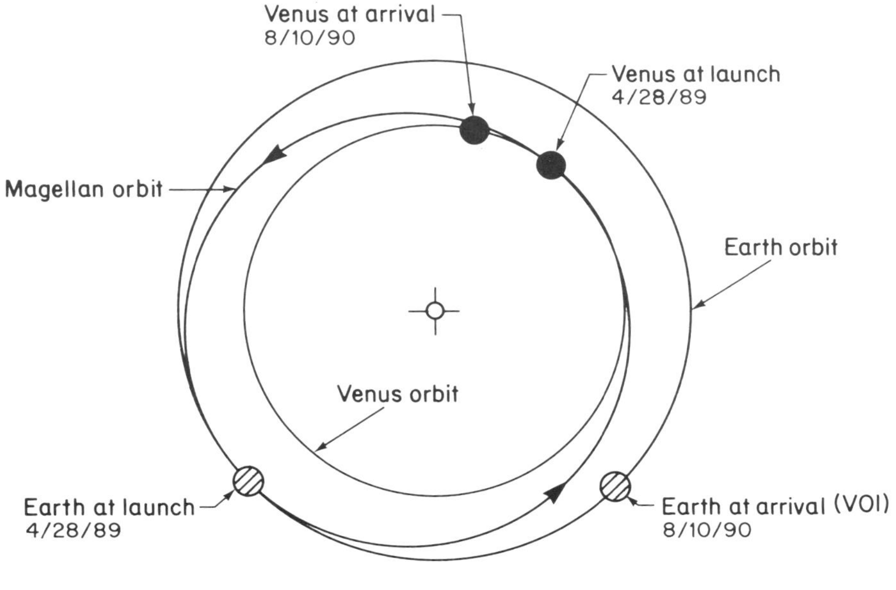

Fig. 15.1. A schematic of Magellan's cruise to Venus.

(a)

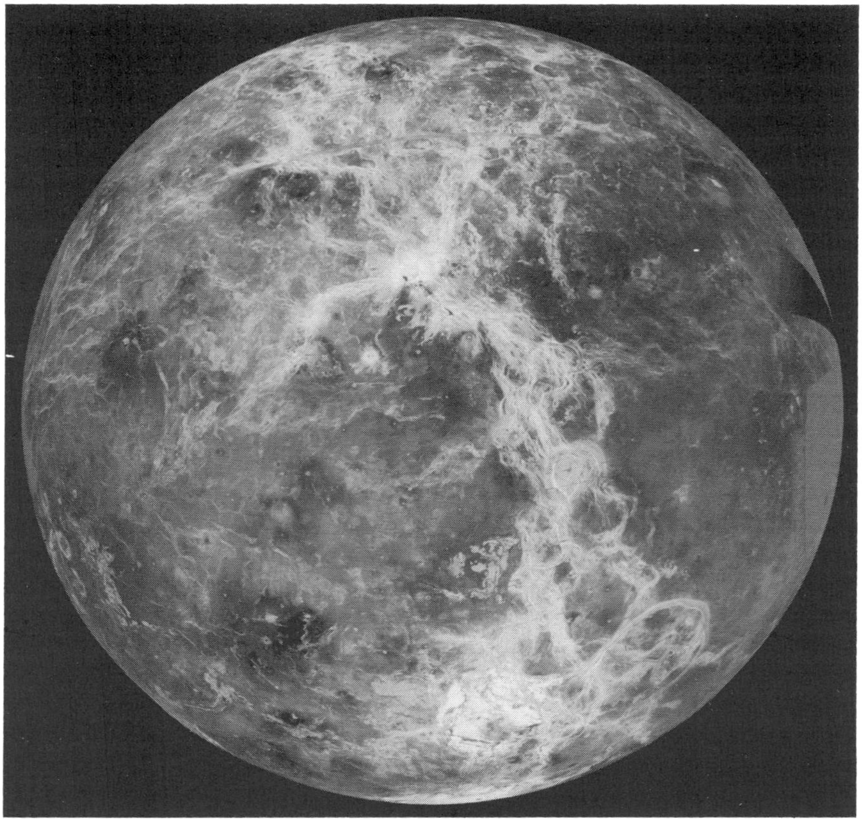

(b)

Fig. 15.2. Magellan: (a) riding an Inertial Upper Stage (IUS) for deployment, and (b) a rendition of one hemisphere of Venus derived from Magellan data (the contrast in this image indicates the extent to which the planetary surface reflects radar energy, the surface is permanently obscured by cloud).

would not reach Venus until August 1990, fully six months after Galileo had passed by. That such a compromise was necessary was a consequence of NASA's post-Challenger reluctance to try to launch two shuttles within a single window. Magellan was lucky to retain this early slot, because the shuttle's return to service had been delayed to September 1988, and there was a queue of high-priority payloads in line for the initial missions.

The IUS stack was released by Mark Lee some six hours after Atlantis reached orbit. The orbiter lingered until Magellan's solar panels were confirmed to have deployed, and then it withdrew. Half an hour later, the IUS fired its two stages in succession. It put Magellan on an extremely accurate trajectory, so only a modest mid-course correction was needed; this was made on 22 May.

When Magellan arrived on 10 August 1990, Ames' Pioneer Orbiter was still operating. Despite concern that it might misfire, the solid rocket motor burned perfectly. A set of small hydrazine thrusters actively damped out the coning effect so as to ensure that Magellan was deposited into the necessary 250 × 8,750-km orbit, and the entire braking package was then jettisoned. Over the next three weeks, data were sampled to check out the radar, which scanned for 18 minutes either side of the lowest point. Its resolution varied with altitude, but it was at its 100-metre best at latitude 10° N. Magellan suddenly fell silent on 16 August, but it came back on line after 14 hours. However it shut down again on 21 August, and this time it took 18 hours to recover. The cause of the problem was a mystery, and if the fault recurred frequently it would disrupt the mapping. A bug was later identified in the software that took care of the navigational star sighting at apoapsis, but because it was an asynchronous fault it struck only intermittently; but when it did it confused the spacecraft's computer, causing it to run through a recovery procedure. When JPL realised what was happening, it revised the programme so that it would not take so long to recover.

MAPPING MISSION

The mapping mission started formally on 1 September. The planet's rotation made the ground-track migrate 18 km between successive mapping passes. Day by day, Magellan's 24 × 15,000-km strips of terrain – called 'noodles' – were joined together by the image processors at JPL to build up a continuous sequence which progressively advanced around the equator. The primary mission goal was to map 70 per cent of the planet in the first rotation (to avoid the confusing term of 'day' in the case of Venus, this period was dubbed a *cycle*), but it achieved 82 per cent coverage! Although the spacecraft had suffered a few problems, JPL secured funding for another cycle to cover the far-southern latitudes that it had missed, and which the Veneras had been unable to map. The second cycle began on 16 May 1991, ran to 15 January 1992, and increased the coverage to 96 per cent. In the third cycle, Magellan added stereoscopic studies of selected terrain objects for detailed 3-D analysis. A problem with the transmitter in July 1992 terminated routine mapping; by this point, 21 per cent of the surface had been covered stereoscopically. For the next few months the radar was switched on only to fill in gaps in the map. At the end of the cycle on 13 September, the overall coverage had increased to 98 per cent.

GRAVITY SURVEY

But a bonus was on offer. On 14 September 1992, Magellan fired a thruster to lower its periapsis from 250 km to 180 km. This made it more sensitive to variations in the density of the planet immediately below, so the spacecraft could be used to make a gravity survey that would allow the internal structure of the planet to be inferred. Magellan now kept its antenna facing the Earth and transmitted only a carrier wave, and monitoring of the Doppler effect on this signal allowed the variations in the way the spacecraft moved to be deduced, and over time, attributed to variations in the gravitational field.

As Magellan completed its fourth cycle on 25 May 1993, it fired its thruster again. This lowered the periapsis from 172 km to 145 km. By skimming the atmosphere, it could use the resulting aerobraking to progressively lower its apoapsis. This was done because the gravity survey worked best at low altitude, and lowering its apoapsis would allow the spacecraft to spend a higher proportion of each orbit in close proximity to the planet. Aerobraking had never been done before, so this represented another experiment. It allowed a major orbital change to be made without using much propellant. (It was so successful that JPL decided to use aerobraking to adjust the orbit of Mars Global Surveyor in 1997, so that it could carry extra instruments instead of propellant.) By 3 August 1993 the apoapsis had been lowered to 540 km, so the low periapsis was lifted out of the atmosphere. The gravity survey continued until the end of the sixth cycle, on 10 October 1994. The data covered 95 per cent of the planet. In effect, as a bonus on top of its original mission, Magellan had surveyed the inner structure of Venus in greater detail than had been done for the Earth. But by this point, Magellan was in pretty bad shape, and the funding had dried up.

LAST ACT

On 12 October, Magellan penetrated so deeply into Venus' dense atmosphere that it was unable to re-emerge, and it burned up. Even in this final act, it performed the 'windmill' experiment in which its solar panels were offset and faced into the wind so that the drag encountered in its final moments would induce a spin, the rate of which measured the atmospheric density. Pioneer Venus had succumbed to a similar fate on 9 October 1992; it had been assigned a single-cycle mission, but had functioned for 21 cycles (14 Earth years). Pioneer Venus and Magellan were totally different, but complementary, spacecraft. Pioneer Venus provided long-term monitoring of Venus' atmospheric structure, and Magellan mapped its surface. The surface pressure of the primarily carbon dioxide atmosphere is 90 times that of the Earth's at sea-level; the air pressure on Venus corresponds to the water pressure at an ocean depth of 1,000 metres on the Earth. Furthermore, the temperature at the surface is a fairly uniform 470 °C. Venus is a 'runaway greenhouse' on an awesome scale. The low-resolution radars on the Venera spacecraft had revealed strange topographic structures on the surface, and when Magellan was dispatched, the nature and origin of these features was still a mystery.

Magellan revealed that Venus is a world dominated by volcanism, with a staggering array of different types of feature, but there was *no* convincing evidence of plate tectonics. Unlike the Earth and Mars, whose surfaces have two general levels, most of Venus is at the same level. Although there are mountain belts, there are no large basins, and although there are craters in profusion, there are few small ones; bodies that would reach the Earth's surface are destroyed by the dense Venusian atmosphere. The crucial clue provided by the cratering record was that a spasm of volcanism 500 million years ago resurfaced the entire planet. Although Venus is dimensionally similar to the Earth, the two planets are certainly not twins.

From a scientific viewpoint, Magellan was clearly a total success. Although the original concept had to be scaled down to release funds to help cover the shuttle's overruns, following revisions of its design, the mission went ahead as planned. It was only slightly delayed by the loss of the Challenger, and easily recovered from the cancellation of the Centaur stage.

In terms of operating the shuttle, the dispatch of this first planetary spacecraft marked a significant milestone, but, from the shuttle's point of view, Magellan was simply a payload in need of transport into low orbit.

16

Home planet

It is sometimes the case that an understanding of a complex system can be attained only when it becomes practicable to stand back and observe the system as a whole. One of the primary benefits of placing satellites into orbit around the Earth is that they can scrutinise their home planet.

But it is not simply a matter of looking down from a new height; it is the combination of the *global perspective* with *time-resolution* that is so important to observing the Earth's processes at work. A satellite circling the globe every few hours can not only survey a wide area; it can do so rapidly, thereby enabling the processes to be observed in fine detail. In 1957–1958, teams of observers were sent far and wide across the globe to undertake the International Geophysical Year (IGY) study, but even this enormous endeavour was unable to reveal much information concerning the Earth as a *system*.

THE ATMOSPHERE

The atmosphere is structured in layers, each with characteristic physical and dynamical properties. It is convenient to classify these layers in terms of their temperature profiles.

The troposphere is the lowest layer. Depending on latitude, it extends to an altitude of between 10 km and 15 km. This layer contains most of the cloud formations, and circulates the weather system. The temperature decreases with increasing altitude, so it was natural to presume that this trend continued all the way out to the boundary with space. However, it was discovered that there is a reversal of the trend at an altitude of about 20 km. Indeed, by 50 km, in the stratosphere, the temperature is back to that at sea-level. Thereafter, the temperature decreases rapidly, and reaches its minimum at 80 km, in the mesosphere. Above, however, in the low-density thermosphere, the temperature soars.

This layering is a stable feature, so there are distinct boundaries. The upper edge of the troposphere is called the tropopause, that of the stratosphere is the stratopause, and that of the mesosphere is the mesopause.

The existence of the stratosphere as a distinct atmospheric feature was discovered in the nineteenth century, when instruments were flown on balloons. At that time, the atmosphere was divided into the troposphere and the stratosphere, which was presumed to be synonymous with the upper atmosphere. This world-girdling region between 20 km and

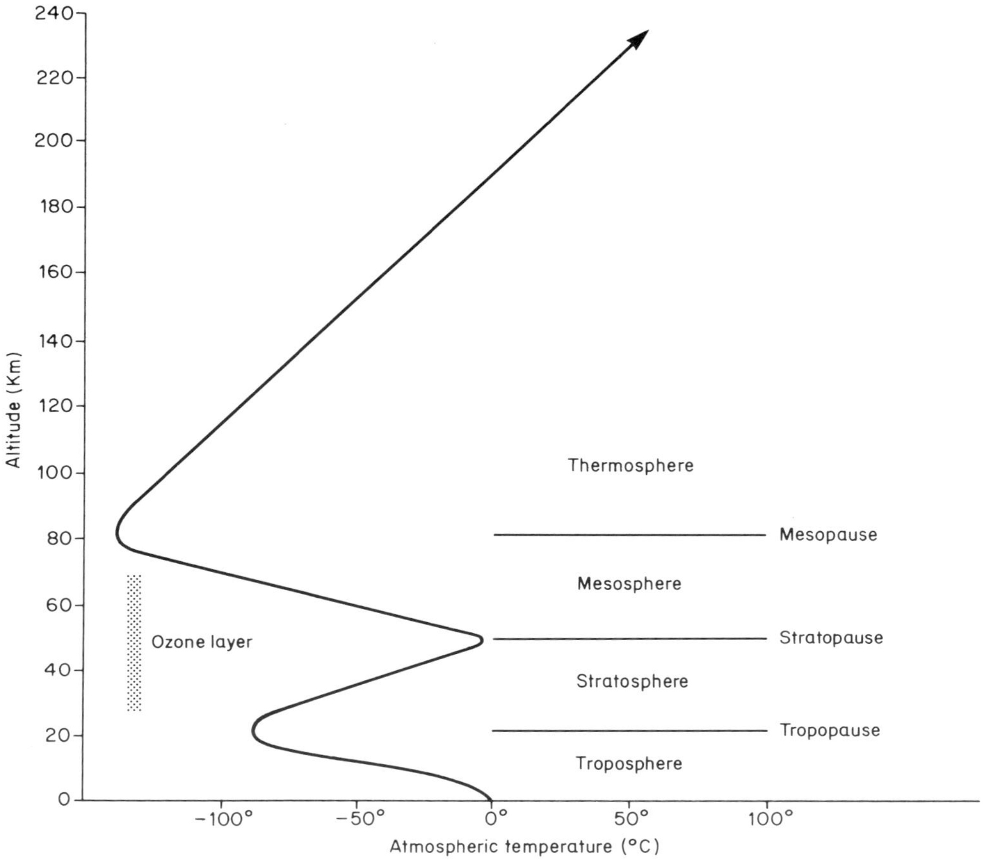

Fig. 16.1. A schematic of the thermal layering of the Earth's atmosphere.

50 km is a true layer, because vertical motion across the tropopause is fairly rare. Stratospheric winds are fierce, however, so it undergoes extensive horizontal mixing. The stratosphere is the site of strong interaction of the radiative, dynamical and chemical processes. In 1930 Sidney Chapman finally explained the mystery of stratospheric warming: it is ozone absorbing the ultraviolet from the Sun. The heating effect is strongest near the stratopause, the altitude at which the ozone is most concentrated. If it were not for the ozone, this thermal structuring would not be present, and the atmosphere *would* have progressively cooled all the way up to the mesopause. It arises because, as the gas density decreases, radiative cooling becomes dominant. Above 80 km, the mesopause, the temperature rises again because the rarefied atmosphere is dominated by atomic oxygen, which is excited by sunlight. Oxygen therefore plays a crucial role in determining the temperature profile of the upper atmosphere. In the lower atmosphere, oxygen exists as undissociated di-atomic molecules (O_2). The intense sunlight at high altitude dissociates the oxygen, but recombination of atomic oxygen with undissociated O_2 leads to the formation of tri-atomic molecules of ozone (O_3). It is a dynamic process, however; the ozone is in equilibrium of formation and dissociation. At even higher altitudes, the recombination rate de-

creases, leaving only the atomic oxygen. A surprising discovery made on an early shuttle flight was that a spacecraft glows as a result of its interaction with atomic oxygen, a phenomenon that was promptly dubbed 'shuttle glow'.

Lightning is a high current electrical discharge that occurs when two or more electrical charge centres in the troposphere become connected through a discharge channel. Because the discharge generates very high temperatures, the surrounding air expands in a shock wave producing thunder. Various mechanisms have been proposed for the electrical charge generation within a thunderstorm, but the process is stimulated when rapid convection transports water droplets, ice particles or specks of dust in a severe updraft. Although lightning discharges occur as cloud-to-ground flashes, they also occur as intracloud flashes or as flashes between neighbouring clouds and, because these are visible from above as well as from below, they can be studied from space.

The Night-day Optical Survey of Lightning (NOSL) project involved a 16-mm film camera with a zoom lens, a diffraction grating on the lens to show spectral information, and an optical sensor for flash timing. It was a hand-held unit, and the astronaut's commentary was recorded on tape. Although the crew of STS-2 were alerted by meteorologists four times, they managed to observe only one storm and to record only one discharge, which was visible on four successive frames. It was a poor return on the time involved, but later flights had greater success. Indeed, on STS-4, in one particularly severe storm complex over South America, the astronauts noted that sympathetic lightning events progressively transferred the locus of activity across distances of 600 km. Such large-scale coordination is not readily observed from the ground because the field of view of any particular observer is so restricted; in fact, this was subsequently defined as 'mesoscale' lightning. NOSL's objective was to derive a correlation between the degree of lightning and the convective state of a storm. A weather satellite which could identify storms likely to give rise to tornadoes would enable specific warnings to be issued to threatened areas.

STS-26 saw the introduction of the Mesoscale Lightning Experiment (MLE). This used a payload bay video camera fitted with a low-light imager, but unlike NOSL this new project did not require crew participation; the camera was remotely-controlled by Houston. It was a *synoptic* programme in the sense that it recorded the weather on the groundtrack while the orbiter flew through the Earth's shadow, but because it was so straightforward it could be done on every flight, and it formed a long-term study. Given knowledge of the orbiter's position, it was possible to determine the size of the lightning storm, as well as measure its intensity. Whenever the orbiter passed near a storm that was within observational range of ground-based lightning-detection stations, if the astronauts were free they would be invited to make direct observations.

The magnetosphere is a consequence of the strong magnetic field created in the Earth's core. It is an electrically-conducting plasma of charged particles, extending down into the mesosphere. Its lower region is the ionosphere, and its upper boundary, which is buffeted by the solar wind, is the magnetopause. A surprising discovery was that in addition to discharging to the ground, thunderstorms can launch intense bolts of lightning *upwards* towards the ionosphere. Several distinct forms of such lightning, dubbed 'sprites', have been noted.

The thermosphere has no specific boundary; it simply extends far out into space with ever-decreasing density. At 250 km, which is the lower edge of the shuttle's operating zone, the pressure is one 10-billionth that at sea-level, but this is still 10 million times that out in interplanetary space.

The gas density at the base of the thermosphere imposes sufficient drag to claw down a satellite in a matter of hours, but Explorer 51, launched in 1973, was manoeuvrable. It spent most of its time above 160 km, but it periodically fired its motor to lower its perigee in order to make a few passes down near the mesopause, and then boosted itself back to safety. It is impracticable to sample the mesosphere by this means, however, and the stratosphere is totally unreachable, so remote-sensing techniques have to be employed. Early results from Explorer 55 in 1975 prompted concern that ozone was being depleted by transient events such as plumes of ash from volcanoes, and interest immediately focused on the aerosols, the fine spray of tiny droplets suspended in the rarefied gas. In 1979 the second Applications Explorer (AEM-2) flew the Stratospheric Aerosol and Gas Experiment (SAGE), and for a decade this photometer observed the Sun during orbital sunrise and sunset. By monitoring the spectral absorption features due to the passage of sunlight through the atmosphere, it measured ozone and aerosols concentrations, and because the light's trajectory was horizontal it was able to measure this concentration as a function of altitude as well as location. This enabled the distribution of aerosols in the atmosphere to be recorded in three dimensions. Furthermore, over time it was possible to monitor the dynamics of the ozone layer. AEM-2 flew in a 55-degree inclination orbit. Another satellite, Nimbus 7, was launched in 1978 to extend the coverage into the polar regions. This far bigger satellite carried a wide range of instruments: the Total Ozone Mapping Spectrometer (TOMS), the Stratospheric Aerosol Monitor (SAM), the Earth Radiation Budget (ERB), and the Solar and Backscattered-UltraViolet (SBUV). The TOMS was really an improved type of SAGE. A third satellite, the Solar And Mesosphere Explorer (SAME), launched in 1981, employed a Sun-synchronous polar orbit in order to study the processes controlling the equilibrium in the ozone cycle.

The British Antarctic Survey based at Halley Bay had been monitoring overhead ozone concentrations since the IGY. In the mid-1970s, it noted the appearance of a progressively more significant depletion during the southern summer. Each year, the ozone concentration fell rapidly during the early spring, remained low throughout the summer, then slowly recovered. Even though the concentration in 1982 fell by 20 per cent, the data were not published because TOMS data did not show anything unusual. In 1984, the ozone concentration fell by 30 per cent. When another site confirmed its reality, the results were written up, and a paper appeared in *Nature* in May 1985. Upon re-examining its TOMS output, NASA discovered that a 'filter' in the analysis programme had been set to reject 'spurious' data, so it had failed to report this unexpectedly pronounced variation in Antarctic ozone. Fortunately, the raw data dating back to October 1978 had been archived, and when reprocessed this not only confirmed the Halley Bay discovery, it revealed the true scale of the phenomenon. The STS-13 mission of October 1984 turned out to have been extremely well-timed.

RADIATION BUDGET

Sally Ride used the RMS to lift the 2,275-kg Earth Radiation Budget Satellite (ERBS) out of the bay. When one of the two 7-metre solar panels jammed, she turned it to face the Sun to warm the stuck mechanism, and then shook the satellite until the panel unfolded and locked into position. It was a three-axis stabilised platform that kept its instruments facing the ground-track, and it had a steerable TDRS antenna to relay data. Over the next

10 hours it used its thrusters to climb to 600 km. At 57 degrees inclination, ERBS covered the same latitude range as the still-operating Applications Explorer. In addition to an advanced form of the Applications Explorer's instrument (this one known as SAGE 2), it had Nimbus 7's ERB instrument. Further ERB sensors followed on NOAA 9 in December 1984, and yet another on NOAA 10 in September 1986. Because these latter satellites were to fly in polar orbit, they had to ride Atlas rockets rather than the shuttle. They all measured the vertical distribution of stratospheric aerosols, ozone, water vapour, nitrous oxides, and other pollutants with an altitude resolution of 1 km. The NOAA satellites were operated by the National Oceanic and Atmospheric Administration. All of these data were made available to the National Center for Climate Control.

For purposes of analysis, the Earth was divided into a matrix of 1,000 km^2 'cells'. The orbits of the mid-inclination satellites precessed round the equator in such a way as to yield a monthly average for each cell within their latitude range. As the Earth rotated under them, the polar satellites were able to monitor the entire atmosphere on a daily basis; and because their orbits were not Sun-synchronous, they passed over each cell slightly later each day, and so yielded detailed diurnal studies every month. Between them, these satellites mapped local, zonal, regional and global variations, and over time they enabled seasonal variations to be monitored. Such long-term worldwide data were basic to achieving a real understanding of the solar–terrestrial relationship.

In the tropics, the Earth absorbs far more energy from the Sun than it radiates into space. At higher latitudes less energy is received, but the leakage continues. The flow is entirely one-way at the poles during the extended periods of darkness. The differences in temperature in different locations drive the weather system in which atmospheric and oceanic circulation try in vain to redistribute this energy in order to establish a state of thermal equilibrium. The oceans serve as a store for this energy; if it were not for the oceans, the Earth would have a very different climate. The 'radiation budget' – the net energy reaching the surface – is the key measure of the Sun's influence on the Earth.

The radiation budget is dependent on variation in solar energy reaching the Earth, on the veiling of the lower atmosphere by dust, ice crystals and aerosols in the upper atmosphere, and upon the amount of carbon dioxide, the greenhouse gas which, by soaking up thermal energy, reduces leakage to space. The first step was to determine the energy spectrum of insolation at the Earth's distance from the Sun, then to establish any cyclic and longer-term divergent trends.

The Earth's orbit is not circular, but slightly eccentric. On average, the Earth is about 150 million km from the Sun, but it is a few million kilometres closer during the northern winter than during the southern winter. This introduces an annular variation in insolation. It is possible that the magnitude of this eccentricity varies over a long timescale, in which case there will be a long-term cyclic variability in insolation.

The output from the Sun is not constant, as sunspots cause the energy emitted by the Sun to vary over an 11-year cycle. And, of course, the Sun is slowly evolving. It may even be a 'variable star' over very long timescales.

The Earth's atmosphere is essentially transparent to visible light, but it absorbs longer wavelengths, and over most of the infrared spectrum the atmosphere is opaque. Some of visible component of sunlight is reflected back into space by stratospheric clouds. That which passes is scattered. Some is reflected but the rest is diffused, which is why the sky appears blue. Much of what reaches the lower atmosphere is selectively absorbed, depending on the chemical composition of the atmospheric gases. Light is absorbed by electrons jumping straight to high states of excitation within individual atoms, but it is not re-radiated as light; the electron drops back to its ground state in a series of steps, so it gives rise to a veritable cascade of low-energy photons. In the infrared region of the spectrum, the energy is absorbed by gaseous molecules, and it makes the component atoms vibrate. Because such molecules are stable, they lock this energy in the lower atmosphere. Water vapour and carbon dioxide are particularly effective at soaking up thermal energy. In fact, the Earth's surface would be 35 degrees cooler if it were not for the 0.3 per cent of carbon dioxide in the atmosphere. Only a small fraction of this thermal energy manages to leak into space by radiative cooling.

To calculate the radiation budget, therefore, it is necessary to determine the insolation, subtract the energy reflected back into space, and then subtract, from the energy that reaches the lower atmosphere, that which leaks into space by radiative cooling. The residual is the energy which serves to heat the Earth, drive its weather system and sustain its biosphere.

While SAGE monitored the concentration of stratospheric agents, ERB measured their effect. Each ERB comprised two instruments, one of which aimed four cavity-radiometers straight at the ground. Two of these nadir-pointing sensors had fields of view sufficiently wide to view the entire disk of the Earth from limb to limb, and the others used narrower fields for improved resolution. One of each pair measured reflected insolation, and the other measured escaping thermal energy. A fifth radiometer was aimed at the Sun to measure broadband insolation. The second instrument scanned narrow-field sensors to either side of the ground-track, and scanned a swath as the satellite travelled along its orbit. Of the three thermistor-bolometers; one sensed reflected insolation, one sensed emitted thermal radiation, and the third recorded across the entire range in order to measure the Earth's total output. Overall, therefore, ERB measured solar insolation, visible and ultraviolet reflected radiation, and infrared emission. Correlating the variations in the radiation budget with the effects observed by SAGE would be crucial to understanding the role of the stratosphere in the solar–terrestrial relationship.

These satellites provided sufficient *hard* data to enable computer models of the weather and climate to be calibrated. It had been difficult to compute the precise effects of cloud and carbon dioxide as factors affecting warming and cooling. Insolation is reflected primarily by clouds and snow, so the extent and location of cloud and snow cover is a crucial factor. It is an interconnected system, because significant warming affects oceanic and atmospheric circulation, which affects the weather, which in turn affects the extent, type and distribution of cloud cover. The observations confirmed that cloud cover is a net cooling factor, so both the extent of cloud cover and the type of cloud are important because different types of cloud have different reflectivity.

In March 1982, STS-3 had tested an instrument to measure the ultraviolet component of insolation. This Solar Ultraviolet Spectral Irradiance Monitor (SUSIM) had been developed by the Naval Research Laboratory. Beginning with the flight of Spacelab 2 in July 1985, SUSIM was flown as often as possible. Its role was to determine whether the energy in this part of the spectrum varied over the solar cycle. Finding out was important, because ultraviolet radiation played a vital role in the ozone cycle. Artificial chlorine compounds form a major constituent of the upper-atmospheric aerosols; the ionising radiation dissociates the compounds, and liberated chlorine then reacts with the ozone. This was suspected of being the active agent in ozone depletion. Although the objective of STS-34 in October 1989 was to dispatch the Galileo spacecraft, it also carried a GAS package with the Shuttle-SBUV (SSBUV), a rebuilt form of the instrument carried by Nimbus 7, NOAA 9 and NOAA 11 (the most recent in this series, added in 1988). This radiometer measured the fraction of the ultraviolet insolation which was reflected by the stratosphere. It flew on the shuttle frequently, not so much to collect data, which was being done by the satellites, but to calibrate those other instruments in order to guard against their falsely identifying long-term trends. Space is a harsh environment, and exposure to solar ultraviolet and to atomic oxygen escaping from the Earth's atmosphere can degrade sensors.

Between them the SAGE instrument on ERBS and TOMS on Nimbus 7 showed that in 1987 the ozone over Antarctica fell to 50 per cent, which was sufficient to define this as a 'hole' in the stratosphere. This hole was less serious in 1988, but grew progressively worse with each subsequent year. The ozone is not a static layer; it is in a state of dynamic equilibrium, with creative and destructive processes in balance. The depletion is the result of fluctuations in the rates of these processes in which the destructive process is stimulated by the presence of the chemically active aerosols. It was ironic that although the industrialised nations were concentrated in the northern hemisphere, their pollution was having an influence at the south pole. In the winter of 1988–89, however, slight depletion of *northern* ozone was noted for the first time, so clearly it was a global problem. Combining data from Nimbus, NOAA, SolarMax and ERBS revealed that the output of the Sun varied by "a few tenths of a percent" on a two-week timescale; that is, half of a solar rotation. Measuring the variability in its output during a full 22-year magnetic cycle is a key objective. It has been calculated that if the Sun varied over a period of hundreds of thousands of years, with a systematic change of 0.5 per cent per century, it could account for all of the Earth's past climate changes. Picking such a trend out from the shorter cycles, and the 'noise' of sunspots, will be difficult, but nothing could have greater long-term significance for the future of humankind.

Although it was crucial to monitor long-term trends, to establish those which were cyclic from those which might be divergent, no scanning system operates indefinitely. NOAA 9's ERB scanner failed in January 1987, NOAA 10's in May 1989, and that of ERBS followed in February 1990. In each case though, the nadir-viewing instruments continued to operate. SolarMax fell back into the atmosphere in 1989, and the Solar and Mesospheric Explorer was crippled by a battery fault in 1989. No satellite had done more to monitor stratospheric ozone depletion than Nimbus 7, which lost both its ERB and TOMS instruments in early 1993. Luckily, another TOMS had been launched on a Russian Meteor satellite in August 1991, and, a month later, NASA had deployed the first satellite of its new Mission To Planet Earth (MTPE) initiative, which took its name from the phrase coined by Sally Ride, in 1987, in writing a far-sighted report for future NASA objectives: *Leadership and America's Future in Space*. (Mission to Planet Earth has recently been renamed by NASA, and is now the Earth Science Enterprise.)

UPPER ATMOSPHERE RESEARCH SATELLITE

At 7 tonnes, the Upper Atmosphere Research Satellite (UARS) was the biggest package devoted specifically to atmospheric research. Discovery climbed to its maximum operating altitude to place the satellite in the required 600-km orbit, which at 57 degrees was as highly inclined as the shuttle could economically attain. The suite of 10 instruments was to carry out the first *systematic* study of the upper atmosphere's physical and chemical processes. After three days of checking out the instruments, Mark Brown hoisted the UARS out of the bay and deployed the single large solar panel and the boom with the TDRS antenna. Based on the experience with the Gamma Ray Observatory, on which the boom had jammed, Sam Gemar and Jim Buchli were standing by in the airlock, ready to overcome any problems. On this occasion though, there were no problems, and the satellite was released as planned.

Fig. 16.2. An artist's impression of the Upper Atmosphere Research Satellite (UARS).

The basic spacecraft systems were carried on the same type of Fairchild bus as was first used by SolarMax. Indeed, the attitude-control controller was that retrieved from SolarMax, refurbished for reuse. However, the massive instrument module on the triangular bus bore little resemblance to the compact form of SolarMax. UARS employed a combination of boxes, cylinders, antennas and booms that projected at all angles. In addition to measuring vertical temperature profiles and mapping the concentrations of ozone, methane, water vapour and aerosols, the instruments were to chart the upper atmospheric winds that shape the global distribution of chemicals which filter up from the lower atmosphere, in order to compile a comprehensive database on the mechanisms controlling the structure and variability of the upper atmosphere. Four of these instruments were specifically devoted to the ozone issue, four were devoted to measuring solar energy, and two measured winds. A large part of the value of this package was its ability to perform *coordinated* observations. UARS' primary objective was to monitor two northern winters. Of course, it was confidently expected that it would operate for at least five years and, with a little luck, possibly until the follow-on MTPE satellites could take over.

The Microwave Limb Sounder (MLS) was a radiometer for detecting thermal emission from vibrating molecules. Its antenna was aimed at the limb in order to integrate energy from the thickest possible air sample. Like occultation measurements at sunrise and sunset, this had the advantage of revealing vertical structure, but because MLS directly sensed molecules by their emissions rather than indirectly by their absorption of sunlight, it was able to take data continuously. In practice, MLS was able to sense chlorine monoxide, hydrogen peroxide and water vapour at altitudes as low as 16 km, with 3-km vertical resolution. Built by JPL, it used a new semiconductor transducer to turn the millimetre range incident energy into radio that could then be readily amplified. It provided, for the first time, a global survey of the key stratospheric aerosol, chlorine monoxide.

The Halogen Occultation Experiment (HOE), supplied by the Langley Research Center, made observations of the Sun's passage over the Earth's limb to measure the distribution of hydrofluoric acid, hydrochloric acid and assorted nitrous agents, as well as methane, water vapour and carbon dioxide.

The National Center for Atmospheric Research built the Cryogenic Limb Array Etalon Spectrometer (CLAES). This infrared emission spectrometer measured the distribution of nitrogen and chlorine compounds, ozone, methane and water vapour.

The Improved Stratospheric And Mesospheric Sounder (ISAMS) was supplied by the University of Oxford's Department of Atmospheric Physics. This infrared radiometer was able to measure carbon dioxide, carbon monoxide, nitrous oxide and nitric acid, as well as methane, ozone and water vapour.

The overlap in detector capabilities facilitated cross-checking to ensure that there was no instrumental bias. Methane and nitrous oxide were measured because, like carbon dioxide, they act as greenhouse agents. In addition to revealing chemical composition, such infrared detectors were able to determine temperature profiles. Water vapour is an excellent indicator of atmospheric motion. The University of Michigan's High-Resolution Doppler Imager (HRDI), and the Wind Imaging Interferometer (WINDII) from Canada's York University, were to monitor wind speed and direction. HRDI measured in the 10–45 km altitude range, while WINDII operated above 80 km. Both instruments used interferometric analysis of Doppler variations in spectral data. The result was the first global survey of upper atmospheric wind patterns.

The TOMS instrument on Nimbus 7 had revealed *rivers* of water vapour hundreds of kilometres long flowing in the lower atmosphere, some with as much water as the Amazon, and it was suspected that other atmospheric constituents were concentrated into such flows. The UARS discovered continent-sized windstorms raging in the mesosphere. Two patterns were discernible: migrating diurnal *tides* sweep around the world once a day, remaining in sunlight, while non-migrating tides are fixed over the major landmasses, and wax and wane with the diurnal cycle. Although these winds rage at 300 km/s, the very low density of the air at that altitude means that they harness very little energy. The upper atmosphere is rather more dynamic than had been believed.

Solar energy is the driving force behind the Earth's weather system, so it was vital that this be monitored in parallel with the atmospheric observations. A SUSIM instrument, carried for the ozone work, measured the ultraviolet component of insolation. The University of Colorado had supplied the SOLar–STellar Irradiance Comparison Experiment (SOLSTICE) to test a technique for checking the calibration of instruments such as SUSIM. Hot blue stars emit strongly in the ultraviolet, and a selection of such stars was measured by SOLSTICE to create a reference against which variations in insolation could be checked. One of the instruments tested on the first Spacelab flight in 1983 was JPL's Active Cavity Radiometer Irradiance Monitor (ACRIM); the UARS carried an improved version to measure the total output from the Sun. The Particle Environment Monitor (PEM) was an X-ray spectrometer supplied by the Southwest Research Institute in Texas to study the direct effects of solar wind particles impinging on the upper atmosphere.

The stabilised UARS flew with its single 10-metre solar panel out to one side so that its chemistry and wind sensors had a clear view of the nadir and of the limb on the far side, and the solar instruments were on a scan-platform on the front of the bus in order to track the Sun. Although the satellite was not in polar orbit, its sensors could survey up to 80° latitude, and it was readily able to follow the Antarctic ozone hole. With each passing year, the dynamics of the processes became better understood. In winter, when insolation is at a minimum, the air over the continent forms an isolated vortex that locks in the aerosols. They are then frozen into droplets in stratospheric clouds, and so serve as platforms for slow chemical reactions. Upon the Sun's return in the spring, photochemical processes tip the ozone equilibrium in the direction of depletion. Chlorine monoxide has a particularly avid affinity for ozone, and its abundance has been increased by human activity. This potentially runaway reaction tails off with the dissipation of the clouds in the summer, at which time the altered atmospheric circulation lets ozone flow in from subpolar latitudes. It is this factor which makes the hole a cyclic feature. The situation in the northern hemisphere is less conducive to the formation of a distinct hole, but depletion can take place if local conditions are conducive. The UARS proved conclusively that it was the *artificial* compounds in the stratosphere that caused the ozone hole, and that, as had been suspected, chlorofluorocarbon (CFC) agents were the real culprits. The Antarctic hole not only grew deeper each year; it developed earlier, lasted longer, and was bigger. In 1992, it was almost as large as the continent itself and extended over the tip of South America, possibly because it was enhanced by particulates injected into the stratosphere by the eruption of Mount Pinatubo in the Philippines in June 1991.

The ultraviolet component of solar insolation in the wavelength range nearest the visible spectrum – the so-called UV-A – is totally absorbed by the atmosphere, and is harmless. It is the UV-B component of insolation (in the range 2900–3200 Å) that is of concern. In areas of the stratosphere with severe ozone depletion it passes straight through, and this is dangerous to life on the surface because even brief exposure can promote skin cancer. The shorter-wavelength ionising radiation (UV-C) would be even more dangerous if it were to penetrate, but this is absorbed by molecular oxygen. The extreme-ultraviolet region (shorter than 1800 Å) is absorbed by thermospheric atomic oxygen. Clearly, although ozone forms only a tenuous layer (only a few parts per million) in the upper atmosphere, it is of vital importance to life on the surface.

The good news was that, as of 1990 the concentration of certain stratospheric aerosols *fell* as a result of the limits placed by the Montreal Protocol on ozone-depleting chemicals.

Methyl chloroform, in particular, had risen 4 per cent per annum since records began in 1978, but starting in 1990 it dropped progressively by 2 per cent per annum. Although this 'reversal' was encouraging, it was only one of many agents. Even if the emission of all destructive agents is terminated, it may take a *century* for the reservoir in the lower atmosphere to deplete.

The situation with global warming is less clear cut. The onset of such a runaway trend diverges only marginally from the norm, so it is easily hidden by natural fluctuations. ERBS gave the first direct evidence that volcanic plumes can cause significant change in the radiation budget. Mount Pinatubo – a particularly energetic eruption – injected a plume of sulphur dioxide to an altitude of 30 km, to form a stratospheric haze of reflective droplets of sulphuric acid. This haze rapidly spread throughout mid-latitudes, and it prevented sufficient light penetrating the atmosphere to lower the global average temperature by 0.3° to 0.6 °C. This global cooling counteracted the ongoing warming for several years. As a rule, carbon dioxide levels have increased annually since measurements began for the IGY in 1958, and the heat capacity of the atmosphere has matched this. Mysteriously, this relentless rise in carbon dioxide tailed off in 1991 and then remained stable until 1994. A similar pause had been seen in the increase of carbon dioxide in the early 1960s, following the injection of a plume by a volcano in Bali, Indonesia. It was not clear though, *how* carbon dioxide was drawn from the atmosphere by a volcanic eruption. A 2 per cent increase in the Earth's radiation budget would be sufficient to melt the polar ice caps, and a similar decrease would restore ice-age conditions; in fact, strictly speaking we are still in an ice age. Whatever Mount Pinatubo's effect on carbon dioxide, the cooling by its stratospheric veil cancelled out a century's-worth of warming by industrialised human civilisation, which puts our effect on the planet into stark context.

1992 was designated the International Space Year (ISY), in order to focus international efforts to study the solar–terrestrial relationship, the Earth's atmosphere and global climate. In particular, it set out to determine the build-up of carbon dioxide likely to result from the depletion of the rain forests. Ironically, in July the UARS was nearly 'knocked out' by a spring which locked the drive of its solar panel, but repeatedly cycling the motor eventually shook the obstruction free, and routine operations resumed. However, a few months later the ISAMS failed; MLS failed in April 1993 and CLAES exhausted its coolant a few weeks later. Although the battery unit is not very healthy, it is sufficient for this reduced payload, so this satellite is still making a significant long-term contribution.

The UARS concentrated on the stratosphere, but the mesosphere's reaction to increased levels of methane and carbon dioxide could point the way for global change. Goddard SFC proposed the Thermospheric, Ionospheric, Mesospheric Energetics & Dynamics (TIMED) satellite for the first comprehensive study of the physical and chemical processes across the 60-km to 180-km altitude range. Unfortunately, funding imposed repeated delays, and this will not now be launched until the turn of the century.

ATMOSPHERIC LABORATORY FOR APPLICATIONS AND SCIENCE

The Atmospheric Laboratory for Applications and Science (ATLAS) was a Spacelab to provide detailed snapshots to supplement the long-term studies by the UARS.

Upon its inception, the ATLAS was known as the Spacelab Earth Observation Mission (SEOM). The first mission, set for September 1986 on flight 61K, was cancelled following the loss of Challenger. In its original form it would have employed a *short* pressurised module and a pallet, but when it was reincarnated as ATLAS, it was without the module. For ATLAS-1, a second pallet was added to accommodate instruments (ISO, AEPI, SEPAC, the Belgian Grille Spectrometer and the FAUST telescope) which had been assigned to the first SEOM, but subsequent missions employed only one pallet, with just the core package, so that a free-flyer could be carried as a secondary payload.

The main atmospheric monitor had three instruments. The Atmospheric Trace MOlecule Spectrometer (ATMOS), evaluated on Spacelab 3 in 1985, measured infrared absorption features at orbital sunrise and sunset to chart the concentration of atmospheric trace constituents. A new instrument was the Millimetre Atmospheric Sounder (MAS) flown for the Max Planck Institut in Germany. Like the MLS on the UARS, it detected microwave emission from vibrating molecules, so it could gather limb data continuously. It measured the concentrations of both ozone and of the chlorine monoxide produced during its depletion, and supplemented these data with temperature profiles. The SSBUV was carried to reassess the long-term calibration of similar instruments on satellites. All of the instruments devoted to measuring the Sun's contribution to the Earth's radiation budget had flown before. An ACRIM was flown to check on that carried by the UARS, so that its data could continue to be used with confidence. To illustrate the scale of this problem, the SUSIM instrument on the UARS lost 90 per cent of its sensitivity during its first three years of operation. The SOLar CONstant (SOLCON) and SOLar SPECtrum (SOLSPEC) instruments had been tested on Spacelab 1. These measured, respectively, the total output of the Sun and its energy spectrum across the infrared to ultraviolet range. The energy delivered in the different parts of the spectrum varies over the solar cycle. SUSIM accurately measured the ultraviolet component to follow its variability over the solar cycle, because many of the photochemical processes in the atmosphere are conditioned by this. All of these instruments had been finely calibrated so that they could test the underlying models. The detailed snapshots served to complement the long-term studies by the UARS and various NOAA satellites.

ATLAS-1 flew three other instruments to study the atmosphere. Atmospheric Lyman-Alpha Emission (ALAE), built by CNES in France, measured absorption in the extreme-ultraviolet to determine the abundance ratio of hydrogen to deuterium, the heavy form of hydrogen, because knowing this ratio would shed light on how water is processed by the atmosphere. The Belgian Grille Spectrometer noted infrared absorption when the Sun was on the limb, in order to chart stratospheric water vapour, methane and nitrogen compounds, and the Imaging Spectrometric Observatory (ISO) measured ultraviolet emission lines resulting from photochemical reactions, to provide information on the energy transfer processes in the upper atmosphere. Both of these instruments had flown on Spacelab 1, as had the instruments for plasma physics. The Atmospheric Emissions Photometric Imaging (AEPI) and Space Experiments by Particle ACcelerators (SEPAC) studied cause-and-effect relationships linking the ionosphere to the upper atmosphere. The Grille Spectrometer was subsequently repackaged as the Mir InfraRed Atmospheric Spectrometer (MIRAS) and sent up to the Mir space station in 1995 to undertake an extended study.

Although the eruption of Mount Pinatubo took place a few months prior to the launch of the UARS, the satellite had been able to monitor the initial phase of the spread of the volcano's stratospheric veil. By the time ATLAS-1 flew, the stratospheric winds had spread what had initially formed as a concentrated stream girdling the globe at the volcano's latitude, into a thin haze that almost covered both hemispheres. ATLAS provided a comprehensive chemical survey of the effects of this unusually massive volcanic plume. But one of the most significant results of the ATLAS mission was the detection by its ATMOS of significantly higher concentrations of chlorine and fluorine, the photochemical decay products of CFCs. The results could be directly compared with those of seven years previously: chlorine was up by 25 per cent, and fluorine was up by 70 per cent. It was no coincidence, therefore, that at its peak six months later, the Antarctic ozone hole covered the entire continent. During the following northern winter, total ozone fell by 10 per cent, so the second ATLAS flight was timed to study its recovery. The third ATLAS was set for November in order to observe the onset of the northern winter as well as the start of the Antarctic ozone hole's recovery. One specific objective was to study the southern vortex as it relaxed and allowed mixing between the ozone-depleted polar and the surrounding stratospheric circulation. The Millimetre Atmospheric Sounder (MAS) indicated that there was a significant day–night variation in ozone concentration at the level of the mesopause, with less ozone on the sunlit side of the Earth. It was in this coldest region of the atmosphere that the ozone was *most* vulnerable. Unfortunately on ATLAS-3, MAS suffered a power spike early on, and it never recovered. SSBUV verified the decrease in northern-hemispheric total ozone between the first two ATLAS flights. Although the Sun was at sunspot maximum in 1991, and was still very active in the years that followed, thereby complicating the task of measuring its total output, its disk was clear of spots throughout the second half of ATLAS-2, so ACRIM and SOLCON were able to make very accurate sustained measurements. The EURECA free-flyer, which was deployed in July 1992 and spent a year in space, carried versions of SOLCON (SOVA) and SOLSPEC (SOSP), so these two sets of data facilitated mutual confirmation. SUSIM showed a significant fall in solar ultraviolet after 1985, when the Sun had been near minimum activity, which was a greater variation than expected. To fully understand the solar–terrestrial relationship, it was clearly necessary to accumulate data over a long period. The initial plan was to fly ATLAS annually, but this proved to be too great an operational load. After its third flight the programme was further scaled-back, in part due to budgetary pressure but also to create opportunities for Shuttle–Mir. By building a *calibrated* database, the ATLAS flights in 1992, 1993 and 1994 made a substantial contribution to preparations for the MTPE satellites which will fly in the coming decades.

ATLAS-3 was combined with the first flight of a new SPAS free-flyer. This carried the Cryogenic Infrared Spectrometer and Telescope for the Atmosphere (CRISTA), which was to undertake the first global survey of atmospheric trace constituents with spatial resolution to show small-scale structures related to previously discovered wind phenomena such as the atmospheric waves and tides structures. CRISTA's detailed three-dimensional map was to enable models of the atmosphere and the processes affecting the Earth's radiation budget to be updated. It followed up on results from stratospheric balloons, which had hinted at the existence of such waves. CRISTA was the first satellite instrument to provide sufficient time resolution to sample with high spatial resolution. With the satellite moving

Fig. 16.3. The release of the Cryogenic Infrared Spectrometer and Telescope for the Atmosphere (CRISTA), an environmental-monitoring package which exploited the heavyweight SPAS free-flyer.

around its orbit at 8 km/s, a fast-acting spectrometer was required, and spectra were taken once a second. To achieve this speed, it was necessary to cryogenically chill the detector. It took a minute to scan the line-of-sight up and down across the altitude range, and this enabled the distribution of 15 gases to be monitored in unprecedented detail during the week that the satellite was in use. The Naval Research Laboratory's Middle Atmosphere High-Resolution Spectrograph Instrument (MAHRSI) flew as a piggy-back payload on the University of Wuppertal's big CRISTA telescope on its first outing. It observed the horizon to measure ultraviolet absorption in sunlight. The hydroxyl free-radical has a tendency to combine with nitrogen compounds to make nitric acid, which has a voracious appetite for ozone. Its main goal was to make the first global three-dimensional survey of stratospheric hydroxyl and nitric oxide, but it also derived thermal profiles up to 120 km (the coldest part of the atmosphere, the mesosphere) so that this could be correlated with the chemistry, because ozone depletion is increased by the formation of stratospheric ice. What was really needed was a truly massive platform in space, with the entire panoply of instruments making coordinated measurements.

Not actually part of ATLAS, but clearly complementary, was an instrument developed by the Langley Research Center. Measurement of Air Pollution from a Satellite (MAPS) measured the global distribution of carbon monoxide, a greenhouse gas that is released by industrial facilities. It was an infrared radiometer that detected the concentration of a gas by its characteristic absorption of the background of thermal emission from the Earth. It did not need to scan the limb, and it was not restricted to observing while the Sun was on the horizon, so it was one of the first instruments able to collect data continuously. It flew four times over a 12-year period (its first outing being on STS-2), and on each occasion it produced a 'snapshot' of the state of the atmosphere. Its data have been used to track the atmosphere's response to efforts to reduce industrial emissions. MAPS was repackaged for installation on Mir in 1997, and a lightweight version is to fly on one of NASA's EarthWatch satellites.

SPACEBORNE LIDAR

Langley Research Center developed the Lidar-In-space Technology Experiment (LITE) that was carried on STS-64 in September 1994. Although a lidar is standard equipment on aircraft engaged in meteorological research, this was the first time that one had been assessed in orbit. A lidar is similar in principle to a pencil-beam radar, except that it uses a laser rather than a microwave system. A conventional weather radar detects reflections from droplets of water or crystals of ice, but it does not directly sense air moisture. LITE sensed molecular reflectance, so it *could* measure water vapour concentration and the moisture content of soil. It complemented instruments such as Langley's SAGE (on the ERBS) which measured stratospheric aerosols. In effect, the LITE produced a three-dimensional map of the structure of the atmosphere in the 10 km to 40 km altitude range. Its thermal profiles could be correlated with the location, movement and concentration of atmospheric constituents. From its vantage point in orbit, this instrument was able to reveal the large-scale structure of clouds and storms in unprecedented detail. For calibration, simultaneous measurements were made by meteorological stations and by airborne lidars.

The laser unit was mounted on a Spacelab pallet alongside a 1-metre diameter telescope to detect the reflected light. The telescope's mirror was actually the engineering test article left over from the Orbiting Astronomical Observatory programme; it had been in storage for 20 years. By the time the laser beam reached the surface it had expanded to cover a circle almost 300 metres wide. Each sample involved emitting a brief pulse, then measuring the return in the narrow-field-of-view telescope. The fact that the laser light was coherent made detecting the reflection relatively straightforward, but a fast-acting solid-state detector was necessary. Lasers for near-infrared, green and ultraviolet were available. Because the laser unit and its boresighted telescope were fixed, staring straight up from the pallet, the orbiter was placed into a fairly dizzying 2 degrees per second roll to scan the laser. Although it weighed 2 tonnes, the LITE instrument was sufficiently lightweight to be converted into a free-flyer, should this be deemed appropriate.

The Department of Defense was also interested in the results, because they offered the possibility of tracking aircraft and missiles by virtue of otherwise invisible vapour trails.

TERRA FIRMA

The Earth is not a sphere; its rotation makes it an oblate spheroid, which bulges at the equator and is slightly flattened at the poles. It is clearly not a perfect spheroid either; the continents and the oceans are not arranged symmetrically, and there is a 20-km difference in elevation between the peak of the highest mountain and the bottom of the deepest trench on the ocean floor. Early gravity-surveys of a century ago, using straightforward mechanical techniques, found that the thin crustal layer varies in thickness and is not homogeneous. Continental rock turned out to be less dense than ocean floor rock. In effect, the continents represent the 'scum' that floated to the surface when the mantle was differentiated into layers by heat escaping from the molten core. The mantle is still in a state of semi-plastic flow; hot plumes rise, and cool plumes sink. It is believed that these flows cause the continents to break apart, creating new oceans which open up as the large crustal plates split and drift apart, only to collide with one another and coalesce at some other point, millions of years later. The different plumes have different densities, so the Earth's inhomogeneity extends deep into its mantle. These variations from a uniformly dense idealised sphere manifest themselves as irregularities in the planet's gravity field.

This differential gravity posed no significant problem until the early 1960s, when it was realised that it would complicate the task of aiming an intercontinental-range ballistic missile crossing the north pole. Consequently, the Department of Defense strongly supported the geodesy project set up by NASA with the launch in November 1965 of Explorer 29 (GEOS-1), the first US Geodetic Satellite. Its main objective was to provide a single reference frame to integrate all the existing geodetic surveys, so that surveyed positions could be located within 10 metres. When it introduced the Atlas missile, the Air Force's uncertainty in Moscow's location was 10 *kilometres*, so this geodetic data was a real improvement. The global frame was based on an accurate determination of the Earth's centre of mass.

Once in orbit, the satellite extended a 20-metre boom towards the Earth, to stabilise its antenna. By transmitting a spiralling beam, the satellite enabled receivers below to precisely track its motion. By working backwards, it was possible to compute the variations of the gravity field. When the first satellite fell silent in 1967, Explorer 36 was sent up to continue the survey. Although this radio-based system initially introduced a dramatic improvement, its capacity for further improvement was soon exhausted. A laser-based system, however, offered great potential. The new satellite, launched in 1976, was the LAser GEOS (LAGEOS). The surface of this 60-cm diamater sphere incorporated a large number of tiny corner-cube reflectors to return a laser irrespective of the angle of incidence. Such a mirror could reflect beams from different sites simultaneously. Furthermore, in its 6,000-km polar orbit this satellite could be seen from sites on different continents at the same time. At such an altitude, LAGEOS was barely affected by variations in the Earth's gravity field, but it took several years to build up a large enough network of laser sites to define its orbit with sufficient accuracy to transform the satellite into a predictable reference point in the sky, against which arbitrary points on the Earth could be located to within 3 cm. A long series of measurements using a surveyed grid enabled the drift-rate of continents to be computed with an accuracy of 5 mm per year, which was sufficient to provide confidence in the observed motions of 2 cm per year. As a surprise result, it was also possible to measure *vertical* motions caused by tides in the crust in response to the Moon's gravitational pull, and even ongoing crustal relaxation after the retreat of the great northern ice sheets, thousands of years ago. In general terms, the polar radius is 21 km less than the equatorial radius, and the Earth is pear-shaped with a 40-metre 'dimple' at the south pole.

In 1984 NASA and the Italian Space Agency agreed a joint programme to launch a second satellite. Two points of reference would enable positions to be determined to within a few millimetres. Although originally assigned to a shuttle mission in 1987, it was October 1992 before LAGEOS-2 was finally rescheduled. Although released into a

Fig. 16.4. Deployment of LAGEOS-2, a joint venture between NASA and the Italian Space Agency. It carried mirrors to reflect laser beams fired to study the dynamics of the Earth's crust.

28-degree inclination orbit, the Italian IRIS two-stage solid rocket raised the plane to 52 degrees as it climbed to 6,000 km to deliver the 400-kg satellite. As its orbit will not decay for millions of years, Carl Sagan seized the opportunity to etch maps inside the satellite, depicting the continents as they are believed to have been 268 million years ago, as they are today and, based on observed drift-rates, as they should appear in 8 million years' time. The immediate objective was to reveal any unsuspected disturbances of its polar orbiting predecessor's orbit. In the longer term, the intention was to characterise the extent to which the axial pole wanders, and to measure the rate at which the Moon is sapping the Earth's angular momentum, lengthening the day and increasing the size of the Moon's orbit. Taken out of context, this tiny sphere was popularly criticised as being unworthy of a shuttle mission, but although it was the only satellite deployed, it was actually the secondary payload.

To map the variations in density in the Earth's upper mantle and crust in fine detail, it is necessary to fly a geodetic satellite at low altitude. It was for this reason that Magellan's

Venus orbit was lowered for the gravity survey phase of its mission. The LAGEOS satellites had been placed high in order to escape perturbations, so as to provide reference points for mapping locations and measuring surface motions. To probe the Earth's internal structure, Germany arranged for GFZ to be deployed from the Mir space station in 1995. This 22-cm sphere was ejected from the airlock. At 480 km, the 20-kg satellite was easily perturbed by variations in the gravity field, so it needed intensive tracking. However, it was soon brought down by air-drag. A few medium-altitude reflectors are also in use. Japan's Experimental Geophysical Satellite (EGS) – a comparatively large 2-metre sphere – was placed at 1,500 km at 50 degrees in 1986, and a small French reflector, called Stella, was put into a slightly elliptical polar orbit at 750 km altitude in 1993.

The Soviet Union placed Cosmos 1989 and Cosmos 2024 in high orbit at 65 degrees in 1989, but these reflectors rode into orbit with GLONASS satellites, the equivalent of NAVSTAR, and their role was to determine the extent to which the Moon's gravity disturbed the orbital spacing required to operate the satellites of a GPS system.

The Department of Defense's GPS satellites could be used to monitor vertical crustal movements. Long-term measurements by the University of Arizona using a network of receivers distributed across the Tucson Basin revealed that the floor of the basin was sinking, indicating that it was losing its capacity to hold a water table. Accurate vertical measurements were feasible with GPS in its *coarse* civilian mode of operation, because the immobile receivers were able to refine their positions by taking continuous satellite fixes. A net of receivers on the slopes of a volcano might be able to detect the distortions caused by movement of magma prior to an eruption. There was seemingly no end to which satellites could help with geophysical research.

MAPPING FROM ORBIT

The mixed bag of instruments tested on the November 1983 Spacelab mission included a metric camera supplied by ESA to assess the potential for high-definition photography for mapping from orbit. It was mounted in the porthole in the roof of the pressurised module, and was operated directly by Robert Parker; this was fortunate because he was able to free the film-feed mechanism when it jammed early on. It produced excellent results, although a microwave radar, intended to take complementary imagery, malfunctioned, and this part of the experiment had to be cancelled. The European Metric Camera was assigned to the first of the SEOM missions, but when this was transformed into ATLAS, and the pressurised module was deleted, it was dropped from that payload.

NASA flew its own Large Format Camera (LFC) on STS-12 in August 1984 and again on STS-13 in October. Although derived from a high-altitude metric stereographic mapping camera, it was bigger and more advanced optically and electronically than aircraft cameras. This bulky 400-kg unit was mounted on an MPESS frame in the bay, and it was remotely operated. Its cassette had film for 2,400 exposures, and each time it took a picture a pair of small cameras recorded orthogonal star fields in order to provide a reference frame. Its first flight was primarily for calibration purposes. In addition to nadir imagery, it took oblique and limb pictures for assessment of its capabilities; and star fields were photographed to prepare for the flight planned for March 1986, when it was to be used to 'capture' Halley's Comet.

The LFC's primary terrestrial application was to be in supporting the Landsat programme in the ongoing search for natural resources. From 260 km, the 30-cm focal length f/6 lens was able to focus a 220 × 440-km field on a 22 × 44-cm negative. Its 20-metre resolution was four times better than in a Landsat image, so it was to assist in identifying features showing interesting spectral characteristics. Its mapping capability is evidenced by the fact that it was able to include the entire state of Massachusetts in a single frame. However, responsibility for running the Landsat system was transferred from NASA to a newly-created commercial organisation (EOSAT), and the LFC's mission to observe Halley's Comet was pre-empted by the loss of Challenger.

SHUTTLE RADAR LABORATORY

On its second test flight, in November 1981, the shuttle carried a sophisticated scientific payload on behalf of NASA's Office of Space and Terrestrial Applications. Part of this was the Spaceborne Imaging Radar (SIR) developed by the Jet Propulsion Laboratory.

The Landsat satellites used visual and infrared imaging to chart the Earth's natural resources, but they required clear weather. A radar can see through cloud, so is capable of all-weather operations. JPL adapted an L-Band side-looking Synthetic-Aperture Radar (SAR) developed for airborne reconnaissance, and flew it on SeaSat in 1978. At 800 km altitude, its 25-metre resolution imagery was sufficient to map the state of the oceans. It was particularly valuable because it was able to monitor the sea beneath tropical storms,

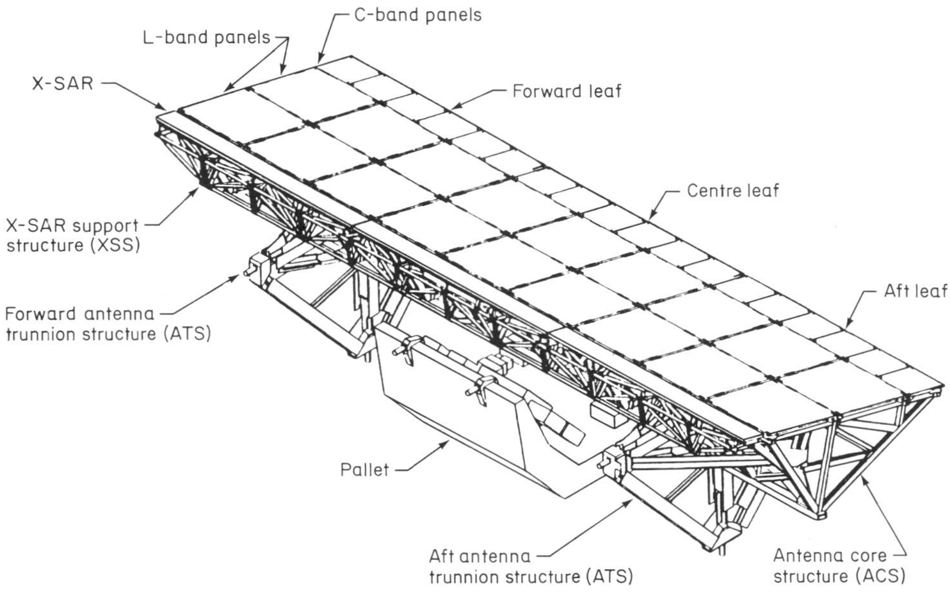

Fig. 16.5. A schematic of the Spaceborne Imaging Radar (SIR) and the X-SAR (X-Band Synthetic Aperture Radar) antennas of the Shuttle Radar Laboratory.

which would otherwise be visible only as an atmospheric structure. An S-Band radar was then under consideration to map Venus through its perpetual cloud cover. The SIR was an L/C-Band dual-frequency system. From the shuttle's 225 km, under optimum conditions, it provided a resolution of 10 metres – sufficient for general terrain-mapping. The 2 × 9-metre planar array system was carried on a Spacelab pallet. The shuttle flew upside down, and the radar viewed 47 degrees off the nadir to scan a swath 50 km wide. The Soviets tested a radar on Cosmos 1076 in 1979, for later use on its Okean sea-survey satellites. And, of course, the Department of Defense was developing its much higher-resolution Lacrosse reconnaissance radar.

On its STS-2 test, the SIR concentrated on areas for which comprehensive Landsat data were available, so that the two techniques could be compared; but when the shuttle overflew Africa, data were taken to test the radar's ability to see through dense vegetation. It proved possible to chart jungle terrain – something that was otherwise impracticable. Surprisingly, the radar could also penetrate 10 metres into the extremely fine dry sand of the eastern Sahara, the Earth's most arid region, and it found previously unsuspected ancient dry river channels in Egypt and the Sudan. Analysis of tracks in the desert, almost undetectable at surface level because of dunes, revealed sites of human activity dating back to the Stone Age.

When the SIR flew again, in October 1984 on STS-13 (the same mission that deployed the ERBS), the archaeologists had set a real challenge: could the radar locate the lost city of Ubar, the hub of the frankincense trade some 4,000 years ago, which lore said was somewhere in the Omani desert? Bertram Thomas, in following ancient desert tracks on his searches in the 1930s, had criss-crossed the Empty Quarter but found nothing. Some argued that there never had been such a city at all. The SIR revealed new tracks, and an expedition set off to follow these new tracks to look for evidence. After a fruitless search, the team took a slight detour south of the most promising track to stop off at the village of Ash Shisr, because there was an old dry well there that, in the light of its proximity to the new track, merited a more detailed study. This proved fortunate, because excavation revealed a walled enclosure which, at that time, would have constituted a city. Furthermore, a geological study revealed that the structures had been built over a limestone cave, and when this had begun to subside in 100 AD the residents had moved on. This contribution to archaeological research was an unexpected bonus. The SIR's primary utility was to complement multispectral imaging for mapping natural resources geomorphology. A natural resource in short supply in a desert is water. The radar could see substructure likely to hold a shallow water table, and later drilling at seven sites in Saudi Arabia tapped water. To a nation which relies on costly desalination for its fresh water, satellite prospecting was attractive. Amongst the morphological features on the target list were tectonic studies of the floor of the African Rift Valley, coastal land forms in the Netherlands and suspected meteor impact structures in Canada. In addition, it charted wave patterns beneath Hurricane Josephine and the propagation of the so-called extreme-waves of ocean currents.

For its second outing, the SIR radar had been upgraded: its planar array was hinged so that it could peer over the starboard sill at different angles. Scanning the same swath on two successive passes would facilitate stereoscopic mapping. Unfortunately, data sampling was severely limited on this occasion by communications problems. First, the orbiter's steerable TDRS antenna malfunctioned and had to be locked in position; which prevented relaying data in real time. As there was only one TDRS satellite at that time, a tape unit had been installed to record SIR data while the shuttle was out of contact, so data from each swath were stored on tape, then the orbiter was reorientated to point its antenna at the satellite for the download. But the radar mission would in any case have been interrupted. Out at 36,000 km, geostationary satellites are more exposed to the solar wind than is a low-orbit shuttle. When the TDRS detected a blast of high-energy particles, it went into hibernation to protect itself, and it was 13 hours before it came back on line. This interruption further complicated the SIR mapping. On this flight, the radar imaging was coordinated with conventional photography by the LFC, to assist with feature identification.

The only problem with the radar occurred at the very end. In its original form the planar array had been mounted on a fixed frame, which had occupied the front half of the bay. To make the array more compact so that the ERBS could be carried further aft, the unit unfolded across the space made available by the deployment of the satellite. Although the panels folded back, the thermal insulation blanket had expanded, and the latches would not engage. Undeterred, Sally Ride pushed the top panel flat with the end-effector of the RMS, and it finally locked into position.

When the SIR flew again a decade later, it was alongside the European X-SAR radar, and it was called the Shuttle Radar Laboratory (SRL). In this SIR-C form, the planar array of JPL's radar was 12 metres long and 3 metres wide and, for the first time, it steered its beam electronically and exploited both horizontal and vertical polarisation for better spectral discrimination. Having increased to 10 tonnes, it was the biggest payload-bay instrument ever carried by the shuttle. The X-SAR – so-called because its SAR operated in the X-Band – was a lightweight 250-kg instrument with a 40-cm wide mechanically-tilting array using a single polarisation. It was under evaluation for a future ESA remote-sensing satellite. Because the two radars used different frequencies, they could run simultaneously and, by scanning the same track, their data could be combined to make three-shade false-colour imagery to bring out a wide variety of detail. This time everything worked flawlessly. The SIR took six times more data than it had during its earlier trial. In addition to mapping 20 per cent of the Earth's land surface, it observed several test sites that had been extensively surveyed by on-site teams, in order to evaluate its ability to discern different types of detail. A battery of video and still cameras recorded these passes to assist the analysis. The orbiter manoeuvred in order to sample sites several days apart, to highlight short-term changes. The crew split into shifts to facilitate 24-hour operations, and took 133 hours of radar data on 400 selected sites. In order to safeguard the radar mapping, this time a high-capacity tape unit was carried and all the data were stored in addition to being periodically downlinked. A follow-up flight was made a few months later to record seasonal changes. Tom Jones flew on both in order to provide continuity in procedure. As a former employee of the Central Intelligence Agency, Jones is believed to have been primarily involved in airborne radar-imaging reconnaissance.

With multiple polarisation, the SIR-C was not only able to *identify* a crop as wheat, it could also say what *type* of wheat and how healthy it was. It could also measure moisture content of soil. These were tasks previously feasible only by multispectral *optical* sensors. The X-SAR could determine the type and water-equivalent content of snow, and so provide advance warning of the thaw run-off. As before, the deserts proved remarkably interesting, but this time, because cameras shadowed the radar, direct comparison confirmed that the radar could see so much more. A chain of three 15-km diameter craters was discovered near Aorounga in northern Chad. As well as following up on previously noted ancient river beds in the Sahara, the radar spotted a previously unsuspected dry channel at a sweeping bend in the River Nile. Eight hours after the shuttle reached orbit on SRL-2, a volcano on the Kamchatka peninsula, Mount Kliuchevskoi, erupted; the radar was able to see through the massive ash cloud, which was spectacular in the optical imagery, to chart the progress of the mud flows beneath. And when a powerful earthquake occurred off the coast of Japan a few days later, the radar looked for tsunami waves, but saw nothing. The archaeological test this time was Angkar Wat in Cambodia. Most of this vast temple complex is within the jungle, but the radar was able to survey its true extent. The Chinese Academy of Sciences had requested mapping of an ancient desert section of the Great Wall, long since buried by sand, and they had also requested a tracing of the Silk Road in the north-western part of the country, where it is very difficult to follow on the ground.

In order to achieve the best possible comparative data, on the last few days of its SRL-2 mission the shuttle used GPS to refine its orbit to fly within 100 metres of the ground-track of its previous mission. The resulting topographic data had sufficient resolution to reveal surface movements between the two flights to an accuracy of a few centimetres. Although the SIR could detect the expansion of a volcano due to the rise of magma preparatory to an eruption, and so could, in principle, provide early warning, in practice this capability would be useful only if the radar data could be processed in real time. However, the SIR produced a prodigious stream of angle, phase and reflectance data which required complex processing to produce an image. It took six months to reduce the SIR data from each SRL flight to imagery, and a *further* nine months to refine this to highlight the specific detail needed by the 52 science teams involved in the demonstration.

When the SRL programme had been proposed in 1993, it had called for one flight each in 1994, 1995 and 1996. Although the second flight had been brought forward, the third had been pushed back indefinitely. At one point, NASA considered flying it in 1997, but it was subsequently squeezed out of the manifest as flight opportunities were switched from science missions to the assembly of the International Space Station.

The Defense Mapping Agency – one of the many and varied organisations in the orbit of the Department of Defense – announced in 1996 that it intended to work with NASA to build an interferometric imaging radar to map the planet's surface topography with *30 times* the accuracy currently available. This Shuttle Radar Topography Mission is an outgrowth of the SIR radar, and many of the systems are to be reused, but in this case two radars are to be carried to make simultaneous measurements. To achieve a baseline, one will be carried on a 60-metre mast while the other will remain in the bay. This 11-day mission is planned for the year 2000. From a 57-degree orbit it should be able to make an excellent three-dimensional map of 80 per cent of the land surface, but because it is to be undertaken for military reasons this is likely to remain classified.

Although JPL had developed the SIR independently of the MTPE programme, its potential was soon noted. After the success of the SRL missions, JPL argued for a long-duration shuttle-deployed SIR free-flyer or, even better, a satellite for a Titan 4 to launch into polar orbit for global ecological surveying. A remote-sensing system is more than a space-based instrument, however; an archival facility will be needed if the data is to be exploited. In the long term the cost of an advanced 'ground element' can far exceed the up-front investment in the space-based asset. With a wide variety of instruments returning data for MTPE on a long-term basis, an integrated facility will be required to process, archive and make available the data. Although the multispectral imagery from the Landsat satellites had initially been processed by NASA, the Reagan administration decided to privatise commercially viable satellites, and Landsat was contracted to the Earth Observation SATellite Corporation (EOSAT), a company formed in 1984 by Hughes and RCA specifically for the task. This venture was undermined by the total loss of Landsat 6 and the postponement of its replacement, but it served as a model for the kind of support that will be required for MTPE to be successful.

EARTH OBSERVING SYSTEM

The ambitious objective of the Earth Observing System (EOS), announced by NASA in March 1988 as the core of the bold MTPE programme, was to integrate multi-disciplinary research to develop a *synergistic* series of studies of the atmosphere, the oceans, and the biosphere in order to determine relationships in terms of the processes which cycle energy, water and chemicals. Defining the state of the Earth as a 'system' is the first step towards developing a model with unprecedented scope for predicting climatic change.

The initial plan called for two huge polar platforms, each with a comprehensive suite of instruments and a life of 15 years. In 1991, however, Congress told NASA to scale down its plan, so it suggested flying a fleet of smaller satellites, based on two buses. Even these would require the service of the Titan 4, however. It was not long, therefore, before NASA was told to split each of these into three UARS-like satellites for launch on the Atlas 2AS, the top of the range of the new commercial system. The advocates of the integrated package argued that simultaneous data acquired by different instruments would help in the understanding of complex interactions. Others insisted that a fleet of smaller satellites would prevent an entire suite of instruments from being wiped out by a single failure. This was very much a case of different points of view being driven by different beliefs, but the hidden agenda for this restructuring was not to minimise the technical risk; it was to facilitate Congressional control of what had started out as a monolithic project which, although laudable in its goal, was ludicrously expensive.

Appointed in 1992 as NASA's Administrator, Dan Goldin's first task was to cut the fat out of his agency. He split the Office of Space Science and Applications (OSSA) into the Office for the Mission To Planet Earth to manage local projects such as EOS, and the Office of Planetary Sciences to coordinate Solar System missions. He was also a great advocate for small narrow-focus satellites, so he supported the proposal to split EOS into a constellation of satellites. This 'small-is-best' strategy also spawned the Earth Probe (EP) programme, which called for a series of single-instrument packages which could ride the Pegasus air-launched rocket into low orbit. Although there was a temptation to use this as a means of testing new instruments, it was decided to give immediate priority to instruments of proven worth, such as TOMS and ACRIM. Although Congress substantially cut the EOS budget, the incoming Clinton administration, while allowing NASA's overall budget to fall, expressed its strong support for this particular project.

The EOS will therefore exploit a variety of satellites, but at its core will be four series of custom-built platforms, which will be launched in sequence over the programme's 20-year life. To ensure that successive satellites are state-of-the-art, each bus will be sufficiently flexible to accommodate instruments drawn in 'pick-and-mix' fashion from a continually updated pool. The EOS programme is therefore not about satellites at all; it is about remote-sensing technology.

In addition to already-flown instruments such as ACRIM, MLS, SAGE, SOLSTICE and TOMS, the initial pool includes:

Dual-Frequency Altimeter and Microwave Radiometer
: An altimeter to measure wave height and the direction and speed of ocean currents, as well as wind speed at sea-level. Its radiometer will provide corrections for the effects of tropospheric water vapour.

Sea-viewing Wide-field Sensor
: This combines the Coastal Zone Colour Scanner flown on Nimbus 7 with the Advanced Very-High-Resolution Radiometer of the NOAA satellites to monitor ocean phytoplankton. It was proposed for, but not installed on, the third-generation Landsat, but it later flew on Orbital Sciences Corporation's SeaStar satellite.

Tropical Rainfall Monitor
: A precipitation radar: in effect, an aircraft weather radar modified for look-down application. It was flown on the Japanese TRMM satellite.

Tropospheric Emission Spectrometer
: A comprehensive instrument to measure the three-dimensional distribution of almost all atmospheric gases that radiate in the infrared.

Atmospheric InfraRed Sounder
: An instrument to map the distribution of gases and water vapour, to map cloud cover, to measure sea and land surface temperature, and to measure vertical thermal profiles with 1 °C and 1-km resolution.

Spaceborne Imaging Radar
: The radar system as flown as part of the Shuttle Radar Laboratory.

Lidar-In-Space
: The lidar system as tested on STS-64.

Lightning Imaging Sensor
: An instrument to clarify the relationship between cloud electrification and the process of precipitation to establish why this is different over oceanic and continental regions. It was flown on the Japanese TRMM satellite.

Measurement of Pollution in the Troposphere (MPT)
: An instrument to measure the three-dimensional distribution of carbon monoxide and methane in the troposphere.

Multi-angle Imaging Spectro-Radiometer (MISR)
: An instrument to measure the effect of clouds and tropospheric aerosols on the radiation budget, and to monitor biomass cycles.

Moderate Resolution Imaging Spectrometer (MODIS)
: An instrument to assess surface temperature, vegetation, deforestation, snow cover and moisture in order to study biological and physical processes with 1 km^2 resolution over

oceanic and land areas, and to measure ocean colour to yield concentrations of phytoplankton and sediment in suspension.

High-Resolution Dynamics Limb Sounder
An instrument to measure the state of the ozone layer and the chemicals which deplete it, correlated with vertical thermal profiles.

Multi-frequency Imaging Microwave Radiometer
An instrument to measure atmospheric water content, rainfall rates, soil moisture, and ice and snow cover. It will be carried by ESA's Metop satellite.

Clouds and Earth Radiant Energy System (CERES)
An instrument to study the Earth's radiation budget by measuring clear-sky radiative flux, radiative forcing and feedback processes, and radiant input to the atmosphere and the ocean. It was flown on the Japanese TRMM satellite.

Earth Observing Scanning Polarimeter
An instrument to measure the net effect of clouds and aerosols in heating and cooling the planet, on a global scale.

Geoscience Laser Altimeter System
An instrument to monitor seasonal variation of the polar ice sheets.

Advanced Spaceborne Thermal Emission and Reflection Radiometer (ASTER)
An instrument to measure surface spectral radiance and surface temperature, to map surface elevation, composition and vegetation, and to monitor cloud, sea ice and polar ice.

These instruments (and there are a great many more) have been listed simply to illustrate the tremendous potential for studying the Earth as a *system* from low orbit.

NASA's first EOS satellite is set for launch by Atlas in 1998. This 5-tonne platform is to use ASTER, CERES, MISR, MODIS and MPT to measure radiative processes, with the objective of characterising the relationships between the continents and the atmosphere. Dry land soaks up heat faster than the ocean, but it also releases its energy more readily, so its diurnal variability is considerably greater; the oceans, in contrast, serve as a vast long-term store of energy. The tropical zone, in which most of the rain falls, is the heat engine of the planet. The MTPE programme is being undertaken in the same spirit of cooperation as the World Weather Watch. It is a joint effort by the US, Europe and Japan. Following NASA, ESA and NASDA decided to fly a series of satellites in sequence rather than a single large satellite. Envisat and Metop, ESA's first two satellites, will be launched in 1999 and 2002. Two smaller Earth Resources Satellites (ERS) were launched to experiment with potential instrumentation. Japan launched its ADvanced Earth Observing Satellite (ADEOS) by H2 in 1996, but it was knocked out by a power failure in June 1997, and the follow-up will not be ready until 1999. Although this was only a small platform, the instruments that it carried were a *pot pourri* of state-of-the-art sensors from around the world, and its loss, together with NASA's Lewis satellite a few months later, served to illustrate that any and all satellites are at risk once they are in space.

AN INTEGRATED PLATFORM

In merging its proposed space station with the Russian effort already underway, NASA agreed to switch to a 51-degree inclination for the International Space Station, a move that offered much better scope than the originally intended 28 degrees for utilising this massive new facility for MTPE research. Although observational instruments cannot be mounted on the pressurised modules, there is ample accommodation for attached payloads on the long truss. It is to be hoped that once the International Space Station is in service, many of the above-mentioned instruments will be mounted on it, because, even though it will not allow Sun-synchronous coverage, it will monitor all but the most extreme polar regions. Furthermore, instruments on the station will be able to be serviced if they malfunction, and replaced when they are rendered obsolete; and the station will be able to serve as a test-bed for completely new instruments. The International Space Station will, therefore, form a vital component of the Mission-To-Planet-Earth programme, or the Earth Science Enterprise, as NASA recently renamed it.

Part 4: Outpost

17

Unexpected opportunity

The early Soviet and the American space programmes were manifestations of the Cold War. It was competition, and space was the arena. Astronauts played out the ancient ritual of single combat, in which the opposing forces lined up and dispatched a single representative to fight to the death in order to decide the *entire* issue. In its modern equivalent, the nation with the most impressive space feats would be shown to have the best technology and, as a result, surely, the best society; a society which the rest of the world would wish to emulate. Nothing less than the future of civilization was at stake. America lost the race to put a man into orbit, but it raised the stakes, and developed the technology required to land on the Moon so seemingly effortlessly that the Soviets later denied that *they* had ever intended to do the same.

Having shown its mettle, America relaxed, took stock of what its achievement had cost, and decided that its interests would be best served by a cost-effective transportation system. Meanwhile, the Soviets flew a succession of space stations. By the time construction of the Mir complex began, the Cold War had spawned Star Wars (the Strategic Defense Initiative), but the two space programmes had so little in common that there was no scope for open competition. Each nation pursued its own objectives, and ignored the other. The sudden collapse of the Soviet Union in 1991 drew the Cold War to its spectacular conclusion, and the situation was transformed overnight. With the Russian economy in free-fall, and its new partners in the new Commonwealth of Independent States claiming ownership of the former Soviet Union's nuclear weapons, the US decided that it needed to *support* Russia, and space became the arena in which to make a public display of cooperation.

COOPERATION

At a Washington summit on 17 June 1992, President George Bush and President Boris Yeltsin agreed to coordinate their efforts in space astronomy, astrophysics, solar–terrestrial physics, and Solar System exploration; and they ordered their respective space agencies "to give consideration to" a joint venture involving their human spaceflight programmes.

A month later, in Moscow, NASA Administrator Dan Goldin and RSA Director Yuri Koptev issued a follow-up memorandum of understanding which called for a cosmonaut to fly on a shuttle in 1994 in return for an astronaut making an extended visit to the Mir space station in 1995, and this was formalised in the Human Spaceflight Cooperation protocol that was signed on 5 October 1992.

With the assumption of office of the Clinton administration in January 1993, NASA's space station plan was ordered back to the drawing board with the objective of reducing the overall cost. At the same time, the climate of cooperation with the Russians blossomed, and in September it was decided to merge their respective plans to form a single structure. As a step towards this, in December the October 1992 protocol was formally extended by Vice President Al Gore and Prime Minister Viktor Chernomyrdin to include a series of visits by the shuttle to Mir, starting in 1995, and the stationing of a succession of astronauts aboard the space station for a total of two years. From the Russian viewpoint, because NASA had agreed to pay $100 million per annum for 1994–1997 (for a total of $400 million), this was another fee-contract to lease time aboard its orbital laboratory; but it was a notable departure for NASA, because it had traditionally cooperated on a no-exchange-of-funds basis with its international partners. The contract for the Shuttle–Mir missions, which were seen as Phase One of the new space station programme, was formally signed in June 1994.

In February 1994, Sergei Krikalev flew as a Mission Specialist on STS-60 Discovery. This flight also carried a Spacehab of experiments, and it suffered a variety of faults which prevented the deployment of the Wake Shield Facility free-flyer. It was ironic that "the last Soviet citizen", as Krikalev had been called by the Western Press, had far more experience in space than all of his colleagues combined. As a precursor for the Shuttle–Mir dockings, it had been agreed that Krikalev's backup, Vladimir Titov, would fly Discovery to rehearse the rendezvous phase of such a mission. STS-63 had been scheduled for June, but the lack of payloads for its Spacehab module had prompted its postponement, and it did not fly until February 1995, which was just a month before Norman Thagard was to lift off on a Soyuz rocket to fulfil the original June 1992 agreement; he and his Russian colleagues were to be retrieved by the shuttle later in the year.

Meanwhile, to prepare for Shuttle–Mir, STS-51 Discovery had evaluated the Trajectory Control System (TCS), which employed a laser rangefinder to augment the orbiter's inertial navigation system in the final phase of its rendezvous; the ORFEUS–SPAS satellite played the part of Mir for this manoeuvring trial. On STS-66, as it retrieved the CRISTA-SPAS satellite, Atlantis tested the procedure it was later to use to approach Mir. This involved approaching from below, rather than from ahead of, the satellite, an approach which had been used just once before in rendezvousing with the HST, because it had minimised contamination of the telescope with efflux. Atlantis also tested the deployment of the recumbent seats to be used by Mir's residents for their return to Earth. Immediately upon landing, Atlantis had been sent back to Rockwell so that it could be refurbished and fitted with the apparatus required to dock with Mir.

CLOSE APPROACH

The most stunning of these rehearsals, however, was STS-63 Discovery's rendezvous, dubbed 'Near-Mir'. Flown by Jim Wetherbee and Eileen Collins, STS-63 was launched on 3 February. In fact, the rendezvous with Mir was a secondary objective, and it was to be undertaken only if doing so would not jeopardise the deployment of a SPARTAN free-flyer. The determinant was *propellant*. If the launch missed its five-minute window, correcting the discrepancy between the two orbital planes would use so much propellant that it would become impracticable to deploy and retrieve the SPARTAN. Based on previous experience, the shuttle had a one-in-three statistical chance of succeeding. Nevertheless, lift-off took place at precisely the moment that Mir's orbital plane intersected the Kennedy Space Center.

However, as soon as Discovery reached orbit, two of its RCS thrusters malfunctioned, and began to leak. This led to concern that the nitrogen tetroxide might coat instrumentation mounted on Mir, so the final go-ahead for the planned close approach was made contingent on the leaks being stemmed. The plan called for Discovery to draw to a halt 10 metres from the androgynous docking port that Atlantis was to use, but if the leak could not be stopped it would be allowed no closer than 125 metres. Fortunately, during the three-day chase the problem was overcome by the expedient of shutting down the leaking thrusters and then letting them drain dry. On 6 February, Discovery approached from below, then, with its nose high and its bay facing Mir, stationed itself directly in front of the complex. This was how a shuttle usually approached a satellite, not how Atlantis was to make its approach; but the objective was to assess the orbiter's station-keeping characteristics, and the line of approach was not crucial to that test. Mir was reorientated so that Kristall's axis was aligned along the velocity vector, with its docking port aimed at the newcomer. Aboard Discovery, Vladimir Titov was responsible for communications with the Mir cosmonauts; they had been talking by VHF radio virtually continuously since the shuttle had established line-of-sight contact.

Discovery paused at 300 metres to await permission to proceed with the final phase of the approach. From this point, the upward-firing thrusters were not to be used, so as not to blast Mir with their efflux (a flight mode referred to as 'low-Z' because only the thrusters aimed obliquely 'up' could be used). Mir's solar panels had been set edge-on to the shuttle to further protect the sensitive transducers. At the moment that permission was granted, Discovery started to reduce the separation. The manoeuvring was done by the orbiter; Mir was to hold its orientation (if it drifted, Discovery was to pause and wait for the complex to be realigned before resuming the approach). During the final phase, Wetherbee flew from the aft station, viewing Kristall's port through the camera in the specially installed window in the roof of the Spacehab (positioned near where a boresighted camera would be when Atlantis flew in carrying the Russian-built docking system), and Collins maintained a running commentary, providing range and closing-rate data.

As Discovery closed in, the television networks relayed the video downlinks from the two vehicles split-screen fashion. The closing rate was so low that it required about forty minutes – almost half an orbit, much of it in darkness – to reach the 10-metre limit. The video showed members of the two crews waving to each other. Closest approach was achieved at 2220 MT. After ten minutes, Wetherbee announced that he hoped their successors would be able to shake hands, and then eased Discovery back to 125 metres so that he could make a slow fly-around to photograph the complex using the large-format IMAX motion-picture camera in the bay.

From passing the 125-metre point on the approach, to finally departing, Discovery had spent about three hours in close proximity to Mir. It had demonstrated that an orbiter could approach Kristall's axial port within an 8-degree cone, and a 2-degree tolerance in orientation, maintain its position at a point 10 metres out, and manoeuvre around the complex. In addition, it had exercised communications and coordination procedures between Houston and Kaliningrad, and between the two crews. "When all was said and done", Wetherbee reflected, "it turned out to be easy". This engineering demonstration had cleared the way for Atlantis to attempt to close that final 10 metres.

FIRST TOUR

On 14 March 1995, Norman Thagard became the first NASA astronaut to be launched on a Russian rocket. His colleagues on Soyuz-TM 21 were spacecraft commander Vladimir Dezhurov and flight engineer Gennadi Strekalov. This spacecraft, in its original form, had been introduced in the mid-1960s. On its test flight, in April 1967, a string of problems that culminated in a twisted parachute had led to the death of its pilot, but since then it had been turned into the space station programme's workhorse. The rocket on which it was launched had an even more impressive heritage. The direct descendant of the Semyorka, the world's first intercontinental-range ballistic missile which Sergei Korolev had developed in the mid-1950s, it had launched Sputnik in 1957, Yuri Gagarin in 1961, and every cosmonaut since. The 7.5-tonne Soyuz spacecraft comprised three modules: the cylindrical service module at the rear contained the engines, the spheroidal orbital module on the front was a living quarters and a cargo hauler, and the descent module was sandwiched in between. At just 2.2 metres in diameter, and 2 metres tall, the bell-shaped descent module was rather cramped for a crew of three pressure-suited cosmonauts; hence the need for the 2.2-metre orbital module. After the standard two-day rendezvous, the ferry automatically docked with the orbital complex, and the newcomers were given a traditional bread-and-salt welcome by the resident crew of Alexander Viktorenko, Yelena Kondakova and Valeri Poliakov.

As the world's first modular space station, Mir was a physical incarnation of NASA's aspirations. Its base block had been launched in early 1986, just weeks after the shuttle had been grounded by the loss of Challenger. Derived from the 20-tonne Salyut, the first of which had been launched in 1971, the Mir base block was the first spacecraft specifically designed to be permanently occupied, and it was fitted out as a habitat. Like its predecessors, it had an axial docking port at either end, but it also had a ring of four radial ports at the front. In 1987 it had been augmented at the rear by an 8-tonne astrophysical laboratory and at the front by two 20-tonne modules – one in 1989 with additional environmental systems and the other in 1990 with a set of furnaces. The assembly of the complex had been far slower than intended, and the final two modules had been indefinitely grounded, but, as the only orbital facility capable of sustaining human life, Mir had found its niche as an international laboratory for microgravity research. NASA had yearned for a permanent space station since its inception. It had mounted three expeditions on Skylab, one of which had lasted three months, and had hoped to use the shuttle to refurbish that station, but it had plunged back to Earth before the shuttle was ready. With its shuttle restricted to a fortnight in space, NASA had seized upon the opportunity to send an astronaut for a tour aboard Mir.

As a physician, Thagard's primary research involved gathering biomedical data on how he adapted to the space environment; on his four shuttle missions he had never had time to adapt to weightlessness. His data were to extend that from the Space Life Sciences missions which had, of necessity, focused on the initial phase of the process. By taking samples of blood, urine and saliva on a regular basis and by a log of food and fluid intake, post-flight analysis would be able to track changes in body fluid and blood chemistry in fine detail. He was also to perform tests to measure cardiovascular, regulatory, metabolic, musculatory, bone and psychological effects. Although the planned three-month tour was insignificant in comparison to the 438-day marathon which Valeri Poliakov, also a physician, brought to a close upon his departure on 22 March, it would match, and indeed exceed, the longest tour of duty aboard Skylab, which was NASA's only experience of long-duration adaptation to space. Previously, fee-paying international researchers had been limited to the period of a crew exchange, although in several cases this had facilitated visits of up to a month. Originally, Thagard was to have visited during an extended handover, but when it had been decided to expand the cooperative programme and to retrieve him when Atlantis made its first visit, his role had been upgraded to membership of the new resident crew. The mounting of the *Stars And Stripes* on the bulkhead alongside the Russian national flag symbolised this status.

The decision to integrate the national space programmes had finally released the funding required to complete the assembly of the complex. The last two modules – incomplete and in storage – were redesigned. Spektr was to be launched first, closely followed by Priroda. To overcome the fact that Mir was forever short of power, it had been decided to remove some of the remote-sensing equipment from Spektr so that a second pair of solar panels could be fitted, and, to enable NASA to undertake a serious programme of research, it was to ferry up some 800 kg of apparatus as internal cargo. Priroda was to have a substantial amount of NASA apparatus built into it. The plan had been to have Spektr fully commissioned prior to Thagard's arrival, but a variety of individually minor delays had conspired to delay its final preparation. As soon as it had become clear that the module would not be launched in time, a Progress ferry – one of the Soyuz-based spacecraft which routinely resupplied the station – had delivered sufficient apparatus to enable him to start his work. Nevertheless, as he settled down to life aboard Mir, Thagard expected the rest of his gear to arrive in early May; but this soon slipped to late May.

Life aboard Mir was completely different from that on a shuttle mission. Whereas NASA mission planners broke each crew member's day into five-minute frames and ordered tasks to make the most productive use of the limited time in orbit, then monitored progress on a continuous basis by way of the TDRS geostationary relay satellite network, the routine on Mir was considerably more relaxed. Although the crew worked through a prioritised list of tasks, the fact that the complex was usually in communication for no more than 20 minutes even on a favourable pass, and was out of contact for up to nine hours at a time on other passes, meant that they worked independently and set their own pace. Furthermore, experience had shown that it was not practicable to sustain the level of activity expected of a shuttle crew, so Mir crews worked a five-day week. Of course, it was not an environment conducive to the pursuance of hobbies, and most cosmonauts spent much of their spare time working on their experiments; but the fact that they had chosen to do so voluntarily made it seem less like work. Of the 200 kg of NASA apparatus that had been flown up previously, the most significant item was the Mir Interface to Payloads (MIPS), a laptop which could control NASA apparatus and send and receive data via Mir's telemetry link. Its burst mode enabled Thagard to make optimum use of the times that Mir was in communication, to keep in touch with his support team using e-mail; but he could also employ Mir's short-wave to relay messages via the ham radio network. Although this radio also enabled him to keep up with US news, Thagard increasingly suffered a sense of cultural isolation.

When Progress-M 27 docked on 11 April, it brought additional supplies for Thagard's programme, and 48 fertilised quail eggs; he was to place them in an incubator, allow them to develop for specific periods, and then freeze them so that post-flight analysis could study the process of embryonic development.

On 12 May, Dezhurov and Strekalov made the first of a series of spacewalks to prepare the complex for Atlantis' arrival. One of the solar panels on Kristall was to be transferred to Kvant 1, the small module at the rear of the complex. After laying cables to the motor

Fig. 17.1. The Mir space station, as viewed by Atlantis on its historic docking mission. Note that the Kristall module is on the complex's forward axis, and a Soyuz is docked at the rear port.

that an earlier crew had set up, they retracted the panel, and on 17 May they dismounted the 500-kg unit and used a crane to swing it to the rear of the complex. It took longer than expected to mount it on the new motor, so they had no time to plug it in. On Mir, though, tasks which cannot be finished can be left for another day. They returned on 22 May, plugged it in and extended its concertina array, then set off to retract Kristall's other panel so that the module could be moved off the lower port to clear the way for Spektr which, having been launched on 20 May, was already on its way. Kristall was first swung up onto the axis, then over to the right-hand port. After docking on the axis on 1 June, Spektr was transferred to the lower port. On 10 June, the last act in reconfiguring the complex was to return Kristall to the axial port so that, as Atlantis closed on the port at the module's far end, the orbiter would be well clear of the rest of the complex. The irony of the androgynous docking port was that it had been installed to mate with the Soviet shuttle, which had long-since been cancelled. The port was about to be used for its intended purpose, but the nationality of the visitor was not as had been expected.

ATLANTIS ... NOW ARRIVING!

After several postponements due to weather, STS-71 Atlantis was launched on 27 June. It was flown by Robert Gibson and Charles Precourt. In addition to Mission Specialists Greg Harbaugh, Ellen Baker and Bonnie Dunbar, it was carrying Anatoli Solovyov and Nikolai Budarin, Mir's next crew, as passengers.

The Orbiter Docking System (ODS) was mounted near the front of the payload bay. Set in a twin-triangular truss, it was basically two interconnected tubes, one leading from the middeck hatch to the tunnel running to the Spacelab mounted further aft, the other running upwards, through an airlock, to the androgynous docking system. Built long ago to enable Buran to dock with Mir, this docking system had been bought from Energiya. It had been hoped to take receipt of it in July 1994, but permission to import the pyrotechnically-fitted

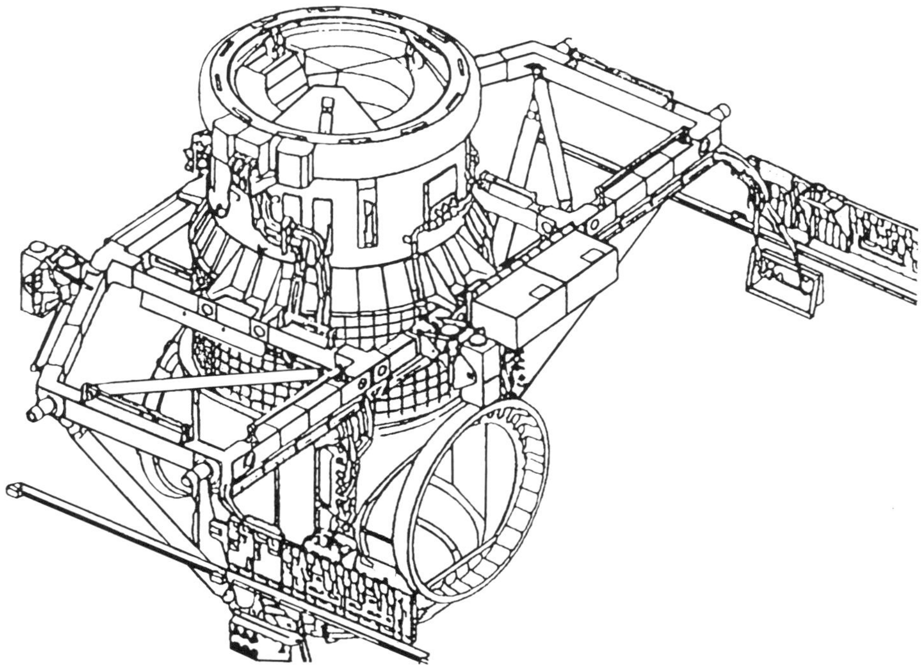

Fig. 17.2. A schematic of the Orbiter Docking System (ODS) that was installed in the shuttle's bay to dock with Mir. The ODS consisted of an airlock and a transfer tunnel, which ran from the orbiter's middeck to the Spacelab module at the rear of the bay. The Russian-supplied Androgynous Peripheral Docking System (APDS) that was mounted on the upper rim of the ODS mated with an identical unit on the end of the Kristall module.

hardware had been delayed, so it had not arrived until September. It had immediately been sent to Rockwell to be mated with the ODS. The TCS rangefinder was on the sidewall of the bay. Of the four-shuttle fleet, Atlantis had been chosen for these Mir missions for the simple reason that at the end of 1993, when the decision had been taken to perform a series of Mir dockings, it had been undergoing a refit at Rockwell, so the modifications needed to accommodate the ODS had been added to the refit. The ODS was to become a permanent fixture in Atlantis' bay, and this orbiter was to be dedicated to the Shuttle–Mir programme. It was pointed out by some disgruntled scientists that Shuttle–Mir had commandeered many flight opportunities that would otherwise have serviced a variety of Spacelabs, and that this political junket had been bought at the expense of the scientific programme. However, for this first docking, Atlantis had a Spacelab fitted as a life-sciences laboratory, and Baker and Dunbar were to study the state of adaptation to weightlessness of the Mir residents. In addition, this criticism failed to acknowledge the potential for research from maintaining a succession of astronauts aboard Mir.

Atlantis rendezvoused with the Mir complex on 29 June. It approached from below, as had Discovery, but it did not pass by and then ascend to dock by drawing back along Mir's velocity vector (the V-bar approach); instead, it ascended the radius vector (R-bar) that came straight up from the centre of the Earth. When Discovery had made its V-bar approach, the low-Z thrusters had consumed much more propellant than if the high-Z thrusters had been used, because the low-angled thrusters directed only a small fraction of their impulse to halt the motion. On the precursor flight, this penalty in propellant had been traded against the familiarity of the line of approach. By using the R-bar approach, Atlantis would genuinely save propellant. It relied on the gravity gradient of one vehicle orbiting beneath the other to brake its very slow climb during the final phase without firing its upward-directed thrusters at all. In fact, to overcome natural forces of orbital dynamics, Atlantis was required to fire its downward-directed thrusters intermittently in order to *maintain* its closure rate. The real attraction of this approach for NASA was that it was fail-safe: if the shuttle was disabled, gravity would draw it down and away from Mir, thereby precluding the possibility of a collision. Mir was reorientated to aim its androgynous port towards the shuttle. Dunbar, who had backed-up Thagard for the Soyuz–Mir mission, and so was fluent in Russian, handled VHF communications with Mir.

At 1450 MT, Atlantis was in position 300 metres directly beneath Mir. After a pause, it closed to 100 metres, then paused again. At 1623 MT, as the two spacecraft passed north of the equator, Atlantis resumed its approach, then paused again at 10 metres. At 1655 MT, with Mir back over Russian territory, and once again in contact with Kaliningrad, Atlantis started the final phase of the approach. The rules required the docking to be achieved before Mir flew out of range of the tracking network, which gave Gibson about 15 minutes to complete the manoeuvre. Stationed at the rear of the flight deck, he initially peered through the overhead window, but for the docking he switched to the monitor displaying the view from the camera on the centreline of the ODS. To ensure that the two docking units were aligned, as he slowly ascended he had to stay in the centre of an ever-narrowing cone, and this was done by maintaining a stand-out bullseye over the target in the centre of Kristall's port. When Atlantis paused just a few metres out, Mir was put in free-drift to ensure that its attitude-control system would not compete with the orbiter's to control the linked structure. Atlantis closed at a rate of 3 cm per second, offset by no more than 2 cm from the axis of the approach cone, and with an angular discrepancy of less than 0.5 degrees. Gibson had flown the 100-tonne orbiter with such skill that he had made the unprecedented manoeuvre appear deceptively straightforward. Exactly on time, at 1700 MT, while over Lake Baikal, the two sets of triple-petals meshed and the capture latches they carried engaged. There was a momentary shudder as the two spacecraft jostled one another, and the springs within the mechanism absorbed these residual motions. After about 15 minutes in this soft-docked state, Atlantis fired its thrusters to push its guide-ring against Kristall so as to ensure that the twelve primary latches around the rim of two collars were in perfect alignment for the hard-dock, then the ring was retracted to form a hermetically sealed tunnel between the two vehicles. An hour later, when Mir flew back into communications range, Dezhurov swung open the Kristall hatch, Gibson opened the ODS hatch, and they greeted one another in the tunnel. In the Kaliningrad control room, Yuri Koptev and Dan Goldin savoured the moment. The first order of business on Mir was for everyone to congregate in the base block for a 'photo opportunity' which amply demonstrated that it had not been built to accommodate *ten* people.

After a well-earned night off, it was time for business. In the Spacelab, Dezhurov, Strekalov and Thagard, who were far advanced with their preparations for return to Earth, were given a thorough medical examination using the life-science equipment that included a treadmill, a veloergometer, and an LBNP chamber. The scientists who had complained that the science programme had been hijacked to facilitate a political junket could hardly criticise the examinations undertaken on these three subjects, who had spent over a hundred days in space. This was the first opportunity for NASA to test long-duration cases whilst still in the space environment. Even with the EDO wafer in its bay, the shuttle is limited to a fortnight in space. Only a permanent orbital complex can provide the time to fully adapt to the space environment, and Mir was the only such facility. Only by having the shuttle visit Mir could the sophisticated biomedical technology at NASA's disposal be applied to space-adapted cases. The political symbolism of this docking was strong, but this first visit also produced tangible scientific results.

While the retiring Mir crew served as guinea-pigs for the biomedical studies, their successors transferred cargo. Atlantis had a great deal of water in reserve as a result of the reaction in its power-generating fuel cells. This 'waste' is normally vented to space, but on Mir, water is a precious resource. On the whole, Mir recycles 60 per cent of its water, but only that condensed from the water vapour is potable. It had been decided to donate the excess water to top up Mir's tanks, but because there were no pipes it had to be pumped into small containers which were transported manually. Nevertheless, half a tonne of water was donated by this means.

The formal handover between the two Mir crews occurred when Solovyov and Budarin put their couch liners and Sokol suits aboard Soyuz-TM 21, and Dezhurov, Strekalov and Thagard offloaded theirs to Atlantis. The cargo transferred to Atlantis included frozen quail chicks, accumulated film and materials and biomedical samples. The Russians exploited the shuttle's payload capacity to return expired items which were usually discarded with empty cargo ferries – including elements of the Salyut 5B computer – so that their service life could be reassessed. This aspect of the joint mission relied on quartermastering skills to track all the items going in each direction and ensure that they were properly loaded for the return to Earth (a laser bar-code reader was under development to assist transhipment in the future).

Late in the preparations for the mission, it had been decided that rather than keep the combined 'stack' orientated as it had been at the moment of docking, with the shuttle below Mir, Atlantis would reorientate it at the start of each day so that Mir's solar panels could generate the maximum power. Because the differential gravity field would tend to restore the most stable position, maintaining this 'solar inertial' attitude would involve Atlantis in making frequent adjustments. The stack was to re-establish the gravity-gradient attitude each evening, and be left so overnight. When it was discovered that much more propellant was consumed doing this than had been expected, it was realised that it was because the shuttle computer's mass model was not able to deal accurately with having such a large attached mass and, as it overcompensated using its fine-control thrusters, it oscillated back and forth, either side of the optimal alignment. The long-term solution was a better mass model, but on this occasion the orientation tolerance was relaxed, and the resulting minor deviations did not seriously degrade Mir's power. The inadequacy of the computer's mass model had never been suspected. With this revelation and a variety of other engineering data, NASA was racing up the learning curve of how to operate a shuttle in conjunction with a massive orbital structure; taking the technical risk out of operations planned for the International Space Station was the primary objective of the Shuttle–Mir programme. On 1 July the hatches were closed to enable Atlantis to perform manoeuvres whilst docked with Mir, to test the integrity of the ODS (it did not lose its hermetic seal), and to observe the dynamics of Mir's solar panels (they wobbled alarmingly), and then the tunnel was reopened.

Late on 3 July, as the spacefarers bade their farewells and then congregated in their respective vehicles, Atlantis pumped up the complex's atmosphere to a pressure of 15.4 psi as a parting gift, thereby obviating the need to send up any air aboard the next cargo ferry. Kristall's hatch was closed at 2332 MT, and the ODS hatch was closed a quarter of an hour later. This must have been a sad moment for Dunbar. Having trained for Mir as Thagard's backup, she had been assigned to the first docking mission to transfer to the complex along with Solovyov and Budarin, to follow-up Thagard's programme until she was retrieved by Atlantis upon its return later in the year. This plan had had to be cancelled, however, when it became clear that leaving an astronaut aboard would conflict with the 'long' visit planned by ESA (the factor determining the size of the resident crew was the capacity of the lifeboat Soyuz). However, NASA had booked a continuous slot thereafter, and the presence of an astronaut would restrict other visitors to the duration of a handover. Only after NASA had used up its assigned time, in mid-1998, would others be able to make extended visits. The demand for access was so great that the problem was fitting everyone in.

On 4 July, Solovyov and Budarin undocked their ferry from Mir's rear port, withdrew 100 metres, then flew out to the side of the complex so that they could photograph Atlantis' departure. When the docking latches released, the spring-loaded mechanism eased the two vehicles apart. Explosive bolts would have been fired if the latches had failed to disengage; and if the bolts failed to fully separate the two components, Atlantis carried tools to enable a spacewalking astronaut to unfasten the 96 bolts that held the docking system on the ODS assembly. If this had proven necessary, however, fouling Kristall's

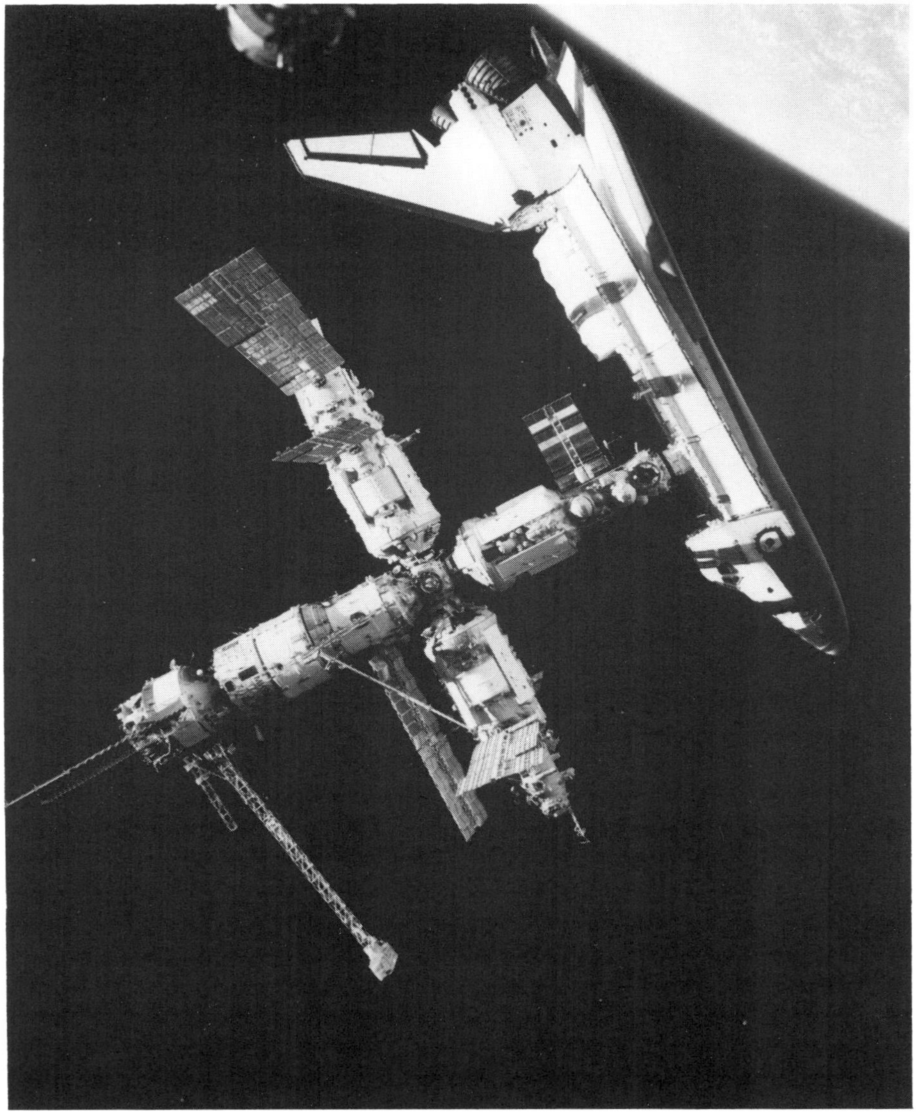

Fig. 17.3. STS-71 Atlantis docked with the Mir space station. This photograph was taken from the Soyuz, which undocked from the orbital complex and flew out to the side to document the event. No such pictures were taken of the subsequent dockings, so this depiction is unique.

port in this way would almost certainly have brought the Shuttle–Mir programme to a premature conclusion.

Atlantis returned to the Kennedy Space Center on 7 July. During the descent, Dezhurov, Strekalov and Thagard wore NASA pressure suits. Reclined couches had been installed on the middeck for them, so that they would not have to endure sitting upright at the start of their readaptation to gravity – the time that they would be most at risk of orthostatic intolerance, and most likely to black out. Contrary to instructions, Thagard climbed out of his couch and walked away to the recovery van. The delay in launching

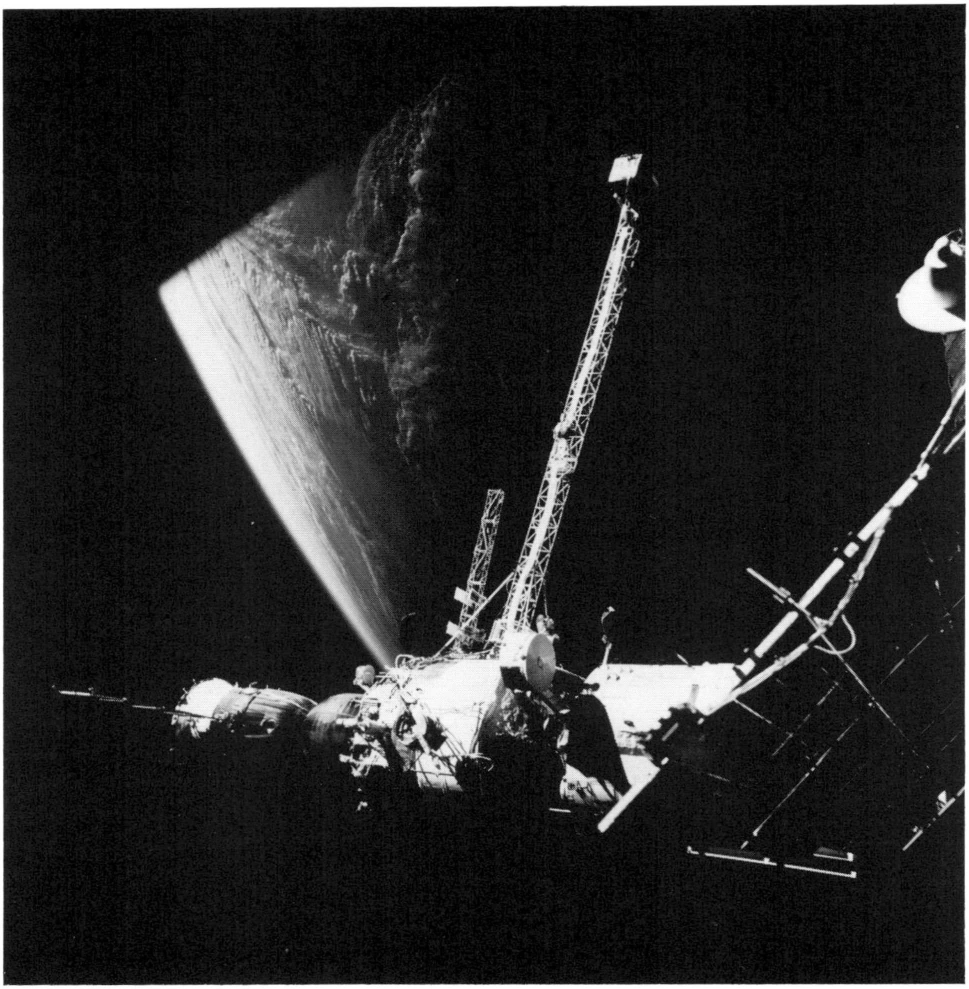

Fig. 17.4. A view of the rear of Mir, showing some of the construction work carried out on the Kvant 1 module, which is docked at the rear of the base block, and has a Soyuz at its rear. The box at the end of the tall Sofora truss is designed to provide more efficient control when rolling the orbital complex.

Spektr, and the resultant delay in sending up Atlantis, had meant that instead of a 90-day tour, he had spent 115 days in space. Despite the problems, though, he had managed to carry out his entire research programme, and NASA now had detailed medical data on how the human body would adapt during the planned three-month tour on the International Space Station.

Thagard's debriefing highlighted many interesting points: Mir was roomy, and very habitable, but it definitely had the look and feel of a locker room that had been lived in for a decade; he had got on well with Dezhurov and Strekalov, but days had passed by without hearing English on the voice link, and he had suffered cultural isolation; he had yearned for his family; being the equivalent of a laboratory rat in such a rigorous biomedical study was no joke; the requirement to log food intake was a disincentive to eating; and he had dreaded the prospect of extending his tour to six months. After four shuttle flights and a tour on Mir, Thagard retired from NASA a few months later, and returned to academia.

Although the primary objectives of the Shuttle–Mir programme were that America and Russia learn how to work together in space and reduce the technical risk in building and operating a joint orbital facility, the opportunity to have astronauts serve tours aboard Mir prompted NASA to review the science programme for the International Space Station. This identified several dozen experiments that could be undertaken early, aboard Mir, at little or no cost-penalty to the individual project budgets, and these were brought forward. Clearly, although the first docking was primarily an engineering test flight, the ongoing Shuttle–Mir programme would enable NASA to build up science activities on Mir.

SPACE TRUCK

On 12 November, STS-74 Atlantis was launched. As there were no long-stay residents to be retrieved, there was no need for biomedical facilities, so the rear of the bay was given over to the Russian-built Docking Module (DM). At 4.2 tonnes, this 4.6-metre long, 2.2-metre diameter compartment was really an extremely stretched Soyuz orbital module, fitted with an androgynous port at either end.

On 14 November, Chris Hadfield used the remote manipulator to unstow the DM and position it directly over the ODS. Opportunity was taken to collect configuration data for the computerised Space Vision System (SVS). Reference dots on the two units were tracked by obliquely-angled video cameras, while Hadfield set the DM above the ODS. The relative displacement of these references was measured, and an animated boresighted-viewpoint was computed. The SVS will be used to mount modules on the ports of the International Space Station, on which direct viewing will sometimes be impracticable. With the DM in position, the arm was placed in 'limp' mode, and Atlantis fired its thrusters to nudge the guide ring of the ODS up against the DM to achieve a soft-dock. As soon as the residual relative motions had been damped, the ring was retracted to hard-dock; only then was the arm withdrawn. If there had been a problem, Jerry Ross and Bill McArthur were ready to make a spacewalk in an attempt to overcome it.

Atlantis rendezvoused with Mir on 15 November, and flew the same R-bar approach as before. Although Kristall was now back on Mir's *radial* port, having the DM on the ODS provided four metres of extra clearance from the solar panels that projected from Kvant 2, Spektr and the base block. To provide further clearance for the DM, the base block's

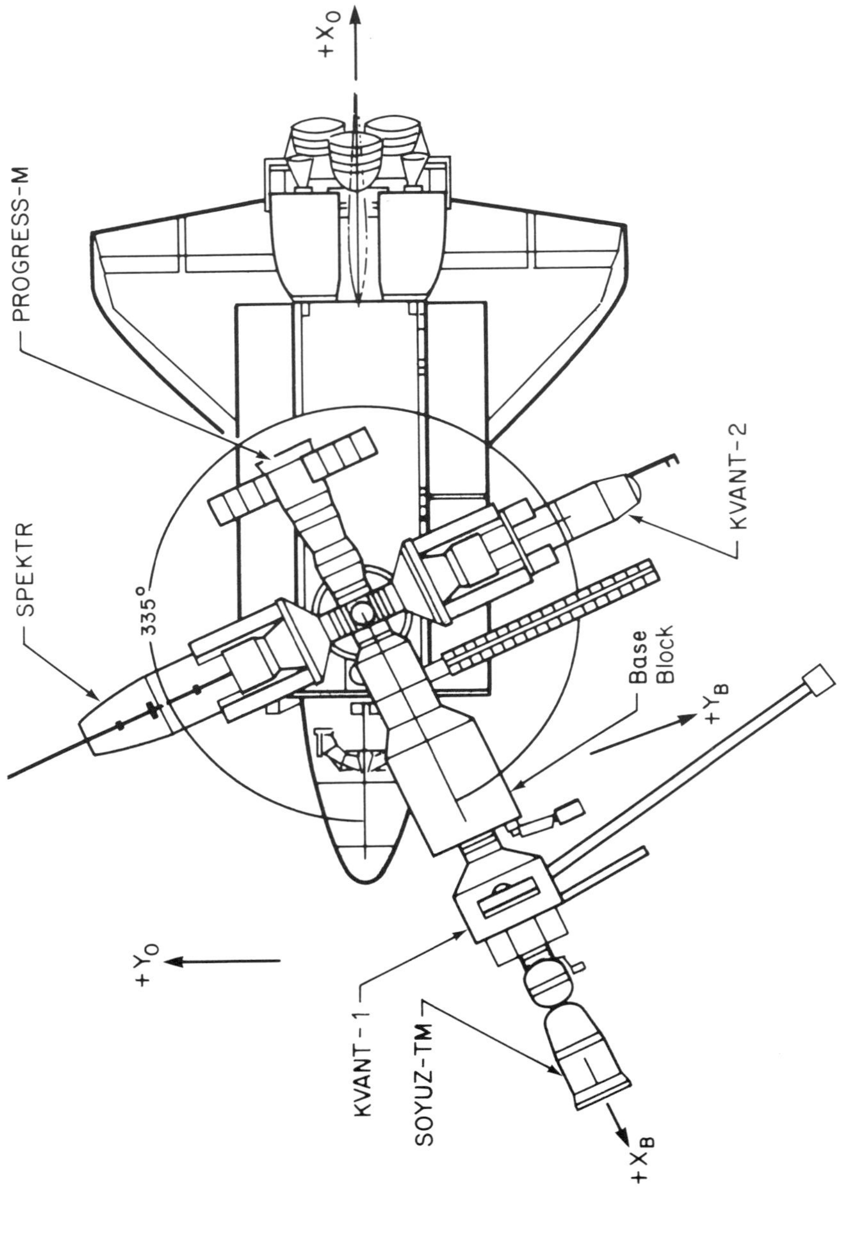

Fig. 17.5. A schematic of the alignment of STS-74 Atlantis once the Docking Module (DM) had been mated with the Kristall module, which by this time had been relocated to a radial port of the Mir complex.

panels had been rotated face-on to the Kristall module. Ken Cameron did not have a direct view of the docking system, so the RMS was fixed in position above and to the side of it to give a side view of the final few metres of the approach. Atlantis had never manoeuvred in such a confined space before, but Cameron made it look easy. Upon opening the hatches, Cameron and Yuri Gidzenko shook hands in the tunnel, then the two crews met in the base block for the ritual photo-opportunity. This time, with national flags of Russia, America, Canada (for Hadfield) and Germany (for Reiter) on display, it was evident that this complex had indeed become an *international* facility. In fact, Yuri Koptev had recently noted that because the RSA had been underfunded by 180 billion roubles in the current financial year, continued Mir operations were possible *only* because of the money earned from leasing it to fee-paying guests.

For the first time, the shuttle really had served as a *space truck* by delivering a module to a space station. The to-Mir cargo comprised 700 kg of food and water and some 300 kg of items for NASA's research. The 375 kg of cargo for return to Earth included processed materials, computer disks of experiment data, biomedical samples, and a variety of expired and broken hardware that was to be returned to the manufacturer for study. The Russians had never been able to return such amounts of cargo from Mir before so, by this simple act, Atlantis was making a significant contribution to Mir operations. Overall, 275 items were taken aboard Mir, and 195 were retrieved.

This time it was not necessary to maintain Mir in solar-inertial attitude, because the Sun angle was better and the power-level was higher. Nevertheless, Atlantis manoeuvred to test its ability to control such an offset centre of mass. On 18 November, when the orbiter departed, the DM was left attached to Kristall. The fact that it had been mated 'in the blind' boosted NASA's confidence that it would be able to add modules to the International Space Station. This tricky flight had "far exceeded expectation", noted shuttle manager Tommy Holloway. Dan Goldin announced that Mir was "proving to be an ideal test site for vital engineering research", and that the first two dockings were "already paying back benefits", by providing "proximity and docking operations" and by "simulating an early construction flight".

COMINGS AND GOINGS

STS-76 Atlantis docked at the DM on 24 March 1996. In its payload bay this time was a Spacehab, full of cargo.

The original plan had been to carry cargo in a Spacelab, but the internal configuration of this laboratory module did not lend itself well to bulk haulage, and Spacehab turned out to be ideal. In July 1995 a contract had been signed to fly one of the existing pair of Spacehab modules on this mission, and to mate the other with an engineering test article to construct a double-length module for use on later flights. Mounted well aft of the ODS, almost on the orbiter's centre of mass, the single-length module could carry 700 kg more than it could in its usual position at the front of the bay, and this took its capacity to 2 tonnes. In this case, it delivered 980 kg of material for the Russians and 740 kg for NASA's programme, and 500 kg was off-loaded. In addition, the module had a Biorack of experiments, some of which were transferred to Mir and the rest run independently, and a freezer in which to return accumulated biomedical samples. The cargo transfers were routine; the highlight of this visit was the spacewalk by Linda Godwin and Rich Clifford.

As this was the first time that astronauts had worked alongside an orbiter that would not be able to chase after them if a tether snapped, they wore backpacks incorporating SAFER manoeuvring units. And because it was the first time that astronauts had worked near Mir, the surface of which was unfamiliar, they were not to go beyond the top of the DM. After dismantling a now-redundant camera, they installed the Mir Environmental Effects Package (MEEP) cassettes on the short module. They then tested a new portable foot-restraint and a tether system intended for use in assembling the International Space Station.

When the hatches were closed on 28 March, Shannon Lucid remained aboard Mir with Yuri Onufrienko and Yuri Usachyov, who had begun their tour in February. The plan was for STS-79 Atlantis to retrieve Lucid upon its return in early August. Her tour of duty started two years of continuous habitation for NASA astronauts aboard Mir; Atlantis was to be the taxi that ferried them up and down.

Priroda was finally launched on 23 April, docked on 26 April, and was swung across to the radial port opposite Kristall the next day, thereby finally completing the Mir complex. It incorporated an array of remote-sensing instruments developed collaboratively. The NASA apparatus it contained was to support successive astronauts; this included the Mir Electric Field Characterisation Experiment (MEFCE), the Microgravity Isolation Mount (MIM), the High-Temperature Liquid-Phase Experiment (HTLPE), the Queen's University Experiment in Liquid-Diffusion (QUELD) experiment, the Glovebox (GBX), and the Optizon Liquid-Phase Sintering Experiment (OLIPSE) that was to be done in one of Kristall's furnaces. A SAMS instrument had been sent up prior to Thagard's tour, and Lucid's first task was to mount it in the new module to assess its microgravity environment. Then she set up video cameras and the MIM in the Glovebox, installed SAMS sensors, and ran an experiment to determine how well the MIM could damp out the ambient vibrations, to assess whether it would allow particularly sensitive microgravity experiments to be performed in the 'noisy' environment of an inhabited spacecraft. Over the following months, Lucid's varied programme included processing samples in the OLIPSE and QUELD furnaces, sampling air and water quality, monitoring ambient radiation, embryological studies, and, of course, gathering data for the biomedical database; but her favourite task was photography for geological, ecological and environmental studies, because it gave her an opportunity to gaze out of Mir's high-fidelity porthole. Houston updated a four-day task-list every night, and Lucid worked through it in her own time, checking off the items as and when she finished them. The only real problem that she encountered during the early part of her tour was when the MIPS died and she had degraded communications until a replacement processor card was dispatched on a Progress ferry.

On 21 June, Onufrienko and Usachyov were informed of the financial problems which had delayed the fabrication of the rocket that was to have delivered their successors in July, and that they would need to extend their tour to late August. This was a disappointment for Lucid, because she had expected to play her part as host to the French cosmonaut who was to accompany the handover. On 12 July, however, she was told that she too would have to extend her tour by six weeks, because it was necessary to replace STS-79's SRBs, which meant that she *would* be present for the handover. In mid-August, in an extensive series of combustion experiments in the Glovebox, she burned 80 candles for the Candle Flame Experiment (CFE) and eight solid fuels for the Forced Flow Flame-spread Test (FFFT); and she processed more QUELD and OLIPSE samples. By this point, she had finished her core programme, so she used her extension to set up the BioTechnology System (BTS), to grow wheat in the Svet cultivator for the Greenhouse experiment, to run the Anticipatory Postural Activity (APA) experiment, and to further characterise the body in its fully-adapted state by measuring muscle stimulation with the Belt-Pack Amplifier System (BPAS). Frenchwoman Claudie Andre-Deshays arrived on 19 August with Valeri Korzun and Alexander Kaleri. As a specialist in rheumatology and neurology, she focused on adaptation to microgravity, and then left with Onufrienko and Usachyov on 2 September.

Lucid spent her brief time with the new residents preparing to return to Earth, but on 7 September she claimed Kondakova's 169-day woman's record, and on 17 September, with Atlantis already on its way, she claimed Reiter's 179-day visitor's record too. This time, Atlantis carried the double Spacehab. The docking on 19 September was complicated by a recently-deployed solar panel which was projecting within three metres of the orbiter's nose. John Blaha had flown with Lucid on two shuttle missions, so they were old friends, and he had come to replace her on Mir. The moment that they transferred their Sokol suits and couch liners marked the first-ever astronaut handover. Just as did the resident crews, Lucid briefed Blaha on where everything was stowed, and suggested how best to work aboard the complex.

Tom Akers, the loadmaster, evaluated a new method of handling the cargo. Rather than stow items in lockers and transfer them individually, everything was packed in locker-sized canvas bags and the shipping was monitored by a bar-code reader. There was nothing like doing something for real for discovering the problems. Stowing the return cargo turned out to be harder than expected, because everything had to be weighed and placed consistent with centre-of-mass requirements. Some 2,250 kg was sent to Mir, and 1,000 kg was retrieved. Lucid's results alone took up twenty canvas bags. The Russians sent back an expired Orlan EVA suit, so that its state of degradation could be assessed. Atlantis left on 24 September. An important test in the Spacehab involved the Active-Rack Isolation System (ARIS) which Boeing had developed for microgravity experiments on the International Space Station. In contrast to the magnetically-levitated MIM, ARIS relied on a mechanical suspension system to damp out micro-accelerations.

Immediately upon landing, the density of Lucid's skeleton and key musculature was imaged by a nuclear magnetic-resonance scanner. To her surprise, she rapidly readapted to gravity. Shuttle–Mir programme manager Frank Culbertson opined that Lucid had "set the standard" for NASA's work aboard Mir. She was subsequently welcomed into the select group of astronauts to have been awarded the Congressional Space Medal of Honor. Upon taking over, Blaha expressed his view that there was no better way for NASA to prepare for the future than to have a cadre of astronauts learn to live and work in space by spending tours on Mir, because everything from the exercise regime to the logistics system was new to the agency.

Blaha's programme built upon Lucid's, but included several new experiments. One of the packages delivered by Atlantis was a 'powered transfer' with mammalian cartilage cells for a BTS experiment involving tissue growth. The Diffusion Crystallisation Apparatus for Microgravity (DCAM) employed a semi-permeable membrane to grow protein crystal over a long period. The Binary-Colloid Alloy Test (BCAT) investigated crystallisation of alloyed colloids. Blaha had a number of technological tests to perform in order to provide information important for fitting out the various modules of the International Space Station. One of these was an instrumented foot-restraint which monitored how much his body moved whilst he worked 'in place', and another used a push-off pad to measure the force imparted in moving around in the complex. The Passive Accelerometer System (PAS) measured low-intensity continuous effects due to air-drag and the gravity-gradient so as to further characterise the microgravity environment. The MIr Structural Dynamics Experiment (MISDE) measured the transient stresses on the complex resulting from thruster firings and docking operations, and thermal effects due to flying into and out of the Earth's shadow. Once installed, these instruments were to record data whenever NASA had microgravity experiments running.

In early December, the wheat that Lucid had planted in the Svet in August completed its cycle and produced grain, so Blaha 'harvested' it. This was the first time that a staple had grown to maturity, and it had important implications for the prospects of making a station self-sufficient in food production. The wheat and most of the grain was frozen for return to Earth for a comprehensive biochemical analysis. During a press conference at the end of the year, Blaha was asked whether he was eager for the shuttle to retrieve him on schedule; he replied laconically, "if it gets here, it gets here". Clearly, like Lucid, he was enjoying his tour. And the shuttle was once again launched on its first attempt; despite the early fears of the programme's critics, no launch attempt had been scrubbed during the final moments of the countdown.

Mike Baker and Brent Jett tested upgraded software as they docked STS-81 Atlantis on 15 January 1997, and the MISDE sensors measured the resultant stresses on the orbital complex. The double Spacehab carried the now routine cargo mix. Marsha Ivins was the loadmaster in control of the cross-loading, which was done mainly by John Grunsfeld and Jeff Wisoff. A treadmill that did not transmit vibrations to the orbiter's frame was assessed, as another step towards minimising the pollution of the microgravity environment by the presence of a crew. Blaha left on Atlantis on 20 January. Upon landing in Florida two days later, he was "absolutely stunned" at the strength of the Earth's gravity, and, in contrast to his predecessors, he obliged the medics and let them carry him off the shuttle.

GOOD TIMES AND BAD

On Mir, Jerry Linenger, a physician, set about a programme that combined biomedical, biotechnology, fluid physics and materials-processing assignments, but the highlight of his tour was to be when he made a spacewalk to retrieve two dust traps set up a year earlier and to deploy the Advanced Materials Exposure Experiment (AMEE). As had Lucid, he played host to a handover between Russian crews. Vasili Tsibliev and Alexander Lazutkin brought Reinhold Ewald with them, but it happened to be a far more eventful handover than was intended. On 24 February, thick smoke suddenly spewed from the module at the rear of the complex. When more than three people were aboard Mir, the Vika chemical burner supplemented the oxygen output of the Elektron regenerative system. The canister had split, released oxygen into the electronics, and ignited a fire. Although the cosmonauts were on the scene within seconds with portable extinguishers, and smothered the system with foam, the combustion was sustained by the direct injection of oxygen for 10 minutes.

The entire unit was reduced to soot-blackened scrap. For the first hour or so, until the air filtration system could extract the worst of the smoke, everyone wore face masks and portable oxygen bottles, and for the next few days they wore small filter-masks so as not to inhale particulates. It was the worst fire yet. On an earlier station it would have resulted in immediate evacuation, but Mir could not simply be abandoned, so the mess was cleared up and work resumed.

But Mir's problems had barely begun. On 4 March – two days after Korzun, Kaleri and Ewald left – Progress-M 33 had to abort its attempt to redock when its automated system failed. Tsibliev tried to steer it in using a remote control system, but this too broke down, so the ferry was ordered to withdraw. The next day, one of the Elektron units shut down because a bubble of air was blocking the flow of water in the electrolysis canal. This was a serious problem, because the other Elektron had been out of commission for some time, and now the only means of making oxygen was the sole remaining Vika. If that failed, there was only a few days worth of bottled oxygen, but this was an emergency reserve.

One clear lesson learned from earlier stations was that an empty station could be disabled by a fault that could easily have been fixed if a crew had been aboard. To leave Mir vacant, even for a short period, would be to invite trouble. As if to underline this danger, on 19 March an orientation sensor failed and the attitude control system tried to correct what it took to be an unexpected motion. By the time the cosmonauts were able to intervene, the flywheels had built up a runaway three-axis roll. The computer-controlled gyrodynes had to be turned off, the rotation cancelled manually by thrusters, and then the complex stabilised in the gravity-gradient while the fault was analysed. It is doubtful that control could have been reasserted from the ground, because problems with the communications link meant that Mir was in communication with Kaliningrad for only 10 minutes on favourable passes. The situation began to improve on 8 April, when Progress-M 34 brought a stock of 60 canisters for the surviving Vika, and parts to repair the transmitter and to service the balky Elektron. With the short-term future of the complex assured, Tsibliev, Lazutkin and Linenger settled down to work, and on 29 April, Tsibliev and Linenger went out to retrieve the PIE and MSRE and to mount the AMEE package. Despite having been drummed into helping with maintenance chores, Linenger had made progress with his science assignments. Indeed, by processing fifty samples for the QUELD experiment, he had not only finished his own programme, but had begun to work his way through the samples intended for his successor, Michael Foale, who arrived on STS-84 on 17 May 1997. Atlantis delivered the heaviest logistics load yet: in addition to 500 kg for NASA, the 1,200-kg Russian cargo included a 120-kg Elektron to replace that which had long ago failed, a 165-kg gyrodyne, and a stock of lithium hydroxide canisters. Also aboard, but only as a visitor, was veteran Mir resident Yelena Kondakova. When Linenger left on 22 May, he took 400 kg of accumulated NASA research results with him.

Foale had been assigned a programme of protein crystallisation, Earth observations, life sciences, materials processing and engineering tests, and was to assist with maintenance tasks, help unload the Progress scheduled for late June, look after Mir while Tsibliev and Lazutkin made two spacewalks in early July, then host the August handover. The repairs to the environmental systems got off to an excellent start, but then, completely unexpectedly, disaster struck in the shape of an errant ferry. Because the redocking of Progress-M 33 had had to be aborted, Progress-M 34 was to recertify the remote-control system. It undocked on 24 June, then returned the following day. As Tsibliev was manoeuvring it, the 7-tonne ferry went out of control and smashed into the Spektr module, badly mangling a solar panel and puncturing a coolant radiator. The twisting of the panel's motor installation, and the force on the hull transmitted through the struts supporting the conformal radiator, combined to break the hermetic seal of that module, and the air inside began to vent to space. Tsibliev ordered a retreat to the Soyuz at the front of the complex. As Foale made his way across the cramped docking adapter he heard "a loud bang". Lazutkin, who was still in the base block with his feet anchored to the floor, felt the shock-wave propagate through the structure. In seconds, there was a faint hiss; air was escaping to space. Tsibliev knew that Spektr had to be the source of the leak; its hatch had to be closed, and *fast*.

Cables had been laid between the modules to form an integrated power grid linking the various solar panels to the storage batteries distributed throughout the complex. There was no time to unplug the cables, so they would have to be *cut*. If the hatch could not be closed, the crew would have no choice but to retreat to the Soyuz, which would mean abandoning the airless station. A heavy cutting tool was stored in the docking adapter for precisely such an emergency. It took several minutes to clear the aperture and fit the hatch, but by that time it was evident that it was a *slow* leak. Later monitoring showed that the integrity of the docking collar had not been breached as it had absorbed the energy of the impact. On Earth, Linenger opined that the two worst accidents that could befall a spacecraft were fire and depressurisation. Mir had suffered both within the space of a few months.

With the station secure, it was time to tackle the immediate consequences. The quartet of solar panels on Spektr had contributed almost half of Mir's power, but, with the cables cut, this power was no longer being fed into the power distribution system. The complex had survived the decompression, but it now faced a power crisis. It was essential that the batteries not be *totally* drained, so each man dived into one of the other radial modules and switched off all the apparatus. Nevertheless, a terrible crisis was in the making, because the complex needed power to be able to continue to generate power. The power output from a solar panel is related to the angle of insolation by a sinusoidal function; it produces its peak output only when it is face on. As Mir flew around its orbit, it generally kept the same orientation with respect to the Earth below, so the panels had to be rotated to track the Sun, and it took power to run these motors; each panel had its own motor. If the batteries were depleted, there would be insufficient power to rotate the panels, which in turn meant that no power would be available to recharge the batteries; it was a runaway process – a spiral to oblivion. The vital thing was to switch off *everything*, to conserve the batteries, and then, with half the generating capacity denied, endeavour to remain on the safe side of the cusp in the power curve. The power situation was complicated by the fact that the orientation of the complex was being controlled by the gyrodynes, and these, too, consumed electrical power; they had to be deactivated. This left the complex in free-drift. It was at about this point that Mir flew back into communications range with Kaliningrad. The first the flight controllers knew of the problem was when the telemetry stream did not materialise, because the transmitter had been switched off together with everything else. In any case, there was nothing the ground team could do; the crew in space were effectively on their own.

The Soyuz had been powered up immediately, just in case it proved necessary to make a hasty escape, and now it was the only part of the complex not crippled by the power loss.

Its thrusters were used to stabilise the complex and then to reorientate it so that the majority of its remaining solar panels faced the Sun. It took several hours of continuous adjustments to fully charge the batteries; only then could Mir's control systems be reactivated. For the next few days, as they inspected the station, the cosmonauts worked by torchlight, and they set up a sleep roster so that they would not be taken by surprise by a sudden deterioration. And there were still the slow-to-develop problems which could, in the end, force abandonment. With power low, the cooling system was ineffective. The soaring temperature was not just uncomfortable for the crew; the Vozdukh carbon dioxide scrubber overheated and had to be turned off, which forced reliance on the limited supply of lithium hydroxide canisters. The Elektron suffered, so waste water could not be electrolysed to make oxygen, which forced reliance on the sole remaining Vika, and this meant having masks and fire extinguishers to hand. The toilet was operational, but its reprocessing system could not be used because the tank for the Elektron was full. And so it went. Nevertheless, the situation *was* improving with each day. It was by no means clear, however, that they would *not* be forced out. One thing was sure: if Mir was vacated, it was unlikely to be reoccupied. The spacecraft and its crew were in a symbiotic relationship; just as the crew relied on it for their survival, it relied on their continued presence. But basic survival was not the objective. Mir was a laboratory, and if it was to have a long-term future, then, at the very least, the undamaged solar panels on the Spektr module would have to be brought back on line.

In the original scheme, after Progress-M 34 had redocked, verifying its remote-control, it was to have been discarded to clear the aft port for its successor, which was already on the pad. Set for 27 June, this new launch was postponed, to give time to work out a way to restore the power, and to fabricate the necessary materials and tools. It was concluded that although it might be feasible to run external cabling from Spektr's panels to sockets outside the base block, it would be simpler to try to reconnect them internally. With Spektr exposed to vacuum, however, the hatch would have to remain closed. But how could new cables be run through a sealed hatch? An ingenious scheme was conceived: one of the Konus drogues in the docking adapter would be modified to serve as an air-tight electrical junction. There were two of these drogues, one of which was kept permanently on the axial port for dockings. The other, which was detachable, had been moved around the radial ports as necessary to facilitate the movements of the add-on modules, which swung themselves around on their short Ljappa arms. This Konus had last been used to accept the Priroda module. It comprised the hollow drogue, which was essentially a conical guide plate, and the bulbous end cap that contained the clamp for the mechanism at the tip of the probe. The two parts were connected by a ring of bolts. The plan was to remove the end cap and the clamp, and to bolt on a new unit with a set of electrical sockets on either side. If this could be mounted on Spektr's hatch, which, like the other modules, was currently sealed by a simple disk-like cover, then the situation should be retrievable. It would require an 'internal spacewalk', however, to redeploy this hatch and to lay the necessary cables within the stricken Spektr module. Trials in the hydrotank by Anatoli Solovyov and Pavel Vinogradov, who were to be Mir's next crew, showed that the task was feasible, so the equipment was loaded aboard Progress-M 35. As this ferry made its final approach on 7 July, the view from its docking camera clearly depicted the extent to which one of Spektr's solar panels had been twisted in the collision, but the other three panels seemed to be undamaged.

Kaliningrad had hoped that Tsibliev and Lazutkin would be able to effect repairs within a week, but when the cosmonauts reviewed the proposed procedure they requested additional time to prepare, so the internal spacewalk was pushed back a week. Fortunately, because the Konus was not already on Spektr's hatch, the preparatory work of exchanging the junction-plate for the drogue mechanism could be done in a shirt-sleeved environment. It was important that Spektr's power be restored prior to the August handover, because the new crew's arrival would place an extra demand on the environmental system, and because power would be required for the experiments to be performed by the visitor. A spacesuited rehearsal without depressurising the multiple docking adapter was scheduled for 15 July, and the real thing was set for 18 July. With power restored to the maximum available, the next crew would be able to concentrate on the external activities required to locate and plug Spektr's leak, so that the module could be repressurised and, with a little luck, much of its apparatus salvaged. The extent to which this proved practicable was to determine NASA's future aboard the complex. This module housed 50 per cent of NASA's science apparatus. It was questionable whether there would be sufficient scientific yield to justify continuing the Shuttle–Mir programme in the event that Spektr had to be written off. If the errant ferry had struck and depressurised Kristall, the loss of its androgynous docking system would have terminated the programme, and Foale would have been obliged to return with Tsibliev and Lazutkin in the Soyuz upon the completion of their tour, because Atlantis would not be able to return in September. A great deal, therefore, was riding on the outcome of this makeshift repair.

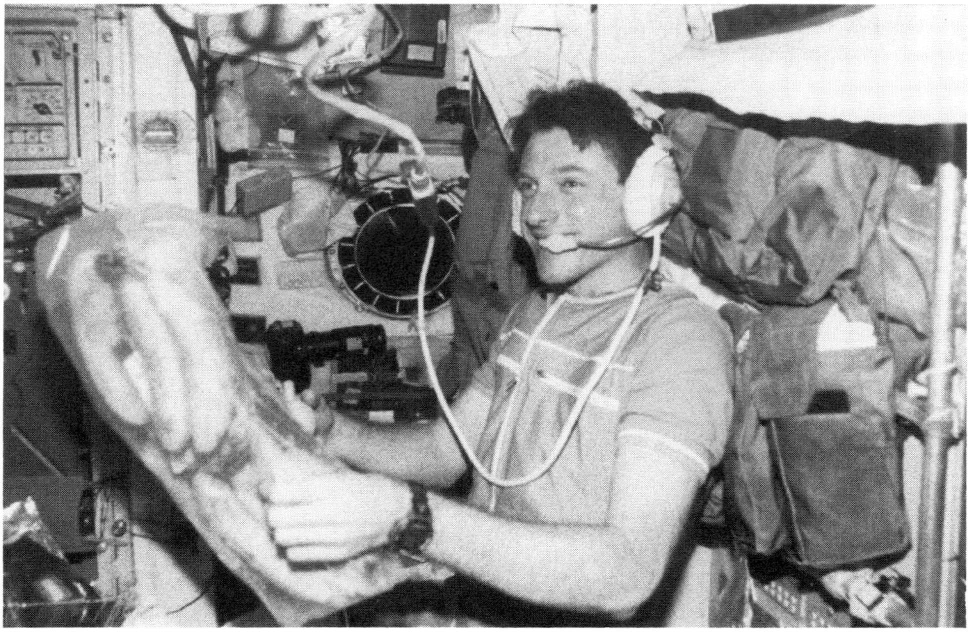

Fig. 17.6. Aboard the Mir base block, Michael Foale inspects a newly delivered package of fruit.

Although the preparations progressed well, the mounting stress evidently took its toll on the commander. On 12 July, Tsibliev reported sensing a heart arrhythmia. When a full test confirmed this the next day, he was instructed to take a combination of heart medication and tranquilisers to help him relax. It was also decided that Foale should support Lazutkin in the difficult repair, which was put back another week to allow Foale time to prepare. Foale had backed up Linenger in the preparations for Linenger's spacewalk, so he was familiar with the basic procedures for using the Orlan suit. On 16 July, while rehearsing the process of disconnecting the cabling within the docking adapter, Lazutkin unplugged the cable that distributed data from the main orientation sensors to the gyrodynes. Unfortunately, he had neglected to switch over to the back-up first, so the system immediately began to drift, with the result that the power output from the solar panels fell. The crew scrambled to switch off apparatus, but the complex was rapidly overwhelmed by another power crisis. In a rerun of the earlier process, the cosmonauts were obliged to retreat to the Soyuz and use its thrusters to reorientate the complex so that the solar panels could feed power and recharge the batteries.

Undeterred, the cosmonauts argued to be allowed to proceed with the repair on 25 July, but they were told that it had been decided to reassign the job to their successors. Furthermore, without power to run his experiments, Leopold Eyharts, the Frenchman who was to have worked through the handover, was reassigned to the next Soyuz, scheduled for early 1998. Anatoli Solovyov and Pavel Vinogradov arrived on 7 August, and a week later Tsibliev and Lazutkin left Foale with the new residents and returned to Earth. The first order of business was to swap the new Soyuz from the rear to the front of the complex, so that the spacecraft would be readily available in the event of trouble during the internal spacewalk to mount the connector plate on Spektr's hatch. Accordingly, on 22 August Foale retreated to the Soyuz descent module, Spektr's hatch was replaced, and then Vinogradov eased through feet-first to hook up the cables which would enable the computer to face the undamaged panels to the Sun and to feed the resulting power into the main grid. This done, he searched for any sign of where the collision had punctured the hull; but he could find nothing, so he retrieved some of Foale's belongings and closed the hatch. But although the panels produced power, they could not be rotated. When Solovyov and Foale went out on 6 September to inspect the damage, their first task was to manually orientate the three surviving solar panels into the optimum position. They did not have time to complete the inspection, so the site of the leak remained a mystery. Several spacewalks had been assigned, so it was a certainty that they would soon return to locate the hole, and hopefully seal it.

Given the recent requirement for spacewalking, NASA decided that David Wolf should supersede Foale in September because, unlike Wendy Lawrence, who was to have had the next tour, Wolf was EVA-trained, and so would be better able to assist in an emergency. It was far from certain that Foale *would* be replaced, however. Since the fire on Linenger's tour, Congressional critics had argued that Mir was dangerous, and had called for NASA to terminate Shuttle–Mir because it must have learnt all that it had hoped from joint operations, and that surely the recent loss of power and half of the US apparatus meant that there could be no scientific justification for stationing more astronauts aboard. NASA, in contrast, maintained that it was still learning valuable lessons in operating a space station, and in any case dealing with the fire and the depression was excellent on-the-job training for situations which could easily hit the International Space Station. In short, NASA saw Mir's problems not as an excuse to bale out but as a learning curve to be climbed, which is precisely why it had seized upon the joint programme in the first place; so, by any measure, the Shuttle–Mir programme had served to give NASA hands-on experience of operating a space station as a long-term facility.

THE CALM AFTER THE STORM

While Atlantis was in place, Scott Parazynski and Vladimir Titov made a spacewalk. A 'cap' had been added to the manifest at the last minute just in case the Russians finally traced Spektr's leak to the motor housing of the damaged solar panel, if so, and if the motor could be removed, the cap was to be used to seal the leak. After retrieving the MEEP, they tied the 'cap' to the DM. Unfortunately, because the damage to Spektr was on the Priroda side of the complex, opposite Atlantis, it was impracticable for the astronauts to inspect the site of the leak, this would have to be left to the station's crew.

Wolf's tour passed off uneventfully. In addition to tending to his experiments, he made a spacewalk with the Mir commander, Anatoli Solovyov, on 14 January 1998, at which time he employed a portable spectrometer to assess the state of the outer surface of the Kvant 2 module. It had been intended to venture down onto the base block, which had been in space for over a decade, but the need to set aside time for maintenance on the airlock hatch had ruled this out. Although Mir continued to suffer computer faults, the fact that there was no real drama for the Press to report meant that it dropped from the headlines, which vindicated the decision to send Wolf to continue the American presence.

The next shuttle arrived on 24 January 1998, but this time it was Endeavour not Atlantis. It brought Andy Thomas who, upon being retrieved by Discovery in June, was to round out the programme.

18

An island in the sky

When NASA submitted its post-Apollo plan to the Nixon administration, it proposed a reusable shuttle to provide routine cheap access to low orbit, and a space station to serve as an orbital laboratory and as a way-station for missions heading for deep space. However, the innovative shuttle's projected development cost was so high that NASA had to sacrifice the station. But as soon as the shuttle was declared operational, it dusted off its station plan and began lobbying for support, and, to its surprise, found key members of the Reagan administration to be very receptive. The President was taken with the idea, and particularly appreciated the suggestion that international participation be sought to build the station, so, in January 1984, in his State Of The Union Address to Congress, he directed NASA to build a space station that would be permanently occupied by eight astronauts, and in an echo of Kennedy's historic challenge, ordered it done within a decade. A few months later, the European Space Agency (ESA) announced that it would build a laboratory to be added to the American space station.

FREEDOM

Although the structure shown to Reagan consisted of a compact cluster of modules with a pair of solar panels, the design selected was Grumman's Power Tower concept. This was a 120-metre keel aligned vertically to use the gravity gradient for stability, a horizontal truss to carry the solar panels, and a cluster of modules at either end to facilitate simultaneous observation of the Earth and the sky. As the design was reviewed, however, doubts began to arise. Would the long narrow keel be sufficiently rigid? How would the astronauts move from one cluster of modules to the other? Other issues concerned the specification of the modules themselves, which were linked together in such a way that it would be difficult to isolate a specific compartment in an emergency. As a result, the Power Tower was rejected in late 1985.

A year later, having evaluated submissions, NASA accepted the Dual Keel concept by Lockheed and McDonnell Douglas. It had two 150-metre vertical trusses linked top and bottom by 45-metre spars for increased structural strength. Like the Power Tower, it would exploit the gravity gradient to maintain a vertical orientation, and it would have Earth-observational and astronomical apparatus at the ends, but in this case these would be remotely operated from a single cluster of modules half-way up and between the two keels, beside the horizontal truss supporting the solar panels. This eliminated the need for a frequent-use transportation system along the keel, and the modules themselves incorporated key safety features. As before, once the station was complete, astronauts would be delivered and retrieved by a succession of shuttles.

Within months of this revision, however, the enormous dual keel had been dropped from the configuration, and with it went the dedicated Earth and astronomical observational sites. This left only the module cluster on the truss with the solar panels. Stripped of its vertical structure, gravity-gradient stability was impracticable. External apparatus would have to be mounted on the truss. The heavy keel, it was argued, could be added later. To give the project a sense of identity, it was decided to name the structure 'Space Station Freedom'. It reflected Reagan's hostility towards what he had so recently dubbed 'the evil empire' of the Soviet Union. So, just as Apollo had been two decades earlier, the space station became a symbol of the Cold War.

Despite having selected the design, the funding to begin construction work was not forthcoming. Instead, the annual Congressional budget review demanded a succession of redesigns to reduce costs. During much of this hiatus, the shuttle fleet was grounded by the loss of the Challenger. In 1990, Congress sent NASA back to the drawing board and told it to come up with a design that would significantly reduce costs. Ninety days later, NASA proposed a configuration that could be crew-tended within five years, and then gradually be upgraded to allow permanent occupancy. To cut the number of shuttle flights required for the assembly sequence, the truss had been shortened by 30 per cent, and the modules had been scaled down by 40 per cent to ensure that they could be delivered fully outfitted. The occupancy level of the habitat had been cut from eight to four astronauts. Already frustrated by NASA's funding delays, ESA now faced not only the prospect that its laboratory would not be launched until the end of the century, but also that there would not be sufficient crew capacity to staff it permanently.

Dan Goldin took over from Richard Truly as NASA administrator in April 1992. One of his first acts was to instigate another redesign of Freedom with the objective of further slashing costs. He also ordered that a 'lifeboat' be added, to ensure that the crew would be able to return to Earth in an emergency.

Following close upon the Human Spaceflight Cooperation protocol of October 1992, NASA began to consider the possibility of purchasing a specially-modified Soyuz ferry for use as the lifeboat, to preclude the cost of building a completely new Assured Crew-Return Vehicle (ACRV), as the lifeboat had been prosaically termed. This lifeboat was to be ferried to the station in the shuttle's bay and attached to the station by using the RMS. Negotiations were instigated to define the modifications which NASA would require.

When the Clinton administration took office in January 1993, the provocative name of 'Freedom' was quietly dropped. In March, Goldin was told to come up with a more cost-effective plan that would be compatible with Clinton's promise to reduce the deficit.

ALPHA

On 7 June, NASA submitted three options. Option-A was to employ proven hardware, including a propulsion, guidance and navigational system developed for spy satellites, and incorporate those systems intended for Freedom that would be most cost-effective. Option-B was essentially what had already been proposed, and so would maximally exploit the work done on Freedom. Option-C was a 'minimalist' facility that called for the Columbia orbiter to be decommissioned and its engine unit mated to a specially modified external tank, then used to put a large one-piece space station into orbit more or less completely fitted out, as a latter-day Skylab. On 17 June the White House selected Option-A, and asked Congress to award the necessary funding, which it duly did on 23 June, but with a majority of a *single* vote. NASA was given three months to 'transition' to this new configuration. Boeing was nominated as prime contractor on 17 August, and the detailed project definition for 'Alpha' was sent to the White House on 7 September.

In parallel with this technical review, in which the RSA – in addition to ESA, Canada and Japan, NASA's formal international partners for the space station – had been included in a 'consultancy' capacity, the White House had been endeavouring to expand the scope of the historic 1992 agreements with Russia, and on 2 September Vice President Al Gore and Prime Minister Viktor Chernomyrdin signed an accord which called for their space station proposals to be merged into a single orbital structure. The fact that this announcement took NASA's partners by surprise highlighted the rapid pace at which the programme was being transformed.

On 1 November, NASA and the RSA announced an addendum to the 'Alpha' plan submitted on 7 September. This 'Space Station Programme Implementation Plan' involved three phases. Phase One was a logical expansion of the October 1992 plan to fly each other's crews, and it called for the shuttle to make a series of dockings with Mir. This would be conducted between 1995 and 1997, and involve NASA leaving a succession of astronauts aboard Mir. Phase Two called for sufficient elements of the new station to be in orbit by 1998 to be able to support continuous habitation. Phase Three involved flights through to 2002 to complete the station.

The three-phase plan was given to the White House on 4 November, and it was eagerly accepted on 29 November. The formal invitation to Russia to join NASA, ESA, NASDA and the CSA in the development of the international facility was issued on 6 December, and the RSA signed with NASA on 16 December. ESA formally reaffirmed its support in January. The first meeting of the full international team, which included prime contractor Boeing and its subcontractors Rockwell, Lockheed and McDonnell Douglas, was held on 24 March to review the overall systems-design of the new proposal.

In June 1994, even as its critics in Congress continued to snipe at the budget, Gore and Chernomyrdin signed the interim cooperation agreement to build what was then informally dubbed 'International Space Station Alpha'. In responding to concerns expressed by some in Congress that the Russians would not be able to deliver, Chernomyrdin assured, tongue-in-cheek, that Russia would continue with the project even if NASA had to pull out!

TWO IN ONE

With the political agreement in place, the two agencies got together to define the detailed configuration and the assembly sequence. NASA took its Option-A (Alpha) configuration, restored certain elements that it had elided to lower the cost, and settled on a design which used 75 per cent of the hardware intended for the 1992 version of Freedom. However, there were significant revisions. One immediate decision was to employ a Russian manoeuvring unit. This tug was to be built by Khrunichev, the company which had made all the Mir modules. It would not only be cheaper; it would be able to be replenished in orbit. The simplest way forward was to order an off-the-shelf tug, and somehow mount it on the scaled-up Alpha, but to have done no more than this would have been to miss the golden opportunity offered by cooperating with the Russians.

In its own concept, NASA could not leave a crew aboard its station until it was virtually complete; residence would not be practicable until the habitat module was installed. During construction, it would be 'tended' only whilst a shuttle was docked. The Mir 2 base block was a habitat. Surely it could be docked at the other end of the tug. By *starting* with these two Russian 'core modules', this unproductive initial period could be eliminated. Although the assembly process would take just as long to complete, the crucial 'permanent habitation capability' could be advanced by two years.

The key to integrating the two technologies was the tug. The propulsion unit was given a multiple docking adapter with an androgynous axial port to mate with the first of NASA's modules, and a pair of radial ports for Soyuz ferries. It would be linked at the far end to the base block, which NASA took to calling the 'Service Module'. As the functionality of the tug was refined, it acquired logistics and power generation capabilities, so it was named the 'functional energy block' (or, by using the Cyrillic acronym, the FGB). These two core modules would provide orbital manoeuvring, attitude control, power supply, logistics and crew accommodation in the initial phase of assembly. While NASA built up one end of the complex using hardware that it had designed for Freedom, the RSA was free to develop the other end, to create a hybrid which integrated both technologies.

As NASA Administrator Dan Goldin put it, following substantial votes in favour of the station in both parts of the Congress in July 1994, the station was "no longer just a design ... we are building hardware". To answer Congress's doubts about Russian participation, NASA agreed to buy the tug outright.

In January 1995, after resolving details of configuration, assembly sequence, schedule and cost, NASA confirmed its contract with Boeing to fabricate its hardware, and Boeing immediately began negotiations with its major subcontractors. Although Boeing would not play a direct role in building European and Japanese modules, it had to ensure that their hardware would integrate into the overall system. The FGB was subcontracted to Lockheed because it had already worked with Khrunichev to commercialise the Proton rocket, and on 8 February a $200 million contract requiring 'on-orbit delivery' of the FGB was agreed. Khrunichev was to build the Service Module too, but under a separate contract to be funded by the RSA.

On 28 September 1995, in an unprecedented move, Congress voted NASA a multi-year budget of $2.1 billion per annum through 1996–2002, thereby guaranteeing the financial commitment required to manufacture and assemble the space station. Unfortunately, despite a pledge by Boris Yeltsin, the RSA did not have the stable cash flow required to construct the Service Module, and funding for the proposed new research modules seemed ever less likely to be forthcoming. The RSA's dilemma was that in order to work on building up the new space station, it would have to terminate further operations on Mir. The worst outcome would be to give up Mir and then be denied the funding necessary to participate in the bold new venture. It made more sense, in the short term, to keep Mir running, so that it could continue to serve as an international research laboratory. In November, therefore, the RSA announced its intention to transfer Spektr and Priroda to the new station. NASA's response was that it had no objection to anything which would increase the science capability of the complex, so long as achieving it did not delay the assembly schedule or increase its cost. On 11 December, however, the RSA tabled a radical change in plan. Surely it would make more sense to dock the FGB at the front of the existing complex. This would make the best use of existing assets, and would allow faster commissioning of the new station because it could support continuous occupancy by larger crews. As a compromise, NASA suggested it help to keep Mir running while the space station was assembled by extending Phase One, by adding two missions to extend the Shuttle–Mir programme well into 1998, and it offered to deliver Russia's Solar Power Platform to relieve pressure on the RSA's finances by reducing the number of Proton rockets required. This agreement was formally ratified in July 1996, by which time NASA had quietly dropped the name 'Alpha', it was now simply the 'International Space Station'.

With Russian participation declining, it became clear that the space shuttle would be the workhorse of the assembly process. The decision to employ a 51-degree orbital inclination meant that the shuttle would not be able to carry the fully-outfitted NASA laboratory, and most of its equipment racks would have to be delivered later. NASA had originally intended to use a laboratory-sized logistics carrier to resupply Freedom, but this was not in Option-A. Rather than try to reinstate it, NASA decided to accept an Italian offer to supply a Multi-Purpose Logistics Module (MPLM), and it struck a deal whereby, in return for Italy donating a cargo module, NASA would fly up a refurbished test article carrying a large centrifuge to serve as a life-sciences module; this was the 'no exchange of funds' principle at work. Assembly of the International Space Station will have to be borne by Atlantis, Discovery and Endeavour, because Columbia, the first of the fleet, which is four tonnes heavier, cannot carry heavy payloads. Of two long-awaited upgrades, the aluminium–lithium Lightweight External Tank was on track for use in 1998 but, after several stops and starts, the Advanced Solid Rocket Motor had been finally cancelled in late 1993.

By the end of 1996, it was clear that cash flow problems were significantly delaying the Service Module, but the FGB, which was being paid for by NASA, was on schedule. As a contingency in case the Service Module could not be launched before the FGB expended its propellant, thereby leaving the complex without attitude control, NASA decided to adapt a manoeuvrable stage used to deploy military satellites to act as an 'interim control module'. In April 1997, NASA decided to put back the launch of the node, which had been set for November, by at least six months, in order to accommodate the slippage in fabricating the Service Module. It hoped, however, that it would be able to make up much of this delay by pursuing a more aggressive schedule thereafter, so that the project would be completed more or less as intended. To launch the FGB and the node on time, at the end of 1997, just to stick to the schedule had been deemed futile. Although basic fabrication of the FGB was finished, NASA ordered modifications so that its propellant tanks could be replenished by a Progress tanker on one of the docking adapter's radial ports, not just by way of the Service Module.

THE INTERNATIONAL SPACE STATION

The Service Module habitat and the FGB power and logistics module – the International Space Station's 'core' elements – are 20-tonne vehicles derived from Mir technology.

The FGB is structurally similar to Mir's Kristall module, in that it has a long cylindrical main body with a small docking adapter at the far end. NASA's first node will be docked at the adapter's axial port and Soyuz ferries will use its radial ports. The main compartment of the FGB is fitted with lockers for dry cargo. The main axial port is for the Service Module. This has the same configuration as Mir's base block, and is to provide environmental support for a crew of three throughout the assembly. The Solar Power Platform (SPP) truss will be mounted on the upper port of the Service Module, and will carry a thermal radiator, a set of gyrodynes for attitude-control, a roll-control thruster unit, and motors for eight solar panels similar to the Cooperative Solar Array that was tested aboard Mir. Each of these will produce 6 kW for NiCd batteries. The two panels on the FGB will add 4 kW, and those on the Service Module will add 9 kW. An ESA-supplied RMS-like arm will run on rails on the SPP. Whereas the Russian modules are autonomous vehicles, NASA's elements are totally passive, and they will initially be reliant on the FGB for navigation, propulsion and power.

As prime contractor, Boeing undertook to develop the pressurised modules and their internal environmental systems. All the modules are cylinders with an external diameter of 4.42 metres and an internal diameter of 4.25 metres. In contrast to the core modules, which comprise variously-sized and variously-shaped compartments, these modules form 2-metre wide, 2-metre high rooms, each with a distinct sense of 'up and down' (the ceiling provides illumination; the dark floor contains utilities and storage lockers; wall-mounted standardised racks contain apparatus), and the shallow conical end-caps have 2-metre diameter ports that incorporate a 1.3-metre square hatch with a centrally-mounted 20-cm window. As well as providing a solid structural link, the rings of these ports, the Common Berthing Mechanism (CBM), facilitate inter-module environmental services. The hatches can be readily closed to isolate adjacent modules, without the need to disconnect utilities. Although it is not required for the basic configuration, connectivity is offered at both ends of a module in order to maximise the potential for future expansion of the complex. In addition to a second node to carry the international modules, an airlock and a 1.4-metre high blister-like cupola – offering panoramic viewing through peripheral windows for supervision of operations with remote manipulators – will be mounted on the first node, together with a back-up docking system for the shuttle.

NASA has two module types: the shorter modules of the nodes and the longer modules of the laboratory and the habitat. In addition to axial ports, the 5.5-metre long node has a ring of four radial ports (offset to one end) to serve as a nexus for attaching other modules. The first node will provide command-and-control and power-distribution functions. It will be linked to the FGB by a semi-conical Pressurised Mating Adapter (PMA). The 8.5-metre laboratory will be connected to the other end of the node. It can accommodate a dozen racks of apparatus. The structurally-identical habitat has six crew cabins, a galley with a freezer, an oven, a hot-and-cold water faucet, a hand washer, a trash compactor, a shower, a toilet, a gymnasium, and a wardroom with a high-fidelity viewport.

A conventional sea-level oxygen–nitrogen atmosphere will be regulated by drawing on supercritical cryogenic tanks that have to be periodically replenished. Air temperature and humidity will be regulated by condensing heat exchangers. Carbon dioxide will be removed by a regenerative molecular sieve. Undesirable trace constituents such as carbon monoxide, methane and ammonia will be removed by charcoal and sorbent beds. A multi-filtration unit using heaters and filters will retrieve potable water from the shower and galley. A separate vacuum filtration unit will extract water from urine. Solid waste will be returned to Earth by shuttle. High-capacity fans will ventilate the pressurised compartments via ducts within the attachment collars. Flame detectors in open areas, and smoke detectors in confined spaces, will detect fires and trigger an automatic carbon dioxide suppressant system. The thermal regulation system will use internal coolant loops that circulate water and then transfer heat to external loops which run ammonia through 23-metre long radiators on the truss; three radiators are mounted in line on pallets on either side, shaded from direct sunlight.

The truss, thermal control, command, communications and navigation systems will be provided by McDonnell Douglas. Signals will be relayed via the TDRS network to maintain continuous high-capacity communications via S-Band uplink and K_u/K_a-Band downlink. Spacecraft in line-of-sight will be able to use UHF. A control-moment gyroscope (CMG) corresponding to Mir's gyrodynes will provide three-axis stability. The GPS network will provide navigation so as not to tie up ground-based tracking systems. Flight operations will be managed by the Space Station Control Center at the Johnson Space Center in Houston.

The structure will fly with its truss horizontal and with its principal axis on the velocity vector. The central section of the truss, which constitutes half of its 94-metre total length, is a 4.3-metre wide hexagonal-section frame mated to the laboratory by a 13-metre spur. This also accommodates the GPS antenna, the mobile transporter and the portable work-platform which will be used by spacewalking astronauts. The port side will contain a UHF antenna and a section of the crew-and-equipment translation assembly, and carry a thermal radiator on its trailing edge. The starboard section is identical, but with the S-Band antenna. The 2-metre square-section port and starboard truss extensions at either end carry the solar panels and their associated storage batteries. The mobile transporter is a 2.5-metre flat-pallet mounted on rails running along the leading edge of the hexagonal truss, and it will carry the remote manipulator's mobile servicing system. The Canadian Space Agency had developed the shuttle's RMS, so it scaled it up to a 17.7-metre long heavy-duty arm for the station. To facilitate installing modules on the station 'in the blind', it will incorporate the Space Vision System. The Mobile Remote-Servicing Platform will carry tools for the arm's end-effector, and mountings, power and communications sockets for apparatus to be installed externally. The Special-Purpose Dexterous Manipulator is a pair of 2-metre long multi-jointed arms set on the mobile platform to manipulate external payloads with dexterity comparable with that of the human hand.

The solar panels, storage batteries, and power distribution system will be supplied by Rockwell. Each 11.8-metre long truss section contains six high-capacity NiH batteries, and has a rotating joint to keep its panels facing the Sun. Each 36-metre long panel frame will expose 287 m^2 of transducers capable of generating 23 kW. As power-loss is proportional to current as well as to cable length, the output from the transducers is converted to high-voltage-low-current for transmission along the cables running along the truss to the module cluster, and converted back by direct-to-direct converter units to low voltage for use (the same principle as used on commercial power transmission lines). All NASA-supported modules will use 160 volts dc, and Russian modules will use 28 volts dc. The shuttle fuel cells generate 30 volts dc. Although half of the 96 kW from NASA's solar panels is to be reserved for station operations, this leaves sufficient margin to ensure that the laboratories will not be starved of power.

All international modules which will be delivered by shuttle will have the same diameter as NASA's pressurised elements. Italy's 6.6-metre Multi-Purpose Logistics Module will have a 9-tonne cargo capacity in 16 racks in a wrap-around configuration. The same shell will be used for Italy's life-sciences module, which will house a 2.5-metre centrifuge. It is also likely to be used to make an eight-rack module for ESA. Japan's 10-metre experiment module is a far more elaborate facility, with two small arms to manipulate apparatus on an unpressurised pallet at its far end, and a collar on its roof to accept a pressurised experiment logistics module.

The sprawling football-field-sized structure, with modules clustered in the centre and massive solar panels mounted at each end of a framework truss, may be a far cry from the elegant wheel-in-space imagined by Wernher von Braun, and the magnificent double wheel depicted by Stanley Kubrick in his classic science fiction film *2001: A Space Odyssey*, but it nevertheless has an elegance all of its own.

ASSEMBLY

The detailed manifesting of the International Space Station assembly might be revised to meet operational exigencies, but the overall process is clear. Assembly is to follow on from Shuttle–Mir; in effect, Phase One of the international programme, which is to conclude in 1998. The sequence involves two further phases. The first major milestone occurs when NASA's laboratory is up and running.

Table 18.1. Phase Two manifest, *c.* mid-1997

Date	Label	Launcher	Cargo
Jun 1998	1A/R	Proton	FGB
Jul 1998	2A	STS-88	Node 1; PMA 1 and 2, and one stowage rack
Dec 1998	1R	Proton	Service Module
Dec 1998	2A.1	US	miscellaneous logistics
Jan 1999	2R	Soyuz	commissioning crew
Jan 1999	3A	US	truss Z1; PMA 3; CMGs
Mar 1999	4A	US	truss P6
May 1999	5A	US	US Lab
Jun 1999	6A	US	MPLM (with lab racks); UHF; SS-RMS
Aug 1999	7A	US	Airlock; h.p. gas pallet
Oct 1999	7A.1	US	miscellaneous outfitting

In June 1998, a Proton – the most powerful operational rocket in the Russian inventory – is to blast off from the Baikonour Cosmodrome in Kazakhstan and arc over to the north-east on a 51-degree inclination. Its payload, the FGB, will be placed into the same plane as Mir, but it will make no attempt to initiate a rendezvous.

Flown by Bob Cabana and Fred Sturckow, STS-88 Endeavour will be launched into the same plane a fortnight later. On the first day of the two-day rendezvous with the FGB, Nancy Currie will use the RMS to lift the primary payload – a node with a pressurised mating adapter on each end – out of the rear of the bay, raise it high above the orbiter, turn it around, and then mate it with the orbiter docking system near the front of the bay. As in the case of rendezvousing with Mir, Endeavour will approach the FGB from directly below. At this point, the FGB will be stabilised in a vertical orientation with its docking adapter facing down. Cabana will manoeuvre Endeavour so that the FGB is stationary a mere three metres above the now vacant rear bay. Although she will not be able to see the FGB, because it will be behind the node, Currie will reach around with the RMS and, guided by the SVS, grab a trunnion pin on the FGB's docking adapter. Although it will be the heaviest payload ever to be manoeuvred by the arm, the FGB is well within the arm's specifications. The final stage in the process will be to raise the FGB, draw it forward, and then mate it with the docking system at the end of the adapter on top of the node.

Fig. 18.1. A computer graphic of how STS-88 Endeavour, with NASA's node already mounted on the Orbiter Docking System (ODS), will employ the Remote Manipulator System (RMS) to capture the previously-launched Russian-built FGB, and mate it with the adaper at the far end of the node.

The next day, Jerry Ross and James Newman will conduct the first of three spacewalks to complete the mating process. Many of the power and data links on the NASA part of the station will employ external, rather than internal, connectors. On their first outing, the two astronauts will first set up work sites in the bay, then run safety wires between the elements of the stack, and finally install cables between PMA 1 and the FGB. The next day, once the command links have been verified, the orbiter's docking system will finally be pressurised, the node's environmental system will be powered up, and first the local and then the remote PMAs will be pressurised. On their next spacewalk, the following day, Ross and Newman will install handrails, worksites, and the S-Band communications antenna on the node. The day after that, the astronauts will progress through the node into the FGB in order to set up communications apparatus. Displacing the node's side ports towards its PMA 2 end made it possible to accommodate a standardised rack on each adjacent internal face. Only one rack will be in place at this stage; it will carry apparatus for the commissioning crew. The other racks will be installed by subsequent shuttles. On their final outing, Ross and Newman will mount a toolkit on the node for future spacewalkers. Endeavour will then leave the node in the care of the FGB.

In December, the Service Module will be launched. The FGB will rendezvous and dock at its front port. Even though the FGB has the node in trail, it is able to manoeuvre because its main engines are mounted peripherally, but once it is docked with the Service Module it will relinquish this role to *that* vehicle. Its integrated propulsion unit (ODU) is the same as installed on Mir: it burns Unsymmetrical DiMethyl Hydrazine (UDMH) in nitrogen tetroxide oxidiser. As these are hypergolic, no ignition system is needed. It will periodically reboost the station to overcome orbital decay. There are pipes in the aft docking collar to enable a Progress ferry to pump propellant right through the Service Module into storage tanks set around the hull of the FGB. With six tonnes of capacity, these tanks can maintain sufficient propellant for a year's-worth of reboost.

Over the next month or so, two shuttles will start the process of erecting facilities on the node, and a Soyuz will dock at the lower node of the FGB's docking adapter to deliver Yuri Gidzenko, Sergei Krikalev and Bill Shepherd, the commissioning crew who will live in the Service Module. From this point on, the station will be permanently inhabited. In addition to PMA 3, which will be mounted on the node's lower port, a stubby, square-section truss will be installed on the upper port. This truss contains high-pressure gas tanks, CMGs and K_u-Band communications, but their addition is opportunistic manifesting because they will not be required until later; it is merely convenient to locate them in this Z1 truss. The next shuttle will mount, on this truss, a Power Module which will incorporate a pair of arrays of solar transducers, a set of high-capacity NiH storage batteries, and a thermal radiator.

If the schedule unfolds as planned, the addition of NASA's laboratory in May will be a major milestone in the assembly process. Because it is to be mounted in line with the node, the shuttle will have to dock at the PMA 3, then use its arm to disconnect PMA 2 and stow it prior to mounting the laboratory on the node; then the adapter will be remounted axially. The shuttle cannot lift a fully-equipped laboratory to a high-inclination orbit, so only half of the complement of racks will be installed for launch; the rest of the racks will be ferried up in MPLMs. These next few missions will also add an airlock for the node, and the station's own remote manipulator (SS-RMS), which be used in later assembly operations.

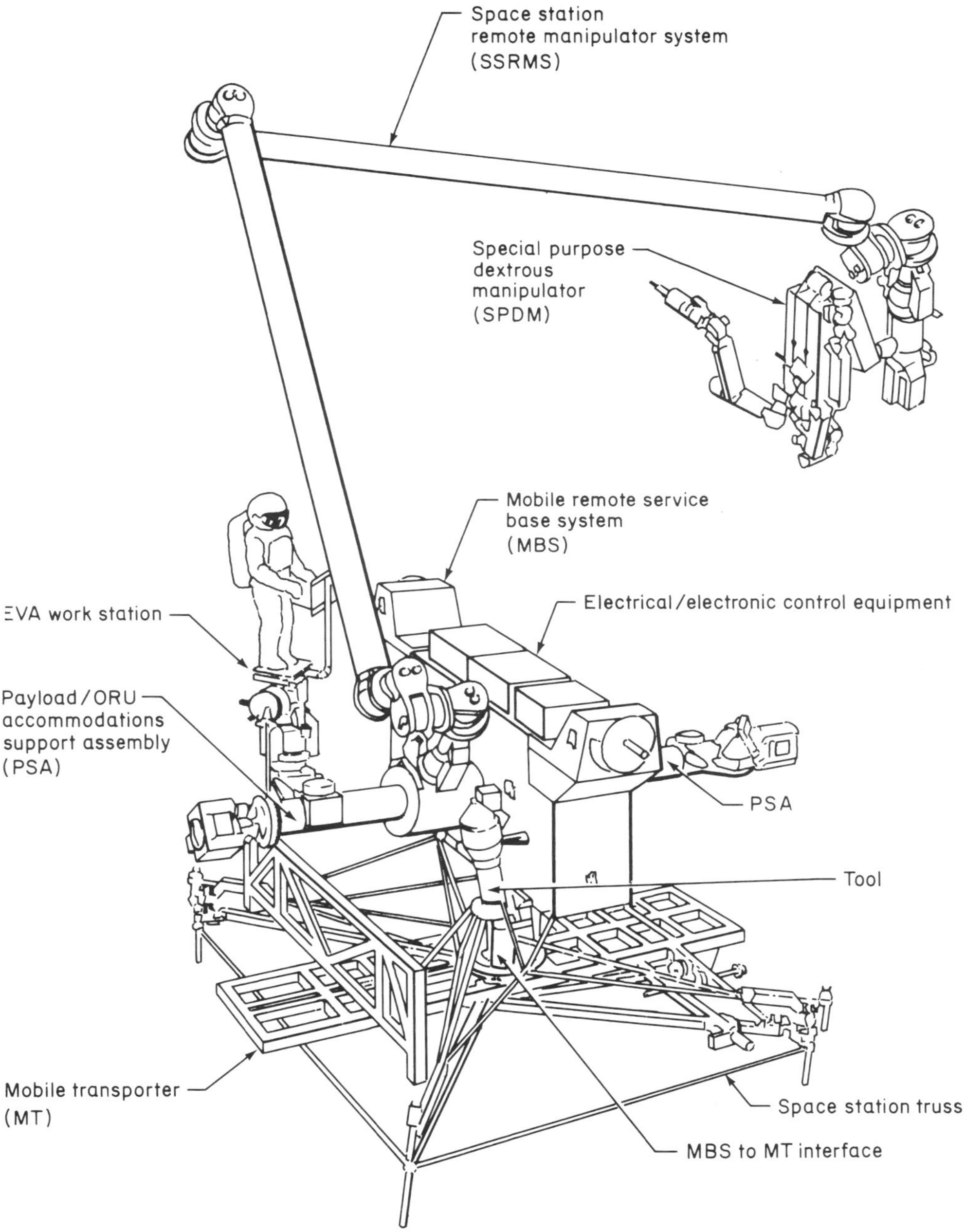

Fig. 18.2. A schematic of the sophisticated Space Station Remote Manipulator System (SS-RMS) which is to be Canada's main contribution to the international complex. For a sense of scale of the apparatus, note the astronaut's work station on the left. The entire assembly will run on rails along the leading edge of the main truss, so the arm will be able to reach any part of the NASA end of the complex.

The early addition of the Power Module will ensure that the laboratory will not have to draw power from the core modules, and this will transform the station into a viable science platform. An operational laboratory marks the end of Phase Two. Some of the Phase Three flights – appropriately dubbed 'utilisation flights' – will linger, after installing apparatus, in order to perform experiments.

Table18.2. Phase Three manifest, *c.* mid-1997

Date	Label	Launcher	Cargo
Jan 2000	UF-1	US	MPLM (ISPRs); solar panel and batteries
Feb 2000	8A	US	truss S0; MT
Mar 2000	UF-2	US	MPLM (ISPRs); MBS, Lab sys
Jun 2000	9A	US	truss S1; CETA-A
Jul 2000	9A.1	US	SPP truss with four solar panels
Oct 2000	11A	US	truss P1; CETA-B
Nov 2000	12A	US	truss P3 and P4
Mar 2001	13A	US	truss S3 and S4
Apr 2001	10A	US	Node 2; N_2 tanks
May 2001	1J/A	US	truss P5; O_2 tanks; JEM's ELM-PS
Aug 2001	1J	US	JEM's PM
Sep 2001	UF-3	US	MPLM (ISPRs)
Jan 2002	UF-4	US	AMS; Express pallet; Spacelab pallet (SPDM; ATA; h.p. gas)
Feb 2002	2J/A	US	solar panels and batteries; JEM's EF; ELM-ES
May 2002	14A	US	four solar panels for SPP, cupola and port rails (on Spacelab pallet)
Jun 2002	UF-5	US	MPLM (ISPRs); Express pallet
–	2E	US	truss S5; MPLM (system stowage and ISPRs); JEM's SFA
–	16A	US	US Hab
–	17A	US	solar panels and batteries; MPLM (Hab outfitting); h.p. O_2 tanks
–	18A	US	ACRV 1
–	19A	US	MPLM (outfitting)
–	15A	US	truss S6; solar panels and batteries
–	UF-6	US	MPLM (ISPRs); attached payloads
–	UF-7	US	Centrifuge
–	1E	US	Columbus Orbital Facility (COF)

Note: This manifest is listed in full in order to illustrate the unprecedented magnitude of the task. Like all previous schedules, the timeline will undoubtedly be stretched and, to some extent, flight assignments will be reassigned.

One theme running through Phase Three is the assembly of the truss structure. The first task is to mount the spur with the central hexagonal-section segment to the laboratory. This will be fitted with thermal radiators, facilities for the manipulator arm, carts for astronauts, antennas and scientific packages as it is progressively extended to port and starboard so that the remaining solar panels can be installed. Once the truss is assembled, the panel mounted on the node will be relocated. Meanwhile, the SPP will be mounted on the upper port of the Service Module. Adding the second node will involve repeating the relocation of the axial adapter. This node opens the way to add the international modules provided by Japan, Italy and ESA. NASA will add its habitat towards the end of the assembly process, in order that the station's crew can be doubled, and at this time a second lifeboat (currently planned to be a Soyuz) will be mounted on the FGB.

Even on the optimistic schedule, Phase Three will not *start* until early 2000, and it will run for several years. Although the final element is unlikely to be in place before 2003, this does not mean that the station will not be productive. Thanks to the introduction of Russian hardware for the core modules, it will be possible to maintain crews aboard within months of starting assembly. Until it installed its habitat, an item which was far down the original manifest, NASA would otherwise have been restricted during these years to the 'tended' phase in which the station would be used only by a succession of utility flights.

By opting to try to continue to operate Mir as an international facility until the end of the century, the RSA postponed its commitment to the International Space Station, but it plans to install a cluster of modules on the Service Module. Although the details of these elements have yet to be defined, their configuration is likely to be similar to that of the Mir

Fig. 18.3. A computer rendition of the International Space Station as it is intended to appear upon completion of Phase Three in 2003. Thereafter, this core structure may well be greatly expanded.

modules, because they will undoubtedly be launched by the same rockets, whose payload shape and mass are standardised.

PREPARATIONS

A mere six months after President Reagan ordered that a space station be built, STS-12 tested the deployment mechanism of a large solar panel on which Lockheed had begun work a decade earlier. Although this frame was a mere 75 mm thick in its stowed form, it unfolded out to 32 metres. The test article carried dummy cells, but an operational panel with gallium arsenide transducers will be able to generate 12.5 kW. Such panels will be installed in pairs on each of the International Space Station's Power Modules, and the completed station will have four modules, with a total capacity of 96 kW.

With the onset of the Shuttle–Mir programme, NASA put its advanced transducers into a proven Russian deployment frame, thereby forming the Cooperative Solar Array, which was mounted on one of Mir's modules. Verification that the cells would withstand the harsh space environment was important, because the International Space Station will, at least initially, be totally dependent on solar panels for energy. It is intended to augment, the transducer arrays, and later to supersede them, with a completely new power system. This will use a large parabolic mirror to focus solar energy on a canister of salts which, upon melting, will drive a turbine that will generate electricity. At the Earth's distance from the Sun, the energy in sunlight is about 1.3 kW. Even the best gallium arsenide transducer is only 14 per cent efficient. A 'solar dynamics' system, on the other hand, should yield 45 per cent of this energy. However, in order to achieve this high efficiency the mirror will have to remain locked within a few degrees of the Sun, so it will need a very stable mount. A solar panel, in contrast, produces power sinusoidally with angular displacement, so is more tolerant of orientation. However, a mirror-based unit will not be as susceptible to degradation as a solar transducer, and so it will provide a better long-term solution. Early in planning for Shuttle–Mir, NASA proposed mounting a prototype of this new power system on Mir once that complex was 'mothballed', for trials, but because of the Russian decision to continue operating its own station this plan had to be shelved.

When NASA started to consider possible configurations for its space station, it realised that one of the most crucial factors involved whether the main truss should be assembled by astronauts, or whether it would be necessary to design self-deploying structures. To assess the ability of astronauts to assemble structures, several rod-and-pin construction kits were developed and tested (as ACCESS and EASE) on spacewalks. In November 1985, Jerry Ross and Sherwood Spring repeatedly assembled and disassembled these frames. Their progress was subjected to time-and-motion analysis, and stereoscopic film was transferred to a computer for further analysis. These tests showed that astronauts could indeed carry out fine-scale construction, so proposals to develop self-deploying frames were shelved.

The continuing operations of the shuttle enabled the devices intended for the station to be tested. Only about 40 per cent of apparatus *designed* to operate in microgravity actually works first time, and the rest has to be refined to overcome unforeseen flaws. The Station Heatpipe Advanced Radiator Experiment (SHARE) was such a case. This prototype element for the radiator, which is to cool the station, vapourised liquid ammonia and transported it along a 15-metre long pipe by thermally-induced convection. A fin assembly at the far end radiated the heat to space and recondensed the ammonia, which then flowed back to the evaporator. The station will have six radiator panels, each employing several dozen such units joined edgewise. One advantage of such a technology is that because its elements are independent, it will provide a highly-redundant thermal control system, and because it has no mechanical apparatus it will not draw any power from the station. Furthermore, it should run for years without needing attention.

The prototype was to have been tested in 1986, but the Challenger accident delayed the first flight to STS-29, when the test was foiled by air bubbles. A sub-scale model incorporating a transparent cover was flown on STS-37, and when, as expected, this too became clogged, the fluid flows were filmed. A redesigned element was tested on STS-43. Although this too developed bubbles, the pipe was able to clear itself. The orbiter turned to put the radiator in the shade, so that different heaters could be applied to verify that the evaporator could deal with the range of predicted thermal loads. Finally, the heatpipe was deactivated and allowed to freeze before being restarted, to show that it would survive an extended period off-line in shadow. The shuttle's continuing operations enabled testing to be conducted 'on the fly' as secondary objectives on already-planned missions. Although such payloads hardly ever hit the headlines, they were vital for the long-term future of the programme, because testing reveals design flaws, and with a structure incorporating as much new technology as the International Space Station, the risk of technological failure was significant.

As in the case of preparations to repair the Hubble Space Telescope, where nothing was left to chance, for the International Space Station even apparently straightforward apparatus was verified. This included new work stations, rigid umbilicals, foot restraints, the trays to carry pipes and cables along the truss, and even tool boxes. Tests were conducted to verify that astronauts wearing spacesuit gloves could manipulate the plugs and sockets to be used to hook up external payloads. Instrumented foot restraints were used to measure the forces imparted by astronauts working on typical maintenance tasks. Astronauts removed thermal insulation from an NiH battery, unplugged it, and then restored it, in order to identify any problems which could impede the replacement of batteries in the Power Modules. Because *time* is a premium resource on a spacewalk, they will use a 'translation aid' to move along the truss, so once STS-37 had fulfilled its primary mission, and its payload bay was clear, Jay Apt and Jerry Ross went out to test several competing CETA designs. But not everything went smoothly; on STS-80, when Columbia's hatch refused to open, the test of the crane which will swing the bulky batteries around within the truss had to be cancelled. It was done a year later though, on STS-87.

The Shuttle–Mir programme provided another way for NASA to minimise technological risk. Although the shuttle often retrieved satellites, it did so using its manipulator. To dock with Mir or the International Space Station, the shuttle is required to manoeuvre precisely down an ever-narrowing cone leading to the docking system. The requirement is to stay within an 8-degree cone, and to maintain a 2-degree tolerance in orientation. When Atlantis made its first approach, it closed at a rate of 3 cm per second, offset by no more than 2 cm from the axis of the cone, and with less than a 0.5-degree angular discrepancy, thereby verifying that the orbiter was indeed a remarkable flying machine. Upon its return, Atlantis performed an even more difficult task by docking 'in the blind', because the pilot's line-of-sight view was blocked by the body of the module which was to be added to the station. This was also accomplished without incident. And as the shuttle resupplied the Russian station with a varied assortment of cargo, NASA learnt many lessons regarding loadmastering that will assist logistical operations with the International Space Station.

OPERATIONS

A crew will typically have a commander, a flight engineer and four research scientists, who will serve tours of three or six months. Each will have a personal laptop that will serve as the main interface to the station's systems. It will be plugged in wherever convenient to undertake a given task, and will retrieve operating manuals from the mainframe as required. The use of this 'soft' interface eliminates the need to provide a distinct physical link to each and every system from a large bank of infrequently-used switches and dials. With a staff of six, the sense of isolation should be considerably less than that experienced by cosmonauts on two-seat Salyut space stations. And with a shuttle docked at the same time as a Russian handover, there could be as many as *sixteen* people aboard the International Space Station for periods of a week or so. With a mix of nationalities, the cultural isolation should also be diminished, especially if communication with family and friends by videolink and e-mail is easy. In effect, therefore, the International Space Station will be more of a community than its Skylab, Salyut and Mir precursors.

The International Space Station will operate a 90-day service cycle providing 30 days of high-quality microgravity, 10 days of maintenance, another 30 days of microgravity, then a period of resupply, reconfiguration and reboost. As a result, shuttle and tanker flights will be scheduled to fit in with experiment cycles, as well as with orbital dynamics. When the station is fully assembled, therefore, its operation will represent a relatively light load for the fleet of shuttles, so they will once again be available for independent missions. Shuttles will fly a V-Bar approach, and dock axially at the leading node. The station's orbit will be boosted after each shuttle visit and immediately prior to the next resupply tanker's arrival.

Although every crew member will have to be sufficiently knowledgeable to fly a Soyuz lifeboat back to Earth, only the Russian crew, who will launch on Soyuz, will require training to perform rendezvous and docking operations. Everyone else will ride on shuttles, which will become logistics and personnel transports.

ROUND THE CLOCK

The fully-operational International Space Station will 'do science' on a 24-hour basis, not just because its crew will work in shifts, but because much apparatus will be capable of being remotely operated, and because the communications via TDRS satellites will facilitate telescience.

One aspect of having a large crew – a benefit that will in all likelihood not be commented upon because it will be taken for granted – will be that while the commander and the flight engineer focus on station operations and perform routine maintenance, the scientists will be able to work on their own programmes. This point is worth noting formally, because on Mir maintenance was undertaken at the expense of the science programme.

ALL YEAR ROUND

In contrast to serving as a way-station to deep space, as NASA had first envisaged, the descoping of Freedom to make it affordable forced the agency to dedicate its laboratory to microgravity research; the Europeans and Japanese followed suit. Consequently, the initial goals of the International Space Station's work will focus on life and materials sciences. A substantial amount of experiment apparatus has already been thoroughly tested on Spacelab shuttle flights, so the station's research is likely to begin energetically. Although the expectation is that the station will secure contracts for materials-processing on a commercial basis, to begin with it will serve the Centers for Commercial Development of Space (CCDS), which NASA set up with industrial and academic co-sponsorship, and which, by use of the Spacehab middeck augmentation module, have completed their development work and, as a result, are ready to utilise the unprecedented opportunities presented by a permanent orbital facility. The laboratories will be fitted with International Standard Payload Racks (ISPR) so that hardware can be upgraded and modified as necessary. A modular approach is also essential, because the laboratories will need to support an integrated research programme, on a continuous basis, over a lifetime of several decades. This approach will enable the scientific focus to evolve and also enable apparatus to keep pace with the state of the art. In addition, plug-in compatibility will permit apparatus from different nations to be distributed to satisfy dynamic power requirements, and be integrated in a 'pick-and-mix' fashion to match specific experimental requirements. Since some apparatus will be required only occasionally, there will also have to be long-term storage facilities.

Even with the EDO wafer in its bay, a shuttle orbiter is restricted to a fortnight in space. The International Space Station will facilitate the slower-growing proteins and larger tissue cultures, for example. It will transform the way in which such experiments are undertaken. Scientists working in a laboratory will repeat their experiment, fine-tuning its configuration in order to hone a crucial measurement, and then vary it, so as to establish the sensitivity of the data to parameters, all of which takes time. But on a shuttle mission, time is a precious resource that must be shared between the many payloads that have to be carried in order to justify the cost of a launch. On a station, however, the economics are different. The cost of the science is essentially decoupled from the shuttle's running cost. Once an experiment is on the station, it can be run for prolonged periods, and rerun, at no extra cost to the shuttle. At long last, therefore, the shuttle, in combination with the station, will cut the cost of *operating* in space, and industrial microgravity materials-processing should, at last, become financially justifiable for those markets which can support the premium required to meet the unique production costs.

So far, it has not been just the *overhead* which has deterred commercial processing in orbit; it has been the reliance of any activity on the vagaries of the shuttle manifest. NASA has historically been able to offer only infrequent flights on a very indeterminate schedule. It is difficult to justify commercial investment in a process that can be invoked for only a fortnight at a time, and twice a year at most. Spacehab Incorporated faced this dilemma from the start; development funding and initial customers were forthcoming only after NASA had reserved slots in its manifest for the facility. The ongoing nature of operations aboard the International Space Station will remove both of these obstacles; it will be possible to run a process on a semi-continuous basis. If raw material is ferried up in bulk, mass production will be practicable, and the output will be collected whenever convenient. Production will have been decoupled from the vagaries of flight operations. If necessary, the process can be run on a free-flyer, such as the Wake Shield Facility (WSF), which returns on a regular basis for servicing. Free-flyers will carry apparatus requiring extremely 'pure' microgravity. For little or no overhead, the station will be able to service an entire *flotilla* of free-flyers. The WSF, for example, would require an output of just 200 kg per annum to yield an estimated turnover of $50 million, and this would provide a sufficient rate of return to recover the development costs over a financially acceptable period. It is to be hoped that other areas of research already investigated on the shuttle and aboard Mir – such as the crystallisation of zeolite and the processing of biological agents for the pharmaceutical industry – will become equally attractive for commercialisation in concert with the International Space Station.

WHAT GOES UP MUST COME DOWN

The work of the Mir space station, and indeed its Salyut predecessors, was hindered by a deficiency in its logistics infrastructure. It was *not* a problem of resupplying the station, because automated Progress ferries docked every few months to replenish its fluids and dry cargo, and expired parts from the environmental system, food packaging and other rubbish could be put into a departing Progress, which then 'ceased to exist', as the Russians expressed it, by burning up during re-entry; the problem that faced Mir was *returning* material to Earth.

The Soyuz spacecraft that ferries crews to and from the station is a cramped bell-shaped capsule just 2.2 metres in diameter and 2 metres tall. Although the current Soyuz-TM model can return three cosmonauts and up to 120 kg of cargo, this is barely sufficient to return the film, crystals and log books that accumulate on long missions. To correct this imbalance in the transportation infrastructure, a small capsule was built for delivery within a Progress. Once it had been loaded with 120 kg of cargo, it was mounted in the ferry's hatch so that it could be ejected as the ferry deorbited over the capsule's recovery zone. The rationale for this Raduga capsule, however, was to enable a crew to return the fruits of their work on an *incremental* basis, and thereby prevent the build-up of more material than they would be able to take away upon the completion of their six-month tour. The root of the problem is that a descent capsule which uses an ablative shield to protect it during re-entry needs to be small. The Soyuz capsule is well-suited to a crew with minimal cargo making the brief trip down from Mir, but it cannot return with bulky items. It was for this reason that, following Apollo, NASA opted for a transportation infrastructure based on a spaceplane incorporating a massive payload bay. As a result, it was not until Shuttle–Mir that it became practicable to return apparatus for refurbishment. And the Russians took advantage of this opportunity by offloading a *tonne* of apparatus each time Atlantis docked. The shuttle's participation made it possible to return gyrodynes from the attitude-control system for refurbishment, to return failed Elektrons from the environmental system for analysis, and to return Orlan spacesuits for inspection to reassess their service life. For NASA, the only issue in transporting these objects was the stowing of them in such a way as not to disrupt the vehicle's centre of mass.

The significance of this observation is that the shuttle will be similarly able to serve the International Space Station. In addition to delivering cargo it will return any items that have reached the end of their service life so that this life can be reassessed; worn out or broken objects will be returned for refurbishment and reuse; free-flyers will be returned to be re-equipped; and accumulated product will be offloaded. If NASA had opted to retain the Apollo capsule and use it to service a space station assembled using pre-equipped Skylab-style modules, then it would have faced the same logistics problem as has hindered Mir operations.

As NASA's initial proposal had argued, the shuttle and the station form an integrated system. As an orbital outpost, the station will facilitate all-year-round operations which will minimise the overhead of scientific and commercial activities, and the shuttle's prodigious logistics capacity will sustain the station by exchanging crews, delivering consumables and new apparatus, and offloading product and expired and unwanted apparatus; either without the other is deficient.

THE END OF THE BEGINNING

Over time, this initial focus on microgravity work will be augmented by instruments for the observational sciences. Two significant advantages derive from orbital flight: firstly, it provides an unprecedented view of the sky; secondly, by orbiting every 90 minutes, it enables the planet to be studied as an integrated system, on a global scale, with a fine time and spatial resolution. The International Space Station will offer significant benefit in both of these observational sciences, and this will not simply be the result of its position in orbit. Its greatest asset will be its human crew.

It is undoubtedly true, as those critical of human spaceflight never miss an opportunity to point out, that the development of an orbital facility capable of human habitation is an *extremely* expensive project. They have argued, rightly, that it would be cheaper to operate automated satellites. But satellites have to be designed to fit specific mass and size constraints: that is, they must not be too heavy to be lifted by their assigned rocket, and they must fit within its shroud. Once in orbit, they are required to operate autonomously. In the case of the larger and more valuable satellites, the development cost is usually offset by a long lifetime. Since satellites usually carry a suite of instruments, if an instrument fails the satellite continues in its degraded state; and if its power or communication systems fails, the entire package generally has to be written off. Even if the satellite performs perfectly for 20 years, towards the end of its mission its sensors will be degraded and its instruments will in all likelihood have been rendered technologically obsolete. What is needed is a platform that combines a very long life with both reliability and flexibility. The most reliable and flexible component that can be built into a system is a human operator.

The International Space Station will, in the fullness of time, make the astronomical and environmental monitoring programmes more cost-effective, not necessarily in terms of their individual disciplines with their specific budgets, but together, as a total package, alongside the ongoing microgravity work and the running of the free-flyers, because it will provide an infrastructure for operating, servicing and upgrading such instruments. So not only will such programmes be decoupled from the operational lifetime of their technologies and from the pace of technological advance; their participants will, for the first time, be able to design instruments more or less free of the traditional limitations of mass, size, power and communications limitations. By being able to rely upon the shuttle to provide transportation, and upon the station to provide utilities, the appallingly long time currently taken to conceive, develop and fly an instrument will be reduced, which will indirectly benefit work in that field. With the International Space Station operational, it would therefore be *negligent* not to exploit it to its fullest. With 40 metres of the main truss available for mounting 'attached payloads', it is clear that a wide variety of automated apparatus could be installed.

Astronomers put telescopes in space not simply to avoid the twinkling that results from air currents, but because the atmosphere absorbs ultraviolet and infrared light. These scintillation regions of the spectrum carry much useful information. Satellites such as the Orbiting Astronomical Observatory and the Infrared Astronomical Satellite made pioneering studies, but they were short-lived.

It is almost a truism to some that the presence of a human crew will effectively rule out any programme that requires the fine pointing of observational apparatus. Yet the Spacelab missions, which had clusters of telescopes on the Instrument Pointing System for solar and astronomical studies, proved the naivety of this remark. And, of course, the more massive the platform, the less significant will be the vibrations induced by the activities of its crew. It will be perfectly practicable to mount an IPS on the International Space Station's primary truss; indeed, there is no reason why there should not be *several* such mounts, each with a different set of instruments. As the ASTRO flights showed, the IPS can be operated by the astronauts, by the ground team, or fully-automatically. Such instruments need not, therefore, impose an additional burden on its crew. The ongoing nature of operations will facilitate comprehensive studies, and not just the snapshots that were so hectically gathered by the few shuttle missions devoted to astronomy. As for looking at the Earth, the instruments used by the Upper Atmosphere Research Satellite, those of the ATLAS package and, in due course, those that are to be carried by the satellites of the Earth Observing System, should also be placed on the truss. By having all of these sensors making simultaneous measurements, it will be possible to make *integrated* studies. With a platform in place that can be augmented and which provides a built-in capacity for repair of attached payloads, it will surely make little sense to continue to construct independent satellites unless they require a significantly different orbit.

The International Space Station should not be seen as a short-term venture. It is actually only the core of a complex that will be progressively expanded. By designing its modules to offer more connectivity than is required in the basic configuration, NASA has expressed its confidence that assembling this cluster of modules will be just the beginning. Once it is up and running, there will be an incentive to expand the station, because the overhead of each new facility will be small. However, as facilities are added, the workload, even with telescience, will exceed the capacity of the crew, so it will be necessary to add another habitat to enable the crew complement to be increased. If it is decided to keep free-flyers in space, and undertake maintenance *in situ*, rather than back on Earth, it will become necessary to add a large airlock that can act as a berthing facility for free-flyers and as a garage for when they are not in use, so that their apparatus can be reconfigured in a 'shirt-sleeved' environment. A sketch of the International Space Station showing it as it will be at the conclusion of Phase Three corresponds, therefore, only to the completion of Stage One; it will mark *the end of the beginning* of humankind's expansion into space.

We might not return to the Moon until far into the next century, and it will undoubtedly be many more years before we follow our automated spacecraft to explore the Solar System; but when this is finally done, the International Space Station, or what it has by then become, will, as von Braun clearly envisaged so long ago, serve as the first step on the journey, the outpost in low orbit, the way-station to destinations beyond. And, as NASA pointed out to Nixon, the station and the shuttle are symbiotic partners that combine to form a system with extraordinary promise.

Part 5: Conclusions

19

The evolving role

Prior to the dawning of the Space Age, the winged spaceplane, as a development of the aeroplane, possibly with rocket assistance at launch, had been the 'obvious' technological route to spaceflight.

In the 1950s and early 1960s, however, a spaceplane was not technologically feasible, due to the shortfall of rocket power to attain orbital speed, the absence of thermal protection to withstand hypersonic re-entry of the atmosphere, and the unsuitability of contemporary electronics technology. The significance of this last obstacle is not so much that electronics based on vacuum tubes would not have been able to implement a fly-by-wire control system, but that in the absence of computers on which to design such a vehicle it would have been necessary to employ the same development methodology as used for contemporary aircraft, in which a succession of prototypes was built and subjected to flight test, a process that would have been inordinately difficult. If the radiative thermal protection system *had* been available in the early 1960s, it could have been used to enable the X-15 or the X-20 to achieve orbital flight once large enough rockets had been developed. By a supreme irony, by the end of the 1960s, when microelectronics had matured and rocket thrust was readily available, the throwaway rocket was no longer deemed to be economically viable.

Although the spaceplane was ordered at the earliest time that it became technologically feasible, its development nevertheless represented a challenge, and it fell behind schedule. In retrospect, was this delay significant?

The rate at which the expendable launchers were to be phased out was linked to the rate of progress in developing the shuttle, so there was no shortage of rockets. In fact, it was the shuttle that lost out, because it lost several commercial satellite contracts to the expendables; even this was insignificant, however.

A more serious consequence of the delay in the shuttle's development, in the long term, was the loss of the Skylab space station.

THE MISSED OPPORTUNITY

When the hope was to perform the first shuttle test flight in 1978, it was intended that STS-3 should rendezvous with Skylab and attach a motor so that the abandoned space station could be refurbished and reoccupied. Fred Haise and Jack Lousma were to fly this mission, which was set for September 1979.

Columbia was to have drawn alongside Skylab so that a remotely-controlled propulsion module could be flown over from the payload bay to dock at the station's axial port. Martin Marietta built the propulsion module in 1978 using off-the-shelf Apollo systems, but by the end of that year it had become clear that the shuttle would not be ready to make even its first test flight in 1979. In December 1978 the rescue was cancelled, because in the run-up to the maximum of solar activity, which was due in 1980, the upper atmosphere became inflated, thereby increasing the drag on the station to such an extent that it became clear that the orbit would decay long before the shuttle could possibly be ready to fly! Skylab duly re-entered in July 1979.

If it had been practicable to refurbish Skylab, it would undoubtedly have served as the core of the space station that NASA was intent upon developing just as soon as it could persuade a President to back it. However, Skylab's availability would have served as no more than a stimulus because, in itself, the station was not a solution. And even if the shuttle could have been introduced in time to rescue the station, and its cargo capacity exploited to fly up a comprehensive renovation kit, installing this would have been tricky, because the station had not really been intended to be serviced. Nevertheless, its availability would have served as a *seed* to provide a sense of operational imperative and a technological focus; without it, the result has been a decade of costly indecision and waste.

Skylab and the early Salyuts, although very different in size, were essentially similar in that they were designed to be launched fully-outfitted, inhabited by a succession of crews, and then discarded when their resources were exhausted. Although the later Salyuts proved able to sustain residents on extremely long flights, besides hosting short-term visitors, Mir was the first space station designed for *ongoing* inhabitation, and its modular philosophy has enabled it to be built up over the years into a sophisticated complex. Mir is, therefore, a high-fidelity precursor to the International Space Station, which will face similar design and operational challenges.

NATURAL SELECTION

As originally conceived, the shuttle was seen as the *universal* spacecraft which was to replace all previous satellite launchers and crew-carrying spacecraft. As the National Space Transportation System, its role was to be all-encompassing, but over the years it has, for a variety of reasons, evolved, primarily by means of the imposition of constraints.

The shuttle was specifically designed to provide access to low orbit, one of its primary roles being to deploy satellites. Those destined for geostationary orbit were either to be fitted with kick motors for insertion into transfer orbit, or be ferried into GSO by the purpose-designed IUS. A related mission, to facilitate *in situ* satellite repair, was demonstrated early on with SolarMax and Leasat-3, as was the retrieval of satellites, with the two HS-376s whose PAM motors had failed. The fitting of a new kick motor to a stranded

Intelsat demonstrated yet another variation on this theme. However, the plan for the shuttle to recover satellites from *high* orbit was undermined by the cancellation of the OMV. This weakness in the shuttle's capability was illustrated by the loss of Leasat-4.

Having designed the Space Transportation System to serve as a truck, to ferry satellites into low orbit for deployment, it would be inappropriate to criticise the shuttle for adding no value to its cargo. Each satellite successfully deployed was an objective accomplished. If it had not been for the shuttle-only policy, many of these satellites could well have ridden expendable rockets. However, some satellites had been designed specifically to exploit the orbiter's capacious payload bay; in addition to most of the classified satellites, this category included the HS-381 and HS-393 wide-bodied geostationary relays, and the Hubble Space Telescope. Although the shuttle made practicable the concept of a free-flyer – a configuration that was welcomed by the scientific community – the interplanetary spacecraft deployed by the shuttle paid a heavy price for their brief ride into low orbit. The IUS–Centaur fiasco was shameful and, for the planetary spacecraft missions, a costly debacle.

During its early years of operating the shuttle, NASA demonstrated a remarkable ability to perform dynamic scheduling, not only by adding mundane tasks left over from one flight to the next, but also by undertaking spacewalks of increasing complexity to rectify faults in satellites. This spontaneity represented a significant operational accomplishment, because it was achieved within the context of a sustained drive to increase the flight rate. The low rate at which the shuttle operated – at least as measured relative to the original plan – represented another constraint. Even with only a dozen flights a year (a rate that has never been achieved) there were simply too many satellites – both government and commercial – for the shuttle to launch. Although the largest single contract was to carry two dozen GPS satellites for the Department of Defense, there were many more commercial satellites booked on the shuttle. The folly of overly relying on the shuttle was amply illustrated by the mad scramble for the few remaining rockets following the loss of Challenger.

The Challenger accident led to the imposition of additional operational constraints. The cancellation of the plan to run the SSMEs at 109 per cent not only restricted the shuttle's capacity in highly-inclined orbits to such an extent that the Air Force withdrew its interest, it affected the assembly of the planned space station. By abandoning its plan to fly the shuttle from Vandenberg AFB, the Air Force denied NASA access to polar orbit, and thereby closed off the missions involving meteorological and environmental satellites, including the imminent repair of the ailing Landsat. One constraint affecting the shuttle's to-orbit payload capacity was the extent to which it jeopardised the safety of a RTLS landing, in which the orbiter's centre-of-mass was a crucial factor. It was the risk of the Centaur disgorging its cryogenic propellant upon a rough landing that led to its being shelved. The orbiter could not carry the maximum five HS-376 or GPS satellites, even though they were to be dispatched in orbit, because a full load would set the centre-of-mass too far forward in a launch abort leading to a RTLS or TAL landing. A Spacelab module represented a significant risk if the orbiter also had a full load of propellant in its forward-RCS cluster. As a result, the shuttle's capacity to-Earth was rather lower than to-orbit.

The offloading of the commercial satellites to expendable launchers pre-empted the hope that the shuttle would generate revenue. However, switching to the mixed-fleet policy was a positive step, which not only struck an effective balance in capabilities by stripping the manifest of straightforward deployment missions, but also went a long way towards relieving the pressure on NASA to increase the shuttle's flight rate.

Upon Return-To-Flight, NASA initially hoped to be able to build up the flight rate, but it soon came to terms with the fact that it could sustain only six to eight flight per year. This low rate further pared the manifest. Although satellites were still to be deployed, these were the ones which could not reasonably be offloaded to expendables. The shuttle was to focus on tasks that only it could undertake, and the multifaceted Spacelab finally rose to the top of the manifest. Although spacewalking became less common, when it occurred it was no less ambitious, as the fitting of a kick motor to the Intelsat satellite and the repair of the Hubble Space Telescope amply demonstrated. In fact, the shuttle proved to be an excellent platform for spacewalking. Although the RMS was invaluable, the most effective item in the toolkit turned out to be the human hand.

Apart from all these *operational* constraints, the most significant factor that constrained the shuttle's role was imposed before the shuttle's development was authorised. NASA had put forward the novel transportation system and the space station as an integrated package. In forcing NASA to choose one or the other, Congress constrained the transportation's role. Although the shuttle was to be a space truck, and was to ferry satellites into low orbit, it was *not* to assemble an orbital base. The shuttle was to launch, achieve its objectives and return to Earth. This was fine for satellite deployments, and for short-run observational and microgravity work using Spacelab, but because the shuttle would have nowhere to go, all such projects would be limited to the time that the orbiter could stay in space, and even with the EDO wafer in its bay the orbiter is limited to a fortnight in space. This constraint on NASA's activities was partially lifted by the decision to install apparatus on Mir, but it will only be banished once the International Space Station becomes available. A fair assessment of the shuttle must, therefore, consider the value of the transportation system in the context of the station, because they are, in reality, elements in an integrated system.

THE HEAVY-LIFTER

When it was decided to develop the shuttle, another significant constraint was placed on NASA's operations. The abandonment of the Saturn 5 denied it the option of sending a 100-tonne fully-integrated payload into orbit. In theory, the so-called Shuttle-C was to have served this role, but this exploitation of shuttle-based technology was always unlikely to be pursued. Without a heavy-lift capability, NASA was obliged to design its station to be built in orbit from modular components that would fit into the shuttle's bay. Using a few larger components to form the core of the station was therefore an option that was denied. If it *had* been decided to retain the Saturn 5, its pad facilities would have to have been rebuilt. By the time the shuttle entered service, the Saturn's old Launch Processing System – which was more of an electro-mechanical mechanism than a computer – had been rendered technologically obsolete. And given that by its very nature the Saturn 5 would have been used very infrequently, retaining a pad for its use would have been prohibitively expensive. Quite apart from anything else, of course, the VAB, the MLP, and the pads on LC 39 were all to be converted to accommodate the shuttle, so the demise of the Saturn 5 was virtually inevitable. The most significant technological heritage of the mighty Saturn 5 is the F1, five of these engines being clustered to form its first stage's power unit. If ever the need arises, this could be integrated into a new rocket, but it would be expensive. It would undoubtedly make better financial sense simply to restart production of the Energiya – the heavy-lifter that was cancelled by Russia in the aftermath of the collapse of the Soviet Union – because it is a more modern technology than the old lunar behemoth. Alternatively, for a modestly sized payload, the recently uprated Titan 4 could be used.

The irony, however, is that although NASA *could* now augment its station to exploit elements far heavier than could be delivered by the shuttle, its design is configured to match the constraints imposed by operating the shuttle at less than its theoretical maximum power, and at the 51-degree orbital inclination at which its performance is further degraded. In the future, though, when it comes time to expand the International Space Station, we may see a switch to much larger modules, and a corresponding increase in both the station's capabilities and its crew.

A MATTER OF COST

The cost-benefit analysis which was used to 'prove' that the shuttle would be a more effective long-term solution than the expendable rocket, relied on a very high flight rate that proved impracticable. Although it is common to cite this as a failure of the shuttle to deliver upon its requirements, it is more realistic to acknowledge that the analysis was flawed, at least to the extent that it did not fully appreciate the factors that would dictate the operational pace of the system as a whole.

The only way that the called-for weekly missions could have been mounted would have been to have a large fleet of orbiters which would individually fly only two or three times a year, but the 'turnaround costs' would have undermined the economics of the analysis. The cost of a shuttle mission is difficult to define, and in a sense it is rather arbitrary. One way to compute cost is to start by dividing the total development cost by the anticipated flight rate in order to work out the amortisation overhead, add in a proportion of the annual operating costs, and then the specific development costs associated with the payload. Doing this, however, makes each flight appear considerably more costly than it really is, because the cost of the shuttle's development is in the past, and nothing will be saved by *not* flying the shuttle. It was the amortisation of the development to the once-a-week flight rate which suggested that the shuttle would, in the end, be more cost-effective than to continue with expendables, so long as the expendables were phased out. A four-orbiter fleet cannot fly at that rate. This increases the amortisation overhead for each mission that is flown, but this is an artificial increase. Typically, it takes three to four months to turn around an orbiter and roll it back out to the pad. The resultant operating overhead represents a genuine increase in the cost of a flight in comparison to the originally projected cost but, as experience over the years has shown, a higher-than-expected operating overhead is unavoidable. Nevertheless, this, too, is misleading, because 80 per cent of the shuttle's annual operating costs are *fixed*. The cost of maintaining the facilities is largely independent of the shuttle's flight rate. Thus, it is rather naive simply to divide up the development cost and the annual operating cost to determine a 'basic cost' for each flight. Rather, the cost of mounting any specific mission has to be computed on the basis of its directly incurred costs. In the extreme, this cost would be precisely the difference in the annual budget between making and not making the flight. But to argue this would also be naive, because the system is operated specifically to make flights possible. It is a matter of balance. There is clearly a flight rate at which the shuttle would be underused, and for which the overhead would be prohibitive. Conversely, a specific annual budget imposes a maximum number of flights. In assessing costs, it must be remembered that the cost of developing and operating a payload is not usually borne by the shuttle; the Hubble Space Telescope, for example, was, and is, a completely separate budget. Also, back in the days when NASA deployed commercial satellites, even though it was obliged to charge an unrealistically low fee, it nevertheless did generate income which offset mission costs.

NASA has settled down to a readily sustainable rate of six to eight missions per annum. With a fleet of four orbiters, it would be difficult to increase the flight rate beyond the once-a-month level. United Space Alliance (USA) – the contractor that turns around shuttles and provides pad services – is currently lobbying to overturn the Reagan administration's ruling that the shuttle not be permitted to carry commercial satellites. If it succeeds in this, USA's intent is to lease the shuttle to mount commercial deployment missions. In effect, it would exploit the 'spare' capacity in its turnaround process, and so operate the system at its peak rate. The company's objective is to have demonstrated the commercial viability of mounting its own missions by the time that its service contract comes up for renewal in 2002. It has been the gradual maturation of the shuttle as an operational system, and the degree to which its turnaround process has become subject to control, both in terms of time and cost, that has made commercial operations feasible at such a low flight rate. This expresses USA's confidence that the shuttle will remain operational well into the new millennium.

RISKY BUSINESS

In authorising the development of the shuttle, President Nixon stated that its goal would be to 'routinise' access to low orbit, and it is clear that the shuttle has indeed done so.

The shuttle is the most successful spacecraft ever built to carry a human crew. It has carried more people, more often, in a shorter overall period, than the Soyuz, the workhorse of the Soviet/Russian programme. It has successfully performed a multiplicity of roles, and is about to start the largest space construction project ever attempted.

Yet in suffering a single accident, the shuttle killed more astronauts than flew during the entire Mercury programme. Spaceflight is an inherently risky activity. But how dangerous? In 1979, one NASA estimate said 1 in 50 shuttles would suffer *catastrophic* failure during the ascent and that 1 in 100 would be lost in landing. All other failure modes were expected to be successfully dealt with by the predefined options. Immediately prior to the Challenger accident, NASA's Public Affairs Office put the risk at 1 in 100,000. At this same time, the Air Force put the risk at 1 in 35, with the most likely cause of *catastrophic* failure being an SRB fault. In 1989, following Return-To-Flight, NASA considered the risk of catastrophic failure leading to the loss of the vehicle (although its crew was expected to be able to bale out) at 1 in 168. If another orbiter is lost, in landing for example, NASA will not have the luxury of a two-year grounding; it will be obliged to learn from the experience and resume operations to continue assembly of the International Space Station.

A MIXED FLEET

In the immediate aftermath of the loss of Challenger, America asked itself whether it was right to risk the lives of its astronauts simply to deploy satellites, and it decided that the answer was 'No'. In the resulting mixed-fleet policy, the expendable rockets *complement* the shuttle. A decade later, the shuttle is routinely ferrying supplies to the Mir space station. Is it worth risking astronauts' lives to deliver food, clothing, spare parts, and water to Mir? This question is distinct from whether it is worth NASA's while visiting Mir. Is the ferrying of supplies a fair use of the shuttle? With the exception of propellant (which is to be delivered by the Progress tanker), the International Space Station is to rely on the shuttle for resupply. With an upper stage capable of rendezvousing and docking, it would be practicable, and perhaps appropriate, to launch much of this cargo on expendable rockets. This Automated Transfer Vehicle, which ESA is developing, will be able to serve this role. NASA's OMV, if it had been built, would have been even better. As a *reusable* vehicle which would remain in space and be serviced at the station, the OMV would have been able to retrieve a payload that had been left in parking orbit by a dumb booster. Such vehicles would complement the shuttle in such niche roles, and free it up for more appropriate work.

There is a lobby in favour of phasing out the shuttle, to replace it with a more reusable spaceplane which will be cheaper to operate. In part, this is an economic argument. But it is also an urge to move on, to build a new vehicle. There is a distinction, however, between an *old* technology and a *mature* technology. Having been refined, the shuttle is operating extremely effectively. Each orbiter is regularly upgraded to incorporate improvements in the flight-control system avionics. The shuttle is like the venerable Soyuz ferry, in that it has been refined over the years and is now highly optimised. From its conception, the shuttle was to be an operational vehicle. The X-33 is being pursued as a technology demonstrator; if it leads to a vehicle to supersede the shuttle, it too will require time to mature, to become cost-effective. If, in 2020, the shuttle is still the primary space transportation system, this will not serve as a reason for junking the shuttle; rather, it will reflect its suitability for its role.

This is not to say that the shuttle is perfect; its role has been progressively constrained, as missions have been offloaded. The continuing use of the Soyuz indicates that, despite the existence of the shuttle, there is a niche for a small passenger ferry. If the X-38 is built, it would be folly to use it simply as a lifeboat that will, hopefully, *never* be used. It should be mated with a rocket, integrated into the International Space Station's infrastructure, and used to deliver as well as to recover crews. In the narrow role of crew transport, the X-38 would complement the shuttle.

Filling these niche roles with specialised vehicles should not be perceived as threatening the shuttle, which has a unique capacity to accommodate bulky objects and operate

Fig. 19.1. An artist's impression of the X-38 which is under development to act as a 'lifeboat' for the International Space Station.

pallets of instruments, and, most crucially, is currently the only vehicle capable of returning bulk cargo to Earth.

The X-33, or similar vehicle, offers the potential to supersede the shuttle; but this is far off, and until then the shuttle will continue to provide a key function in an expanded mixed fleet.

THE CHALLENGER LEGACY

If NASA had already started to build its orbital complex before the collapse of the Soviet Union, there would probably not have been a move to merge the Russian and American programmes. With the two facilities at different orbital inclinations (NASA would have used 28 degrees), there would have been no way to link them together; and this would have been so even if it had been decided that cooperation in space should symbolise the end of the Cold War.

If the Challenger accident had not occurred, it is highly likely that NASA would have commenced assembling some suitably constrained form of its Freedom station by the time of the Soviet Union's demise. In a very real sense, therefore, the fact that the two national space programmes have been merged to create a truly international space station is a legacy bequeathed by the crew of Challenger.

Fig. 19.2. An artist's impression of the X-33 about to dock with the International Space Station. Its payload bay doors are open. This vehicle is to evaluate a possible configuration to replace the shuttle.

Shuttle mission log

The following table details each shuttle mission in chronological order, with its sequence number, international designation, orbiter and flight count, launch and landing, landing site and its use count, duration of flight, orbital inclination, crew, with their flight counts aboard the shuttle (prior experience excepted), payloads and mission activities.

#	Designation	Orbiter	Launch	Landing	Runway	Days	Incl
STS-1							
1	1981-34A	Columbia (1)	0700 EST 12 Apr	1021 PST 14 Apr	EAFB (1)	2.26	40

CDR: John Young (1)
PLT: Bob Crippen (1)
Payload: OAST

#	Designation	Orbiter	Launch	Landing	Runway	Days	Incl
STS-2							
2	1981-111A	Columbia (2)	1010 EST 12 Nov	1323 PST 14 Nov	EAFB (2)	2.26	38

CDR: Joe Engle (1)
PLT: Dick Truly (1)
Payload: OAST, IECM, PDP, NOSL, HBT, OSTA-1

#	Designation	Orbiter	Launch	Landing	Runway	Days	Incl
STS-3							
3	1982-22A	Columbia (3)	1100 EST 22 Mar	0905 MST 30 Mar	WSTR (1)	8.00	38

CDR: Jack Lousma (1)
PLT: Gordon Fullerton (1)
Payload: OAST, IECM, OSS-1, MLR, HBT, EEVT (CFES), GFVT (GAS), SSIP, PGU(LIGNIN)

#	Designation	Orbiter	Launch	Landing	Runway	Days	Incl
STS-4							
4	1982-65A	Columbia (4)	1100 EDT 27 Jun	0909 PDT 4 Jul	EAFB (3)	7.05	28

CDR: Ken Mattingly (1)
PLT: Hank Hartsfield (1)
Payload: OAST, DOD82-1 (CIRRIS, UHS), MLR, CFES, IECM, SSIP, GAS, NOSL

#	Designation	Orbiter	Launch		Landing		Runway	Days	Incl

STS-5

5	1982-110A	Columbia (5)	0719 EST	11 Nov	0633 PST	16 Nov	EAFB (4)	5.09	28

CDR: Vance Brand (1)
PLT: Bob Overmyer (1)
MS1: Joe Allen (1,EV)
MS2: Bill Lenoir (1,EV)
Payload: OAST, GAS, SSIP
Satellites: Anik C-3/PAM, SBS-3/PAM

1982-110B	SBS-3	1517 EST	11 Nov
1982-110C	Anik C-3	1524 EST	12 Nov

STS-6

6	1983-26A	Challenger (1)	1330 EST	4 Apr	1053 PST	9 Apr	EAFB (5)	5.00	28

CDR: Paul Weitz (1)
PLT: Karol Bobko (1)
MS1: Donald Peterson (1,EV)
MS2: Story Musgrave (1,EV)
Payload: CFES, MLR, RME, NOSL, GAS
Satellite: TDRS-A/IUS
Spacewalks: Peterson and Musgrave

1983-26B	TDRS-A	– EST	4 Apr

STS-7

7	1983-59A	Challenger (2)	0733 EDT	18 Jun	0657 PDT	24 Jun	EAFB (6)	6.10	28

CDR: Bob Crippen (2)
PLT: Rick Hauck (1)
MS1: John Fabian (1,EV)
MS2: Sally Ride (1)
MS3: Norman Thagard (1,EV)
Payload: OSTA-2, MLR, CFES, GAS, OAST
Satellites: Anik C-2/PAM, Palapa-B1/PAM, SPAS-1

1983-59B	Anik C-2	1701 EDT	18 Jun
1983-59E	Palapa-B1	0936 EDT	19 Jun
1983-59F	SPAS-1	0405 EDT	22 Jun

STS-8

8	1983-89A	Challenger (3)	0232 EDT	30 Aug	0041 PDT	5 Sep	EAFB (7)	6.05	28

CDR: Dick Truly (2,EV)
PLT: Dan Brandenstein (1)
MS1: Dale Gardner (1,EV)
MS2: Guion Bluford (1)
MS3: Bill Thornton (1)
Payload: PFTA, CFES, MLR, ICAT, ISAL, GAS, AEM, RME
Satellite: Insat-1B/PAM

1983-89B	Insat-1B	0349 EDT	31 Aug

#	Designation	Orbiter	Launch			Landing		Runway	Days	Incl

STS-9

9	1983-116A	Columbia (6)	1100	EST	28 Nov	1547 PST	8 Dec	EAFB (8)	10.32	57

CDR: John Young (2,EV)
PLT: Brewster Shaw (1)
MS1: Owen Garriott (1,EV)
MS2: Bob Parker (1)
PS1: Byron Lichtenberg (1)
PS2: Ulf Merbold (1,ESA)
Payload: Spacelab 1

STS-10/41B

10	1984-11A	Challenger (4)	0800	EST	3 Feb	0716 EST	11 Feb	KSC (1)	7.97	28

CDR: Vance Brand (2)
PLT: Hoot Gibson (1)
MS1: Bruce McCandless (1,EV)
MS2: Ron McNair (1)
MS3: Robert Stewart (1,EV)
Payload: MMU, IEF, RME, ACES, MLR, IRT, SSIP, GAS
Satellites: Westar-6/PAM, Palapa-B2/PAM, SPAS-1A
Spacewalks: McCandless and Stewart

1984-11B	Westar-6	1559	EST	3 Feb
1984-11D	Palapa-B2	1137	EST	4 Feb

STS-11/41C

11	1984-34A	Challenger (5)	0858	EST	6 Apr	0538 PST	13 Apr	EAFB (9)	6.98	28

CDR: Bob Crippen (3)
PLT: Dick Scobee (1)
MS1: George Nelson (1,EV)
MS2: James van Hoften (1,EV)
MS3: Terry Hart (1)
Payload: RME, IMAX, Cinema 360, SSIP
Satellite: LDEF
Activity: SolarMax (1980-14A) was retrieved 10 Apr, repaired, and released 12 Apr
Spacewalks: Nelson and van Hoften

1984-34B	LDEF	–	–	7 Apr

STS-12/41D

12	1984-93A	Discovery (1)	0842	EDT	30 Aug	0637 PDT	5 Sep	EAFB (10)	6.04	28

CDR: Hank Hartsfield (2)
PLT: Michael Coats (1)
MS1: Judith Resnik (1)
MS2: Steven Hawley (1,EV)
MS3: Mike Mullane (1,EV)
PS: Charles Walker (1)
Payload: OAST-1, LFC, CFES, RME, CLOUDS, IMAX, SSIP
Satellites: Leasat-2, SBS-4/PAM, Telstar-3C/PAM

1984-93B	SBS-4	1640	EDT	30 Aug
1984-93C	Leasat-2	0913	EDT	31 Aug
1984-93D	Telstar-3C	0920	EDT	1 Sep

#	Designation	Orbiter	Launch		Landing		Runway	Days	Incl

STS-13/41G

#	Designation	Orbiter	Launch		Landing		Runway	Days	Incl
13	1984-108A	Challenger (6)	0703 EDT	5 Oct	1226 EDT	13 Oct	KSC (2)	8.22	57

CDR: Bob Crippen (4)
PLT: Jon McBride (1)
MS1: Kathryn Sullivan (1,EV)
MS2: Sally Ride (2)
MS3: David Leestma (1,EV)
PS1: Marc Garneau (1,Canada)
PS2: Paul Scully-Power (1)
Payload: OSTA-3, CANEX, RME, TLD, APE, IMAX, GAS
Satellite: ERBS
Spacewalks: Sullivan and Leestma

1984-108B	ERBS	1530 EDT	5 Oct

STS-14/51A

#	Designation	Orbiter	Launch		Landing		Runway	Days	Incl
14	1984-113A	Discovery (2)	0715 EST	8 Nov	0700 EST	16 Nov	KSC (3)	7.99	28

CDR: Rick Hauck (2)
PLT: David Walker (1)
MS1: Anna Fisher (1)
MS2: Dale Gardner (2,EV)
MS3: Joe Allen (2,EV)
Payload: DMOS, RME
Satellites: Anik D-2/PAM, Leasat-1
Activity: Palapa-B2 was retrieved 12 Nov and Westar-6 was retrieved 14 Nov
Spacewalks: Allen and Gardner

1984-113B	Anik D-2	1630 EST	9 Nov
1984-113C	Leasat-1	0700 EST	10 Nov

STS-15/51C

#	Designation	Orbiter	Launch		Landing		Runway	Days	Incl
15	1985-10A	Discovery (3)	1450 EST	24 Jan	1623 EST	27 Jan	KSC (4)	3.06	28

CDR: Ken Mattingly (2)
PLT: Loren Shriver (1)
MS1: Ellison Onizuka (1,EV)
MS2: James Buchli (1,EV)
SFE: Gary Payton (1)
Payload: ARC
Satellite: Magnum/IUS

1985-10B	USA-8	1200 EST	25 Jan

#	Designation	Orbiter	Launch		Landing		Runway	Days	Incl

STS-16/51D

16	1985-28A	Discovery (4)	0859 EST	12 Apr	0854 EST	19 Apr	KSC (5)	6.99	28

CDR: Karol Bobko (2)
PLT: Don Williams (1)
MS1: Rhea Seddon (1)
MS2: Jeff Hoffman (1,EV)
MS3: David Griggs (1,EV)
PS: Charles Walker (2)
CO: Senator Jake Garn (1)
Payload: CFES, AFE, PPE, SSIP, GAS
Satellites: Anik C-1/PAM, Leasat-3
Spacewalks: Griggs and Hoffman

1985-28B	Anik C-1	1839 EST	12 Apr
1985-28C	Leasat-3	1000 EST	13 Apr

STS-17/51B

17	1985-34A	Challenger (7)	1202 EDT	29 Apr	0911 PDT	6 May	EAFB (11)	7.00	57

CDR: Bob Overmyer (2)
PLT: Fred Gregory (1)
MS1: Don Lind (1,EV)
MS2: Norman Thagard (2,EV)
MS3: Bill Thornton (2)
PS1: Taylor Wang (1)
PS2: Lodewijk van den Berg (1)
Payload: Spacelab 3, GAS
Satellite: GAS (Nusat)

1985-34B	Nusat	1614 EDT	29 Apr

STS-18/51G

18	1985-48A	Discovery (5)	0733 EDT	17 Jun	0611 PDT	24 Jun	EAFB (12)	7.07	28

CDR: Dan Brandenstein (2)
PLT: John Creighton (1)
MS1: Shannon Lucid (1)
MS2: John Fabian (2,EV)
MS: Steven Nagel (1,EV)
PS1: Patrick Baudry (1,France)
PS2: Prince Sultan Salman Abdul Aziz al-Saud (1)
Payload: FPE, ADSF, GAS, HPTE
Satellites: Morelos-1/PAM, Arabsat-1B/PAM, Telstar-3D/PAM, SPARTAN-101

1985-48B	Morelos-1	– –	17 Jun		
1985-48C	Arabsat-1B	– –	18 Jun		
1985-48D	Telstar-3D	– –	19 Jun		
1985-48E	SPARTAN-101	1401 EDT	20 Jun	1100 EDT	22 Jun

#	Designation	Orbiter	Launch			Landing			Runway	Days	Incl

STS-19/51F

19	1985-63A	Challenger (8)	1700	EDT	29 Jul	1245	PDT	6 Aug	EAFB (13)	7.95	50

CDR: Gordon Fullerton (2)
PLT: Roy Bridges (1)
MS1: Story Musgrave (2,EV)
MS2: Anthony England (1,EV)
MS3: Karl Henize (1)
PS1: Loren Acton (1)
PS2: John-David Bartoe (1)
Payload: Spacelab 2, SAREX, PGU(LIGNIN)
Satellite: PDP

1985-63B	PDP	– –	1 Aug	– –	2 Aug

STS-20/51I

20	1985-76A	Discovery (6)	0658	EDT	27 Aug	0616	PDT	3 Sep	EAFB (14)	7.10	28

CDR: Joe Engle (2)
PLT: Dick Covey (1)
MS1: James van Hoften (2,EV)
MS2: Mike Lounge (1)
MS3: Bill Fisher (1,EV)
Payload: PVTOS
Satellites: Aussat-1/PAM, ASC-1/PAM, Leasat-4
Activity: Leasat-3 was repaired
Spacewalks: Fisher and van Hoften

1985-76B	Aussat-1	1333 EDT	27 Aug
1985-76C	ASC-1	1808 EDT	27 Aug
1985-76D	Leasat-4	0645 EDT	29 Aug

STS-21/51J

21	1985-92A	Atlantis (1)	1115	EDT	3 Oct	1000	PDT	7 Oct	EAFB (15)	4.07	28

CDR: Karol Bobko (3)
PLT: Ron Grabe (1)
MS1: Dave Hilmers (1,EV)
MS2: Robert Stewart (2,EV)
SFE: William Pailes (1)
Satellites: DSCS III-2 and DSCS III-3 (both on the same IUS)

1985-92B	USA-11	– –	4 Oct
1985-92C	USA-12	– –	4 Oct

STS-22/61A

22	1985-104A	Challenger (9)	1200	EST	30 Oct	0945	PST	6 Nov	EAFB (16)	7.03	57

CDR: Hank Hartsfield (3)
PLT: Steven Nagel (2)
MS1: James Buchli 2,EV)
MS2: Guion Bluford (2,EV)
MS3: Bonnie Dunbar (1)
PS1: Reinhard Furrer (1, Germany)
PS2: Ernst Messerschmid (1, Germany)
PS3: Wubbo Ockels (1, ESA)
Payload: Spacelab D1
Satellite: GAS (GLOMR)

1985-104B	GLOMR	0034 EST	31 Oct

#	Designation	Orbiter	Launch		Landing		Runway	Days	Incl

STS-23/61B

23	1985-109A	Atlantis (2)	1929 EST	26 Nov	1334 PST	3 Dec	EAFB (17)	6.88	28

CDR: Brewster Shaw (2)
PLT: Bryan O'Connor (1)
MS1: Mary Cleave (1)
MS2: Sherwood Spring (1,EV)
MS3: Jerry Ross (1,EV)
PS1: Rodolfo Neri Vela (1,Mexico)
PS2: Charles Walker (3)
Payload: EASE, ACCESS, CFES, IMAX, GAS, DMOS
Satellites: Morelos-2/PAM, Aussat-2/PAM, Satcom-K2/PAM-D2
Spacewalks: Ross and Spring

1985-109B	Morelos-2	0247 EST	27 Nov
1985-109C	Aussat-2	2020 EST	27 Nov
1985-109D	Satcom-K2	1650 EST	28 Nov
1985-109E	OEX	2200 EST	29 Nov

STS-24/61C

24	1986-3A	Columbia (7)	0655 EST	12 Jan	0558 PST	18 Jan	EAFB (18)	6.08	28

CDR: Hoot Gibson (2)
PLT: Charles Bolden (1)
MS1: Franklin Chang-Diaz (1,EV)
MS2: Steven Hawley (2)
MS3: George Nelson (2,EV)
PS: Robert Cenker (1)
CO: Congressman Bill Nelson (1)
Payload: MSL, IR-IE, IBSE, CHAMP, SSIP, GBA, HH-PCG
Satellite: Satcom-K1/PAM-D2

1986-3B	Satcom-K1	1626 EST	12 Jan

STS-25/51L

25	1986	Challenger (10)	1138 EST	28 Jan	– –	– –	–	–	28

CDR: Dick Scobee (2)
PLT: Mike Smith (1)
MS1: Judith Resnik (2)
MS2: Ellison Onizuka (2,EV)
MS3: Ron McNair (2,EV)
PS: Greg Jarvis (1)
CO: Christa McAuliffe (1)
Payload: CHAMP, RME, FDE, TISP, SSIP, PPE
Satellites: TDRS-B/IUS, Halley/SPARTAN-203

STS-26

26	1988-91A	Discovery (7)	1137 EDT	29 Sep	0937 PDT	3 Oct	EAFB (19)	4.04	28

CDR: Rick Hauck (3)
PLT: Dick Covey (2)
MS1: Mike Lounge (2)
MS2: George Nelson (3)
MS3: Dave Hilmers (2)
Payload: PVTOS, PCG, IRCFE, ARC, IFE, MLE, PPE, ELRAD, ADSF, SSIP, OASIS
Satellite: TDRS-C/IUS

1988-91B	TDRS-C	1750 EDT	29 Sep

#	Designation	Orbiter	Launch	Landing	Runway	Days	Incl

STS-27

#	Designation	Orbiter	Launch	Landing	Runway	Days	Incl
27	1988-106A	Atlantis (3)	0930 EST 2 Dec	1536 PST 6 Dec	EAFB (20)	4.38	57

CDR: Hoot Gibson (3)
PLT: Guy Gardner (1)
MS1: Mike Mullane (2)
MS2: Jerry Ross (2,EV)
MS3: Bill Shepherd (1,EV)
Satellite: Lacrosse

	Designation	Satellite	Launch
	1988-106B	USA-34	1630 EST 2 Dec

STS-29

#	Designation	Orbiter	Launch	Landing	Runway	Days	Incl
28	1989-21A	Discovery (8)	0957 EST 13 Mar	0635 PST 18 Mar	EAFB (21)	4.98	28

CDR: Michael Coats (2)
PLT: John Blaha (1)
MS1: James Bagian (1)
MS2: James Buchli (3)
MS3: Robert Springer (1)
Payload: IMAX, SHARE, OASIS, PCG, PGU(CHROMEX), SSIP, AMOS, AEM
Satellite: TDRS-D/IUS

	Designation	Satellite	Launch
	1989-21B	TDRS-D	1600 EST 13 Mar

STS-30

#	Designation	Orbiter	Launch	Landing	Runway	Days	Incl
29	1989-33A	Atlantis (4)	1447 EDT 4 May	1243 PDT 8 May	EAFB (22)	4.04	28

CDR: David Walker (2)
PLT: Ron Grabe (2)
MS1: Norman Thagard (3,EV)
MS2: Mary Cleave (2)
MS3: Mark Lee (1,EV)
Payload: MLE, FEA, AMOS
Satellite: Magellan/PAM/IUS

	Designation	Satellite	Launch
	1989-33B	Magellan	2101 EDT 4 May (entered Venus orbit on 10 Aug 1990)

STS-28

#	Designation	Orbiter	Launch	Landing	Runway	Days	Incl
30	1989-61A	Columbia (8)	0837 EDT 8 Aug	0637 PDT 13 Aug	EAFB (23)	5.04	57

CDR: Brewster Shaw (3)
PLT: Dick Richards (1)
MS1: Jim Adamson (1)
MS2: David Leestma (2)
MS3: Mark Brown (1)
Payload: L3, EDS, SILTS, "the Head"
Satellites: a pair of DoD satellites (the main one possibly deployed by the SPDS)

	Designation	Satellite	Launch
	1989-61B	USA-40	– – – –
	1989-61C	USA-41	– – – –

#	Designation	Orbiter	Launch			Landing			Runway	Days	Incl

STS-34

#	Designation	Orbiter	Launch			Landing			Runway	Days	Incl
31	1989-84A	Atlantis (5)	1254	EDT	18 Oct	0933	PDT	23 Oct	EAFB (24)	4.99	34

CDR: Don Williams (2)
PLT: Mike McCulley (1)
MS1: Franklin Chang-Diaz (2,EV)
MS2: Shannon Lucid (2)
MS3: Ellen Baker (1,EV)
Payload: IMAX, GAS(SSBUV), GHCD, PM, STEX, MLE, IMAX, SSIP, AMOS
Satellite: Galileo/IUS
1989-84B Galileo 1915 EDT 18 Oct (entered Jupiter orbit on 7 Dec 1995)

STS-33

#	Designation	Orbiter	Launch			Landing			Runway	Days	Incl
32	1989-90A	Discovery (9)	1923	EST	22 Nov	1630	PST	27 Nov	EAFB (25)	5.00	28

CDR: Fred Gregory (2)
PLT: John Blaha (2)
MS1: Story Musgrave (3)
MS2: Manley Carter (1)
MS3: Kathryn Thornton (1)
Satellite: Magnum/IUS
1989-90B USA-48 – – – –

STS-32

#	Designation	Orbiter	Launch			Landing			Runway	Days	Incl
33	1990-02A	Columbia (9)	0735	EST	9 Jan	0135	PST	20 Jan	EAFB (26)	10.87	28

CDR: Dan Brandenstein (3)
PLT: Jim Wetherbee (1)
MS1: Bonnie Dunbar (2,EV)
MS2: David Low (1,EV)
MS3: Marsha Ivins (1)
Payload: IMAX, PCG, FEA, AFE, CNCR, L3, MLE, AMOS, LBNP, SILTS
Satellite: Leasat-5
Activity: LDEF was retrieved on 12 Jan
1990-02B Leasat-5 0819 EST 10 Jan

STS-36

#	Designation	Orbiter	Launch			Landing			Runway	Days	Incl
34	1990-19A	Atlantis (6)	0250	EST	28 Feb	1009	PST	4 Mar	EAFB (27)	4.43	62

CDR: John Creighton (2)
PLT: John Casper (1)
MS1: Mike Mullane (3)
MS2: Dave Hilmers (3)
MS3: Pierre Thuot (1)
Satellite: a DoD satellite (released by SPDS)
1990-19B USA-53 – – 1 Mar

#	Designation	Orbiter	Launch			Landing			Runway	Days	Incl

STS-31

35	1990-37A	Discovery (10)	0833	EDT	24 Apr	0649	PDT	29 Apr	EAFB (28)	5.05	28

CDR: Loren Shriver (2)
PLT: Charles Bolden (2)
MS1: Steven Hawley (3)
MS2: Bruce McCandless (2,EV)
MS3: Kathryn Sullivan (2,EV)
Payload: IMAX, APM, PCG, RME, PMP, SSIP, AMOS
Satellite: HST
1990-37B HST 1537 EDT 25 Apr

STS-41

36	1990-90A	Discovery (11)	0747	EDT	6 Oct	0657	PDT	10 Oct	EAFB (29)	4.09	28

CDR: Dick Richards (2)
PLT: Bob Cabana (1)
MS1: Bill Shepherd (2)
MS2: Bruce Melnick (1,EV)
MS3: Tom Akers (1,EV)
Payload: GAS(SSBUV), ISAC, PGU(CHROMEX), VCS, SSCE, PMP, PSE/AEM, RME, SSIP, AMOS
Satellite: Ulysses/PAM/IUS
1990-90B Ulysses 1348 EDT 6 Oct (now in heliocentric orbit)

STS-38

37	1990-97A	Atlantis (7)	1848	EST	15 Nov	1643	EST	20 Nov	KSC (6)	4.91	28

CDR: Dick Covey (3)
PLT: Frank Culbertson (1)
MS1: Robert Springer (2,EV)
MS2: Carl Meade (1,EV)
MS3: Sam Gemar (1)
Satellite: a DoD satellite (not on an IUS)
1990-97B USA-67 – – – –

STS-35

38	1990-106A	Columbia (10)	0149	EST	2 Dec	2154	PST	10 Dec	EAFB (30)	8.96	28

CDR: Vance Brand (3)
PLT: Guy Gardner (2)
MS1: Jeff Hoffman (2,EV)
MS2: Mike Lounge (3,EV)
MS3: Bob Parker (2)
PS1: Samuel Durrance (1)
PS2: Ronald Parise (1)
Payload: ASTRO-1, BBXRT, SAREX, AMOS

#	Designation	Orbiter	Launch				Landing				Runway	Days	Incl

STS-37

39 1991-27A Atlantis (8) 0922 EST 5 Apr 0655 PST 11 Apr EAFB (31) 5.98 28

CDR: Steven Nagel (3)
PLT: Ken Cameron (1)
MS1: Jerry Ross (3,EV)
MS2: Jay Apt (1,EV)
MS3: Linda Godwin (1)
Payload: CETA, CLIP, APM, SAREX, PCG, BIMDA, RME, AMOS, SHARE
Satellite: GRO
Spacewalks: Ross and Apt

1991-27B	GRO	1737	EST	7 Apr

STS-39

40 1991-31A Discovery (12) 0733 EDT 28 Apr 1455 EDT 6 May KSC (7) 8.30 57

CDR: Mike Coats (3)
PLT: Blaine Hammond (1)
MS1: Guion Bluford (3)
MS2: Greg Harbaugh (1,EV)
MS3: Richard Hieb (1)
MS4: Donald McMonagle (1,EV)
MS5: Lacy Veach (1)
Payload: CIRRIS, FUV, HUP, URA, QINMS, CIV, STP (UVLIM, ALFE, SKIRT, APM, DSE), RME, CLOUDS
Satellites: SPAS-2/IBSS, CRO, MPEC (USA-70)

1991-31B	SPAS-2-01	1017	EDT	1 May	1825	EDT	2 May
1991-31C	USA-70	–	--	6 May			
1991-31D	CRO-C	–	–	1 May			
1991-31E	CRO-B	1403	EDT	1 May			
1991-31F	CRO-A	–	–	3 May			

STS-40

41 1991-40A Columbia (11) 0925 EDT 5 Jun 0839 PDT 14 Jun EAFB (32) 9.09 39

CDR: Bryan O'Connor (2)
PLT: Sid Gutierrez (1)
MS1: Rhea Seddon (2)
MS2: James Bagian (2,EV)
MS3: Tamara Jernigan (1,EV)
PS1: Drew Gaffney (1)
PS2: Millie Hughes-Fulford (1)
Payload: SLS-1, GBA, MODE, AEM, ARE, RAHF, GPWS

#	Designation	Orbiter	Launch			Landing			Runway	Days	Incl

STS-43

#	Designation	Orbiter	Launch			Landing			Runway	Days	Incl
42	1991-54A	Atlantis (9)	1102	EDT	2 Aug	0823	EDT	11 Aug	KSC (8)	8.89	28

CDR: John Blaha (3)
PLT: Mike Baker (1)
MS1: Shannon Lucid (3)
MS2: Jim Adamson (2,EV)
MS3: David Low (2,EV)
Payload: GAS(SSBUV), SHARE, OCTW, GAS(TPCE), APE, PCG, BIMDA, PMP, SAMS, SSCE, UVPI, AMOS, LBNP
Satellite: TDRS-E/IUS
1991-54B TDRS-E 1714 EDT 2 Aug

STS-48

#	Designation	Orbiter	Launch			Landing			Runway	Days	Incl
43	1991-63A	Discovery (13)	1911	EDT	12 Sep	0039	PDT	18 Sep	EAFB (33)	5.35	57

CDR: John Creighton (3)
PLT: Ken Reightler (1)
MS1: James Buchli (4,EV)
MS2: Sam Gemar (2,EV)
MS3: Mark Brown (2)
Payload: AMOS, APM, MODE, SAM, CREAM, PARE, PCG, PMP, RME, ESC
Satellite: UARS
1991-63B UARS 0023 EDT 15 Sep

STS-44

#	Designation	Orbiter	Launch			Landing			Runway	Days	Incl
44	1991-80A	Atlantis (10)	1844	EST	24 Nov	1434	PST	1 Dec	EAFB (34)	6.95	28

CDR: Fred Gregory (3)
PLT: Terry Henricks (1)
MS1: Story Musgrave (4)
MS2: Mario Runco (1,EV)
MS3: Jim Voss (1,EV)
PS: Tom Hennen (1)
Payload: IOCM, MODE, AMOS, M88-1 (MMIS), CREAM, SAM, RME, VFT, UVPI, BFPT, EDO-MP, Terra Scout
Satellite: DSP-16/IUS
1991-80B USA-75 0103 EST 25 Nov

STS-42

#	Designation	Orbiter	Launch			Landing			Runway	Days	Incl
45	1992-02A	Discovery (14)	0952	EST	22 Jan	0807	PST	30 Jan	EAFB (35)	8.05	57

CDR: Ron Grabe (3)
PLT: Stephen Oswald (1)
MS1: Norman Thagard (4,EV,PC)
MS2: Dave Hilmers (4)
MS3: Bill Readdy (1,EV)
PS1: Roberta Bondar (1,Canada)
PS2: Ulf Merbold (2,ESA)
Payload: IML-1, IMAX, GBA, SSIP, GOSAMR, PMP, RME

#	Designation	Orbiter	Launch			Landing			Runway	Days	Incl

STS-45

46	1992-15A	Atlantis (11)	0814	EST	24 Mar	0623	EST	2 Apr	KSC (9)	8.93	57

CDR: Charles Bolden (3)
PLT: Brian Duffy (1)
MS1: Kathryn Sullivan (3,PC)
MS2: David Leestma (3)
MS3: Mike Foale (1)
PS1: Byron Lichtenberg (2)
PS2: Dirk Frimout (1)
Payload: ATLAS-1, GAS(SSBUV), STL, PMP, SAREX, VFT, RME, CLOUDS, GAS, APE

STS-49

47	1992-26A	Endeavour (1)	1940	EDT	7 May	1357	PDT	16 May	EAFB (36)	8.88	28

CDR: Dan Brandenstein (4)
PLT: Kevin Chilton (1)
MS1: Pierre Thuot (2,EV)
MS2: Kathryn Thornton (2,EV)
MS3: Richard Hieb (2,EV)
MS4: Tom Akers (2,EV)
MS5: Bruce Melnick (2)
Payload: ASEM, PCG, UVPI, AMOS
Activity: Intelsat 603 (1990-) was retrieved 2000 EDT 13 May, fitted with an Orbus motor, and released 0053 EDT 14 May
Spacewalks: Thuot, Thornton, Hieb and Akers

STS-50

48	1992-34A	Columbia (12)	1212	EDT	25 Jun	0742	EDT	9 Jul	KSC (10)	13.81	28

CDR: Dick Richards (3)
PLT: Ken Bowersox (1)
MS1: Bonnie Dunbar (3,PC)
MS2: Ellen Baker (2)
MS3: Carl Meade (2)
PS1: Larry DeLucas (1)
PS2: Gene Trinh (1)
Payload: USML-1, EDO-1, PMP, SAREX, UVPI, AERIS

STS-46

49	1992-49A	Atlantis (12)	0957	EDT	31 Jul	0911	EDT	8 Aug	KSC (11)	7.97	28

CDR: Loren Shriver (3)
PLT: Andy Allen (1)
MS1: Jeff Hoffman (3,EV)
MS2: Frankin Chang-Diaz (3,EV)
MS3: Claude Nicollier (1,ESA)
MS4: Marsha Ivins (2)
PS: Franco Malerba (1,Italy)
Payload: TSS 1, LDCE, UVPI, IMAX, EOIM/TEMP, CONCAP, PHCF, IMAX, AMOS
Satellite: EURECA-1

	1992-49B	EURECA-1	0307	EDT	2 Aug

#	Designation	Orbiter	Launch			Landing			Runway	Days	Incl

STS-47

50	1992-61A	Endeavour (2)	1023	EDT	12 Sep	0853	EDT	20 Sep	KSC (12)	7.94	57

CDR: Hoot Gibson (4)
PLT: Curt Brown (1)
MS1: Mark Lee (2,PC)
MS2: Jan Davis (1)
MS3: Jay Apt (2)
MS4: Mae Jemison (1)
PS: Mamoru Mohri (1,Japan)
Payload: Spacelab J1, GEF, SAMS, PCG, CHF, GBA, ISAIAH, SSCE, SAREX, AMOS, UVPI

STS-52

51	1992-71A	Columbia (13)	1309	EDT	22 Oct	0905	EST	1 Nov	KSC (13)	9.86	28

CDR: Jim Wetherbee (2)
PLT: Mike Baker (2)
MS1: Lacy Veach (2)
MS2: Bill Shepherd (3)
MS3: Tamara Jernigan (2)
PS: Steven MacLean (1)
Payload: USMP 1, CANEX, CMIX, SPIFEX, GAS(TPCE), CVTE, HPP, ASP, PCG, PSE
Satellites: LAGEOS-2/IRIS, CTA

1992-71B	LAGEOS-2	0956	EST	23 Oct
1992-71C	CTA	–	–	31 Oct

STS-53

52	1992-86A	Discovery (15)	0824	EST	2 Dec	1243	PST	9 Dec	EAFB (37)	7.30	57

CDR: David Walker (3)
PLT: Bob Cabana (2)
MS1: Guion Bluford (4)
MS2: Jim Voss (2)
MS3: Rich Clifford (1)
Payload: GAS(ODERACS), GLO, CRYOHP, MIS, STL, VFT, CREAM, RME, FARE, HERCULES, BLAST, CLOUDS
Satellite: a DoD satellite

1992-86B	'DOD-1'	1500	EST	2 Dec

STS-54

53	1993-03A	Endeavour (3)	0859	EST	13 Jan	0838	EST	19 Jan	KSC (14)	5.97	28

CDR: John Casper (2)
PLT: Donald McMonagle (2)
MS1: Mario Runco (2,EV)
MS2: Greg Harbaugh (2,EV)
MS3: Susan Helms (1)
Payload: DXS, CGBA, PGU(CHROMEX), PARE, SAMS, SSCE
Satellite: TDRS-F/IUS
Spacewalks: Runco and Harbaugh

1993-03B	TDRS-F	1515	EST	13 Jan

#	Designation	Orbiter	Launch		Landing		Runway	Days	Incl

STS-56

54	1993-23A	Discovery (16)	0129 EDT	8 Apr	0737 EDT	17 Apr	KSC (15)	9.25	57

CDR: Ken Cameron (2)
PLT: Stephen Oswald (2)
MS1: Mike Foale (2)
MS2: Ken Cockrell (1)
MS3: Ellen Ochoa (1)
Payload: ATLAS-2, SAREX, GAS(SUVE, SSBUV), CMIX, PARE, STL, CREAM, HERCULES, RME, AMOS
Satellite: SPARTAN-201-1

1993-23B	SPARTAN-201	0211 EDT	11 Apr	0320 EDT	13 Apr

STS-55

55	1993-27A	Columbia (14)	1050 EDT	26 Apr	0730 PDT	6 May	EAFB (38)	9.98	28

CDR: Steven Nagel (4)
PLT: Tom Henricks (2)
MS1: Jerry Ross (4,PC)
MS2: Charles Precourt (1)
MS3: Bernard Harris (1)
PS1: Ulrich Walter (1, Germany)
PS2: Hans Schlegel (1, Germany)
Payload: Spacelab D2 (Anthrorak, Biolabor, Baroreflex, MEDEA, HOLOP, MAUS), ROTEX, SAREX

STS-57

56	1993-37A	Endeavour (4)	0907 EDT	21 Jun	0852 EDT	1 Jul	KSC (16)	9.99	28

CDR: Ron Grabe (4)
PLT: Brian Duffy (2)
MS1: David Low (3,PC,EV)
MS2: Nancy Sherlock/Currie (1)
MS3: Jeff Wisoff (1,EV)
MS4: Janice Voss/Ford (1)
Payload: Spacehab 1, SHOOT, CONCAP, GBA, FARE, BLAST, SAREX
Activity: EURECA-1 was retrieved on 24 June
Spacewalks: Low and Wisoff

STS-51

57	1993-58A	Discovery (17)	0745 EDT	12 Sep	0356 EDT	22 Sep	KSC (17)	9.84	28

CDR: Frank Culbertson (2)
PLT: Bill Readdy (2)
MS1: Jim Newman (1,EV)
MS2: Dan Bursch (1)
MS3: Carl Walz (1,EV)
Payload: IMAX, PCG, PGU(CHROMEX), APE, PMP, RME, LDCE, AMOS
Satellites: ACTS/TOS, ORFEUS-SPAS-1
Spacewalks: Walz and Newman

1993-58B	ACTS	1713 EDT	12 Sep		
1993-58C	ORFEUS-1	1106 EDT	13 Sep	0750 EDT	19 Sep

#	Designation	Orbiter	Launch	Landing	Runway	Days	Incl

STS-58

58 1993-65A Columbia (15) 1053 EDT 18 Oct 0706 PST 1 Nov EAFB (39) 14.00 39

CDR: John Blaha (4)
PLT: Rick Searfoss (1)
MS1: Rhea Seddon (3,PC)
MS2: Bill McArthur (1)
MS3: David Wolf (1)
MS4: Shannon Lucid (4)
PS: Martin Fettman (1)
Payload: SLS-2, EDO-2, OARE, SAREX, PILOT, LBNP

STS-61

59 1993-75A Endeavour (5) 0427 EST 2 Dec 0026 EST 13 Dec KSC (18) 10.83 28

CDR: Dick Covey (4)
PLT: Ken Bowersox (2)
MS1: Kathryn Thornton (3,EV)
MS2: Claude Nicollier (2)
MS3: Jeff Hoffman (4,EV)
MS4: Story Musgrave (5,PC,EV)
MS5: Tom Akers (3,EV)
Payload: HST apparatus (COSTAR, WF/PC-2), IMAX
Activity: The HST (1993-75B) was retrieved 0348 EST 4 Dec, serviced and released on 10 Dec
Spacewalks: Hoffman, Musgrave, Akers and Thornton
1993-75C HST solar array – – 6 Dec

STS-60

60 1994-06A Discovery (18) 0710 EST 3 Feb 1419 EST 11 Feb KSC (19) 8.30 57

CDR: Charles Bolden (4)
PLT: Ken Reightler (2)
MS1: Jan Davis (2)
MS2: Ron Sega (1)
MS3: Franklin Chang-Diaz (4,PC)
MS4: Sergei Krikalev (1,Russia)
Payload: Spacehab 2, WSF-1, SAREX, APE, GBA (CAPL)
Satellites: GAS(ODERACS, BREMSAT)
1994-06B/G ODERACS 0954 EST 9 Feb
1994-06H Bremsat 1423 EST 9 Feb

STS-62

61 1994-15A Columbia (16) 0853 EST 4 Mar 0810 EST 18 Mar KSC (20) 13.97 39

CDR: John Casper (3)
PLT: Andy Allen (2)
MS1: Pierre Thuot (3)
MS2: Sam Gemar (3)
MS3: Marsha Ivins (3)
Payload: USMP-2, OAST-2, EDO-3, DEE, GAS(SSBUV), LDCE, APCG, CPCG, PSE, CGBA, BDS, MODE, AMOS, LBNP, PILOT

#	Designation	Orbiter	Launch			Landing			Runway	Days	Incl

STS-59

#	Designation	Orbiter	Launch			Landing			Runway	Days	Incl
62	1994-20A	Endeavour (6)	0705	EDT	9 Apr	0954	PDT	20 Apr	EAFB (40)	11.24	57

CDR: Sid Gutierrez (2)
PLT: Kevin Chilton (2)
MS1: Jap Apt (3)
MS2: Rich Clifford (2)
MS3: Linda Godwin (2,PC)
MS4: Tom Jones (1)
Payload: SRL-1 (SIR-C, X-SAR), MAPS, CONCAP, SAREX, STL, VFT, GAS, TUFI

STS-65

#	Designation	Orbiter	Launch			Landing			Runway	Days	Incl
63	1994-39A	Columbia (17)	1243	EDT	8 Jul	0638	EDT	23 Jul	KSC (21)	14.75	28

CDR: Bob Cabana (3)
PLT: James Halsell (1)
MS1: Richard Hieb (3,PC)
MS2: Carl Walz (2)
MS3: Leroy Chiao (1)
MS4: Don Thomas (1)
PS: Chiaki Naito-Mukai (1,Japan)
Payload: IML-2, EDO-4, EDO-MP, APCF, CPCG, AMOS, OARE, MAST, SAREX

STS-64

#	Designation	Orbiter	Launch			Landing			Runway	Days	Incl
64	1994-59A	Discovery (19)	1823	EDT	9 Sep	1413	PDT	20 Sep	EAFB (41)	10.95	57

CDR: Dick Richards (4)
PLT: Blaine Hammond (2)
MS1: Jerry Linenger (1)
MS2: Susan Helms (2)
MS3: Carl Meade (3,EV)
MS4: Mark Lee (3,EV)
Payload: LITE, SAFER, ROMPS, TCS, GBA, SSCE, BRIC, RME, MAST, SAREX, AMOS, SPIFEX
Satellite: SPARTAN-201-2
Spacewalks: Lee and Meade

#	Designation	Orbiter	Launch			Landing			Runway	Days	Incl
	1994-59B	SPARTAN-201	1730	EDT	13 Sep	1645	EDT	15 Sep			

STS-68

#	Designation	Orbiter	Launch			Landing			Runway	Days	Incl
65	1994-62A	Endeavour (7)	0716	EDT	30 Sep	1002	PDT	11 Oct	EAFB (42)	11.24	57

CDR: Mike Baker (3)
PLT: Terry Wilcutt (1)
MS1: Steven Smith (1)
MS2: Dan Bursch (2)
MS3: Jeff Wisoff (2)
MS4: Tom Jones (2,PC)
Payload: SRL-2 (SIR-C, X-SAR), MAPS, CPCG, BRIC, PGU(CHROMEX), CREAM, MAST, GAS

#	Designation	Orbiter	Launch			Landing			Runway	Days	Incl

STS-66

66 1994-73A Atlantis (13) 1159 EST 3 Nov 0734 PST 14 Nov EAFB (43) 10.94 57

CDR: Donald McMonagle (3)
PLT: Curt Brown (2)
MS1: Ellen Ochoa (2,PC)
MS2: Joe Tanner (1)
MS3: Jean-Francois Clervoy (1,ESA)
MS4: Scott Parazynski (1)
Payload: ATLAS-3, GAS(SSBUV), ESCAPE 2, NIH-R, PCG-TES, PCG-STES, NIH-C, SAMS, HPP
Satellite: CRISTA-SPAS-1
1994-73B CRISTA-1 0750 EST 4 Nov 0805 EST 12 Nov

STS-63

67 1995-04A Discovery (20) 0022 EST 3 Feb 0650 EST 11 Feb KSC (22) 8.27 51

CDR: Jim Wetherbee (3)
PLT: Eileen Collins (1)
MS1: Bernard Harris (2,PC,EV)
MS2: Mike Foale (3,EV)
MS3: Janice Voss/Ford (2)
MS4: Vladimir Titov (1,Russia)
Payload: Spacehab 3, CONCAP, GLO, CSE, CRYOSYS, IMAX, SSCE, AMOS, MSX, TCS
Satellites: SPARTAN-204, GAS(ODERACS)
Activity: The closest point of approach (10 metres) of Near-Mir was 1420 EST 6 Feb
Spacewalks: Foale and Harris
1995-04B SPARTAN-204 0726 EST 7 Feb 0633 EST 9 Feb
1995-04C/H ODERACS 2350 EST 4 Feb

STS-67

68 1995-07A Endeavour (8) 0138 EST 2 Mar 1347 PST 18 Mar EAFB (44) 16.63 28

CDR: Stephen Oswald (3)
PLT: Bill Gregory (1)
MS1: John Grunsfeld (1)
MS2: Wendy Lawrence (1)
MS3: Tamara Jernigan (3,PC)
PS1: Samuel Durrance (2)
PS2: Ronald Parise (2)
Payload: ASTRO-2, EDO-5, GAS, PCG-TES, PCG-STES, SAREX, CMIX, MSX, MACE

STS-71

69 1995-30A Atlantis (14) 1532 EDT 27 Jun 1054 EDT 7 Jul KSC (23) 9.80 51

CDR: Hoot Gibson (5)
PLT: Charles Precourt (2)
MS1: Ellen Baker (3,PC,EV)
MS2: Greg Harbaugh (3,EV)
MS3: Bonnie Dunbar (4)
UP: Anatoli Solovyev (1), Nikolai Budarin (1)
DN: Vladimir Dezhurov (1), Gennadi Strekalov (1), Norman Thagard (5)
Payload: Spacelab, IMAX, SAREX, ODS
Activity: Shuttle/Mir Mission-1 (SMM-1) docked 0900 EDT 29 Jun and undocked 0710 EDT 4 Jul

#	Designation	Orbiter	Launch			Landing			Runway	Days	Incl

STS-70

70	1995-35A	Discovery (21)	0942	EDT	13 Jul	0802	EDT	22 Jul	KSC (24)	8.93	28

CDR: Tom Henricks (3)
PLT: Kevin Kregel (1)
MS1: Nancy Sherlock/Currie (2)
MS2: Don Thomas (2)
MS3: Mary Ellen Weber (1)
Payload: MSX, NIH-R, BDS, CPCG, NIH-C, BRIC, SAREX, VFT, MIS, AMOS, HERCULES, WINDEX, RME, MAST
Satellite: TDRS-G/IUS

1995-35A TDRS-G 1555 EDT 13 Jul

STS-69

71	1995-48A	Endeavour (9)	1109	EDT	7 Sep	0738	EDT	18 Sep	KSC (25)	10.85	28

CDR: David Walker (4)
PLT: Ken Cockrell (2)
MS1: Jim Voss (3,EV,PC)
MS2: Jim Newman (2)
MS3: Michael Gernhardt (1,EV)
Payload: GBA(CAPL, IEH, UVSTAR, GLO, CONCAP)
Satellites: SPARTAN-201-3, WSF-2
Spacewalks: Voss and Gernhardt

1995-48B	SPARTAN-201-3	1142 EDT	8 Sep	–	–	10 Sep	
1995–48C	WSF-2	0725 EDT	11 Sep	–	–	14 Sep	

STS-73

72	1995-56A	Columbia (18)	0953	EDT	20 Oct	0745	EDT	5 Nov	KSC (26)	15.92	39

CDR: Ken Bowersox (3)
PLT: Kent Rominger (1)
MS1: Kathryn Thornton (4,PC)
MS2: Catherine Coleman (1)
MS3: Michael Lopez-Alegria (1)
PS1: Fred Leslie (1)
PS2: Albert Sacco (1)
Payload: USML-2, EDO-6

STS-74

73	1995-61A	Atlantis (15)	0730	EST	12 Nov	1201	EST	20 Nov	KSC (27)	8.18	51

CDR: Ken Cameron (3)
PLT: James Halsell (2)
MS1: Chris Hadfield (1,Canada)
MS2: Jerry Ross (5,EV)
MS3: Bill McArthur (2,EV)
Payload: SVS, DM, ODS, GLO
Activity: SMM-2; the DM was mounted on the ODS 0217 EST 14 Nov, docking with Mir 0128 EST 15 Nov; undocking ODS from DM 0316 EST 18 Nov; the DM was left on Mir's Kristall module

1995-61B DM – – 18 Nov

#	Designation	Orbiter	Launch	Landing	Runway	Days	Incl

STS-72

#	Designation	Orbiter	Launch	Landing	Runway	Days	Incl
74	1996-01A	Endeavour (10)	0441 EST 11 Jan	0242 EST 20 Jan	KSC (28)	8.91	28

CDR: Brian Duffy (3)
PLT: Brent Jett (1)
MS1: Leroy Chiao (2,EV)
MS2: Winston Scott (1,EV)
MS3 Koicho Wakata (1)
MS4: Daniel Barry (1,EV)
Payload: GAS(SSBUV), PCG, PARE, STL, CPCG
Satellite: OAST/SPARTAN-206
Activity: the Japanese SFU (1995-11A) was retrieved 0557 EST 13 Jan
Spacewalks: Chiao, Scott and Barry

Designation	Satellite	Launch	Landing
1995-01B	SPARTAN-206	0632 EST 14 Jan	– – 16 Jan

STS-75

#	Designation	Orbiter	Launch	Landing	Runway	Days	Incl
75	1996-12A	Columbia (19)	1518 EST 22 Feb	0858 EST 9 Mar	KSC (29)	15.73	28

CDR: Andy Allen (3)
PLT: Scott Horowitz (1)
MS1: Franklin Chang-Diaz (5,PC)
MS2: Jeff Hoffman (5)
MS3: Maurizio Cheli (1)
MS4: Claude Nicollier (3,ESA)
PS: Umberto Guidoni (1)
Payload: USMP-3, TSS-2, EDO-7, MGBX, CPCG
Activity: the tether of the TSS broke, inadvertently turning the payload into a satellite

Designation	Satellite	Launch	Landing
1996-12B	TSS-2	– –	25 Feb

STS-76

#	Designation	Orbiter	Launch	Landing	Runway	Days	Incl
76	1996-18A	Atlantis (16)	0313 EST 22 Mar	0529 PST 31 Mar	EAFB (45)	9.22	51

CDR: Kevin Chilton (3)
PLT: Rick Searfoss (2)
MS1: Ron Sega (2,PC)
MS2: Rich Clifford (3,EV)
MS3: Linda Godwin (3,EV)
UP: Shannon Lucid (5)
Payload: Single Spacehab, ODS, KidSat, SAREX, MEEP
Activity: SMM-3 docked with Mir 2134 EST 23 Mar and undocked 2008 EST 28 Mar
Spacewalks: Clifford and Godwin

STS-77

#	Designation	Orbiter	Launch	Landing	Runway	Days	Incl
77	1996-32A	Endeavour (11)	0630 EDT 19 May	0709 EDT 29 May	KSC (30)	10.03	28

CDR: John Casper (4)
PLT: Curt Brown (3)
MS1: Andrew Thomas (1,PC)
MS2: Dan Bursch (3,EV)
MS3: Mario Runco (3,EV)
MS4: Marc Garneau (2,Canada)
Payload: Spacehab 4, GBA, TEAMS (GANE, VTRE, LMTE), BETSCE, ARF, BRIC, TPCE/RFL
Satellites: IAE/SPARTAN-207, TEAMS (PAMS/STU)

Designation	Satellite	Launch	Landing
1996-32B	SPARTAN-207	0729 EDT 20 May	1158 EDT 21 May
1996-32C	IAE	1115 EDT 20 May	(jettisoned by SPARTAN carrier)
1996-32D	PAMS/STU	0518 EDT 22 May	

#	Designation	Orbiter	Launch			Landing			Runway	Days	Incl

STS-78

#	Designation	Orbiter	Launch			Landing			Runway	Days	Incl
78	1996-36A	Columbia (20)	1049	EDT	20 Jun	0837	EDT	7 Jul	KSC (31)	16.91	28

CDR: Tom Henricks (4)
PLT: Kevin Kregel (2)
MS1: Richard Linnehan (1,EV)
MS2: Susan Helms (3,PC,EV)
MS3: Charles Brady (1)
PS1: Jean-Jacques Favier (1)
PS2: Bob Thirsk (1,Canada)
Payload: LMS (APCF, AGHF, BDPU, AEM, PGF, STL, OARE, SAMS, MMA, TRE, HGD, TVD, LFE, COIS); EDO-8

STS-79

#	Designation	Orbiter	Launch			Landing			Runway	Days	Incl
79	1996-57A	Atlantis (17)	0454	EDT	16 Sep	0813	EDT	26 Sep	KSC (32)	10.13	51

CDR: Bill Readdy (3)
PLT: Terry Wilcutt (2)
MS1: Jay Apt (4,EV)
MS2: Tom Akers (4)
MS3: Carl Walz (3,EV)
UP: John Blaha (5)
DN: Shannon Lucid (5)
Payload: Double Spacehab, ODS
Activity: SMM-4 docked with Mir 2315 EDT 19 Sep and undocked 24 Sep

STS-80

#	Designation	Orbiter	Launch			Landing			Runway	Days	Incl
80	1996-65A	Columbia (21)	1456	EST	19 Nov	0649	EST	7 Dec	KSC (33)	17.67	28

CDR: Ken Cockrell (3)
PLT: Kent Rominger (2)
MS1: Tamara Jernigan (4,EV)
MS2: Tom Jones (3,EV)
MS3: Story Musgrave (6)
Payload: EDO-9, SVS, CMIX, PARE, BRIC, VIEW-CPL, SEM, CCM, SVS
Satellites: WSF-3, ORFEUS-SPAS-2

Designation	Satellite	Launch			Landing		
1996-65B	ORFEUS-2	–	EST	20 Nov	0326	EST	4 Dec
1996-65C	WSF-3	2038	EST	22 Nov	2101	EST	25 Nov

STS-81

#	Designation	Orbiter	Launch			Landing			Runway	Days	Incl
81	1997-01A	Atlantis (18)	0427	EST	12 Jan	0922	EST	22 Jan	KSC (34)	10.21	51

CDR: Mike Baker (4)
PLT: Brent Jett (2)
MS1: Jeff Wisoff (3)
MS2: John Grunsfeld (2)
MS3: Marsha Ivins (4)
UP: Jerry Linenger (2)
DN: John Blaha (5)
Payload: Double Spacehab, ODS, KidSat, CREAM, SAMS, TVIS
Activity: SMM-5 docked with Mir 2255 EST 14 Jan and undocked 2115 EST 19 Jan

#	Designation	Orbiter	Launch	Landing	Runway	Days	Incl

STS-82

82	1997-4A	Discovery (22)	0355 EST 11 Feb	0332 EST 21 Feb	KSC (35)	9.99	28

CDR: Ken Bowersox (4)
PLT: Scott Horowitz (2)
MS1: Mark Lee (4,PC,EV)
MS2: Joe Tanner (2,EV)
MS3: Steven Hawley (4)
MS4: Greg Harbaugh (4,EV)
MS5: Steven Smith (2,PC,EV)
Payload: NICMOS, STIS
Activity: HST was retrieved 0334 EST 13 Feb, upgraded and released 0141 ESR 19 Feb
Spacewalks: Tanner, Harbaugh, Lee and Smith

STS-83

83	1997-13A	Columbia (22)	1420 EST 4 Apr	1433 EDT 8 Apr	KSC (36)	3.97	28

CDR: James Halsell (3)
PLT: Susan Leigh Still (1)
MS1: Janice Voss/Ford (3,PC)
MS2: Michael Gernhardt (2,EV)
MS3: Don Thomas (3,RV)
PS1: Roger Crouch (1)
PS2: Gregory Linteris (1)
Payload: MSL-1, EDO-10

STS-84

84	1997-23A	Atlantis (19)	0407 EDT 15 May	0928 EDT 24 May	KSC (37)	9.22	51

CDR: Charles Precourt (3)
PLT: Eileen Collins (2)
MS1: Jean-Francois Clervoy (2,ESA,EV)
MS2: Carlos Noriega (1)
MS3: Edward Lu (1,EV)
MS4: Yelena Kondakova (1,Russian)
UP: Mike Foale (4)
DN: Jerry Linenger (2)
Payload: Double Spacehab, ODS
Activity: SMM-6 docked with Mir on 16 May and undocked 21 May

STS-94

85	1997-32A	Columbia (23)	1402 EDT 1 Jul	0647 EDT 17 Jul	KSC (38)	15.70	28

CDR: James Halsell (4)
PLT: Susan Leigh Still (2)
MS1: Janice Voss/Ford (4,PC)
MS2: Michael Gernhardt (3,EV)
MS3: Don Thomas (4,RV)
PS1: Roger Crouch (2)
PS2: Gregory Linteris (2)
Payload: MSL-1R, EDO-11

#	Designation	Orbiter	Launch			Landing			Runway	Days	Incl

STS-85

#	Designation	Orbiter	Launch			Landing			Runway	Days	Incl
86	1997-39A	Discovery (23)	1141	EDT	7 Aug	0708	EDT	19 Aug	KSC (39)	11.85	57

CDR: Curt Brown (4)
PLT: Kent Rominger (3)
MS1: Jan Davis (3)
MS2: Robert Curbeam (1,EV)
MS3: Steven Robinson (1,EV)
PS: Bjarni Tryggvason (1,Canada)
Payload: TAS, MFD/SFA, IEH, SWUIS, MIM, BDS, SVS/RSAD, SEM, PCG-STES, BRIC, SSCE, MSX, SIMPLEX
Satellite: CRISTA-SPAS-2

	Designation	Orbiter	Launch			Landing		
	1997-39B	CRISTA-2	1835	EDT	7 Aug	1113	EDT	16 Aug

STS-86

#	Designation	Orbiter	Launch			Landing			Runway	Days	Incl
87	1997-55A	Atlantis (20)	2234	EDT	25 Sep	1755	EDT	6 Oct	KSC (40)	10.80	51

CDR: Jim Wetherbee (4)
PLT: Mike Bloomfield (1)
MS1: Vladimir Titov (2,EV,Russia)
MS2: Scott Parazynski (2,EV)
MS3: Jean-Loup Chrétien (1,France)
MS4: Wendy Lawrence (2)
UP: David Wolf (2)
DN: Mike Foale (4)
Payload: Double Spacehab, ODS
Activity: SMM-7 docked with Mir at 1538 EDT 27 Sep and undocked at 1328 EST 3 Oct
Spacewalks: Titov and Parazynski

STS-87

#	Designation	Orbiter	Launch			Landing			Runway	Days	Incl
88	1997-73A	Columbia (24)	1446	EST	19 Nov	0720	EST	5 Dec	KSC (41)	15.66	28

CDR: Kevin Kregel (3)
PLT: Steven Lindsey (1)
MS1: Winston Scott (2)
MS2: Kalpana Chawla (1)
MS3: Takao Doi (1,Japan)
PS: Leonid Kadenyuk (1,Ukraine)
Payload: USMP-4, EDO-12
Satellite: SPARTAN-201-4, Sprint

	Designation	Orbiter	Launch			Landing			Runway
	1997-73B	SPARTAN-201-4	–	EST	21	Nov	2109EST	24	Nov
	1997-73C	Sprint	–	–	3 Dec	–	–	3 Dec	

Spacewalks: Scott and Doi

#	Designation	Orbiter	Launch			Landing			Runway	Days	Incl

STS-89

89	1998-3A	Endeavour (12)	2148	EST	22 Jan	1736	EST	31 Jan	KSC (42)	8.82	51

CDR: Terry Wilcutt (3)
PLT: Joe Edwards (1)
MS1: Bonnie Dunbar (5,PC)
MS2: Mike Anderson (1)
MS3: Jim Reilly (1)
MS4: Salizhan Shapirov (1,Russia)
UP: Andrew Thomas (2)
DN: David Wolf (2)
Payload: Double Spacelab, ODS
Activity SMM-8 docked with Mir at 1510 EST 24 Jan and undocked at 1155 EST on 29 Jan

STS-90

–	1998-	Columbia (25)	–	–	– Apr	–	–	– –	–	–	28

CDR: Rick Searfoss (3)
PLT: Scott Altman (1)
MS-: Richard Linnehan (2)
MS-: Dave Rhys Williams (1,Canada)
MS-: Kathryn Hire (1)
PS: Jay Buckey
PS: James Pawelczyk (1)
Payload: Neurolab, EDO-13

STS-91

–	1998-	Discovery (24)	–	–	– May	–	–	– –	–	–	51

CDR: Charles Precourt (4)
PLT: Dominic Gorie (1)
MS-: Wendy Lawrence (3)
MS-: Franklin Chang-Diaz (6)
MS-: Janet Kavandi (1)
MS-: Valeri Ryumin (1,Russia)
DN: Andrew Thomas (2)
Payload: Single Spacehab, ODS
Activity: SMM-9; final Shuttle–Mir mission

STS-88

–	1998-	Endeavour (13)	–	–	– Jul	–	–	– –	–	–	51

CDR: Bob Cabana (4)
PLT: Fred Sturckow (1)
MS-: Nancy Sherlock/Currie (3)
MS-: Jerry Ross (6,EV)
MS-: James Newman (3,EV)
Payload: ODS, Node with two PMAs
Activity: SSAF 01-2 is to attach the Node to the FGB to start assembly of the International Space Station
Spacewalks: Ross and Newman

Glossary

AADSF (Advanced ADSF) See entry for ADSF later in Glossary.

AAEU (Aquatic Animal Experiment Unit) An aquarium for newts and small fish such as the Japanese medaka, to study spawning, fertilisation, embryology and vestibular reactions in microgravity.

abort The abandonment of a mission due to a malfunction sufficiently serious to require an emergency procedure, leading towards a landing. In terms of a shuttle, there are abort options for the various phases of the ascent. In order of availability, they are the RTLS, the TAL, the ATO and then the AOA. It should be noted that only the first two result in an emergency landing *instead of* reaching orbit (in the case of RTLS, at the launch site, and in the case of TAL at an appropriate landing site on the far side of the Atlantic), because an ATO is an 'abort' to a low orbit that can then either be improved by an OMS burn or cancelled by an OMS burn to descend after a single orbit (the 'once around' abort). The ATO option has been invoked only once, on STS-19/51F. (See RTLS for details of that procedure.)

ACCESS (Assembly Concept for Construction of Erectable Space Structures) A 15-metre tall triangular-section truss that was erected in the payload bay to test the ability of spacewalkers to assemble a framework structure using tubular rods and tongue-in-groove nodes with sliding-sleeve locks.

ACIP (Aerodynamic Coefficient Identification Package) Instruments carried on Columbia during its test flights to record the vehicle's structural stresses at different phases in the flight, for comparison with computer predictions.

ACRIM (Active Cavity Radiometer Irradiance Monitor) An instrument to measure total output across the solar spectrum from the ultraviolet to the infrared, in order to determine the energy input to the Earth's radiation budget. After testing on Spacelab 1, it was installed on the UARS, then flown on the ATLAS missions for calibration checks. Its data will be used for climate studies. The Sun's output will have to be monitored throughout a solar cycle before it will be feasible to distinguish long-term trends from cyclic fluctuations, to reveal the extent to which climate change on the Earth can be attributed to solar variability.

ACRV (Assured Crew-Return Vehicle) The means of escape from the International Space Station in an emergency, when a shuttle is not docked. Initially, this role will be served by the Soyuz ferry, but several longer-term alternatives are under active consideration.

ACTS (Advanced Communications Technology Satellite) A NASA-built communications satellite that was developed to demonstrate the advantages of using the K_a-Band. It was awarded funding by Congress specifically to prompt the US service providers to follow up by exploiting this very-high-capacity technology before the market could be captured by foreign operators.

ADSEP (ADvanced SEparation Process) An apparatus to separate and purify biological pharmaceuticals, such as the recombinant haemoglobin on which it was tested.

ADSF (Automated Directional Solidification Furnace) An apparatus to study the process of directional solidification in a multi-zone furnace. It was used to make semiconductors, particularly cadmium telluride, and to make high-performance magnetic composites.

AEM (Animal Enclosure Module) A cage that provided thermal regulation, food and water for a small research animal, such as a rodent. Although it was meant to be leak-proof, in its initial form it tended to leak particulate debris, and it failed to contain the aroma of its occupant. Once it had been redesigned, it functioned satisfactorily.

AEPI (Atmospheric Emissions Photometric Imager) An instrument to study aurorae and other air glow phenomena.

AERIS (American Echocardiograph Research Imaging System) An instrument which was used to measure blood volume during LBNP sessions.

AFE (American Flight Echocardiograph) An off-the-shelf commercially available ultrasonic transducer which recorded the heart's action in standard video format. It was dubbed 'American' in order to distinguish it from the French model that was sometimes carried. It was generally used in conjunction with the LBNP.

AGHF (Advanced Gradient Heating Furnace) A directional-solidification metal-processing furnace capable of operating at high temperatures.

air glow A thermospheric emission phenomena caused by interactions between the various species of atoms and molecules which are present in that tenuous but hot outer layer of the atmosphere. The most significant features are the red and green atomic oxygen lines (at 6300 Å and 5577 Å respectively), but ionised oxygen (7329 Å), molecular oxygen (7640 Å) and hydroxyl (7370 Å) have also been observed. Although air glow is readily seen by astronauts as a thin band of light just above the horizon (actually at an altitude of 100 km) during orbital darkness, it can sometimes be seen, by observers on the surface, as a glow all around the horizon.

AKM (Apogee Kick-Motor) A rocket motor, usually a small liquid rocket, which is fired when a satellite reaches the first apogee of a GTO in order to circularise the orbit at a level just below that of a 24-hour orbit, so that it can slowly drift to its assigned operating station, at which time the motor will be fired again to ease the satellite into GSO, so that it will remain stationary above the equator at that longitude. The motor is usually part of the satellite itself, rather than a bolt-on motor, and its services will be called upon from time to time to ensure that the satellite is not pulled too far off station by the perturbations caused by the passage of the Moon.

ALAE (Atmospheric Lyman-Alpha Emission) An instrument to measure deuterium (heavy hydrogen) in the upper atmosphere by measuring emission at the primary feature of hydrogen's spectral signature, the α line in the Lyman series, at 1215 Å, in the ultraviolet.

ALFE (Advanced Liquid Feed Experiment) A TV camera recorded the motion of a fluid as it passed through a series of pumps, tanks and valves in microgravity.

ALT (Approach and Landing Test) The first phase of the orbiter's flight test programme, for the subsonic phase of the descent.

AMEE (Advanced Materials Exposure Experiment) A sophisticated exposure cassette with the Optical-Properties Monitor that could rotate its sample tray to enable a video camera and a spectrometer to examine each sample and provide real time data. It was mounted outside Mir.

AMOS (Air Force Maui Optical Site) The Air Force used an electro-optical system based at Mount Haleakala, Maui, Hawaii, on a regular basis to acquire sample imagery as shuttles passed overhead, and the orbiter made RCS firings and dumped excess water to enable AMOS to calibrate its sensors.

Anthrorack An apparatus to study human physiology in weightlessness. Built by ESA, it took the form of a double rack on Spacelab. It represented a significant step forward by simultaneously measuring cardiovascular, cardiopulmonary, metabolic and endocrinic activities for coordinated whole-body studies. A bicycle ergometer was included so that data could be collected while the body was in a state of exertion.

AOA (Abort Once Around) See *abort*.

APA (Anticipatory Postural Activity) An experiment to study posture whilst undertaking various activities in weightlessness.

APCF (Advanced Protein Crystallisation Facility) A versatile ESA-built furnace which was capable of running three crystallisation processes (liquid–liquid diffusion, dialysis, and vapour diffusion) and providing continuous photodocumentation in order to study how nucleation initiated and progressed.

APCG (Advanced *PCG*)

APDS (Androgynous Peripheral Docking System) The docking system on the Kristall port of the Mir space station. A compatible mechanism was supplied to NASA for mounting on the ODS, so that the shuttle could dock with the station.

APE (Auroral Photography Experiment) An Air Force-sponsored project involving the use of a 35-mm camera fitted with an image intensifier to record auroral displays and various other 'glow' phenomena.

APM (Ascent Particle Monitor) This studied the particulate matter that polluted the payload bay during the ascent to orbit. Although the bay doors were hermetically sealed to prevent contamination from the atmosphere, it was thought that the vibration might shake loose particulates from the structure, and there was some concern that this might contaminate sensitive payloads.

Arabidopsis A wallcress plant, commonly regarded as a weed. It is widely used in genetic experiments due to its rapid growth (barely 40 days per generation) and comparatively simple genome (only about 20 genes). It was the first plant to be successfully taken from seed to seed in space.

ARC (Aggregation of Red blood Cells) An experiment to test whether microgravity would offer any benefits in existing clinical research and testing for medical diagnosis.

It gave data on the formation rate, structure and organisation rate of red blood cells. Blood cells had been provided by donors suffering a variety of medical conditions, including heart disease, hypertension, diabetes and cancer.

ARF (Aquatic Research Facility) An apparatus to facilitate detailed study of the adaptation to microgravity, including fertilisation and embryo formation, of small aquatic species such as sea urchins.

Ariane A family of rockets with differing capabilities developed by the French corporation Arianespace to launch satellites as a business.

ARIS (Active Rack Isolation System) An apparatus which employed an active mechanical system to isolate an experiment from ambient vibrations. It was one of several isolation mounts being considered for use on the International Space Station. Although evaluated on a Shuttle–Mir flight, it was tested within the Spacehab.

ASE (Airborne Support Equipment) The general term for the various tilt-table cradles used to carry satellites in the shuttle's payload bay and to deploy them in space.

ASEM (Assembly of Station by EVA Methods) A time-and-motion study of techniques for assembling a truss structure in space, performed on STS-49

ASP (Attitude Sensor Package) A Hitchhiker package mounted on the side wall of the bay to evaluate the performance of new attitude sensors for future spacecraft.

ASTP (Apollo-Soyuz Test Programme) The joint effort that led to the docking between an Apollo and a Soyuz spacecraft on 17 July 1975.

ASTRO A series of Spacelab flights that focused on ultraviolet astronomy for the simple reason that because most ultraviolet light does not penetrate the atmosphere there was scope for a suitably instrumented small telescope to make significant discoveries. The HUT, WUPPE and UIT were mounted on the IPS. Originally, a series of flights was planned, but when it became clear that the shuttle could not fly as frequently as predicted, the project was cut back to three, and then to just one, but the results of the first flight prompted a reflight. No more flights are planned, but it is possible that these telescopes might eventually be mounted on the International Space Station.

Astroculture A hydroponics plant cultivator apparatus. A heavily instrumented version was flown on the shuttle to verify that it could regulate the supply nutrient solution, because this was technically demanding. It was developed for use on the International Space Station. During its final test (on STS-73) it showed that potatoes could be grown in space by nourishing tubers, the swollen ends on sub-surface stems containing buds.

Atlantis This orbiter was named after the ketch which undertook extensive oceanographic research for the Woods Hole Oceanographic Institute in the period 1930–1966.

ATLAS (ATmospheric Laboratory for Applications and Science) A pallet-based Spacelab for the study of the Earth's atmosphere and the way that it is influenced by the Sun. Its data complemented the ongoing monitoring by UARS. This exceptionally detailed database on the physical and chemical state of the atmosphere will serve the MTPE programme and, by sampling throughout a solar cycle, set the scene for the EOS that which will fly in the next century. The 'core' instruments were ATMOS, MAS, ACRIM, SOLCON, SOLSPEC, SUSIM and SSBUV.

Atlas A rocket originally developed as the first US intercontinental-range ballistic missile, but soon converted for use as a satellite launcher. In considerably upgraded form,

Atlas is still in use launching medium-to-heavy satellites. In its most powerful form, it uses a Centaur upper stage.

ATMOS (Atmospheric Trace MOolecule Spectroscope) An instrument to measure the solar spectrum at orbital sunrise and sunset – when the Sun was on the limb and its light was passing horizontally through the Earth's atmosphere – so that the trace constituents could be identified from their characteristic infrared absorption, their vertical distribution measured and their geographical concentrations mapped.

ATO (Abort To Orbit) See *abort*.

ballistic capsule A low lift-to-drag (typically 0.25) configuration which, for a small mass, offers a more effective solution than a spaceplane based on a lifting body to the problem of surviving the frictional heat associated with re-entering the atmosphere. This is due to the fact that a small conical capsule that re-enters blunt-end first needs only its base to be protected. Such spacecraft were not meant to be reusable, so an ablative shield could be used. The conical shape maximised the utility of internal volume, but it did not permit bulky cargo to be carried.

BATSE (Burst And Transient Source Experiment) A set of eight scintillation detectors, one on each corner of the rectangular platform of GRO. Between them, they were able to monitor the whole sky, yet, by coordinating their output, immediately pinpoint a specific source to within about 2 degrees. Although its detector was sensitive up to 100 MeV, it maintained a watch on the 20 keV to 2 MeV range and had sufficient time-resolution to trace a light curve for a transient. Some bursts had been seen to last for only a few milliseconds, most for a few seconds, and a few for several minutes. The fast ones formed a single sharp pulse, but the sustained ones, after the initial flash, tended to degrade into a gradually diminishing flicker. BATSE was to measure these light curves in unprecedented detail. It 'focused' on the lower end of the energy range for the simple fact that the rate of events decreased with increased energy.

BBXRT (Broad-Band X-Ray Telescope) This grazing-incidence reflecting X-ray telescope was developed for SHEAL (which was cancelled), but added to ASTRO-1 specifically to study SN1987A. It sat on a separate Spacelab pallet, and incorporated its own a two-axis pointing system. It focused the X-rays onto a solid-state spectrometer which was an improved form of that built for the Einstein Observatory (HEAO-2). This cryogenic detector was sensitive in the energy range 0.3 to 12 keV. BBXRT provided better resolution across a wider spectral range than any previous instrument, and it was the first to yield images at energies above 4 keV. It was later adapted for installation on an automated satellite.

BCAT (Binary-Colloid Alloy Test) An apparatus sent to Mir to study the crystallisation of alloyed colloids.

BDPU (Bubble, Drop and Particle Unit) A generic apparatus with a variety of instruments to study the physics of bubbles, drops and particles in order to improve understanding of fluid processes, particularly how evaporation and condensation affect bubble formation and how bubbles affect solidification. It broke down on STS-78, and had to be repaired in flight. An improved version of this ESA facility is to be installed on the International Space Station.

BDS (Bioreactor Demonstration System) An apparatus in which to develop individual cells into organised tissue. In microgravity, it is possible to grow three-dimensional cultures, a capability which had greatly enhanced the potential for growing transplantable tissue. It was flown on the shuttle primarily for engineering verification, and is to be used on the International Space Station.

BETSCE (Brilliant Eyes Ten K Sorption Cryocooler Experiment) An apparatus to evaluate a system for cooling sensors to 10 K by a reversible hydrogen sorption reaction. If this proved to be viable, it held out the prospect of a revolution in the way that instruments are built to work at near-zero temperatures, by offering a *regenerative* alternative to the current generation of use-once cryogenic dewars, thereby decoupling the service life of an infrared telescope from the amount of coolant it can accommodate. The motivation for this development though, was the SDI's Brilliant Eyes missile-tracking programme, so this astronomical benefit is only a spin-off.

BFPT (Bioreactor Flow and Particle Trajectory) A fluid-dynamics experiment to validate predictions for the action of cell cultures in the STLV.

BIMDA (Bioserve ITA MDA) BioServe was ITA's first customer, and the result was an integrated four-MDA middeck package (dubbed BIMDA). It performed a wide range of experiments involving growing both organic and inorganic crystals, germinating seeds and fixing live cells. A few tests were aborted by mechanical failures on each occasion, but most were completed successfully.

biological research Spaceflight provides an opportunity to study fundamental processes of biology, and the basic functions of living organisms, in the absence of gravity. Life's processes are strongly conditioned by gravity; in effect, life on this planet has adapted to its gravitational field. This applies from basic cellular processes all the way up to fully developed plants and animals. Specifically, in low gravity, the rate of biological processes is slowed. Not only can this fact be exploited to study the mechanisms which drive such processes, similarities of the effects with certain terrestrial bodily diseases is leading to a deeper understanding of the cause of diseases such as osteoporosis. It was only by observing the manner in which the balance between production and destruction of cells was disturbed during spaceflight that the significance of the equilibrium that is in effect on Earth was fully appreciated. Studies of fish produced insights into how the vestibular organs provide a sense of balance. Studies of tissue cultures are revealing how cancers grow. Studies of protein crystals are showing how the body's basic processes operate. A basic technique of experimental science is to study the effect of varying each parameter influencing a system in isolation. Gravity is one factor we have thus far not been able to vary. Our previous knowledge base was derived from data obtained within the Earth's gravitational field, so it should come as no surprise that almost every aspect of biological research in space is producing interesting data. A deeper understanding of biology should facilitate improvements in the pharmaceutical and agricultural industries.

bioreactor A container used to grow (or to 'culture') a sample of biological tissue.

biotechnology Techniques for rearranging and manufacturing biological molecules, tissues and living organisms.

Biotechnology Facility The double ISPR rack in NASA's laboratory on the International Space Station which will contain thermostats for growing organic materials, such as protein crystals and cell and tissue cultures.

BLAST (Battlefield Laser Acquisition Sensor Test) A test of a sensor designed to be used to detect a laser beam from a pilot shot down over hostile territory. Such a rescue aid would transmit the GPS-derived location to the satellite, which would relay the information to a Search and Rescue team; unlike a radio beacon, the laser could not be intercepted by the enemy.

BPA (Back Pain in Astronauts) An SPE study of the back pain which many astronauts had reported. This was thought to be due to the spine's relaxation once unloaded of gravity, as the vertebrae absorb fluid, expand and move apart. For this experiment, an astronaut had a set of markers precisely positioned on the shoulder blades, along the spine and on the hips, and was photographed by a stereoscopic camera while in specific positions, for subsequent analysis.

BPL (Bioserve Pilot Laboratory) A fluid-mixing apparatus in which to conduct biological, pharmaceutical and agrichemical experiments.

BR (BioRack) A generic biological facility built by ESA. It could accommodate a wide variety of experiments and be reconfigured for each mission.

BREMSAT (BREMen SATellite) A small cube-shaped satellite built by the University of Bremen, Germany, to study the near-Earth environment, particularly atomic oxygen concentration and the flux of interplanetary dust. It was ejected from a GAS can.

BRIC (Biological Research In a Canister) A study of the biological effects of microgravity, using a wide variety of small samples.

BSK (BioStacK) An ESA experiment to investigate the effects of microgravity and cosmic radiation on life. The biological samples were sandwiched between radiation monitors which enabled the track of a cosmic ray particle to be traced, and the biological material that it struck identified so that the study could be focused. Bacteria, cress seeds, shrimp eggs and fungal spores were carried.

BTS (BioTechnology System) An apparatus incorporating a rotating vessel for the study of biological materials over extended periods. It was to perform protein crystal and cell culture experiments.

CANEX The name for the mixed programme of experiments conducted aboard the shuttle by a succession of Canadian astronauts. It included CTA, MELEO, QUELD, SPEAM, OGLOW, SATO and SVS.

CAPL (CApillary Pumped Loop) A test of the ammonia-flow cooling system developed for the EOS satellites.

CCAFS (Cape Canaveral Air Force Station) The launch facility immediately to the south of NASA's Kennedy Space Center. It included LC 41, from which Titan 3 rockets were launched, the Satellite Assembly Building and the Shuttle Payload Integration Facility, this latter being located within the Solid Motor Assembly Building in which the Titan 3 rockets were assembled. All of these facilities were secure, so could be used to prepare classified payloads in advance of driving them up the coast for loading into the payload bay of a shuttle orbiter already on the pad and checked out. It had proved impossible to maintain security using NASA's payload processing facility.

CCD camera An electronic camera exploiting Charge-Coupled Device technology to produce an array of solid-state pixels.

CCDS (Center for the Commercial Development of Space) To promote commercialisation, NASA teamed up with private industry to co-sponsor a number of Centers, often on a university campus, each focused on a given technology. These included BioServe at the University of Colorado at Boulder; the Center for Macromolecular Crystallography at the University of Alabama at Birmingham; the Consortium for Materials Development in Space, at the University of Alabama at Huntsville; and the Center for Space Vacuum Epitaxy at the University of Houston.

CCM (Cell Culture Module) An apparatus to study the effects of microgravity on muscle, bone and various other cells.

Centaur A very high performance upper stage burning cryogenic hydrogen and oxygen. It is used on Atlas and Titan rockets. A version was developed for carriage in the shuttle's payload bay, but this was scrapped for safety reasons before it had been used. It would have been able to boost a 4,500 kg satellite to GSO. NASA's primary role for it was in dispatching interplanetary spacecraft.

CETA (Crew and Equipment Translation Aids) A system to enable astronauts to transport themselves and equipment along the length of (and within) the truss of the International Space Station. Four techniques were tested on STS-37, one involving simply moving along a tether in a hand-over-hand fashion, and three with different types of hand-operated cart. For an astronaut unencumbered with equipment, the tether was perfectly adequate, but it was difficult to haul a bulky package with just one hand free, so some sort of cart was clearly essential. All the carts involved the astronaut riding a foot restraint. On the manual cart, the astronaut lay prone and was pulled along the track just as with the tether. On the mechanical cart the astronaut operated an up-and-down lever in the same manner as a railroad buggy. On the electrical cart a rotary crank ran a dynamo that drove a motor which moved the cart. Of the carts, the tests favoured the manual, then the electrical and then the mechanical systems; mechanical advantage was not necessary in microgravity. In the production version, equipment will be able to be carried on the rear of the cart. Such a system will be necessary on the station because, with EVA time limited to about six hours, it will save time in transporting ORUs to and from the worksites at the far end of the truss, thereby making more time for on-site work.

CFCs A family of chlorine compounds (ChloroFluoroCarbons) artificially manufactured for use in cooling systems and aerosol sprays. Once released into the air, these compounds migrate into the stratosphere, where they are the primary cause of ozone depletion.

CFE (Candle Flame Experiment) An apparatus to investigate the process of combustion by observing the flame of a burning candle. For safety, these tests were done in the GBX.

CFES (Continuous Flow Electrophoresis System) A major programme by McDonnell Douglas to develop a system to separate biological materials in solution by using electrophoresis.

CFLSE (Critical Fluid Light Scattering Experiment) Also called Zeno, this apparatus used a laser to study xenon at its critical point. In the absence of gravitational convention, the temperature could be precisely controlled, and the transition observed on a macroscopic scale, across the entire sample.

CFZF (Commercial Floating-Zone Furnace) An apparatus to make semiconductor crystals.

CGBA (Commercial Generic Bioprocessing Apparatus) An apparatus in which to conduct fluid-mixing biological experiments. It could accommodate a large number of chambers to facilitate batch processing for statistically valid testing.

CGF (Crystal Growth Furnace) A furnace for producing metals, alloys, and monocrystals of semiconductors at temperatures up to 1,600 °C, by using the directional solidification technique. It was heavily instrumented in order to make detailed studies of the process of crystallisation. It was a boiler-shaped chamber that occupied an entire Spacelab rack. A similar furnace will be installed on the International Space Station.

Challenger This orbiter started out as a structural test article. It was upgraded for flight. It was named after the ship on which Charles Thomson explored the Atlantic and Pacific Oceans in the 1870s, and for Apollo 17's lunar module. To date, it is the only orbiter to be lost on a mission. This loss, however, was in no way attributable to a flaw in the orbiter's design.

Challenger accident The loss of the orbiter Challenger, launched on 28 January 1986.

CHAMP (Comet Halley Active Monitoring Programme) The use of a low-light camera on STS-24 to photograph Halley's Comet. Unfortunately, the battery was found to be flat, so only standard photography was possible.

Charlotte An autonomous robot mounted on wires. It was tested on a Spacehab by stringing the wires in front of a bank of experiments, and it tended the apparatus by positioning itself so that it could use its arm to flick switches. It was a prototype for a system under consideration for the International Space Station.

CHF (Continuous Heating Furnace) A materials furnace tested on Spacelab-J1 preparatory to installation in the Japanese laboratory of the International Space Station.

CHROMEX A study of plant growth. It involved determining how the genetic material in the cells responsible for root-growth in flowering plants responded to the absence of gravity. It used the PGU cultivator.

Cinema-360 A film camera with an ultra-wide-angle lens which was mounted in the bay to record the spacewalk to capture SolarMax. The '360' was a reference to the fact that the film was to be projected onto the a planetarium dome to generate an all-round 360-degree field of view.

CIRRIS (Cryogenic InfraRed Radiance Instrument in Space) This cryogenically-cooled telescopic infrared sensor was flown on STS-4 for the Air Force, to collect signature data on cruise missiles within the atmosphere, but was frustrated by a stuck cover plate. On STS-39 it made high resolution spectral measurements of aurorae and air glow phenomena on the Earth's limb to help develop sensors designed to detect and track ballistic missiles against this irradiant background. In this case, it was flown for the SDIO. The detector was cooled by liquid helium for extreme sensitivity in the 2 to 25 micron range. The telescope employed the same type of two-axis mount as was used for the BBXRT instrument. As events transpired, on STS-39 CIRRIS was presented with a spectacular southern auroral display. The objective was to identify the 'coldest' sections of the infrared spectrum, in which there would be the least glare to distract a missile-tracking sensor.

CIV (Critical Ionisation Velocity) This payload bay experiment used four canisters, each with a gas (xenon, neon, carbon dioxide and nitric oxide). These gases were released in turn so that they could be observed by the IBSS sensors. It was an SDIO experiment to verify that because the gas would share the orbiter's kinetic energy, it would exceed its critical ionisation velocity as it moved within the Earth's magnetic field, even though it had not been vented at high speed, and would therefore be induced to radiate at its own characteristic wavelength. These gases were selected because their ionisation potentials covered a wide range, so some radiated and others did not. This significance of this test was that sensors could be developed to track combustion products leaking from a spent rocket in space even after it had ceased thrusting.

CLAES (Cryogenic Limb Array Etalon Spectrometer) An infrared emission spectrometer on the UARS. It mapped the geographical and vertical distribution of ozone and the chemicals in the upper atmosphere that deplete it, namely water vapour and various compounds of chlorine and nitrogen.

CLIP (Crew Loads Instrumented Pallet) A workstation mounted in the payload bay to test the structural loads imparted to the structure by an astronaut standing on a foot restraint and performing a variety of typical maintenance tasks. This empirical data was to be fed into the development of the workstations for the International Space Station.

CLOUDS (Cloud Logic to Optimise Use of Defence Systems) High-resolution pictures of clouds, particularly those associated with severe weather and wispy cirrus, were recorded over a wide range of viewing angles in order to quantify *apparent* cloud cover as a function of the angle at which clouds of different types are viewed, so as to calibrate the sensors used by the Defence Meteorological Satellite Programme (DMSP).

CMIX (Commercial MDA ITA Experiments) An apparatus to mix batches of fluid samples for a wide variety of biotechnology experiments.

COIS (Canal and Otolith Integration Studies) An apparatus to make a detailed investigation of the effect of microgravity on the neurovestibular system by wearing a pair of goggles to stimulate the eye, and sensors to measure the involuntary action of the eyes and head, both while at rest and while exercising on the bicycle ergometer. It built upon previous studies, but this time focused on the coupling between the various sensory systems.

Columbia This orbiter, which made the historic first shuttle test flight, was named for the US Navy frigate which had circumnavigated the globe 150 years earlier, and for Apollo 11's command module.

COMPTEL (COMpton TELescope) The GRO's primary imaging system. It used two planar arrays of scintillation crystals mounted 1.5 metres apart to achieve an angular resolution of 2 degrees, which was exceptionally good for a gamma ray telescope. It was sensitive in the 1–30 MeV range, and could distinguish gamma rays from cosmic rays by the manner in which they affected the detector. Since it had a wide field of view, its primary function was to produce a sky survey of unprecedented clarity and detail.

combustion research Although fire was one of the first technologies mastered by primitive humans, like many other processes with which we are familiar, combustion is strongly influenced by the presence of gravity. In space it is possible to study those aspects of burning that are masked by gravity-induced fluid phenomena. This research is yielding new insight into a process that is fundamental to our technological society. As 85 per cent of the world's energy production involves combustion, any breakthrough will yield a very significant return. A one per cent increase in burner efficiency would result in a US saving of $4 billion per year. Combustion can also be used as a synthesis process, for fullerene and fibrous carbon, and the yield of this process is significantly higher in microgravity. As a result of research conducted on the shuttle, the Lewis Research Center has applied for a patent for a device that improves air quality by stabilising fuel-lean flames as well as by reducing nitrogen oxides by-products. Another significant result was the verification of a theory of gas jet diffusion proposed in 1928 by Burke and Schumann. Some 10 per cent of the research papers presented at the 25th International Symposium on Combustion dealt with microgravity research.

CONCAP (CONsortium for materials development in space, Complex Autonomous Payload) An ongoing study of the interaction of materials to the space environment.

COSTAR (Corrective Optics Space Telescope Axial Replacement) An optical system built specifically to correct the spherical aberration in the HST's primary mirror on behalf of the FOC, GHRS and FOS axial instruments (the WF/PC-2 had its own corrective optics). It was installed on the first servicing mission, in December 1993, in place of the HSP (the fourth axial instrument) which was sacrificed.

CPCG (Commercial *PCG*)

CPF (Critical Point Facility) An apparatus to study materials at their critical points, as they change from one phase state to another.

CREAM (Cosmic Radiation Effects and Activation Monitor) A set of both active and passive monitors in the orbiter's cabin (the same sampling points as used for SAM) to measure neutron fluxes, induced radioactivity and the energy spectrum of cosmic rays within the vehicle. The data was analysed by the Department of Defense.

CRISTA (CRyogenic Infrared Spectrometer and Telescope for the Atmosphere) A free-flying SPAS platform which carried, amongst other instruments, the MAHRSI.

critical point The conditions of temperature, pressure and density at which a substance simultaneously exists in its liquid and gaseous states. At the critical point, a fluid fluctuates back and forth in small volumes from the liquid to the gaseous phase.

CRO (Chemical Release Observation) An experiment to collect infrared, visible light and ultraviolet time-resolved signatures associated with the release of liquid propellant from rockets in space. Three CRO subsatellites were ejected from the shuttle and, when safely away, in turn released their contents (small amounts of rocket fuel and oxidiser) so that the IBSS sensors, the CIRRIS instrument on the orbiter, and surface and airborne sensors could record the effect. The CRO satellites were actually 'lightsats' built by Defense Systems Incorporated.

CRYOHP (CRYOgenic HeatPipe) A test of a heatpipe that used cryogenic oxygen as its heat flow agent.

Cryostat A German protein crystal growth experiment on the International Microgravity Laboratory.

CRYOSYS A test of a cryogenic system for cooling focal-plane instruments on spacecraft.

CSA (Cooperative Solar Array) A solar panel flown up to Mir by STS-74 as cargo on the DM, and subsequently mounted on Kvant 1 by the space station's resident crew during a spacewalk. It was 'cooperative' in that it combined the proven Russian deployment mechanism with the gallium arsenide solar transducer technology developed by NASA. The 6 kW which it delivered not only significantly ameliorated the longstanding power crisis on the station; it allowed NASA to monitor the degree of degradation, because this type of transducer is to be used on the International Space Station.

CTA (Canadian Target Assembly) A small satellite specifically marked up to test the SVS.

CTPE (Cryogenic Two-Phase Experiment) A test of a liquid nitrogen-flow heat pipe and a thermal storage system based on the energy involved in a solid–liquid phase transition.

CVTE (Chemical Vapour Transport Experiment) A furnace to heat an ampoule of cadmium telluride to 850 °C to evaporate and dissociate the compound into gaseous cadmium and gaseous tellurium. A temperature gradient caused the material to vaporise at one end of the tube, be transported to the other end, then recrystallise. The objective was to study the process, not just exploit it to make crystal, and the apparatus had a window so that an astronaut could observe the crystallisation process and adjust the set-up to test different operating conditions.

DATE (Dynamic, Acoustic and Thermal Environment) A suite of sensors installed in the bay on the shuttle's test flights to characterise the environment in which instruments would later be required to operate.

DCAM (Diffusion Crystallisation Apparatus for Microgravity) An apparatus which used a semi-permeable membrane to grow protein crystal over a long period, with the growth being filmed.

DDU (Data Display Unit) The basis of the flight deck interface for operating instruments in the payload bay.

DEE (Dexterous End-Effector) A test of an advanced teleoperator arm with an end-effector which used an electromagnetic grapple, and incorporated torque sensors to provide its operator with a sense of tactile feedback that was absent in the RMS. It was developed by Spar Aerospace for use on the International Space Station. This dexterous arm will allow astronauts to perform routine maintenance on external apparatus without the need to undertake a spacewalk. It was carried in the payload bay, hoisted by the standard RMS, tested by inserting variously sized pins into sockets – with this activity being videotaped for detailed analysis – and then restowed.

Delta A rocket that began life as the Thor intermediate-range ballistic missile, and was then upgraded to launch satellites. The last form of the Thor was the 'Delta' variant, and this designator became so universal that the original name fell into disuse. A family of Delta rockets was developed to launch small-to-medium satellites. Its commercialised form is the Delta 2.

DFI (Development Flight Instrumentation) A suite of instruments which were carried on a pallet in Columbia's payload bay during its first few test flights, to record temperatures, pressures and accelerations at various points on the orbiter.

dialysis A process in which the material to be crystallised and its fixing agent are separated by a membrane through which the rate of flow (and hence the rate of crystallisation) can be controlled.

direct ascent On most shuttle missions, the initial orbit is defined by its energy state at the moment of SSME shutdown, and a variety of OMS burns are subsequently performed to manoeuvre into the desired operating orbit. A direct ascent involves augmenting the SSMEs with the OMS engines in order to set the initial apogee at the altitude of the final orbit, at which point a circularisation burn is made. This ascent profile has the virtue of saving time during a rendezvous.

Discovery This orbiter was named after the ship on which, in 1610, Henry Hudson, while searching for the Northwest Passage, discovered the bay that bears his name; and for the ship in which Captain James Cook discovered the Hawaiian Islands in 1779.

DM (Docking Module) A Russian-built 5-tonne module with an APDS port on either end which was flown up by Atlantis and docked at the end of the Kristall module in order to serve as an airlock between a shuttle orbiter and the Mir space station.

DMOS (Diffuse Mixing of Organic Solutions) An apparatus to crystallise organic material. Each of its six stainless-steel reactors held three teflon-coated chambers which remained isolated until electrically-controlled gates released to enable their contents to mix. These gates were opened very slowly (over a period of five hours) so that their action would not to disturb the contents.

DOD The US Department of Defense; the civilian interface to the unified commands of the Army, Navy, Air Force and Marine Corps.

DOD82-1 The Department of Defense's classified payload on STS-4 which contained the CIRRIS and UHS instruments.

DPM (Drop Physics Module) An ultrasonic 'containerless' crucible which incorporated a set of cameras to record the results of sensitive materials-processing experiments. On its first flight, its power supply had to be rewired, but because its designer (Taylor Wang) had accompanied it for the test he was able to carry out the repair immediately. It was subsequently used to study the injection of one drop into another and the behaviour of a multiple-drop system.

DSCS (Defense Satellite Communications System) A geostationary relay system operated by the Department of Defense to integrate its command centres around the world.

DSE (Data Systems Experiment) An assessment of how erasable optical-disk technology withstood space radiation.

DSP (Defense Support Programme) A network of satellites in GSO to provide continuous overwatch for ballistic missile launches. In effect, this was the second generation of the IMEWS system. Most were launched by Titan 3 rockets, but one was launched on the shuttle (STS-44). A 4-metre infrared telescope equipped with a focal-plane array of 6,000 sensors was skewed from the satellite's Earth-pointing axis so that, as it spun at 6 rpm, the field of view would perform a circular scan of the entire visible hemisphere for any sign of the thermal plume of a missile exhaust trail.

DVOS (Direct-View Optical System) A small telescope which could be mounted on a flight deck window and could automatically track a target on the ground as the orbiter passed overhead.

DXS (Diffuse X-ray Spectrometer) An instrument to observe the shell of shock-heated gas which swept up the interstellar medium in the immediate neighbourhood of the Sun and the nearest stars to create a void termed the 'Local Bubble'.

EASE (Experimental Assembly of Structures in EVA) A 2-metre triangular truss-section which was erected in the bay to test the ability of spacewalkers to assemble framework structure.

ECLIPSE (Equipment for Controlled LIquid-Phase Sintering Experiments) An apparatus to assess the properties of materials believed to be likely to have strong, lightweight and yet extremely durable metals.

ecliptic The plane in which the Earth orbits the Sun. With the exception of distant Pluto, all of the planets orbit in almost this plane. The fact that the Sun's equator is not coincident with the orbital plane of the planets complicates the way in which the solar wind blows through the planetary system.

ECT (Emulsion Chamber Technology) A stack of photographic emulsions which recorded the passage of high-energy particulate radiation in order to further characterise shielding requirements for the International Space Station.

EDO (Extended Duration Orbiter) A suite of apparatus to extend the time that an orbiter can remain in orbit to a maximum of 16 days (not including a 2-day margin). It comprised a pallet (a wafer) of cryogenic supplies mounted vertically at the rear of the payload bay, a closed-cycle carbon dioxide scrubber (RCRS) installed in the compartment forward of the middeck, a middeck locker containing a Trash Compactor (TC) and an updated toilet (WCS).

EDO-MP (EDO Medical Programme) A programme of biomedical experiments conducted in conjunction with the 'long duration' missions facilitated by the EDO. Most of these tests involved using the LBNP.

EDS (Energy Deposition Spectrometer) A sophisticated radiation monitor built by Battelle. By measuring radiation in 16 energy bands with high time-resolution, it was able to characterise geographical and temporal variation in dosage. Radiation affects solistate electronics, besides humans. One particular region of study was the South Atlantic Anomaly. As an added bonus, its test on STS-28 coincided with a solar flare.

EES (Energy Expenditure in Spaceflight) An SPE study.

EEVT (Electrophoresis Equipment Verification Test) See entry for CFES.

EGRET (Energetic Gamma Ray Experiment Telescope) An instrument on the GRO, able to sense gamma rays from 20 MeV all the way to the observable limit. Any process capable of radiating at extreme energies is likely to occur so rarely that most of the time there will be nothing to see but black sky, but if anything was active EGRET would detect it. Its spark chamber was an order of magnitude larger and more sensitive than any gamma ray detector previously flown. Its primary function was to complement COMPTEL, and map the sky, but it also extended BATSE's coverage of the bursters to the most extreme energies.

EISG (Experimental Investigation of Spacecraft Glow) A study of the correlation between the intensity of 'shuttle glow' and temperature variation due to moving in and out of the Earth's shadow. The atomic oxygen reaction was enhanced when the skin of the orbiter was baked in the Sun.

ELRAD (Earth Limb RAdiance Experiment) Twilight glow on the horizon (the limb) in the transition to and from sunlight was photographed. Pictures taken at 10-second intervals enabled this irradiance in the upper atmosphere to be determined in terms of the Sun's angle as it dropped below the horizon – data that would enable better horizon sensors to be developed for satellites.

EMU (Extravehicular Mobility Unit) The pressure suit used by astronauts working outside the shuttle; a simple-to-don two-part garment comprising the UTA and the LTA.

Endeavour An Australian ultraviolet telescope that was installed in a pair of GAS cans and flown on STS-42 for trials in advance of being used on an automated satellite. The test was foiled when the lid of the can refused to open.

Endeavour This orbiter was named for Captain James Cook's first ship, and for Apollo 15's command module. It was built to replace Challenger.

Enterprise This orbiter was originally intended to be the first flightworthy vehicle, and be called Constitution, but in 1977, following a letter-writing campaign by *Star Trek* fans, it was renamed for the fictional starship. Although it was used for the ALT programme, this particular orbiter was not upgraded for space flight.

EOIM (Evaluation of Oxygen Interaction with Materials) A study of the extent to which the atomic oxygen in the thermosphere affected several materials. To accelerate the erosion, the shuttle lowered its perigee to 225 km, where the gaseous concentration was higher.

EORF (Enhanced Orbiter Refrigerator and Freezer) An apparatus to store biological samples during an astronaut's tour of duty on the Mir space station.

EOS (Earth Observing System) The constellation of satellites which will serve as the main space-based element of the MTPE programme.

EOS (Electrophoresis Operational System) McDonnell Douglas' long-term goal for CFES was an automated system to process materials on an industrial scale; this apparatus was to be carried on an MPESS in the bay. Unfortunately, the loss of Challenger shortly before the first test led to the project being put into abeyance (and ultimately cancelled).

epitaxy A technique for growing a crystal by depositing material atom by atom on an initial substrate, to expand upon the original structure.

ERB (Earth's Radiation Budget) An instrument carried on a variety of satellites, including Nimbus 7, ERBS and several of the NOAA series (including NOAA's 9 and 10). Each comprised two detectors. One aimed four cavity-radiometers at the ground. Two of these nadir-pointing sensors had fields of view sufficiently wide to see the entire disk of the Earth from limb to limb, the others used narrower fields for higher resolution. One of each pair measured reflected insolation, and the other measured escaping thermal energy. A fifth radiometer was aimed at the Sun to measure broadband insolation. The second instrument scanned narrow-field sensors to either side of the ground-track, and scanned a swath as the satellite travelled along its orbit. Of the three thermistor-bolometers, one sensed reflected insolation, one sensed emitted thermal radiation, and the third recorded across the entire range in order to measure the Earth's total output. Overall, therefore, ERB measured insolation, visible and ultraviolet reflected radiation, and infrared emission. Correlating the variations in the radiation budget with the effects observed by SAGE would be crucial to understanding the role of the stratosphere in the solar–terrestrial relationship.

ERBS (Earth's Radiation Budget Satellite) A satellite deployed by the shuttle in 1984 to study solar–terrestrial relationships. It carried two instruments: ERB and SAGE.

erythropoiesis The process by which erythrocytes (red blood cells) are produced by the body. A significant milestone was passed when cosmonauts spent more than 120 days in space, because this is the time constant of the erythropoietic cycle; after that time their blood stream contained only erythrocytes produced in space. It turns out that, like most biological processes, erythropoiesis proceeds at a slower rate in space than on Earth, so red blood cell count is significantly reduced by prolonged exposure to weightlessness.

ESC (Electronic Still Camera) A 35-mm camera with a one-million pixel focal-plane CCD array. Its output was stored on a dedicated hard disk, electronically processed, and then downloaded to the ground. After testing on STS-48, it was carried on a routine basis. Many of its pictures were automatically posted on the Internet.

ESCAPE A GAS canister with an instrument built by the University of Colorado. It was an imaging telescope with a spectrometer for extreme-ultraviolet solar studies, a region of the spectrum in which relatively little research had been done.

ESOC (European Space Operations Centre) A facility established by ESA at Darmstadt in Germany.

ET (External Tank) The structure which feeds the liquid oxygen and the liquid hydrogen to the orbiter's SSMEs during the ascent to orbit. It is discarded within seconds of MECO, and burns up in the atmosphere.

ETR (Eastern Test Range) The facilities established by the Air Force to track rockets fired out over the Atlantic.

ETTF (Extreme-Temperature Translation Furnace) The first furnace capable of operating at temperatures as high as 1,650 °C in space. It was used to study gravitational influences on various types of material and processing methods.

EURECA (EUropean REtrievable CArrier) An ESA free-flyer to supplement Spacelab missions with extended operations (typically for periods of 6–9 months). Although designed in a modular fashion in order to be reusable, it has so far flown only once. After release by STS-46, it raised its altitude to 500 km to clear the worst of the atomic oxygen at lower altitude, then dropped back down for retrieval by STS-57. It carried a wide selection of materials and technology experiments, many of which tested apparatus intended for use on the International Space Station.

eV (electron Volt) The energy which is imparted to an electron when it is subjected it to an electric potential of one volt. All forms of electromagnetic radiation can be described by their characteristic wavelengths and, by using the speed of light, in terms of frequency; and by using Planck's constant, in terms of corresponding energy. For visible light this energy is just a few eV. Radiation in the gamma ray end of the spectrum, however, equates to a thousand (keV), a million (MeV), a billion (BeV), or even a million million (GeV) times greater energy, so X-rays and gamma rays are described in terms of their energy rather than their equivalent frequency or wavelength.

EVA (ExtraVehicular Activity) NASA vernacular for a spacewalk.

FARE (Fluid Acquisition and Resupply Experiment) A tank to enable the dynamics of fluid transfer to be studied, to assess the design of the system that will replenish fluids on the International Space Station.

FAUST (FAr Ultraviolet Space Telescope) A telescopic camera to study regions of star formation in galaxies by imaging in the far-ultraviolet, a part of the spectrum in which massive young stars radiate most of their energy. It flew on Spacelab 1, and then on ATLAS-1.

FEA (Fluids Experiment Apparatus) Rockwell built this multipurpose support apparatus to facilitate fluid physics microgravity research. It could accommodate samples in gaseous or liquid states, and could heat, cool, mix, stir and centrifugally separate samples in a variety of containers, including floating-zone crucibles. Its instrumentation suite could measure a sample's temperature, pressure, viscosity or other physical parameters, with the data being fed to a handy laptop computer; and the processing could be recorded on film for follow-up analysis. It was installed in a middeck locker, and tested on STS-30.

FES (Fluids Experiment System) An apparatus for conducting fluid physics experiments.

FES (Flash Evaporator System) When a shuttle orbiter is in space, it sheds excess heat via a set of radiators which are mounted conformally on the inner face of the payload bay doors. When safely in orbit, most of the systems are shut down. In contrast, during the ascent, all the redundant systems are running in standby, ready to take over should their primary fail, and the bay doors are shut. A flash evaporator is used to shed the copious amounts of heat produced when all the systems are operating. Spraying water on a heat exchanger is a very efficient way of rapidly drawing off energy; the resulting steam was subsequently vented. The flash evaporator was often used on Spacelab flights to dump the heat produced by the laboratory apparatus. The system was sometimes temporarily disabled by a build-up of ice.

FFEU (Free-Flow Electrophoresis Unit) A Japanese investigation of electrophoresis as a means of separating biologically active substances. On its first flight it became clogged with bubbles, but it subsequently functioned well, and was used to study chromosome separation.

FGB This 'functional energy block' is to be one of the two Russian-built modules which will form the core of the International Space Station. Although built by Khrunichev and launched by a Proton, it was paid for and will be owned by NASA. It will form the link between the Russian and non-Russian parts of the orbital complex.

FGBA (Fluid Generic Bioprocessing Apparatus) An apparatus to study two-phase liquids in microgravity. The Coca Cola Company sponsored its initial test, with an experiment to find out whether a carbonated drink could be made by using separately stored water, syrups and carbon dioxide gas.

FGS (Fine Guidance Sensors) A set of small telescopes on the HST through which guide stars were monitored in order to hold the spacecraft still during extended exposures by the primary instrument.

FILE (Feature Identification and Location Experiment) A CCD camera system with on-board processing capability to test whether a 'smart' sensor could be programmed to look for specific environmental features, and then produce data, but otherwise stay 'off the air'. This experiment was one option for alleviating the problem of environmental satellites streaming in data at ever greater rates, the processing of which was not only expensive, but also often fruitless because it did not produce any results relevant to features under investigation.

floating-zone A technique for growing crystals in microgravity which does not involve the crystal coming into contact with (and hence being influenced by, either chemically or by way of transmitted vibrations) its container.

Fltsatcom (Fleet satellite communications) A geostationary communications satellite system operated by the Navy. Although intended to be superseded by Leasat, it actually proved to be rather longer-lived, so it continued in service alongside the new system. Both are currently being phased out in favour of the UFO system.

fluids and combustion facility A trio of double ISPR racks in NASA's laboratory on the International Space Station. The combustion rack will house a generic chamber with viewing ports for a range of data acquisition systems, and will be configured differently for each specific investigation. The fluid rack is on the far side of the central rack, which will provide common support systems. It will be loaded with apparatus specific to each experiment. A number of containerless crucibles have already been proved on Spacelab missions, and these will be used as required by each investigation. Knowledge of fluid behaviour is essential to a wide range of industrial activities, and although some of the more subtle, yet fundamental, properties of fluids do make their influence felt on Earth, they can be studied effectively only in the absence of gravity.

FME (Foil Microabrasion Package) A micrometeoroid experiment.

FOC (Faint Object Camera) An HST instrument supplied by ESA, to image in the visible and near-ultraviolet regions of the spectrum.

FOS (Faint Object Spectrograph) An HST instrument to record spectra in the visible and near-ultraviolet regions of the spectrum.

FPE (French Posture Experiment) A study of posture in the absence of gravity, in order to provide information on how the load-bearing muscles, with no work to do, adapt.

FSDCE (Fibre-Supported Droplet Combustion Experiment) A study of liquid fuel ignition and burning processes, not only to yield basic data on combustion, but also to increase understanding of soot production. A liquid fuel burns most effectively when it is a fine aerosol spray, so individual drops of fuel were suspended on fibres. In microgravity, it proved possible to achieve burn-times 10 times longer than in experiments on Earth.

FSS (Flight Support Structure) A rotating and tilting frame that was installed in the aft bay to hold a satellite so that it could be worked on by spacewalkers.

fuel cell An apparatus to generate electricity by combining hydrogen and oxygen and, as a by-product, produce potable water. First developed for the Gemini spacecraft, fuel cells were the primary source of power for Apollo (they were considered preferable to arrays of solar transducers). A shuttle orbiter has three fuel cells. Mission rules call for a flight to be cut short if one fails (this has occurred only twice; STS-2 and STS-83). The fuel cells are started several hours before launch and are kept operating throughout a flight. However, a shuttle is to be powered down when it is docked at the International Space Station. To test whether this was practicable, on the last day of some missions, starting with STS-54, a fuel cell was switched off and restarted. A number of problems showed up in these tests; on STS-57, the hydrogen supply value would not close so the fuel cell could not be shut down, and sometimes it would not restart. Although fuel cell technology is an effective way to power a spacecraft on a brief flight, it consumes cryogenic reactants at such a rate that it will be necessary to run the space station using solar energy transducers.

furnace facility Part of NASA's laboratory on the International Space Station. It is a trio of double ISPR racks customised to accommodate the large thermal-isolation chambers, control systems and instrumentation required for materials-processing research.

FUV (Far-UltraViolet) An SDI experiment consisting of a pair of imaging cameras and a low-light camera to record aurorae and day and night air glow phenomena. The cameras sampled in the 1,000–2,000 Å range.

Galileo A shuttle-deployed spacecraft that, following a long roundabout route, became the first artificial satellite of Jupiter.

GANE (GPS Attitude and Navigation Experiment) A test to show that if GPS antennas are distributed around a spacecraft, as is to be done on the International Space Station, it can use the navigation signals to determine orientation as well as position and velocity.

GAS (Get-Away-Special) An oil-drum-sized container for packaging experiments which can be carried in the shuttle's payload bay. Two variants are available: one that is hermetically sealed, and the other with a lid that opens in space to expose the apparatus. These experiments must be provide their own power, environmental control, control logic and data processing. They are flown on a first-come-first-served basis. GAS cans can be individually bolted to the bay sidewall or mounted together on a GBA. The first GAS experiment flew on STS-4.

GBA (GAS Bridge Assembly) A cross-bay structure that can accommodate up to a dozen GAS canisters. If fewer than 12 experiments are carried, ballasted cans will be added to make up the full complement and balance the structure.

GBX (GloveBoX) An isolation chamber into which apparatus for hazardous experiments could be loaded and then manipulated by way of a pair of sealed rubberised glove slots. Several versions were built by ESA, each fitted with filters to deal with different types of possible contaminant. In addition to being used on various Spacelab missions, one was installed on the Mir space station.

GEF (Gas Evaporation Facility) An apparatus to study the process by which evaporating metal forms particles in microgravity in a gaseous atmosphere.

GEO (GEostationary Orbit) See next entry.

geostationary orbit The altitude at which a satellite orbits determines the time that it takes to make a complete revolution: the lower the altitude, the shorter the period. At an altitude of 36,000 km, the period is 24 hours. In an orbit with zero degrees inclination, a geosynchronous orbit, a satellite will appear to remain stationary above one spot on the Earth's equator. Any other orbital inclination will cause such a satellite to 'nod' north/south by a corresponding arc. Some arcs above the equator are densely packed with communications and meteorological satellites.

geostationary transfer orbit A highly elliptical Hohmann orbit used to move a satellite from LEO up to GEO. The manoeuvre is complicated by the fact that it is usually necessary to combine the steep ascent with a significant plane change in order to cancel the orbital inclination.

GFFC (Geophysical Fluid Flow Cell) An apparatus to model oceanic flows and scenarios for different atmospheres.

GFVT (GAS Flight Verification Test) A GAS can containing only instruments to monitor internal conditions. It was flown on STS-3 to certify the canister to carry experiments.

GHCD (Growth Hormone Concentration and Distribution) An experiment to determine the effect of microgravity on growth hormone distribution and concentration in a variety of plants. A pair of plant chambers, a temperature monitor and a nitrogen-gas freezing unit were installed in a middeck locker. The plant chambers held seeds embedded in a filter paper soaked in nutrient. They were loaded a few hours before launch, kept in darkness throughout the flight, and then, just before returning to Earth, the chamber was flooded with nitrogen gas to arrest growth and freeze the plants for post-flight analysis. The aim was to determine the concentration of the auxin growth hormone, using shoots of corn.

GLO A study of the way in which the orbiter glows as a result of its passage through the atomic oxygen plasma in LEO.

GLOMR (Global Low Orbiting Message Relay) A satellite deployed for the Department of Defense to relay bursts of data transmitted periodically by buoys in open ocean.

GOSAMR (Gelation Of Sols, Applied Microgravity Research) An experiment to study the influence of microgravity on the processing of gelled sols.

GPPF (Gravitational Plant Physiology Facility) A generic greenhouse facility flown on the International Microgravity Laboratory. It could accommodate a variety of specific plant experiments.

GPPM (Gas-Permeable Polymeric Membrane) An apparatus to make an advanced polymer with improved gas-permeable characteristics, in order to set the standard against which improvements in the terrestrial manufacturing process could be measured.

GPS (Global Positioning System) A constellation of satellites (NAVSTAR) were regularly distributed around the planet so that analysis of the timing signals which they broadcast would enable the geographical position of a receiver to be precisely calculated. Four satellites were distributed 90 degrees apart in orbits inclined at 55 degrees to the equator, with six sets of satellites in orbital planes spaced at 60-degree intervals girdling the equator. In addition to a geographical fix, the system can calculate the receiver's altitude, and because it can produce a continuously updated reading it is increasingly being used by satellites to track their own orbits, thereby reducing the need for ground-based radar tracking. Originally, the satellites of the operational system were to have been deployed by the shuttle, but in the aftermath of the Challenger accident they were offloaded to the Delta 2. The reason for the system's universal adoption is that a battery-powered hand-held receiver is sufficient to make use of it. As an aside, it is interesting to reflect that GPS is one of the few uses of a satellite radio transmitter *not* to have been specifically predicted by Arthur C. Clarke.

GPWS (General Purpose WorkStation) A Spacelab rack, similar to the Glovebox in that it was a generic facility providing an isolation compartment into which apparatus can be inserted and then manipulated by way of access gloves. It was tested on STS-40 using the RAHF cages, to demonstrate that it would indeed contain particulate debris that had tended to leak from such cages, polluting the Spacelab habitat. It is intended for use on the International Space Station.

Gravitational Biology Facility A trio of double ISPR racks in NASA's laboratory on the International Space Station which will have support systems for transportable modular habitats (such as the AEM and RAHF) for specimens participating in developmental studies of cell, plant and animal biology.

Greenhouse An experiment which involved upgrading the Svet cultivator on Mir in order to study plant reproduction, metabolism and biochemistry.

Great Observatories A name adopted early in the shuttle programme to express the significance of the series of astronomical satellites that the shuttle was to deploy. There were to be four such observatories, together spanning the electromagnetic spectrum all the way from the far-infrared, through the visible, ultraviolet and X-ray bands to the highest energy gamma rays. Only the HST (visible light) and the GRO (Gamma Ray Observatory) have been launched to date, but AXAF (X-ray) is to be launched in 1998, and SIRTF (infrared) will follow at the turn of the century.

GRO (Gamma Ray Observatory) The largest satellite ever put in space specifically to study high-energy astrophysics. It comprised four instruments: COMPTEL, EGRET, OSSE and BATSE. It was the second satellite in NASA's Great Observatories programme, and was subsequently renamed the Compton Observatory.

GS (Grille Spectrometer) An instrument to measure the solar spectrum at orbital sunrise and sunset – when the Sun was on the limb and its light was passing horizontally through the Earth's atmosphere – so that the concentrations of trace constituents at that location could be derived from their characteristic infrared absorption. Following trials on the Spacelab 1 and ATLAS-1 shuttle missions, the Belgian instrument was repackaged and installed on the Mir space station to conduct a year-long survey.

GSC (Grab Sample Container) A vial in which to sample the air aboard Mir for post-flight analysis of trace contaminants and biological fauna which might pose a long-term threat to human habitation.

GSO (GeoSynchronous Orbit) See *geostationary orbit.*

GTO (Geostationary Transfer Orbit) See *geostationary transfer orbit.*

gyrodyne A single-axis momentum wheel to store and release angular momentum so as to rotate the Mir space station. A cluster of gyrodynes are required for complete three-axis control. The advantage of this attitude control system was that it did not consume any propellant, although the gyroscopic units required frequent maintenance. Although it was possible to send up new ones, it was impracticable to return the bulky 165-kg units to Earth for maintenance until the shuttle began to make regular visits.

HALOE (HALogen Occultation Experiment) An instrument on the UARS that measured infrared absorption features in sunlight as it passed horizontally through the upper atmosphere at orbital sunrise and sunset to map the geographical and vertical distribution of methane, hydrofluoric acid, hydrochloric acid and various nitrogen compounds.

handover As the shuttle clears the tower at launch, KSC 'hands it over' to JSC. The PAO at KSC has time to mark the successful conclusion of the count by stating the mission's primary objective (e.g. "Liftoff! On a mission to service the Hubble Space Telescope"), before the PAO at JSC announces that, "Houston is now controlling."

HBT (Hyflex Bioengineering Test) An experiment to test a plant-growth chamber to find out how it responded to different degrees of moisture, in order to establish the optimum set-up for a subsequent experiment conducted on Spacelab 1.

'Head' A donated human skull which was sliced into horizontal sections, fitted with 125 sensors and then filled with tissue-equivalent material to study the way in which space radiation penetrates the head.

HERCULES (Hand-held Earth-oriented Real-time Cooperative User-friendly, Location-targeting and Environmental System) A system for locating the geographical location of any photographed object to within a radius of 2 km. It used an inertial platform that had to be calibrated against reference stars. It was considerably easier to use than the

L3, in that it did not require a sequence of pictures to be taken bracketing a target, so it could be used spontaneously.

HGD (Hand Grip Dynamometer) An apparatus to measure the strength of the muscles in the hand.

HH-DTC (Hand-Held Diffusion Test Cell) An apparatus to enable protein crystallisation by the liquid-diffusion process to be directly observed and recorded for subsequent study.

HH-PCG (Hand-Held PCG) A small unit flown to assess the viability of growing protein crystals in microgravity. Although it proved that usable crystals could be grown, it also demonstrated that the product would have to be protected from the vibrations during the ascent and landing. Larger and more regular protein crystals could be grown in space. The structure was then studied by X-ray crystallography. Discovering the structure of a protein was the first step towards designing a drug to control its action. Crystals did not all grow at the same rate.

Hitchhiker A packaging system for the orbiter's payload bay which could hold a variety of small packages in GAS-like canisters on an MPESS. This programme was managed by GSFC on behalf of a variety of customers.

HOL (Holographic Optics Laboratory) An apparatus to investigate Marangoni convection phenomena.

HPP (Heat Pipe Performance) A continuation of the heat pipe tests, in which the apparatus was spun at different rates to determine whether its function was sensitive to its state of motion and to test different pipe configurations.

HPTE (High-Precision Tracking Experiment) The ability to illuminate an object in space with a laser beam was a key SDI objective. In June 1985, AMOS fired its laser at the shuttle to test its ability to lock the beam on target. The crew had mounted a 20-cm wide corner-cube reflector in the porthole of the orbiter's hatch to reflect the beam back to the ground site.

HRDI (High-Resolution Doppler Imager) A visible-light and near-infrared interferometer on the UARS to measure stratospheric winds.

HRS (High-Resolution Spectrograph) An HST instrument similar to FOS but restricted to the ultraviolet in order to provide higher spectral resolution in this crucial part of the spectrum.

HSP (High-Speed Photometer) An HST instrument to monitor the variations in brightness of objects with a high time resolution.

HST (Hubble Space Telescope) The most important scientific payload that the shuttle has ever carried.

HTLPE (High-Temperature Liquid Phase Experiment) An experiment that was installed on the Mir space station.

Human Research Facility The double ISPR rack in NASA's laboratory on the International Space Station which will be fitted to study how the human body responds and adapts to weightlessness, to assess the health of the crew (including heart and lung function, the maintenance of muscle and bone, regulatory physiology, perception, and cognition and balance) and to test the effectiveness of countermeasures to the undesirable side-effects of living and working in space.

HUP (Horizon Ultraviolet Programme) An SDI experiment which used a spectrometer to observe ultraviolet phenomena on the Earth's limb to characterise the background against which a cold missile body, rising above the horizon following its boost phase, would be viewed by a satellite sensor.

HUT (Hopkins Ultraviolet Telescope) A 0.9-metre aperture f/2 reflecting telescope incorporating a spectrograph optimised to study far- and extreme-ultraviolet wavelengths (in the range 425–1850 Å) with a spectral resolution of 3 Å. Its solid-state detector generated real time data. It was built by Johns Hopkins University specifically to be flown on the shuttle.

IAE (Inflatable Antenna Experiment) A package carried on a SPARTAN free-flyer in order to test a system for inflating an antenna. Upon completion of the inflation process, which took five minutes, the antenna comprised a 15-metre wide reflector dish and a 28-metre tall tripod on which to mount focal-point instruments.

IBSS (Infrared Background Signature Survey) An SDIO sensor package flown on the SPAS-2 free-flyer. Its cryogenically cooled infrared sensor, multispectral sensor (the Arizona Imaging Spectrograph) and image-intensified TV cameras observed the orbiter firing its OMS engines, and clouds of gases releases by the shuttle and by several sub-satellites. Its role was to gather signature data to assist in the development of satellite sensors to detect and track ballistic missiles manoeuvring during their mid-course phase.

ICE (Interface Configuration Experiment) A study of the way in which fluids adhere to the walls of a container in microgravity. The results did not match computer models, which indicated that there were subtle hitherto unsuspected forces at work.

IDGE (Isothermal Dendritic Growth Experiment) An apparatus to study the tiny crystalline forms (dendrites) which, under specific conditions, develop when a crystal solidifies in microgravity. A camera displayed the process in real time to scientists so that they could observe its progress, because it was the process of growth, not the final result, that was under investigation.

IECM (Induced Environment Contamination Monitor) A package of instruments carried on Columbia during its test flights to assess the extent to which outgassing and propellant-leakage polluted the payload bay. It was swung around by the RMS, to seek out 'hot spots' in terms of emissivity.

IEH (International Extreme-ultraviolet Hitchhiker) A multinational package of instruments (SEH, USAR and GLO) to study the Sun, the stars and Solar System bodies at extreme-ultraviolet wavelengths.

IFE (Isoelectric Focusing Experiment) An electrophoresis apparatus which used an electric field to separate proteins in solution by using their different surface electrical charges. The electric field creates a pH gradient in the sample – alkaline at one side and acidic at the other – and the biological material migrates to a point giving zero net charge, hence the term isoelectrical focusing.

IFM (In-Flight Maintenance) NASA vernacular for any routine repairs conducted in space.

igloo The small pressurised chamber which contained the control systems for the Spacelab pallet.

IGY (International Geophysical Year) Sputnik was launched to support this international study of the Earth as a global system. The study actually ran for 18 months from mid 1957 through 1958.

IMAX A large-format high-definition camera which used 70-mm film. It was adapted for use on the flight deck, in the payload bay and on a free-flying satellite. IMAX cameras were often carried, and their footage included memorable films such as *The Dream Is Alive* and *The Blue Planet*. The IMAX Systems Corporation was co-sponsored by NASA and the Smithsonian Institution's National Air & Space Museum to develop this camera specifically to record space activities, but it is now also used commercially for a wide range of other topics.

IMEWS (Integrated Missile Early Warning System) A follow-on from the MIDAS and Vela satellites that integrated the roles of both systems in a single package. When the second generation was introduced the name was changed to DSP.

IML (International Microgravity Laboratory) A major Spacelab subprogramme. Its many experiments included SPE, MVI, MWPE, GPPF, OCGF, FES, VCGS, BR, Cryostat, CPF, BSK, MIGF, IMAX, SAMS, AAEU, SCM, LBNP, MAS, PAWS, NIZEMI, RRMD, TEI, RAMSES, BDPU, TEMPUS, FFEU, LIF, QSAM and VIBES.

IMMUNE A study of the response of the immune system to microgravity.

IMU (Inertial Measurement Unit) The gyroscope which provided the orbiter's navigational system with an inertial frame of reference against which to measure attitude and motion.

Intelsat 603 An HS-393 communications satellite that was launched on a Commercial Titan rocket, but left stranded in LEO when its perigee kick-motor failed to separate from the Titan's upper stage. STS-49 rendezvoused with it, and astronauts retrieved it and fitted a new Orbus motor so that it could resume its journey up to GSO. This particular satellite was also known as Intelsat 6-3.

IOCM (Interim Operational Contamination Monitor) A payload bay sensor to follow on from the APM.

IPS (Instrument Pointing System) The high-precision (1-arcsec) telescope mount flown on Spacelab's pallet missions.

IRCFE (InfraRed Communications Flight Experiment) Built by Joseph Prather of JSC, this was tested on STS-26. George Nelson wore a belt pack incorporating infrared-emitting and infrared-sensitive diodes and linked into the standard communications headset. Similar packs were installed on the flight deck. By relaying through wall-mounted transceivers, he was able to maintain two-way communication with the ground while moving around without trailing a wire link. A 24-channel version is under development for use on the International Space Station. During a spacewalk, this system would offer the advantage of immunity from radio-frequency interference.

IRIS (InfraRed Imagery of the Shuttle) An infrared telescope carried on an aircraft that flew along Columbia's track as it penetrated the atmosphere at Mach 25, to record the distribution of thermal stress across the orbiter's belly. After several failed attempts, the data were eventually secured.

IRIS (Italian Research Interim Stage) The rocket stage that was used to deliver the shuttle-deployed LAGEOS up to its operating orbit.

IRT (Integrated Rendezvous Target) An inflatable balloon intended to serve as a radar target to rehearse shuttle rendezvous procedures. It inflated irregularly, and ripped itself to shreds.

ISAC (Intelsat Solar Array Coupon) Prior to making the final commitment to retrieve and repair the Intelsat 603 communications satellite, Hughes flew a 'coupon' of transducers on a shuttle to determine the extent to which its satellite's solar array had been degraded through prolonged exposure to the atomic oxygen present in LEO. This was carried out to assess its subsequent lifetime if it were to be successfully repaired and sent up to its operating altitude, where the gaseous environment is not so corrosive. This small coupon of cells was held out from the orbiter by the RMS in such an orientation as to maximally expose it. Analysis of its deterioration was extrapolated to predict the state of the satellite itself. The results indicated that it would be worthwhile rescuing the satellite.

ISAIAH (Israeli Space Agency Investigation About Hornets) An apparatus containing 18 chambers of various sizes, to study how hornets construct nests in microgravity. The hornets did not seem to be disorientated, but their activities were disorganised.

ISAMS (Improved Stratospheric And Mesospheric Sounder) An infrared emission radiometer on the UARS. It mapped the geographical and vertical distribution of water vapour, carbon monoxide, carbon dioxide, nitric oxide, nitrous oxide, methane and ozone in the upper atmosphere.

ISIS (Infrared Spectral Imaging Radiometer) An instrument to measure radiation from the tops of clouds in the Earth's atmosphere. It was used to test advanced sensors that will be used to measure cloud and aerosol layers.

ISO (Imaging Spectrometric Observatory) An instrument to study air glow phenomena. It measured emission features across a wide part of the spectrum from the ultraviolet to the infrared in order to identify the chemical composition of the radiating material. In addition, the geographical and temporal variations of the air glow were monitored.

ISS (International Space Station)

ISY (International Space Year) An international programme conducted in 1992 to study the solar–terrestrial relationship, the Earth's atmosphere and global climate change. It encompassed a wide variety of investigations (such as destruction of the rain forests in Africa and South America and the effect of their depletion on the build-up of atmospheric carbon dioxide) as well as satellite studies of the upper atmosphere.

ITA (Instrumentation Technologies Associates) A BioServe industrial affiliate. It built the MDA for biotechnology experiments employing fluid mixing.

IUE (International Ultraviolet Explorer) A small astronomical satellite launched in 1978 for ultraviolet spectroscopy. It proved tremendously long-lived, and was only finally switched off in 1996.

IUS (Inertial Upper Stage) The two-stage solid-propellant rocket motor built by Boeing to ferry heavy shuttle-deployed satellites up to GSO. The first stage performed the GTO insertion, and the second circularised the orbit at apogee. This two-stage configuration obviated the need for the payload to make its own circularisation manoeuvre. Although made extremely fault-tolerant so as to ensure that it would be able to deliver its payload, the second stage of the first IUS released by the shuttle suffered a malfunction, and left its satellite to struggle into orbit on its own. Since that inauspicious start, however, the IUS has successfully completed all of its missions. In addition to delivering satellites to GSO, it has dispatched several interplanetary spacecraft. A three-stage variant planned for planetary missions was cancelled early on.

JPL NASA's Jet Propulsion Laboratory in Pasadena, California. Run by Caltech, JPL initially undertook missile work for the Department of Defense, but is now primarily involved in developing and operating scientific spacecraft.

JSC (Johnson Space Center) NASA's centre for the design, development and testing of manned spacecraft located in Houston, Texas, incorporating, among many other facilities, the Mission Control Center (MCC), and the Space Station Control Center (SSCC).

Kaliningrad The facility in a northern suburb of Moscow in which the control room which manages Mir space station operations is situated. It was recently renamed in honour of Sergei Korolev. Incorporating the Yuri Gagarin Cosmonaut Training Centre, it is the RSA's equivalent of NASA's JSC in Houston.

kick-motor A rocket motor that is 'bolted onto' a satellite to undertake a major manoeuvre on its way to its operating station. See PKM and AKM.

KidSat A payload-bay camera which was assigned targets to satisfy requests submitted by school children via the Internet, and whose pictures were downloaded and immediately posted on the Internet.

KSC (Kennedy Space Center) NASA's site for processing and launching the space shuttle and its payloads, located on Merritt Island adjacent to the USAF launch facilities of the Cape Canaveral Air Force Station.

L3 (Latitude and Longitude Locator) It had proven difficult to pinpoint phenomena observed in oceanic regions, and even when the location had been known, recording the location of each photograph taken from space was laborious. An automated system was needed. Developed jointly by NASA and the Department of Defense, L3 was the first attempt. It was tested on STS-28. A Hasselblad camera with a wide-angle lens, it was linked up to a computer. The procedure for locating a target involved taking pairs of photographs of the target 15 seconds apart – one just before and the other after passing over a target – and once the graphics system had digitised and analysed the pictures it was able to work out an approximate latitude and longitude. It worked, but it was cumbersome and did not lend itself to spontaneity.

LACE (Laser Atmospheric Compensation Experiment) An SDI satellite orbiting at 500 km altitude to act as a laser target. By using a cruciform of detectors, it was able to provide the feedback required to adjust the beam so that it struck the 'bullseye' at the centre. It was used to study the way that passage through the atmosphere disturbed a laser beam fired from the ground at a target in space. It was launched in 1990, and carried UVPI as a secondary payload.

Lacrosse A class of high-resolution radar-reconnaissance satellite designed to be launched on the shuttle.

LAGEOS (LAser GEOS) A series of 0.6-metre diameter satellites incorporating 426 corner-cube laser reflectors, for geophysics research.

Landsat A series of Earth-resources satellites. Landsat-4, which suffered a power failure, was to have been repaired in orbit but, because the satellite was in polar orbit, when the plan to launch shuttles from California was cancelled this rescue became impracticable.

launch window The usually short period of time during which circumstances are favourable for dispatching a rocket on a particular mission. A wide variety of factors can determine the ideal moment. One of the most basic constraints is the amount of propellant that will be available, either at launch, or later, to perform the plane-change necessitated by not launching at the precise moment that the required orbital plane intersects the launch site. In some cases, the launch window can last for only a few minutes.

LBNP (Lower Body Negative Pressure) The LBNP fitted around an astronaut's legs and was sealed at the waist. When the pressure was lowered, it would pull blood from the upper to the lower torso. It was 'worn' for up to an hour at a time during the initial phase of adaptation to weightlessness to counter the pooling of blood in the upper torso, and to reacclimatise the heart to an increased load preparatory to the return to Earth. A similar unit had been used on Skylab, but that had been a bulky rigid structure. Space was a premium in the shuttle's cabin, so a collapsible 'sleeping-bag' unit had been developed. It was tested on STS-32, to prepare for the 16-day flights that the EDO would make possible.

LC 39 (Launch Complex 39) A pair of launch pads (39A and 39B) at KSC originally built to launch Saturn 5 rockets for Apollo, but now converted for the shuttle.

LCC (Launch Control Centre) A facility at KSC, close alongside the VAB, from which the process of launching a shuttle can be supervised.

LDCE (Limited Duration space environment Candidate materials Exposure) A series of GAS canisters to expose, for a given period, a variety of candidates for materials under consideration for use on space structures, to assess the extent to which they degraded in the atomic oxygen plasma present at orbital altitude.

LDEF (Long Duration Exposure Facility) A 12-sided cylinder that was deployed in orbit in a stabilised gravity-gradient orientation so that one end was maintained facing the Earth and the other facing space, and with one specific side facing the direction of flight and another protected in the wake. Trays mounted on its surface contained materials which were exposed to the space environment in order to determine how they degraded. The 86 trays had seven samples from NASA, 21 from universities, 33 from private companies, nine from the Department of Defense and six from foreign parties. The topics under investigation included optics, electronics, materials, power systems, and propulsion systems, as well as basic science. Although originally intended to be left in orbit for about six months, it was almost six years before it was finally recovered, this delay being attributable to the Challenger accident. The LDEF framework was designed to be reusable, but it has so far flown only once.

Leasat The operating name for the Syncom-4 series of communications satellites, so called because they were owned by Hughes and leased to the Navy. This HS-381 model was the first Hughes satellite designed to take advantage of the shuttle's wide payload bay. It was large enough to incorporate its own perigee kick-motor, which was adapted from the upper stage of a Minuteman missile.

LEMZ (Liquid Encapsulated Melt Zone) An apparatus to study the growth of exceedingly pure crystals with the floating-zone technique. It was tested by making semiconductors.

LEO (Low Earth Orbit) Generally speaking, a near-circular orbit below an altitude of about 1,000 km, as opposed to GEO.

LFC (Large Format Camera) Although derived from a high-altitude metric stereographic mapping camera, this remotely-operated 400-kg assembly was more advanced optically and electronically. Its cassette contained film for 2,400 exposures, and each time it took a picture a pair of small cameras recorded orthogonal star fields in order to provide a reference frame.

LFE (Lung Function Experiment) An apparatus to study lung function in microgravity.

lidar A laser-based radar.

LIF (Large Isothermal Furnace) A generic apparatus to melt, uniformly mix and solidify in a controlled manner a wide variety of materials.

Life Sciences Facility The gravitational biology research on the International Space Station will be significantly enhanced by the addition of this facility. It incorporates a 2-metre diameter centrifuge able to impart accelerations of up to 2 g, and an isolation glovebox, as well as additional storage for modular habitats.

lifting body A wingless aircraft that uses the shape of its fuselage to provide the lift needed to fly. Such a configuration has a very high lift-to-drag ratio. Following the decision to build the shuttle, the M2F2 (1966–1967), the HL-10 (1966–1970), and the X-24A (1969–1971) were supplemented by studies with the M2F3 (1970–1972) and the X-24B (1973–1975). In the case of the M2, the 'M' indicated manned, and it ran through flight models F1, F2 and F3.

light year Light travels at 300,000 km/s. A light year is the distance that light travels in a year. To appreciate the vastness of the Universe, consider that it takes eight-and-a-half *minutes* for light from the Sun to reach the Earth, four-and-a-quarter *years* to reach us from the nearest star beyond, and a couple of *million* years from the nearest large external galaxy; yet it takes *billions* of years for light from the galaxies that are of most interest to astronomers to reach us.

LIGNIN (Gravity-Influenced Lignification In Higher Plants) Pine, mung-bean and oats were grown to investigate the formation of lignin, the agent which stiffens the stem of a plant so that it can withstand the pull of gravity. The objective was to test whether a wood-plant develops with less of this agent in microgravity. The human gut cannot digest a lignin-cellulose cell so, even though such a plant is rich in carbohydrate and protein, it cannot serve as a food; if a lignin-free variety could be grown, it would broaden the range of foods available. The experiment was conducted in the PGU.

LiOH (Lithium hydroxide) A substance that soaks up carbon dioxide. Unfortunately, it is soon saturated. It is a viable method of cleansing the cabin air on a short flight, but a stock of canisters must be carried and swapped regularly, and since the canister weighs about 20 kg, a long flight with a large crew employs the RCRS closed-cycle system.

liquid–liquid diffusion A crystallisation process in which the material to be crystallised and its fixing agent are separated by a buffer until in space, and then allowed to mix in order to prompt crystallisation.

LITE (Lidar-In-space Technology Experiment) This lidar was carried alongside a telescope on a Spacelab pallet to detect the reflection of the laser by the atmosphere. By the time the laser beam reached the Earth's surface, it had expanded to cover a circle almost 300 metres wide. Each sample involved emitting a brief pulse, then measuring the return in the narrow-field-of-view telescope. Lasers for near-infrared, green and ultraviolet were available. It produced a three-dimensional map of the structure of the atmosphere in the 10–40 km altitude range. Its thermal profiles could be correlated against the location, movement and concentration of atmospheric constituents. From its vantage point in orbit, this satellite was able to reveal the large-scale structure of clouds and storms in unprecedented detail. For calibration, simultaneous measurements were made by meteorological stations and by airborne lidars.

LMC (Large Magellanic Cloud) A small satellite galaxy of our own Milky Way, lying at a distance of about 160,000 light years.

LMTE (Liquid Metal Thermal Experiment) A test of a heat pipe that used a liquid metal for its coolant, a system with the advantage of being able to run its 'hot loop' at a temperature as high as 1,000 °C. For this evaluation, three different configurations were tested.

LPE (Lambda Point Experiment) An apparatus to observe the behaviour of helium at 2 K, the temperature at which it entered the superfluid state in which viscosity became zero (helium is unique amongst the elements in possessing this state). On Earth, thermal and gravitational gradients made the process difficult to study in detail, but in microgravity it was possible to measure the lambda point, as it is called, extremely accurately.

LRU (Line Replacement Units) Components of the shuttle which can be routinely replaced during part of the 'flow' in the between-mission turnaround.

LSRF (Life Sciences Refrigerator and Freezer) An apparatus to store biological samples on life sciences missions. It was used to facilitate the return of the samples of saliva, urine and blood that were taken on a daily basis by the NASA astronauts who spent tours of duty on the Mir space station.

LTA (Lower Torso Assembly) The waist-down part of the EMU suit.

M88-1 A multifaceted Department of Defense assessment of the value of a 'military-man-in-space' (another name for the test). It incorporated Air Force, Army and Navy objectives. An important feature of the experiment was the (Nightmist) UHF secure radio link between the astronaut and the test conductor at each ground site. A picture of each target was taken by a 35-mm camera fitted with a CCD image system, and the imagery was processed in real time to enhance militarily useful detail; then the astronaut's analysis was radioed to the test conductor before the shuttle flew out of line-of-sight communications range. Ships were observed at sea and in port, as were armoured forces manoeuvring in the field, river crossings, artillery barrages, missile launches and airfield operations, and the manoeuvres of an aircraft in flight. A secondary objective was to gain an insight into what the cosmonauts on the Mir space station might have observed over the years.

MACE (Middeck Active Control Experiment) An apparatus to investigate active control of structures in space. It comprised a long flexible polycarbonate structure with a two-axis gimballed payload package at each end, a three-axis acceleration sensor in the middle, and sets of actuators. It measured the forces transmitted through the structure and tested the ability of the active control elements to damp them out. The truss of the International Space Station will carry 'attached payloads'. This experiment would help to determine whether active control mounts would be necessary.

Magellan A shuttle-deployed spacecraft which entered orbit around Venus, and proceeded to use its SAR to make a high-resolution map of the surface, which is otherwise hidden by the perpetual dense cloud cover.

Magnum A class of electronic-reconnaissance satellite.

MAHRSI (Middle-Atmosphere High-Resolution Spectrograph Instrument) An instrument to measure ultraviolet radiation emitted and scattered by the Earth's atmosphere in order to study hydroxyl distribution in the middle and upper atmosphere.

MAPS (Measurement of Air Pollution from Space) An infrared radiometer which measured the concentration of carbon monoxide in the upper troposphere. It detected this gas by its characteristic absorption of the background of thermal emission from the

Earth. It flew four times over a 12-year period, on each occasion producing a snapshot of the state of the atmosphere to assess efforts to reduce industrial emissions of this greenhouse gas. It has since been repackaged to be sent up to the Mir space station, and for installation on one of NASA's EarthWatch satellites.

Marangoni convection A surface tension effect in a fluid that is difficult to study in gravity, but is readily observed in space, where gravitationally-induced convection is absent. It stimulated a weak type of convection in which currents move from low-pressure to high-pressure zones. It was studied not just out of scientific curiosity, but also to determine the extent to which it could disrupt microgravity materials-processing operations, and to assess whether it could be used to *drive* other applications.

MAS (Microbial Air Sampler) A simple experiment in which samples were taken of the air in the orbiter's cabin for later analysis, to test whether there was a build-up of biological contamination which could pose a threat on an extended mission.

MAS (Millimetre Atmospheric Sounder) An ATLAS instrument to measure the concentration of ozone in the upper atmosphere and the trace species that play a part in its depletion. It also produced thermal profiles so that the temperature dependency of these reactions could be studied.

MAST (Military Applications of Ship Tracks) The use of very high-resolution imagery of ship wakes.

materials research The objective of materials research is to study how materials form, and how the process influences the resulting properties, because by controlling the process of formation it is possible to design a material with a specific property. Observing such processes under microgravity is especially productive because on Earth gravity-induced sedimentation and convection degrade the homogeneity of the product. Shuttle studies have produced significant insights into the structure of various metals, and these are being applied to create more corrosion-resistant alloys. Other investigations have focused on forces such as surface tension effects which can be utilised in space, but which cannot be used in isolation on Earth because they are overwhelmed by gravity. One significant realisation was that although surface tension is masked by gravitational disturbances, it nevertheless makes its effect felt as secondary effects in terrestrial manufacturing.

MCC (Mission Control Centre) The flight control room at JSC; radio call-sign 'Houston'.

MDA (Materials Dispersion Apparatus) An apparatus developed by ITA for biotechnology experiments by fluid mixing. It simultaneously processed a block of 100 samples, so it was meant for experiments requiring a great deal of data, rather than limited proof-of-concept studies. Although conceived as a high-capacity protein crystal growth unit, it was suitable for any process involving liquid–liquid diffusion, and the fact that it could be set up to mix as many as four fluids in each container made it a very flexible tool. It could be used for growth of alloy crystals, membrane casting and cell growth.

MEA (Materials Experiments Assembly)

MECO (Main Engine Cut-Off) The point on the ascent-to-orbit timeline at which the main engines (the SSMEs) of a shuttle are shut down; it generally occurs about 8.5 minutes after launch.

MEDS (Multifunction Electronic Display System) The latest generation of 'glass cockpit' technology installed on the shuttle orbiters. These colour matrix liquid-crystal displays superseded the original cathode-ray tubes. Columbia was the first to be fitted, during its 1994–1995 upgrade. The other orbiters will be brought up to this standard as and when their upgrade cycle takes them back to Rockwell.

MEEP (Mir Environmental Effects Package) Four exposure cassettes that were affixed to the DM by spacewalking astronauts on STS-74. Two sampled the ambient particulate matter (one passively, but the other noted the time, size and trajectory of material impacting it), the others exposed construction materials (including insulation, paints, and optical glass coatings) that are to be used on the International Space Station. The entire package was retrieved (on STS-86) for analysis.

MEFCE (Mir Electric Fields Characterisation Experiment) An apparatus to record the state of the ambient energy in the 400 MHz to 18 GHz part of the radio spectrum inside and around the Mir space station.

MELEO (Materials Exposure in LEO) Several materials samples were mounted on the RMS, and the arm was held out from the orbiter to expose these coupons to atomic oxygen erosion.

Mephisto (Materiel Pour d'Étude Des Phenomenes Interessant La Solidification Sur Et En Orbit) A directional solidification furnace. It incorporated a set of video cameras so that scientists on the ground could follow its progress. It involved two furnaces (one fixed, the other movable) and three identical cylindrical samples (two solid, the other liquid). The electrical resistance at the solid–liquid interfaces was measured as the solid samples were driven together.

MET (Mission Elapsed Time) The timeline upon which a flight plan is based; although this allows for the uncertainty in the actual moment of launch, many events are keyed by real time.

Meteor-Priroda A class of Soviet/Russian weather and environmental monitoring satellites.

MFD (Manipulator Flight Demonstration) An in-orbit engineering evaluation of the SFA in advance of its installation on the International Space Station.

MFR (Manipulator Foot Restraint) A foot restraint that can be gripped by the RMS so that an astronaut can be swung on the arm to a work site. Although the basic form is a simple platform with boot grips, a more sophisticated version incorporating a lift-and-tilt foot-activated mechanism was also developed (and used in the delicate work on the HST). A stand-alongside workstation carrying tools can also be fitted. The ability to use the RMS to position astronauts during spacewalks is one of the most under-played successes of the shuttle programme.

MGBX (Middeck GBX) A refinement of the GBX to fit in a middeck locker.

MGME (Mechanics of Granular Materials Experiment) An apparatus to study the physical properties of cohesionless granular materials at low confining pressures in both dry and saturated states.

MICG (Mercury Iodide Crystal Growth) A French crystal experiment on the International Microgravity Laboratory. Because of its high atomic weight, mercury iodide offered the prospect of a very high quantum efficiency in detecting gamma and X-rays. The aim of this experiment was to grow larger and more regular crystals in microgravity.

microgravity Strictly speaking, this corresponds to a 10^{-6} g acceleration, but it is generally used synonymously with weightlessness. The best that can be achieved aboard a shuttle is about 5×10^{-5} g. A free-flying satellite such as EURECA can attain a genuine state of microgravity (so long as it does not carry any 'noisy' apparatus). The use of isolation mounts to damp out ambient vibrations should facilitate the attainment of even lower microgravity levels.

MIDAS (Materials In Devices As Semiconductors) An apparatus to measure the electrical properties of high-temperature semiconductors.

MIDAS (MIssile Defence Alarm System) A series of low-orbiting satellites launched in the 1960s to test sensors capable of detecting ballistic missile launches. These sensors were subsequently operationally deployed in the IMEWS satellites.

MIM (Microgravity Isolation Mount) An apparatus which was sent to Mir to test its ability to use a magnetic field to isolate an experiment from ambient vibrations.

Mir ('Peace', 'Community', 'New World') The first space station to be assembled using independently launched modules. Its 20-tonne habitat (the base block) was launched in February 1986 and the 8-tonne astrophysics module, Kvant 1, docked at the rear in 1987. A cluster of 20-tonne modules was added radially around the multiple docking adapter at the front: Kvant 2 in 1989, Kristall in 1990, Spektr in 1995 and Priroda in 1996. The shuttle mounted the 5-tonne DM on Kristall in 1995. The station is resupplied by the Progress-M ferry, and the Soyuz-TM serves as a crew ferry. The crew can evacuate at any time, because there is always a Soyuz attached. Residency is restricted to a crew of three, but the environmental system can support six people for short periods. The tour of duty is typically six months.

MIS (Microencapsulation In Space) A test to determine whether microcapsules created in space were more regular than those manufactured on Earth. Such biodegradable polymer microcapsules are used to effect timed-release of pharmaceuticals.

MISDE (MIr Structural Dynamics Experiment) A sensor package set up on Mir to measure the forces transmitted through the space station's structure by attitude control, docking and reboost activities.

mixed-fleet policy Following the Challenger accident, the Reagan administration decided that commercial satellites should be offloaded to the 'expendable' launcher which had previously been ordered to be phased out in keeping with the shuttle-only policy, so this new strategy was dubbed the mixed-fleet policy.

MLE (Mesoscale Lightning Experiment) A project to gain night-time imagery of lightning, in an effort to understand the way in which lightning in one area affects neighbouring areas, nearby storms and wind patterns. Payload-bay cameras fitted with image intensifiers were used. They could be controlled from the ground, so the observations could be conducted largely independently of crew activities, although additional studies were sometimes specifically requested. It was possible to monitor a 300-km wide track, as the orbiter flew over an active thunderstorm. Since it was possible to study lightning on a much larger scale than was practicable from the ground, this experiment was called 'mesoscale'. As with cloud-cover studies, this lightning project demonstrated the benefit of being able to view the Earth from space.

MLP (Mobile Launch Platform) The large horizontal slab-like platform on which a shuttle stack is assembled and, when lifted by the crawler, transported to the launch pad. In the days of Apollo, this platform included the tall service tower, but for the shuttle this was eliminated.

MLR (Monodisperse Latex Reactor) An apparatus to make large numbers of latex spheres all of a specific size (i.e. of a single dispersion in scale) between 1 and 100 μm, in order to calibrate apparatus.

MLS (Microwave Limb Sounder) A microwave emission radiometer on the UARS which specifically mapped the geographical and vertical distribution of chlorine monoxide, the primary agent of ozone depletion.

MMA (Microgravity Measurement Assembly) An ESA-built apparatus to monitor the state of the microgravity environment in an instrument module and present its results in real time.

MMIS (Military-Man-In-Space) See M88-1.

MMS (Multiple Mission Spacecraft) The modular bus developed by Fairchild Industries to be used on a variety of scientific satellites. It was the first spacecraft bus built to be able to be serviced by astronauts, which was fortunate because the first satellite to employ it, SolarMax, was the first satellite to be repaired in space. If the shuttle had flown in polar orbit, the Landsat-4 bus would have been repaired. One of the three modules on the bus used for UARS was that retrieved from SolarMax and refurbished for re-use.

MMU (Manned Manoeuvring Unit) The backpack that was used by astronauts requiring to fly free of the shuttle to retrieve satellites. It used nitrogen-gas thrusters to effect control over all six degrees of freedom. It combined exceptional stability and manoeuvrability, but it had limited duration. It will not be used on the International Space Station because it represents too great an operational overhead; astronauts will instead employ EMU suits fitted with the SAFER unit. Although the development of the MMU had begun during the Gemini programme, testing it was beyond the EVA state-of-the-art, but it was later tested *inside* the Skylab orbital workshop. The MMUs are preserved in storage, just in case they should ever be needed again.

MODE (Middeck 0-gravity Dynamics Experiment) A technology experiment to assess the vibration modes of high-fidelity models of truss structures which will be used on the International Space Station. Once the structure had been fitted with strain gauges, it was stimulated by a computerised shake-table which imparted calibrated impulses. It is not possible to fully determine the dynamics of a truss structure in ground trials, because it has to be supported against gravity, and the observed vibration modes are a convolution of both the truss and its support frame. Although testing a model in space did not suffer this complexity, care had to be taken to ensure that the results could be extrapolated to a full-scale truss. Two truss segments were tested: one a 1.6-metre long straight section, and the other incorporating a model of the 'alpha joint' which will rotate the station's solar panels. In other tests, the structures were monitored while floating freely. Another MODE experiment involved mounting containers of silicone oil to the shake-table to record the movements of the fluid inside.

MOMS (Modular Opto-electronic Multispectral Scanner) A camera system which could be operated in various modes, including searching for natural resources and making high-resolution topographic maps. It carried its own GPS-based package so that the position of the orbiter (within about 5 metres) and its orientation (to within 10 arcsec) could be assigned to each image. It was subsequently sent up to the Mir space station.

MPEC (Multipurpose Experiment Canister) This was essentially a GAS canister modified by the Air Force. It was used on STS-39, at which time it deployed a small satellite of a classified nature.

MPESS (Mission-Peculiar Equipment Support Structure) An open framework mounted in the payload bay using keel and sill trunnion pins, supporting a cross-bay bridge which can be reconfigured from one mission to the next to accommodate different payloads. It is often referred to as a 'multi-purpose' structure, but it usually has to be reconfigured to accommodate different payloads.

MPLM (Multi-Purpose Logistics Module) An Italian built pressurised module some 6.6 metres long which will ferry logistics up to the International Space Station.

MSL (Materials Science Laboratory) An MPESS carrying microgravity experiments which flew on STS-24. Although reflights with a variety of experiments had been planned for 1986, the loss of Challenger pre-empted this project, and it did not resume following Return-To-Flight.

MSL (Materials Science Laboratory) A Spacelab mission flown in 1997. When the orbiter had to return early because of a faulty fuel cell, MSL became the first payload to be kept in the bay during turnaround and immediately reflown with the same crew.

MSRE (Mir Sample Return Experiment) An instrument installed outside Mir's Kvant 2 module for micrometeoroid research. It was subsequently retrieved.

MSX (Midcourse Space Experiment) A satellite designed to test sensors for observing ballistic missiles manoeuvring above the atmosphere. It regularly monitored the shuttle, which fired its thrusters specifically for the purpose.

MTPE (Mission To Planet Earth) A comprehensive study of the Earth which considers the planet to be a single dynamic system. The first priority is to define the current state of the system, then to understand the various processes at work and their interactions, and ultimately to unambiguously identify any climate change phenomena. (Recently renamed Earth Science Enterprise by NASA.)

multiplexer–demultiplexer The shuttle orbiter has a fly-by-wire control system in which the pilot's command inputs are issued by moving a stick: movements which the computer digitises, interprets, and then relays to the appropriate systems to produce the specified effect. Actuators and valves require the digital signals to be converted to analogue form. Analogue–digital–analogue conversion is performed by multiplexer–demultiplexer pairs. The use of such a system enabled a single wire to be used to convey a variety of commands and instrumentation readings at different times. Nowadays, such linkage to payload-bay apparatus often includes dedicated processing elements in the chain; in the vernacular these are called 'smart flexibles'.

multi-zone A technique for growing crystals in microgravity in which a sample was moved through several thermal zones in such a way that the length of the sample was subjected to the complete temperature profile as it was processed. It was a very slow method of growing a crystal, but it produced an extremely pure result.

MVC (Measurement of Venous Compliance) An SPE study of fluid movements within the body as it adapts to weightlessness.

MVI (Microgravity Vestibular Investigations) A NASA study that used the Vestibular Sled to study the interaction of the sensory organs which provide the brain with information on the body's orientation and motion. In space, visual and vestibular cues conflict, and cause 'space sickness'. The objective of this study was to identify the precise nature of the coupling between the two systems.

MWPE (Mental Workload and Performance Experiment) A study to define the requirements for a computer workstation for use on the International Space Station. A set of interface tools for a laptop were arranged in different relative positions to assess their ergonomic merit when used in weightlessness. The ISS will be an *embedded system*, in that all of its apparatus will be run via computerised interfaces; so identifying the optimum configuration is important.

NASA (National Aeronautics and Space Administration) The coordinating agency that manages the US space programme. When it was formed in 1958, a number of already existing facilities were reassigned to it, and others were added later. Its facilities include the Ames Research Center, Moffett Field, California; the George C. Marshall Space Flight Center, Huntsville, Alabama; the Langley Research Center, Hampton, Virginia; the Hugh L. Dryden Flight Research Center at Edwards AFB in California; the Lewis Research Center, Cleveland, Ohio; the Robert H Goddard Space Flight Center, Greenbelt, Maryland; the Lyndon B. Johnson Space Center, Houston, Texas; and the John F. Kennedy Space Center on the Atlantic coast of Florida, which adjoins the Air Force's Eastern Test Range based at CCAFS.

Near-Mir The rendezvous demonstration in February 1995, in which STS-63 Discovery closed to within 10 metres of the axial docking port on the end of the Kristall module of the Mir space station, to prove procedures required for the Atlantis docking mission that was to follow.

NICMOS (Near-Infrared Camera and Multi-Object Spectrometer) This extremely-high-resolution imaging system was installed on the HST in 1997. It suffered problems with one of its three cameras, and consumed coolant at such an alarming rate that it was unlikely to run through until the next service call.

NIH (National Institutes of Health)

NIH-C An NIH experiment using the STL apparatus to develop cell cultures using chicken embryos, to observe muscle and bone cells for the effects of microgravity on musculo-skeletal development at the cellular level.

NIH-R An NIH study of rodent embryological growth. The rats were carried in an AEM.

Nimbus A series of weather and environmental monitoring satellites flown in polar orbit. Nimbus 7 – the last in the series, launched in 1978 – revealed the scale of the ozone hole phenomenon that forms over Antarctica each year, following the discovery by ground-based observers of stratospheric depletion.

NIZEMI A centrifuge ('slow-rotating', to impart a force in the range 0.001–1.5 g) fitted with a microscope so that the behaviour of small samples of protein crystals, protozoa, fungi, and plant and animal tissue cells could be studied. A centrifuge in space enabled 'control' tests for experiments studying adaptation to microgravity to be conducted *in situ*, and subject to all aspects of the space environment except weightlessness, rather than in the very different environment on the ground.

NOAA (National Oceanic and Atmospheric Administration)

NOSL (Night-day Optical Survey of Lightning) An early project to observe lightning from space, to test whether it would be practicable for weather satellites to distinguish storms most likely to give rise to tornadoes and lightning strikes.

NSTS (National Space Transportation System) The name for the shuttle while the shuttle-only policy was in effect. Following the switch to the mixed-fleet policy, the shuttle – no longer the national launcher – was down-graded simply to the STS.

OAIMT (Oxygen Atom Interaction with Materials Test) A study of how various materials were degraded by exposure to the atomic oxygen present in low orbit.

OARE (Orbiter Acceleration Research Equipment) This measured the accelerations created as the orbiter performed programmed roll, pitch and yaw manoeuvres, to characterise the disturbance to microgravity experiments when the orbiter manoeuvred, to contrast with the environment when the orbiter exploited the gravity-gradient for stability. It was also used to characterise the residual aerodynamic forces acting on the vehicle, because these can also degrade sensitive experiments (the force imparted by air drag is extremely low, but it is ever-present).

OASIS (Orbiter experiments Autonomous Supporting Instrumentation System) A system to collect and record a variety of environmental data during various phases of a flight. It was mounted in the aft part of the payload bay. It measured strain, sound, temperature and pressure.

OAST (Office of Aeronautics and Space Technology) In addition to instrumentation with which to monitor the performance of the shuttle during its first few flights, this NASA department flew several MPESS-mounted packages on the shuttle, and leased a SPARTAN free-flyer, for a broad programme for the development of space-related technologies. It supplied the instrumentation which monitored the performance of the orbiter during the initial test flights; namely ACIP, DFI, DATE, and OAIMT. Its technological payloads included PEP, SAMPIE, EISG, SKIRT, ECT and CTPE.

O&C (Operations & Checkout) The building at KSC in which payloads are integrated and tested prior to being transferred to the OPF for loading into a shuttle orbiter. It also provides living accommodation for crews in the last few days before a flight.

OCE (Ocean Colour Experiment) A test of a sensor optimised to identify chlorophyll in the ocean, to locate 'blooms' of algae and plankton upwelling from nutrient-rich cold water into the warmer surface layers in which most fish live. Cosmonauts on the Salyut space stations had routinely steered trawler skippers into rich fishing grounds, but by making visual observations.

OCGF (Organic Crystal Growth Facility) A Japanese experiment on the International Microgravity Laboratory.

OCTW (Optical Communication Through Window) This technology experiment tested the use of fibre-optic couplers for controlling payload bay apparatus by passing commands through the aft flight deck windows. Apart from demonstrating that the system worked, NASA wanted to test whether the radiation environment posed a source of interference.

ODERACS (Orbital DEbris RAdar Calibration Spheres) Small radar reflectors were ejected from GAS cans to calibrate the Haystack Near-Earth Assessment Radar which NASA introduced in 1992 to characterise the threat posed to the International Space Station by orbital debris of size greater than 1 cm. The attempt to deploy the first set of reflectors, on STS-53, was foiled by a flat battery in the ejector system.

ODS (Orbiter Docking System) A cylindrical airlock carried upright in the payload bay just aft of the cabin, with a short tunnel from the middeck airlock, and a Russian-built

APDS mounted on top, to enable it to dock at the Mir space station. Upon the completion of the Shuttle–Mir programme, it will be modified to serve the same role with the International Space Station.

OGLOW (Orbiter GLOW) A Canadian study of the way that the orbiter glows as a result of its passage through atomic oxygen.

OLIPSE (Optizon LIquid Phase Sintering Experiment) A NASA experiment that used the Optizon furnace in Mir's Kristall module to process metals by the sintering technique.

OMS (Orbital Manoeuvring System) The two large rocket motors contained within the pods on either side of the shuttle orbiter's tail assembly. They are gimballed so that they can be fired individually or as a pair. They are used to effect major orbital manoeuvres, and then to perform the deorbit burn. In addition, they are used to augment the SSMEs in the event of an ATO abort, and to achieve a high initial apogee for a direct-insertion ascent. The burn immediately after MECO (which is often not employed) is designated the OMS-1 burn; that at first-apogee to circularise the orbit is the OMS-2 burn.

OMV (Orbital Manoeuvring Vehicle) This was essentially a toroidal propellant tank around an axial motor, in a disk-shaped 'tug' that was designed to slot into the orbiter's bay like a wafer in a storage rack so that it would not take up much space. It would use a grapple similar to the end-effector of the RMS to mate with its payloads. The OMV was a priority development, because it was expected to play a vital role in servicing satellites in space. It would have been ideal for deploying satellites in 'high' Sun-synchronous polar orbits, at up to 1,500 km altitude, and also for retrieving such satellites so that the shuttle could replenish them. An upgraded OMV would have been able to deploy small communications satellites in geostationary orbit, and to retrieve them for refurbishment. However, following the Challenger accident, it was decided not to launch commercial satellites on the shuttle, and to abandon the plan to fly from Vandenberg, which denied the shuttle access to polar orbit; so the need for the OMV withered, and it was cancelled in 1990.

OPF (Orbiter Processing Facility) The building at KSC in which orbiters are refurbished following one flight and prepared for the next.

orbiter The shuttle orbiter vehicle (OV). In terms of serial designators, they are: OV-099, Challenger; OV-101, Enterprise; OV-102, Columbia; OV-103, Discovery; OV-104, Atlantis; OV-105, Endeavour.

ORFEUS (Orbiting Retrievable Far- and Extreme-Ultraviolet Spectrometer) A SPAS free-flyer. On its first flight it carried, in addition to its astronomical payload, samples of mirrors and detector materials to assess the effect of the orbiter as it manoeuvred nearby to test a laser rangefinder for use during rendezvous operations. On this occasion it also carried an IMAX camera which recorded the orbiter's departure following the satellite's release and, following retrieval, with the satellite being swung around on the RMS, to make an all-round photographic survey of the orbiter.

Orlan ('Eagle') The Russian EVA suit. This one-piece semi-rigid garment has an integral helmet and backpack. One side of the backpack is hinged, so it can be swung aside to form a hatch to enable the user to climb in and out. Because it must serve a succession of crews, the joints can be adjusted within certain narrow parameters to fit different users. A system of constant-volume joints make it flexible, and the gloves have proved to be sufficiently dexterous to facilitate delicate construction work. The Russians have accumulated a great deal of experience in spacewalking, most of it working outside the Mir space station.

ORS (Orbital Refuelling System) A test conducted early in the programme to prepare to replenish the Landsat-4 satellite, which was in need of extensive repairs and would use up all of its propellant in descending to rendezvous with the shuttle. However, because the shuttle facility at Vandenberg AFB was never commissioned, the shuttle was denied access to polar orbit.

orthostatic intolerance The blackout which follows the drain of blood from the head, upon standing upright immediately after returning to Earth. Regular exercise while in space, stamina enhancing drugs, cardiovascular reconditioning and inflatable positive-pressure g-suit leggings all help combat this brief danger. It is particularly important that shuttle crews be able to stand upright and move about in the immediate post-re-entry period just in case they need to effect an emergency egress.

ORU (Orbital Replacement Unit) Any piece of equipment that has to be replaced in space. On the International Space Station, the ORUs will include the chemical storage batteries for the solar power system.

OSS (Office of Space Sciences) This NASA department flew a mixed package of scientific instruments on the shuttle, most of which were subsequently flown again individually. Its payloads included PGU, PDP, VCAP, FME and SUSIM.

OSSE (Oriented Scintillation Spectrometer Experiment) An instrument on the GRO. It comprised four phoswich detectors that operated in pairs. Being a scintillation device, it measured gamma rays indirectly by the optical flash which is generated within a crystal upon absorbing a gamma ray. Since it was sensitive in the range 0.1–10 MeV, it could see gamma rays released by atoms undergoing radioactive decay, and since this process creates emission lines, OSSE could determine chemical composition. Unfortunately, by its nature such a detector does not offer very high spatial resolution, so its 3.8×11.4-degree field of view set the limit to which it could pinpoint sources. In addition to being used to study energetic processes in discrete objects such as pulsars and quasars, it was a powerful tool for investigating the enrichment of the interstellar medium by supernovae ejecting nucleosynthesised elements.

OSTA (Office of Space and Terrestrial Applications) This NASA department flew several packages on the shuttle, in the context of a very broad programme for the development of space-related applications. The payloads included SIR, MAPS, SMIRR, FILE, OCE, LFC, ORS and MEA.

otolith organs The body's acceleration sensors; small lumps on the end of tiny hairs in the inner ear whose inertia is translated into a sense of movement.

oxygen in the atmosphere The diatomic (O_2) molecular species is concentrated in the lower atmosphere, the troposphere, but the proportion of the triatomic (O_3) molecular species increases in the stratosphere, then decreases in the mesosphere, above which atomic oxygen (O) is predominant (it is this free oxygen that produces the effect called 'shuttle glow').

ozone The tri-atomic form of molecular oxygen (O_3).

ozone depletion The process by which ozone is destroyed. The primary cause of depletion is CFCs which, once they reach the stratosphere, are broken down by the ionising radiation. The free chlorine produced readily breaks down ozone to produce chlorine oxide and diatomic oxygen. Atomic oxygen then 'steals' the oxygen from the chlorine oxide, forming another diatomic oxygen and freeing the chlorine to attack another ozone molecule. A chlorine atom can thus act catalytically to break down a great deal of ozone before it becomes bound up in a compound that is not so easily dissociated. A similar catalytic reaction involves nitric oxide (NO). This breaks down ozone to create nitrogen dioxide, but atomic oxygen then steals back an oxygen atom, recreating the nitric oxide, which can then attack another ozone molecule, the overall result being loss of ozone. The nitric oxide itself results from the breaking down of nitrous oxide, which drifts up into the stratosphere from the troposphere. Another catalyst is hydroxyl (OH), which results from the dissociation of water vapour. What these agents have in common is that they are free radicals with sufficient affinity for oxygen to break a triatomic oxygen bond. The significance of the transformation of triatomic to diatomic oxygen is that whilst ozone strongly absorbs ultraviolet radiation, diatomic oxygen does not. So ozone-depleted zones allow the ultraviolet to reach the ground. The great irony, of course, is that it is the ultraviolet that breaks down the aerosols to release the catalytic agents. The variation in this component of the solar spectrum is therefore the driving factor of the ozone cycle.

ozone hole By definition, a 'hole' develops when the concentration of stratospheric ozone falls below 50 per cent of its nominal level.

ozone layer A layer of ozone in the stratosphere at a height of 20 to 50 km. It absorbs the ultraviolet component of sunlight, particularly at wavelengths shorter than about 3100 Å; this energy heats up this part of the atmosphere, thereby reversing the tropospheric trend in which air temperature decreases with altitude. By absorbing this ionising radiation, ozone played an important part in the development of life on the Earth.

PAM (Payload Assist Module) A McDonnell Douglas packaged form of the Star 48 solid rocket motor. It was designated PAM-D because it was compatible with the company's Delta rocket. It was also made suitable for release from the shuttle's payload bay so that customers would be able to switch payloads booked on the Delta to the shuttle without modification. By selling the kick-motors to such satellites, the company gained a source of revenue from shuttle operations. The 2,200-kg motor was able to send a satellite of 1,275 kg up to GSO. As soon as it had completed the 83-second GTO insertion burn, it released its payload, which had to perform its own circularisation burn upon reaching apogee. Originally, a variety of PAM models were planned, with larger motors to suit heavier payloads, but only one, the PAM-D2, which could send 1,900 kg to GSO, was actually marketed.

PAMS (Passive Aerodynamically-stabilised Magnetically-damped Satellite) A laser-based Attitude Measuring System (AMS) which was used to monitor the efforts of the STU sub-satellite to stabilise itself.

PARE (Physiological and Anatomical Rodent Experiment) An NIH study of muscle atrophy in the hind legs of rodents – muscles that are normally very active but unloaded in space. The rats were housed in an AEM.

PAS (Passive Accelerometer System) An apparatus installed on the Mir space station to measure low-level but continuous accelerations on the complex (as opposed to transient shocks transmitted through the structure), to assess the extent to which factors such as air drag decelerate the complex in its orbit, and how differential gravity drew items inside towards the Earth. It involved placing a small metal ball within a reference frame and recording how it drifted over time. These measurements helped to characterise the effects influencing microgravity experiments. It complemented the MISDE and SAMS instrumentation.

PAWS (Performance Assessment Workstation) An apparatus to study how an astronaut's cognitive skills were affected by the stresses of spaceflight.

Payload Systems Inc The first US private company to send an experiment to the Mir space station. The company negotiated a commercial contract to fly six experiments over a period of three years. The first was a protein crystal growth experiment in which two enzymes (hen egg white lysozyme and D-amino transferase) were processed by vapour diffusion and boundary-layer diffusion crystallisation processes. It was returned after two months. The experiments were packaged by the company on behalf of anonymous pharmaceutical clients.

PCG (Protein Crystal Growth) The production follow-on to the HH-PCG. In its various forms, PCG became one of the most frequently flown applications. Crystalline protein was subjected to X-ray crystallographic analysis to reveal its structure, as a first step towards making a drug to alter its activity; and as the crystals grown in microgravity were larger and their structure was more regular they were more readily studied. Many proteins were tested but because they grew at different rates only those that crystallised rapidly were practicable for shuttle flights which, even using the EDO pallet, are limited to 16 days. Independent experiments on the Mir space station showed that, given time, the slower-growing proteins do yield usable crystals, so this will be a major activity on the International Space Station.

PDE (Particle Dispersion Experiment) An apparatus to study the behaviour of particulates in a gaseous medium. It was used to test theories concerning how material injected into the atmosphere by a meteor impact would behave, how dust storms form on Mars, and how planetary nebulae form. It confirmed that static electricity plays a major role in dust cloud formation.

PDP (Plasma Diagnostics Package) A small sensor package which the RMS deployed just above the bay. It then sampled the contamination of the local environment as the orbiter manoeuvred clear, and returned to retrieve it.

PEM (Particle Environment Monitor) An electron, proton and imaging X-ray spectrometer on the UARS. It measured the flux of energetic particles in the solar wind impinging on the Earth's upper atmosphere.

PEP (Power Extension Package) This MPESS-mounted experiment tested the deployment mechanism for a solar panel. Although in its stowed form it was a mere 75 mm thick, it unfolded to create a panel 32 metres long. The test was important, because the frame was so flimsy that on Earth it could not be tested unsupported, and the support system influenced deployment. The test also enabled the frame's dynamics to be determined. In this case, only a few of the solar cells contained active transducers, and these had been installed so that their power output could be determined; but an operational panel would deliver 12.5 kW. It is intended to mount such a panel on either side of a segment of the International Space Station's truss, thereby to deliver 23 kW, and there will be four pairs on the completed station.

Perseids Often the most reliable meteor stream, it is associated with the orbit of periodic comet Swift–Tuttle. Although activity extends over several weeks, most of the activity is confined to 11/12 August. After the comet made its long-awaited return in late 1992 a higher-than-usual level of activity was predicted for the following year's meteor shower; so the launch of STS-51, which had been set for 4 August, was postponed. The merit of this precaution appeared to be verified when the cosmonauts on the Mir space station spacewalked to inspect the state of their vehicle, and discovered a 10-cm diameter hole punched right though one of the panels.

PFR (Portable Foot Restraint) A simple foot restraint platform that can be inserted into any convenient mount to allow an astronaut to work with both hands free.

PFTA (Payload Flight Test Article) An open-framework ballasted payload representing the size and mass of a large satellite. It was flown early in the programme to verify that the RMS could manipulate a typical satellite.

PGBA (Plant Generic Bioprocessing Apparatus) Although flown for testing on the shuttle, this apparatus is intended for long-term plant growth studies on the International Space Station.

PGF (Plant Growth Facility) A modular form of the PGU that was used to grow seedlings of pine and fir trees, which were fixed to halt their growth for post-flight examination.

PGU (Plant Growth Unit) The cultivator that was used for the LIGNIN and CHROMEX experiments, and later upgraded into the PGF.

phoswich A scintillation detector comprised of two crystals which were optically coupled to form a phosphor sandwich (phoswich). One crystal served as the X-ray detector, while the scintillator acted as a transducer whose visible light was enhanced by a photomultiplier.

PIE (Particle Impact Experiment) An instrument that was installed outside Mir's Kvant 2 module for micrometeoroid research, and subsequently retrieved.

PILOT (Portable In-flight Landing Operations Trainer) When not in space, shuttle pilots tend to fly several times a week to maintain their proficiency, if not actually to travel, so not flying when in space represents anomalous behaviour. Upon the introduction of the EDO, the possibility of a pilot's flying skills significantly deteriorating after 16 days without flying in weightlessness became a matter of some concern. It is especially important that shuttle pilots be 'sharp' during the descent, because the shuttle makes a gliding final approach under manual control and has very unforgiving flight characteristics. In order to enable pilots to run through simulated landings whilst still in orbit, to revitalise their hand–eye coordination, a laptop computer was set up to take inputs from the orbiter's control stick and to display a view representing what the pilot would see in the HUD. It not only allowed pilots to sharpen their skills; it also identified the extent to which their skills had deteriorated while in space. Norm Thagard, who was a pilot, but not a shuttle pilot, used it regularly during his tour aboard the Mir space station to produce a longer run of data.

PKM (Perigee Kick-Motor) A rocket motor – usually a solid rocket – used to boost a satellite from the 'parking' orbit in which its launcher released it, into the 'transfer' orbit which will take it up to GSO. The motor is jettisoned after use.

PM (Polymer Morphology) An experiment devised by the 3M Company to determine the effects of microgravity on morphological formation of polymers in physical transition.

PMP (Polymer Membrane Processing) A study of the formation of polymer membranes in space. Selectively porous membranes are used as filters, to separate fluids and gases, to desalinate water and to release encapsulated drugs, but the force of gravity can corrupt the structure of the membrane. This experiment was undertaken for the CCDS operated by the Battelle Laboratory.

POCC (Payload Operations Control Centre) This facility at MSFC (call-sign 'Huntsville') superseded the much smaller POCC at JSC. The first mission to make full use of this facility was STS-35, with its ASTRO-1 telescope cluster.

PPE (Phase Partitioning Experiment) This used two immiscible liquids to separate (demix) biological agents. The apparatus was repeatedly photographed so that the process could subsequently be studied in detail. It focused on the separation of biological cells and pharmaceutical materials, and was flown many times.

Progress-M The current form of the cargo ferry used to resupply the Mir space station. A 7.5-tonne vehicle derived from the Soyuz spacecraft, it can carry 1,500 kg of dry cargo and 1,000 kg of fluids. The dry cargo is unloaded manually, but the fluids are pumped through pipes in the docking collar into tanks in the base block. The International Space Station's Service Module is derived from Mir's base block, and will be replenished in a similar manner.

protein research X-ray crystallography can reveal the structure of a protein molecule. The best crystalline protein is that grown in the absence of the disturbing effects of gravity. Once a protein's structure is thoroughly understood, drugs can be designed to influence its behaviour. Samples of proteins which crystallise rapidly could be grown on shuttle flights, but large crystals of the more slowly-growing proteins will become practicable only once the International Space Station enters service. Microgravity research will also enable the physics involved in the crystallisation of such a complex molecule to be studied to yield insight into how gravity degrades the process, with a view towards improving the terrestrial process.

PSE (Physiological Systems Experiment) An experiment involving rodents housed in an AEM.

PSN (Positional and Spontaneous Nystagmus) An SPE study of the involuntary eye motion that occurs when the otolith organs are stimulated.

PVTOS (Physical Vapour Transport Organic Solid experiment) An experiment provided by the 3M Company to make organic thin film with a crystalline structure with specific electrical and chemical properties. An organic sample was vaporised and the transported through a buffer gas to crystallise on a shaped surface.

QINMS (Quadropole Ion Neutral Mass Spectrometer) An instrument that accompanied the CIRRIS telescope. It measured contamination in the immediate vicinity of the orbiter, to indicate when the outgassing dropped to a level sufficiently clean to run the telescope. It also directly sampled the gaseous environment when the orbiter flew through an auroral display.

QSAM (Quasi-Steady Acceleration Measurement) A set of sensors to measure the state of the microgravity environment.

QUELD (Queen's University Experiment in Liquid Diffusion) An isothermal furnace set up on the Mir space station to study boundary-layer processes to measure the diffusion coefficients of semiconductors, binary-metals and glasses.

rad The unit of absorbed radiation dose, equivalent to 0.01 Joule/kg of energy absorption.

RAHF (Research Animal Holding Facility) A rack-sized facility for a Spacelab module that provided support for a batch of AEMs.

RAMSES A French investigation of electrophoresis as a means of separating biologically active substances. Unfortunately, it suffered problems with its pump, and testing had to be curtailed.

R-Bar In terms of rendezvous, this is an approach in which the active spacecraft moves on a line linking the centre of the Earth to the passive target. If the approach is from above, it is dubbed 'negative R-Bar'. Prior to Shuttle–Mir, in which the orbiter approached the space station from below, only the HST had been approached in this way. LDEF had been approached from above. A significant advantage of a 'positive R-Bar' approach is that, in the final phase, the differential gravity resulting from the vertical separation slows the rate of closure, so, if the active spacecraft loses its braking thrusters, gravity will, if the closure rate is low, draw the rising vehicle to a halt and then pull it back down, thereby preventing a collision. It was this fail-safe which made R-Bar attractive for Shuttle–Mir, because an impact between the two 100-tonne vehicles would probably be catastrophic to both. The 'R' signified the use of the Earth's radius vector.

RCRS (Regenerative Carbon dioxide Removal System) A closed-cycle system to remove carbon dioxide from the cabin air during EDO missions, because the non-regenerative LiOH system would require a prohibitive number of canisters to be carried. The RCRS was based on the unit which was used on Skylab, and is to be used on the International Space Station. It incorporated a pair of adsorbing cells, only one of which was active at any given time, because while one was active the other was exposed to vacuum to purge it; the two cells swapped roles every 20 minutes. On its first trial (STS-50) it required maintenance, but thereafter proved trouble-free. Although it may eventually be carried by default, the LiOH processor will have to be retained and a few of its canisters carried to provide a contingency in case of a fault in the regenerative system.

RCS (Reaction Control System) The low-thrust rocket motors in the nose and around the tail of the shuttle orbiter to effect fine control over the vehicle's motion, and to adjust its orientation. They draw propellant from the same tanks as the OMS engines, and there is an extensive crossover network to ensure that the system is multiply-redundant, and that the RCS will be able to make a prolonged burn to effect deorbit should the OMS be disabled.

Return-To-Flight The STS-26 mission in September 1988. It marked the end of the shuttle's grounding in the aftermath of the loss of Challenger.

RME (Radiation Monitoring Equipment) A hand-held instrument that measured the ionising radiation within the orbiter's cabin. It sensed gamma rays, electron, proton and neutron fluxes and produced a running rate of exposure in tissue-equivalent rads, for post-flight analysis. Several generations of this apparatus were tested, but it was RME-3 that flew most often.

RMS (Remote Manipulator System) The 'robotic arm' built by Spar Aerospace of Canada for use on the shuttle. An upgraded version is to be installed on the International Space Station.

ROMPS (Robot Operated Materials Processing System) A tele-operation system to operate experiments from the ground. It was tested by extracting samples from storage, loading them into a furnace, and then retrieving and storing the results.

ROSAT (ROentgen SATellite)A multinational facility named for the discoverer of X-rays, Wilhelm Roentgen. A wide-field detector made an improved sky survey, and follow-up high-resolution observations were made of given objects. It was to have been deployed by the shuttle, but the loss of Challenger so delayed this that it was offloaded onto a Delta rocket and launched in 1990.

ROTEX (RObоT EXperiment) A German robotic arm, tested on STS-55. It could be operated either by the astronauts or remotely from the ground. In addition to testing the arm's articulation and end-effector mechanisms, opportunity was taken to assess the suitability of a standard joystick controller in weightlessness.

RRMD (Real-time Radiation Monitoring Device) An apparatus to measure, in real time, the high-energy radiation that penetrated the orbiter's cabin.

RSA (Russian Space Agency) The headquarters, cosmonaut training and mission control facilities are located at Kaliningrad in a northern suburb of Moscow.

RSAD (RMS Situational Awareness Display) A computer display that drew on data from a variety of sensors in order to present the RMS operator with a better appreciation of the arm's position in a confined space in which direct vision is impaired; it is intended to use such an aid on the International Space Station, in conjunction with the SVS.

RSS (Rotating Service Structure) An enclosure on the launch pad that protects the shuttle from the weather. It contains a clean room so that a vertically-integrated payload can be loaded into the orbiter's bay (as was often done by the Air Force). It swings back 120 degrees on rails at an early stage in the countdown.

RTG (Radioisotope Thermoelectric Generator) A container of plutonium 238 that radiated heat as a consequence of the radioactive decay process. If the heat was transformed into electricity by a thermocouple, it could be used to power a spacecraft. Although initially developed to power military satellites, RTGs are now used only to provide power on spacecraft that venture into the outer regions of the Solar System, where the energy in sunlight is insufficient to operate a conveniently-sized array of solar transducers. No fission is involved, so the RTG is not a nuclear reactor; it is simply a container of plutonium dioxide that undergoes natural radioisotopic decay. Although, by everyday standards, the power level of 0.5 kW is minuscule, it is sufficient to run a spacecraft. The only shuttle-launched spacecraft to employ RTGs were Galileo and Ulysses.

RTLS (Return To Landing Site) The first abort mode available to a shuttle during the long climb to orbit. If one or more SSMEs shuts down during the first two minutes, a RTLS will become necessary, but no action will be taken until the SRBs have run their course and have been discarded. The abort involves the rest of the stack climbing as high as it can, the OMS engines augmenting the thrust of the remaining SSMEs if necessary, with a pitch-over so that it can cancel its horizontal speed. It then heads back to KSC, with this large vertical looping trajectory calculated so that the ET can be jettisoned just as the descending orbiter intersects a viable gliding-descent trajectory. For a launch to go ahead, the weather at the SLF must not exceed 50 per cent cloud cover at or below the 8,000-foot level. The manoeuvre has never been tested in reality, but all shuttle pilots rehearse it in a simulator.

S4 (Spacehab Soft Stowage System) A set of colour-coded canvas holdalls in which cargo to and from the Mir space station is carried in the Spacehab. The adoption of this scheme greatly simplified logistics management.

SAFER (Simplified Aid For EVA Rescue) A small nitrogen-gas propulsion unit that was developed to fit underneath the EMU backpack in order to enable an astronaut who had suffered a broken tether to return to safety. During a spacewalk in the orbiter's bay, the shuttle would be able to chase a drifting astronaut; but an astronaut working outside the International Space Station would not be able to be retrieved, so this self-rescue pack is to be carried as standard. It was tested on STS-64, and then used operationally during spacewalks outside Mir, because the shuttle could not have undocked rapidly enough to chase after a drifting astronaut.

SAGE (Stratospheric Aerosol and Gas Experiment) One of the first instruments to monitor stratospheric ozone. Its photometer observed the Sun during orbital sunrise and sunset. By monitoring the spectral absorption features due to the passage of sunlight through the atmosphere, this instrument measured ozone and aerosol concentrations; and because the light's trajectory was *horizontal* it was able to measure this concentration as a function of *altitude* as well as location. This enabled the distribution of aerosols in the atmosphere to be recorded in three dimensions. Furthermore, over time it was possible to monitor the dynamics of the ozone layer.

Salyut ('Salute') The series of Soviet space stations which preceded Mir.

SAM (Shuttle Activation Monitor) A set of sensors to measure the flux of gamma rays as a function of time at various points in the shuttle's cabin. The data were recorded on tape for later analysis by the Department of Defense. It complemented the measurements made by the CREAM sensors.

SAMPIE (Solar Array Module Plasma Interaction Experiment) An evaluation of how the atomic oxygen present in LEO affected samples of the solar cell arrays and high-voltage systems to be used on the International Space Station. The electrical state of the plasma in the immediate environment was monitored in order to calibrate the exposure results.

SAMS (Space Acceleration Measurement System) An instrument that incorporated a set of sensitive accelerometers to measure the microscopic forces which result from the orbiter firing its thrusters, the movements of its crew, and the vibrations caused by its various apparatus. By providing a running record of these forces, it measured the extent to which experiments had been 'polluted'. In addition to being carried routinely on microgravity Spacelab missions, a SAMS instrument was sent up to the Mir space station.

SAR (Synthetic Aperture Radar) The linear resolution of an imaging radar is dependent on the size of the antenna as well as on its frequency. If a spacecraft has only a small dish, it can exploit the SAR processing to improve resolution. This technique exploits the fact that the spacecraft is moving to 'synthesise' a larger antenna. To construct an image, the real antenna has to record the phase of the radar reflection at each point as well as its intensity. The penalty of this technique is that each radar scan generates a vast stream of data.

SAREX (Shuttle Amateur Radio EXperiment) A handheld frequency-modulated transceiver operating at 145 MHz that is routinely carried on shuttle missions; its fold-up antenna is deployed in a convenient window facing the ground. Astronauts use it to keep in touch with 'hams' around the globe (many astronauts are themselves registered stations). On some missions, when orbital dynamics permit it, communication with the cosmonauts on the Mir space station is possible for brief periods. Many of the ground stations host student parties. But SAREX is far more than an experiment; it serves as an assured back-up communications capability in the event that the orbiter's primary system is knocked out. The system on Mir proved its worth when a fire broke out in 1997; the station was out of direct contact with Kaliningrad at the time, so news of the incident was relayed to the flight controllers via the amateur radio network.

SAS (Space Adaptation Syndrome) The 'space sickness' that afflicts about half of humans on initial exposure to weightlessness, due to a combination of vestibular confusion and the pooling of blood in the head and upper torso.

SASE (SAS Experiment) An SPE study of the otolith organs which involved moving back and forth on the Vestibular Sled to provide linear acceleration, while sensors on the legs measured reflex responses to study 'space sickness'.

SATO (Space Adaptation Tests and Observations) A Canadian study of the initial phase of 'space sickness'.

SBUV (Solar Backscatter UltraViolet) An instrument carried on a variety of environmental studies satellites, including NOAA 9 and NOAA 11 Measuring the amount of ultraviolet in sunlight reaching the Earth and in the reflection from its atmosphere enabled ozone concentration in the upper atmosphere and its vertical distribution to be calculated. Ozone absorbs most ultraviolet radiation (which is why the upper atmosphere, the stratosphere, is hot). SBUV took data in 12 discrete spectral bands in the ultraviolet range to measure the ozone absorption feature. It took its measurements when looking vertically down at the nadir. Other instruments operated by measuring absorption of sunlight when the Sun was on the limb.

SCA (Shuttle Carrier Aircraft) A pair of Boeing 747s modified to transport a shuttle orbiter on their upper fuselage.

SCE (Solar Constant Experiment) An instrument to measure the total energy in the insolation.

SCM (Spinal Changes in Microgravity) A refinement of the BPA experiment, but using an echograph in addition to stereophotography.

SDI (Strategic Defense Initiative) A national programme announced by President Ronald Reagan on 23 March 1983. It was to develop a 'peace shield' to protect America from intercontinental-range ballistic missiles in an early stage in their trajectory. Research for techniques to incapacitate an incoming strategic missile was funded. Many space-based approaches were pursued, including kinetic-kill interceptors, X-ray lasers and particle beams, but no 'Star Wars' systems were actually deployed. The programme was later scaled back and refocused to intercept tactical ballistic missiles – a somewhat simpler and more achievable task – as the Ballistic Missile Defence Organisation (BMDO).

SDOC (Space Defense Operations Center) This facility continuously monitored a shuttle's trajectory to provide warning of possible collision hazards. A close encounter is defined as an object in an orbit that would cause it to pass within three kilometres of the orbiter. Whenever such an encounter is predicted, the shuttle alters its orbit slightly to increase the margin of clearance. On STS-48, for example, the upper stage of the rocket that had launched Cosmos 955 would have cut across Discovery's path, some 2 km ahead of it, so the shuttle manoeuvred to open the gap to 16 km. And on STS-72, Endeavour was warned that it would pass near to the Air Force's MISTI satellite, so it adjusted its orbit to open the point of closest approach.

SeaSat A satellite developed jointly by JPL with the Navy. A side-looking SAR which had originally been used for airborne reconnaissance was adapted to study the ocean. Although it was crippled by a power failure a few months after launch in 1978, the quality of the data was so impressive that JPL secured funding to place a similar radar on a spacecraft to be sent to Venus to map its surface through the planet-girdling cloud cover, a mission which flew as Magellan.

SEF (Space Experiment Facility) A generic facility that, on its first test, had two furnaces: one to grow crystals of cadmium telluride, and the other to process metals by sintering.

SEH (Solar Extreme-ultraviolet Hitchhiker) An instrument to observe emissions from the Sun and stars in the extreme-ultraviolet.

SEM (Student Experiment Module) A renewal of the shuttle's educational programme. In effect, it was a GAS can divided up to accommodate 10 compact packages. On its first outing it carried a variety of fluid, crystal and biological demonstrations.

SEPAC (Space Experiments with a Particle ACcelerator) An electron beam emitter that was used to stimulate the ionosphere (in effect, to create a small artificial auroral display) in order to trace the lines of force of the Earth's magnetic field. The electron pulses were matched by ionospheric glows. The experiment had the advantage that the source of the stimulation was calibrated. After test on Spacelab 1, it flew as part of the first ATLAS mission, on which occasion a piece of payload-bay debris shorted out the electron emitter. Low-light monochrome imagery had already been acquired for 60 such displays, so the experiment was deemed to have been successful.

SEPRE (Shuttle Electric Potential and Return Experiment) An instrument to measure the electron density around the orbiter during SETS operations, to monitor charge build-up which might affect the vehicle.

SETS (Shuttle Electrodynamic Tether System) Italy's test satellite for the TSS programme, designed to generate electricity by dragging a wire through the Earth's magnetic field.

SFA (Small Fine Arm) The dexterous remote manipulator arm built to service experiments mounted on the exposed facility which will form part of the Japanese laboratory of the International Space Station. The 1.5-metre long arm had a shoulder roll-and-pitch joint, an elbow pitch joint and a wrist yaw-and-pitch joint.

SFU (Space Flyer Unit) A 4-tonne Japanese satellite that was launched by H-2 rocket and retrieved by STS-72.

SHARE (Station Heatpipe Advanced Radiator Experiment) This heat pipe was a prototype of the unit to be joined edgewise to form the large radiator panels of the International Space Station. Liquid ammonia coolant, vaporised by heat at one end, moved along the 15-metre long pipe by thermally-induced convection. A fin unit at the far end radiated the heat to space, which recondensed the ammonia. The outer vapour transport pipe contained an inner pipe which branched out like a three-dimensional tuning fork at the end that absorbed heat – the evaporator end. This inner pipe transported the liquid. In the evaporator unit, a fine wire mesh functioned in the same manner as an oil lamp's wick, and drew out the liquid ammonia, which then vaporised. Circumferential grooves at the far end enabled the condensate to return to the reservoir. It was installed running the length of the starboard sill of the payload bay. The International Space Station will have six radiator panels, each incorporating several dozen SHARE elements. One advantage is that because its elements are independent, it will provide a highly redundant thermal control system, and because it has no mechanical apparatus it will not draw any power from the station. Furthermore, it will not wear out, so it should be able to run for years. By using surface tension – a force that dominates in microgravity – it is ideally suited to its environment. For trials, the evaporator was driven by an electrical heater. The first test had been set for 1986, but the loss of Challenger delayed it to STS-29; however it was foiled when air bubbles formed. A sub-scale model with a plexiglass cover to facilitate filming was used on STS-37, and it too became clogged, as expected. On STS-43, an updated design was tested. Although the bubbles formed, the heat-pipe was able to clear them and work on. With the system operating, the orbiter turned its tail to the Sun, to chill the radiator, and then a more powerful heater was applied to the evaporator in order to demonstrate that it could shed the highest predicted thermal load. Finally, it was deactivated and allowed to freeze before being restarted, to demonstrate that it would survive an extended period in shadow.

SHEAL (Spacelab High Energy Astrophysics Laboratory) A series of Spacelab flights that had been planned to start in the late 1980s, but which was cancelled following the Challenger accident. One of the instruments, BBXRT, was later incorporated into the first ASTRO mission.

SHOOT (Superfluid Helium On-Orbit Transfer) Apparatus to test the replenishment of a tank of superfluid helium in space. It was a Hitchhiker package consisting of two large dewars – one full and one empty – so that the cryogenic fluid could be moved back and forth from one to the other. At one point, the orbiter was made to perform a series of rapid turns in order to record the movements of this zero-viscosity fluid in its container. Superfluid helium at 2 K is used to chill the detectors of the most sensitive infrared telescopes (such as ISO and the yet-to-be-launched SIRTF), but it rapidly leaks through the pores of its container, so the lifetime of such a telescope is limited by its supply of coolant. SHOOT secured a step towards the capacity to replenish such an instrument.

Shuttle–Mir The series of shuttle rendezvous and docking missions that served to integrate the NASA/RSA space programmes. In effect, this was Phase One of the development of the International Space Station.

shuttle-only policy When the Nixon administration authorised the shuttle's development, it ordered that 'expendable' rockets be phased out and that satellites be launched on the shuttle. This 'shuttle-only' policy manifested itself in the shuttle being designated as the National Space Transportation System (NSTS).

SIMPLEX (Shuttle Ionospheric Modification with Pulsed Local Exhaust Experiment) The shuttle did not actually carry any apparatus specifically for this study of the ionosphere; it simply fired its OMS engines at times designed to enable ground radars to observe the effect on the plasma within the ionosphere (the plumes created substantial voids in it).

sintering The process by which a material can be fused in a high-temperature and high-pressure environment without invoking melting. Unlike melting, mixing and resolidification, the sintering process preserves the structure of the raw material.

SIR (Spaceborne Imaging Radar) Derived from the SeaSat SAR, the SIR scanned a 50-km wide swath with, at best, 10-metre resolution, so it was ideal for terrain mapping; and as it was a dual-frequency L/C-Band system, it also had a capability for resource mapping. The Landsat satellites had pioneered multispectral imagery, but they required clear weather for their optical scanners. The radar was all-weather capable.

SIR (Standard Interface Rack) A modular instrument rack tested on STS-58. This allowed apparatus to be easily reconfigured in flight without jeopardising the mechanical, power and data processing interfaces involving adjacent apparatus.

SKIRT (Spacecraft Kinetic InfraRed Test) A study of the glow phenomena caused by the atomic oxygen plasma in LEO. The orbiter was put through a specific series of manoeuvres to calibrate the sensors.

SLA (Shuttle Laser Altimeter) Used to demonstrate laser altimeter operations in Earth orbit in support of the development of sensors capable of measuring the height and the structure of clouds and the layering of aerosols in the upper atmosphere.

SLF (Shuttle Landing Facility) The 5,000-metre long, 45-metre wide concrete strip built at the Kennedy Space Center to enable an orbiter to land either after a RTLS abort or after a return from orbit. It was initially heavily grooved to prevent aquaplaning but, when it was realised that the grooves increased tire wear at the moment of contact, the grooves were scraped off at the touchdown point (at both ends, because an approach can be made from either end).

SMIRR (Shuttle Multispectral InfraRed Radiometer) An experiment to determine the best spectral bands for remote sensing of natural resources. Its 18-cm telescope was a spare from the Mariner 10 mission to Mercury. It used a detector sensitive in the 0.5–2.5 micrometer infrared spectrum. Even in this trial, the instrument revealed details in the test sites that had been missed by field geologists.

SN1987A The supernova that was seen to explode in the LMC on 23 February 1987. As it was the nearest such event to be observed since the invention of the telescope, it was intensely monitored by a variety of ground-based and satellite-borne telescopes.

Sokol ('Falcon') The Russian launch-and-entry pressure garment. Each suit is tailored for a specific user, so each astronaut left on Mir by shuttle has to take a suit, just in case an emergency descent by Soyuz becomes necessary. Also, because the Soyuz flight couch has a liner which is precisely shaped to fit a specific user, each astronaut must also take this aboard Mir. The point of handover between NASA residents on the Russian space station occurs when the newcomer's suit and couch liner are transferred to the station.

solar constant A measure in the visible-light component of sunlight at the Earth's distance from the Sun; some 1.3 kW/m^2. Although referred to as a 'constant', this is a actually a significant misnomer, because the output of the Sun varies during the roughly 11-year sunspot cycle, and possibly differently over the roughly 22-year magnetic cycle; and it may well vary on an even longer-term basis. In addition, because the Earth's orbit is slightly eccentric, it does not maintain a constant distance from the Sun; and the energy per unit area decreases as the square of the distance, so even this minor variation produces a significant seasonal variation.

SOLCON (SOLar CONstant) An instrument to measure the total output of the Sun across the spectrum from the ultraviolet to the infrared. It is not possible to make this measurement from the ground because these parts of the spectrum are absorbed by the atmosphere. Actually, it is *because* this radiation is absorbed that it is important to measure it, because it is *this* energy which drives atmospheric processes.

SOLSPEC (SOLar SPECtrum) An instrument to measure the energy density across the solar spectrum from the ultraviolet to the infrared. It is necessary to make measurements throughout the solar cycle to identify any cyclic variations in the Sun's spectrum, because the chemical reactions occurring in the upper atmosphere are dependent on the intensity of sunlight at specific wavelengths.

SOLSTICE (SOLar–STellar Irradiance Comparison Experiment) An ultraviolet spectrometer on the UARS to compare the Sun's ultraviolet output with that from a selection of stars which radiate strongly in that part of the spectrum, to establish a system of celestial references against which the solar output can be calibrated. Enabling satellite sensors to be self-calibrating will obviate the need to fly pristine instruments on the shuttle to monitor the degradation of the satellite sensors.

sorption cooling A cooling technique that operates with essentially no vibration, is efficient, and can operate reliably for periods of many years. It uses a powdered metal alloy that absorbs hydrogen refrigerant by a reversible chemical reaction. The powder is cooled to absorb the gas, and the sudden decrease in pressure prompts the temperature to plummet to near-zero, which freezes the remaining hydrogen into a solid block which can then be used to absorb residual heat from an apparatus by sublimation. If the cycle is closed, the sublimated hydrogen can be rechilled at a later date once the device is no longer able to act as an efficient cooler.

SOSP (SOlar SPectrum) A SOLSPEC instrument which flew on the EURECA free-flyer and made extended observations.

South Atlantic Anomaly Located off the coast of Argentina, south of the Brazilian border, this is an 'anomalous' region where the innermost of the Earth's radiation belts – a torus around equatorial latitudes at an altitude of 1,000–5,000 km – dips several hundred kilometres towards the surface. A spacecraft passing through this zone is exposed to a more intense dose in a few minutes than it accumulates during the rest of the orbit.

SOVA (SOlar VAriability) A SOLCON instrument which flew on the EURECA free-flyer and made extended observations.

Soyuz ('Union') The three-person spacecraft which is used to ferry cosmonauts up to and back from the Mir space station.

Spacehab A middeck augmentation module developed to accommodate middeck lockers. Its payloads included BPL, LEMZ, 3DMA, Astroculture, CGBA, SRE, CPCG, ECLIPSE, IMMUNE, SEF, SAMS, Charlotte, BR, LSRF, HTLPE, QUELD, MEFCE, CFZF, ADSEP, PGBA, FGBA, HH-DTC, GPPM, NIH-C, ARIS, ETTF, MGME, EORF, BTS, MIDAS, TEHM and IMAX. Although designed to facilitate commercialisation of microgravity, it is now used as a cargo-hauler.

Spacelab Soon after NASA announced that it intended to develop a reusable space shuttle, ESA decided to build a multifunction laboratory facility to be carried in its payload bay. Spacelab is actually a system of components that can be combined to suit requirements. Its two basic elements are a pressurised module and an unpressurised pallet. Although a 'short' module was developed as the basic unit, this configuration has never been used; the utilities section has always been flown mated to an experiment section as a 'double' module, a 7-metre long cylindrical chamber that fitted snugly within the payload bay. Each module segment can accommodate a roof hatch for either an airlock or a porthole for instruments. The module is connected to the orbiter's middeck airlock by a tunnel, but because the module's hatch is mounted axially, and the airlock is below this level, the tunnel has to be kinked. The 3-metre long U-shaped load-bearing orthogrid was covered by an aluminium-sheet skin, and was mounted in the bay by keel and sill trunnion pins. Because the pallet was originally devised as an unpressurised instrument carrier, a pressurised adjunct (the igloo) was built to carry the necessary control systems. The U-shaped pallet was designed so that five could be laid in train along the length of the bay, but the longest configuration thus far flown employed three pallets. Sometimes the pressurised module is accompanied by a pallet; on occasion, pallets have been used simply to carry cargo. The pressurised module does not contain any built-in hardware; the mission-specific rack-mounted instrumentation is loaded as a single structure during payload integration. Spacelab was engineered to be able to fly a dozen times a year, but the shuttle was not able to meet NASA's early predictions in this respect. Nevertheless, Spacelab has undoubtedly been a major success. The major Spacelab sub-programmes include the Atmospheric Laboratory for Applications and Science (ATLAS), the International Microgravity Laboratory (IML), the Space Life Sciences (SLS), the Life and Microgravity Science (LMS), and the Materials Science Laboratory (MSL).

Spacelab Missions

Name	Mission	Year	EDO	Focus
Spacelab 1	STS-9	1983	x	systems test; general research demonstration
Spacelab 3	STS-17	1985	x	microgravity and life sciences
Spacelab 2	STS-19	1985	x	solar physics
Spacelab D1	STS-22	1985	x	microgravity and life sciences
ASTRO-1	STS-35	1990	x	astronomy
SLS-1	STS-40	1991	x	space life sciences
IML-1	STS-42	1992	x	microgravity
ATLAS-1	STS-45	1992	x	atmospheric studies
USML-1	STS-50	1992	1	microgravity
Spacelab J1	STS-47	1992	x	microgravity and life sciences
ATLAS-2	STS-56	1993	x	atmospheric studies
Spacelab D2	STS-55	1993	x	microgravity
SLS-2	STS-58	1993	2	life sciences
IML-2	STS-65	1994	4	microgravity
ATLAS-3	STS-66	1994	x	atmospheric studies
ASTRO-2	STS-67	1995	5	astronomy
Shuttle-Mir	STS-71	1995	x	life sciences (long-duration Mir crew)
USML-2	STS-73	1995	6	microgravity
LMS-1	STS-78	1996	8	life and microgravity sciences
MSL-1	STS-83	1997	9	materials science
MSL-1R	STS-93	1997	10	materials science
Neurolab	STS-90	1998	13	neurological life sciences

spacesuit See *EMU*, *Orlan*, *Sokol*

SPARTAN (Shuttle Pointed Autonomous Research Tool for AstroNomy) A free-flyer that was built to carry instruments already developed and flown on sounding rockets, in order to enable them to gather data for periods of hours or days, rather than minutes. The initial bus was subsequently superseded by a much heavier platform capable of operating for up to a week.

SPAS (Shuttle Pallet Applications Satellite) This free-flyer was built by Germany to carry small self-contained microgravity packages. The original platform flew twice, although it was unable to be released on its second mission due to a fault in the RMS. It was later sold to the US Department of Defense, which flew it as the IBSS. Germany then built a heavier platform to carry observational instruments for longer periods (up to a week) as CRISTA-SPAS and ORFEUS-SPAS.

SPDS (Stabilised Payload Deployment System) A payload dispenser built by Rockwell for the Air Force to enable it to release certain of its satellites from the shuttle payload bay. It comprised a pair of articulated hinges, one at either end of the payload, which rotated the satellite out of bay and then held it over the port sill while it was given a final check prior to release. It could also be used to retrieve a suitably equipped satellite. It was built to deal with more massive satellites than could be handled by the RMS. Its first use is believed to have been on STS-28, but this remains to be confirmed, although it was acknowledged to have been used on STS-36.

SPE (Space Physiology Experiments) A programme of five Canadian experiments carried out as part of the International Microgravity Laboratory.

SPEAM (Sun Photospectrometer Earth Atmosphere Measurement) A study of the way that sunlight is reflected by the Earth's atmosphere at different wavelengths.

SPIFEX (Shuttle Plume Impingement Flight EXperiment) On the first use, a small package of sensors was retrieved by the RMS and positioned to measure the pressure waves and the composition of the plumes of efflux from the RCS thrusters. As a refinement, the sensors were set on the end of a 10-metre boom that was retrieved and manipulated by the RMS for extended reach. The data was used to define procedures for manoeuvring in close proximity to the Mir space station which, because of its solar panels, was very sensitive to the corrosive effects of efflux (Soyuz ferries had long-ago been restricted in the manoeuvres that they could undertake close to the complex). In the longer term, the data will determine the effects that a shuttle will have on the International Space Station, and be fed into the requirements of the materials which will be used in its construction.

SRB (Solid Rocket Booster) The rockets that are 'strapped onto' the shuttle to augment its thrust at launch are the largest solid-propellant rocket motors every built. In effect, their two-minute cycle lifts the shuttle up out of the atmosphere. After being jettisoned, they deploy parachutes and are recovered from the ocean for examination, refurbishment and reuse.

SSAS (Solid Sorbent Air Sampler) An instrument on Mir to monitor the quality of the air. It sampled the air over a 24-hour period and recorded the presence of trace components.

SSBUV (Shuttle SBUV) This instrument was carried in a pair of GAS cans and mounted on the side wall of the payload bay. This instrument was flown several times a year in order to provide a series of calibration checks of similar instruments running on other satellites providing long-term monitoring. Data were collected when the orbiter passed below such satellites, so that their measurements could be compared. The sensors tend to degrade with prolonged exposure to the space environment so, to have confidence in the fine detail noted in ultraviolet levels, it was necessary to check the calibration on a regular basis. The two-can SSBUV package weighed in at about 550 kg, but a large part of this was the battery pack and data storage system – facilities which were provided by the host spacecraft in the case of the SBUV instrument carried on automated satellites. It was sensitive in the 1,600–4,000 Å range. SSBUV was completely independent of the orbiter and its power supply was sufficient for 40 hours of operation. Although a rather low-profile payload, it served a crucial role in the MTPE programme.

SSCE (Solid Surface Combustion Experiment) A study of the behaviour of combustion in microgravity. The experiment was conducted in a pressurised chamber, and videotaped for subsequent analysis of the manner in which the flame propagated. An ashless filter paper and small pieces of plexiglass were ignited by a hot wire filament. After several shuttle missions, it was sent up to the Mir space station to conduct a further programme of tests.

SSCP (Small Self-Contained Payload) See entry for GAS.

SSIP (Shuttle Student Involvement Programme) Several dozen experiments were selected competitively from proposals submitted by High School students through the National Science Teachers' Association. The first experiment flew on STS-3, and the last flew on STS-42. Their principal investigators had long since left school by the time that their experiments finally flew. It was one of several programmes set up by NASA early in the development of the shuttle to promote public involvement.

SSME (Space Shuttle Main Engine) The trio of hydrogen-burning engines set at the rear of the orbiter. This engine was the first large liquid-propellant rocket motor to be designed to be reusable. It can be throttled between 65 per cent and 106 per cent of a nominal 'rated' thrust. It had been intended to extend the thrust to 109 per cent but, following the Challenger accident, this was cancelled.

SS-RMS (Space Station Remote Manipulator System) The heavy-duty RMS to be installed on the International Space Station. It will employ SVS 'blind-positioning' and will incorporate the DEE for delicate work.

STA (Shuttle Training Aircraft) A Gulfstream aircraft modified to fly the descent profile of an orbiter during the subsonic phase of its gliding approach. To reproduce the steep rate of descent, the aircraft's engines were run in reverse. The cockpit carried an instrument display identical to that of an orbiter.

stack The slang term for the bolted-together shuttle vehicle, comprising the orbiter, the ET and the two SRBs. On the pad, the stack's weight is borne by the SRBs. In flight, the structural load is borne by the ET.

STALE (Suppression of Transient Accelerations by Levitation Experiment) An isolation mount that used electromagnets to damp out ambient vibrations so that they would not disturb an experiment apparatus.

Star 48 The compact spherical solid rocket motor used for the PAM geostationary transfer stage. Two one-off configurations of the motor were also used: one to provide an extra kick after the Ulysses spacecraft was dispatched towards Jupiter by an IUS, and the other to brake the Magellan spacecraft so that it could enter orbit around Venus.

STDCE (Surface Tension Driven Convection Experiment) An apparatus to study the effect of a surface temperature gradient on materials processing making use of surface tension driven convection, which is practicable only in microgravity. To observe the motions in the oil samples, the fluid was seeded with small particles of alumina, which a laser was able to illuminate. It produced data on unstable flows, and revealed surface oscillations never seen in terrestrial experiments.

STIS (Space Telescope Imaging Spectrograph) A multi-function camera installed on the HST in 1997.

STL (Space Tissue Loss) An experiment that exposed samples of human bone and tissues to microgravity to study the extent to which they degraded, and to test a pharmaceutical designed to overcome it. This agent had been developed to reduce tissue loss in a patient whose limb is immobilised in a cast, but it could also be of benefit to astronauts on long missions.)

STLV (Slow-Turning Lateral Vessel) A bioreactor developed for use on the International Space Station. It grows its cell culture in a horizontal cylindrical container which slowly rotates to emulate microgravity and to keep the cells continuously suspended whilst also bathed in nutrients and oxygen. Its cultures would be suitable for replacement of human tissue that is diseased or damaged.

stomach awareness The astronauts' informal name for SAS.

STP (Space Test Programme) A generic programme operated by Department of Defense to test a wide variety of space-based apparatus developed for a multiplicity of purposes.

stratospheric winds There is very little large-scale vertical mixing between the troposphere and the stratosphere, so chemicals released at the surface only slowly migrate into the stratosphere. These emerging gases and aerosols are soon dispersed globally, however, by the horizontal upper atmospheric winds. Although moving at very high speed, these winds are not violent, because the air density is too low to impart much force.

STS (Space Transportation System) The space shuttle.

STScI (Space Telescope Science Institute) The facility in Baltimore which operates the HST on behalf of the astronomical community, in effect running the telescope much like any other high-technology observatory.

STU (Satellite Test Unit) A satellite which was ejected from a Hitchhiker GAS canister in order to participate in the PAMS experiment. Its release mechanism set it rotating, but its role was to test using the residual aerodynamic and magnetic forces at orbital altitude to stabilise itself while being observed by the PAMS laser on the orbiter, in order to demonstrate that a spacecraft could stabilise itself without using thrusters.

SUSIM (Solar Ultraviolet Spectral Irradiance Monitor) A solar ultraviolet spectrometer, one of which was carried on the UARS.

SUVE (Solar UltraViolet Experiment) An instrument developed by students at the University of Colorado to measure the extreme-ultraviolet component of sunlight – a region of the spectrum that is blocked by the Earth's atmosphere – to identify the processes by which the radiation is absorbed, and to study ionospheric effects. The intensity of extreme-ultraviolet is increased by sunspot and flare activity, which is why the outermost fringe of the Earth's atmosphere is heated and inflated when the Sun is at its peak of spot activity.

SVS (Space Vision System) A video processing system that tracked a set of markings on a satellite to calculate its position and orientation with respect to the camera. The objective was to be able to place a payload mounted on the RMS at a specific point, even though doing so would block the line of sight of the astronaut controlling the arm. It was tested with a small sub-satellite (CTA) on STS-52. When the DM was mounted on the ODS on STS-74, opportunity was taken to test the system with large structures, and it was then used to assist with the 'blind' docking in which the DM was connected to the Mir space station. The system was developed to enable modules to be attached to the International Space Station, on which an attachment ring will not necessarily be directly visible to the arm's operator.

SWUIS (SouthWest Ultraviolet Imaging System) A small hand-held telescopic camera that was used to image comet Hale–Bopp.

3DMA (Three-Dimensional Microgravity Accelerometer) Apparatus to measure the tiny accelerations aboard a spacecraft which 'pollute' the microgravity environment it makes available to sensitive experiments, so that their results can be properly interpreted.

TAGS (Text And Graphics System) A fax machine to uplink text and photographs to the shuttle. When tested on STS-29, it initially suffered paper jams.

TAL (Transatlantic Abort Landing) See *abort*.

TAS (Technology Applications and Science) A multidisciplinary package which operated in the bay. On its first use, it had instruments to provide data on the Earth's topography and atmosphere (SLA and ISIR), solar energy (SCE), a study of the critical viscosity of xenon at temperatures closer to its liquid-vapour critical point than was attainable on Earth, an evaluation of a heat-pipe using a two-phase coolant capillary flow, and a new cryogenic cooling system.

TC (Trash Compactor) This manually-operated mechanical compressor was installed in a middeck locker as part of the EDO facility to compress waste for more efficient storage. Before being used operationally on STS-50, it was tested on STS-35.

TCDT (Terminal Countdown Demonstration Test) A full dress rehearsal for the astronauts and the launch team.

TCS (Trajectory Control Sensor) A laser rangefinder that was fixed to the side wall of the payload bay on the early missions to rendezvous with the Mir space station to supply more precise data on separation and closure rate in the final approach than was available from the orbiter's radar. It was superseded on later missions by a hand-held laser which was fired out of one of the overhead windows, and offered the advantage of being able to be aimed at a specific part of the station for consistency.

TDRS (Tracking and Data Relay System) The network of geostationary relay satellites established over the years by NASA. In its final configuration it had two pairs of satellites, at the 'East' and 'West' stations as viewed from the control station at White Sands in New Mexico, stationed 120 degrees apart to provide typically 85 per cent coverage for a satellite in low orbit. Each satellite has two high-capacity steerable dishes for dedicated high-capacity S/K_u-Band bent-pipe relay. The system is used by the shuttle, various scientific satellites including the HST, and the Department of Defense's satellites (the system is capable of operating in a secure mode).

TEAMS (Technology Experiments for Advancing Missions in Space) A Hitchhiker Bridge with an assortment of test apparatus, including the PAMS laser.

TEHM (Thermo-Electric Holding Module) A refrigerator/freezer which could be carried on the middeck to transfer chilled materials to and from the Mir space station.

TEI (ThermoElectric Incubator) A portable incubator used to maintain biological samples at specific temperatures.

telescience The use of bidirectional high-capacity data/video links to enable research teams on the ground to work with astronauts on experiments in space. It was assessed using a small package of materials-processing apparatus on STS-52. The objective was not just to have the astronauts talk the scientists through the experiment, but to transmit video of its process and have the scientists operate the apparatus by remote control. The system was refined on later Spacelab missions, and is to be used routinely on the International Space Station, on which astronauts will physically prepare an experiment and then leave it to the scientists to run it, thereby making the most effective use of both teams.

TEMPUS A containerless crucible employing a radio-frequency coil to electromagnetically manipulate metallic samples. An electromagnetic levitator offered the advantage over an acoustic levitator that it could process the sample in a vacuum.

TEPC (Tissue-Equivalent Proportional Counter) A radiation monitor.

Terra Scout A Department of Defense test of an astronaut's ability to make militarily useful observations from space. It was undertaken by a specialist in interpreting imagery from reconnaissance satellites. The DVOS telescope was used to observe a variety of ground targets, to compare what could be seen visually with what satellite imagery could show. This experiment was undertaken on behalf of the Army's Intelligence Center.

TIROS (TV InfraRed Operational System) The low-altitude polar-orbiting meteorological satellite network operated by NOAA to complement its geostationary GOES network. Although the original generation of satellites in this class were listed as TIROS, the most recent is listed as a NOAA (e.g. NOAA 9).

tissue research Growing cell tissue is one of the fundamental goals of biomedical research. The bioreactors used to grow tissue were tested with cancer tumours that can successfully be developed outside the body. In addition to growing tissue for study, a bioreactor can be used to test the response of anti-cancer drugs. Several bioreactors have been built for the International Space Station, and NASA-built bioreactors have already given rise to a breakthrough in the quality of cancer cultures and have produced significantly better lung and intestine cultures, so their installation on the International Space Station is eagerly awaited. In the long-term, tissues grown in bioreactors will be able to be integrated into the body, to replace damaged or extracted tissue.

Titan Although developed to deliver a heavyweight nuclear warhead, this missile was soon converted to launch satellites. NASA used the Titan 2 to launch its Gemini spacecraft, but the Air Force augmented it with a pair of massive solid rockets and, as the Titan 3, it became the mainstay of the military space programme. The Titan 3 was to have been used to launch the Manned Orbiting Laboratory, which would have carried a Gemini on top. Although the shuttle-only policy obliged the Air Force to relinquish its Titans, once the mixed-fleet policy was adopted following the Challenger accident, it ordered a fully upgraded version, called the Titan 4, to carry payloads that had been built to exploit the shuttle. The latest variant, the Titan 4B, can lift even heavier payloads, and its shroud is even more expansive than the shuttle's cavernous payload bay.

TLD (ThermoLuminescent Dosimeter) A hand-held dosimeter to measure the flux of cosmic rays in the orbiter's cabin.

TOMS (Total Ozone Mapping Spectrometer) An instrument that was carried on Nimbus 7 to map the spatial and temporal variations in the concentration of stratospheric ozone.

TOS (Transfer Orbit Stage) An upper stage employing an Orbus solid rocket motor. It was developed by Orbital Sciences Corporation as a commercial venture to cater for heavier satellites than could be carried by the PAM kick-motor, but were too small to require an IUS. Its first use was to boost the Mars Observer spacecraft out of Earth-orbit (it was to have ridden a shuttle, but had been offloaded to ease the pressure on the manifest in the aftermath of the loss of Challenger); the second was used to place the ACTS into GTO. The TOS was conceived to exploit the shuttle-only policy which was implicit in operating a NSTS, but when NASA was prohibited from carrying commercial satellites on its shuttle the market for the TOS evaporated. However, on both of its missions it performed flawlessly.

TPAD (Trunnion Pin Attachment Device) This was essentially an RMS end-effector which could be mounted on the front of the MMU so that a free-flying astronaut could grapple a satellite. Its first employment was the capture of SolarMax, but an unexpected bolt on the satellite prevented its mechanism from locking. It was to have been used to capture Landsat-4, but this was cancelled. So few satellites have trunnion pins fitted to facilitate retrieval that there has been little call for the TPAD.

TPCE (Tank Pressure Control Experiment) A test of a system to control the pressure in a cryogenic fuel tank.

TPS (Thermal Protection System) In order to be reusable, the shuttle could not employ an ablative shield to protect it from the thermal stress of re-entry, so a radiative system was developed. This initially took the form of a set of 30,000 individually contoured tiles glued onto the aluminium airframe, but the tiles in the less-stressed areas were later replaced by thermal blankets. The tiles were made of extremely pure foamed silica with a coating of borosilicate glass, and the most stressed areas were coated by a carbon–carbon shield. The tiles initially proved susceptible to damage during launch, but later became straightforward to maintain.

transponder The receiver/transmitter on a satellite is defined by its up/down frequencies. A C-Band which receives at 6 GHz and relays at 4 GHz is a 6/4 GHz system.

Band	Range	User	Function
VHF	30-300 MHz	Shuttle/Soyuz/Mir	voice between spacecraft
		tactical military	voice via satellite relay
		ham	voice spacecraft-ground
UHF	300-4,000 MHz	Soyuz/Mir	telemetry
L-Band	1–2 GHz	land mobile and maritime	voice relay

Band	Range	User	Function
S-Band	2–4 GHz	Shuttle many kinds of satellite	voice spacecraft-ground commands and telemetry
C-Band	4–8 GHz	GSO comsats	voice, TV and data relay
X-Band	8–12 GHz	GSO comsats	voice, data and DBS TV
K_u-Band	12–18 GHz	GSO comsats	voice, data and DBS TV
K_a-Band	18–30 GHz	GSO comsats	voice and data

TRE (Torso Rotation Experiment) An upper-body harness carrying sensors to measure the movements of the upper body during typical tasks in space, to complement a vestibular study.

TSS (Tethered Satellite System) A programme to develop the technology for tethered satellite operations. Italy developed a spherical satellite which was mounted on an extensible boom which was carried on a Spacelab pallet. In addition to testing the deployment system, the satellite was to test the viability of inducing an electrical current in a wire cutting across the lines of force of the Earth's magnetic field. A 20-km tether with a copper core was expected to generate a 1 amp current at 5,000 volts. However, on STS-46 the unreeling operation had to be abandoned when the tether jammed after a few hundred metres, so the satellite was reeled back in. On STS-75, the tether broke. A reflight in the foreseeable future is unlikely. One application of a tethered satellite would be to trawl an instrument package in the mesosphere, to make *in situ* measurements far above the upper limit of a high-altitude balloon and far below that of a sustainable orbit, so an operational TSS could make a significant contribution to the MTPE programme.

TUFI (Toughened Unipiece Fibrous Insulation) An improved form of the stick-on tiles of the TPS, made tougher to resist the scratches which all too often damaged the original tiles. To start with, on STS-59, they were applied only at points that had proven most susceptible to damage.

TVD (Torque Velocity Dynamometer) A device to measure the strength of the muscles in the arms and legs.

TVIS (Treadmill Vibration Isolation and Stabilisation) A restraint to allow the treadmill to be used without polluting the microgravity environment. The ergometer had already been isolated by a spring-loaded mount, so isolating the treadmill was simply another step in an ongoing process.

UARS (Upper Atmosphere Research Satellite) The satellite with a comprehensive suite of instruments to investigate the processes which deplete stratospheric ozone and lead to global warming. It provided the first global measurement of the winds that rage in the stratosphere. In effect, it was the first major element of the MTPE programme.

UFO (UHF Follow-On) The Navy's most recent geostationary communications satellite system. It superseded the Fltsatcom and Leasat systems.

UHS (Ultraviolet Horizon Scanner) A telescope that measured the far-ultraviolet spectral characteristics of the horizon to give basic data for a system to provide early warning of ballistic missiles by detecting them as they came into view across the horizon.

UIT (Ultraviolet Imaging Telescope) This instrument was built by GSFC. It used an f/9 telescope with a 0.5-metre diameter mirror and a pair of interchangeable cameras, one sensitive in the range 1,200–1,700 Å and the other in the range 1,250–3,200 Å, both of which recorded their image on film. It not only served as a survey instrument for the HST; by imaging relatively nearby galaxies in the ultraviolet it provided signature data for the way that normal galaxies in the farthest reaches of the Universe will appear once their light has been heavily redshifted down into the visual range in which the HST was operating, so that such galaxies could be distinguished from the other strange objects in the 'deep field' views.

Ulysses A shuttle-deployed spacecraft in a solar orbit perpendicular to the plane of the ecliptic. It made a 'polar passage' over the Sun at the minimum of the solar cycle, and, if all goes well, it will do so again at the turn of the century at the next maximum in solar activity.

URA (Uniformly Redundant Array) An X-ray telescope built by the Los Alamos National Laboratory. It formed an image by standing a plate incorporating 26,000 pinholes some 3 metres in front of the detector. It photographed randomly-selected star fields, in order to test its ability to discriminate point sources, complex collections of point sources and extended sources, but its eventual role will be to assist in battle management, following a nuclear exchange, by precisely locating detonation sites. The test programme had to be curtailed because the power system overheated.

USAR (Ultraviolet Spectrograph for Astronomical Research) An instrument to study Solar System bodies in the extreme-ultraviolet. Specific attention was paid to Jupiter and, as a target of opportunity, comet Hale–Bopp.

US Hab NASA's habitat module on the International Space Station. It will accommodate six astronauts.

US Lab NASA's laboratory module on the International Space Station. It will provide six generic facilities to support apparatus that will be installed as and when necessary to conduct specific research within certain target domains.

Facility	ISPRS
Biotechnology Facility	2
Combustion and Fluids Facility	6
Furnace Facility	6
Gravitational Biology Facility	6
Life Sciences Facility	6
Human Research Facility	4
Total	30

USML (US Microgravity Laboratory) The US-sponsored equivalent of the IML Spacelab. Its payloads included CGF, DPM, STDCE, GBX, SAMS, CGBA, Astroculture, SSCE, EDO-MP, CFE, WIFE, OARE, ZCGF, PCG, ICE, FSDCE, GFFC, PDE, STALE, 3DMA, LBNP and APCF.

USMP (US Microgravity Payload) A pallet of microgravity experiments mounted on an MPESS. Its payloads included LPE, MEPHISTO, SAMS, CFLSE, OARE, IDGE and AADSF.

UTA (Upper Torso Assembly) The rigid upper section of the EMU suit. It incorporated an integrated life-support backpack.

UVLIM (UltraViolet LIMb) A pair of spectrometers to measure air glow phenomena to identify the chemical composition of the ionosphere. A boresighted video film camera recorded the viewing angle in order to assist with data interpretation. It was a Department of Defense experiment.

UVPI (UltraViolet Plume Instrument) A sensor package on the SDI's LACE satellite to assess the feasibility of detecting the thruster plumes from ballistic missiles manoeuvring in space. On several occasions, a shuttle orbiter fired its OMS engines specifically to test whether the UVPI was able to detect it.

UVSTAR (UltraViolet Spectrograph Telescope for Astronomical Research) A GAS package that suffered problems with its gimbal mount.

VAB (Vehicle Assembly Building) The enormous cube-shaped building at KFC in which shuttle stacks are assembled.

VAFB The General Hoyt S Vandenberg AFB at Lompoc, just north of Los Angeles. The Western Test Range.

vapour diffusion A crystallisation process in which the crystals grow in a drop of solution, as solvent from the drop evaporates and diffuses though ambient vapour to a reservoir, thereby increasing the concentration in the remaining solution to prompt crystallisation.

V-Bar In terms of rendezvous, this is an approach in which the active spacecraft assumes a position directly ahead of its passive target, and then slowly closes the gap. This is the standard procedure for approaching a satellite, and will be used to rendezvous with the International Space Station. The 'V' signified the use of the vehicle's velocity vector.

VCAP (Vehicle Charging and Potential) An electron beam employed to stimulate the ambient plasma environment of the ionosphere, while a low-light camera recorded the effect.

VCGS (Vapour Crystal Growth System) A generic facility for growing crystals in space.

VCS (Voice Command System) A control system tested on STS-41 using the payload bay cameras as the subject of the experiment. It used a voice recognition system which had been programmed to obey certain simple commands relating to camera activities. The system functioned well on Earth, but because the intonation of the human voice changed in space due to physiological adaptation to microgravity, it was necessary to reload the unit's phrase memory. It was only a preliminary test. Such a control system might eventually prove invaluable to an astronaut riding a 'smart' RMS, facilitating verbal control over its movements while working on the International Space Station.

Vela A series of satellites in extremely high orbit designed to detect a nuclear detonation by its optical flash and electromagnetic effects. They policed the treaties prohibiting nuclear tests in the atmosphere and in space. This system was supplemented, then superseded by, the IMEWS satellites.

Vestibular Sled A chair mounted on a sled that ran along a track running down the aisle of the first German-sponsored Spacelab. The chair was gimballed to provide rotations and the sled provided translations. These degrees of freedom could be arbitrarily combined to stimulate vestibular effects. It had been planned to carry the sled on Spacelab 1, but it had to be deleted because the total mass exceeded the permitted limit.

VFT (Visual Function Tester) An experiment to measure changes in vision parameters due to microgravity. It used a hand-held binocular eyepiece which could display a variety of visual stimuli. The tests were performed at different times in the process of adapting to microgravity, to investigate any changes in perception of contrast, direction or pattern recognition. In the initial phase of adaptation to microgravity, the eyeball changes shape and its resolving power is slightly impaired, but once this distortion is relieved, visual acuity becomes exceptional, both in terms of linear resolution and the ability to distinguish subtle hues.

VIBES (Vibration Isolation Box Experiment System) An apparatus to measure and damp out disturbances in the microgravity environment due to crew activities. It employed a visco-elastic damper.

VIEW-CAPL (Visualisation In Experimental Water CAPL) An apparatus for directly observing the flow of coolant within a water-based CAPL system.

VTRE (Vented Tank Resupply Experiment) A test of a procedure for replenishing a partly-used tank of propellant, using a capillary vane structure to vent the excess vapour as the liquid was added. The Mir space station is the only satellite ever to have been routinely replenished, but because its tank uses a pressurised-gas bladder to pump liquid to the engine, the gas does not comes into contact with the liquid, and when the tank is topped up the gas is forced back into its supply bottle instead of being vented. The service module of the International Space Station will use this system, but the NASA part of the complex is to use a capillary system.

WAMDII (Wide Angle Michelson Doppler Imaging Interferometer) An optical instrument which observed faint emissions of air glow phenomena, measured the Doppler effect, and computed the speed of high-altitude winds.

WCS (Waste Containment System) The orbiter's toilet. The original unit fed waste straight into a tank that could be emptied only after landing. For EDO, it was modified to feed a bag which could be replaced in flight. The recovered waste was compacted for storage.

WF/PC (Wide Field/Planetary Camera) A pair of cameras fitted into a single instrument on the HST, one camera providing a field of view much wider than usual for such a large telescope, and the other optimised to study the planets of the Solar System. The first WF/PC was replaced on the first servicing mission by the WF/PC-2, which incorporated corrective optics in order to overcome the spherical aberration in the telescope's primary mirror.

WIFE (Wire Insulation Flammability Experiment) An assessment of the conditions under which overloaded wires ignite various types of insulation, to set standards for the design of the International Space Station on which, because the laboratory will be reconfigured from time to time, the power systems will be called upon to operate different loads. The data from this test contributed to the ongoing programme of combustion experiments.

WINDEX (WINdow EXperiment) An investigation of how RCS thruster firings influenced the glow produced by the orbiter's interaction with the ambient atomic oxygen plasma. This highly-calibrated experiment was conducted for the Department of Defense.

WINDII (WIND Imaging Interferometer)A visible-light and near-infrared interferometer on the UARS to measure stratospheric winds.

WSF (Wake Shield Facility) A free-flyer designed to manufacture semiconductor crystal in the 'low pressure' region in its 'wake'.

WUPPE (Wisconsin Ultraviolet Photo-Polarimeter Experiment) An f/10 reflecting telescope with a 0.5-metre diameter mirror, incorporating a detector that recorded both the spectral and the polarisation characteristics of the source. It operated in the 1,400–3,200 Å range with a resolution of 6 Å. Its solid-state detector generated real-time data.

X-20 A spaceplane designed by the Air Force to be launched on a rocket, enter low orbit, make a single pass over a target, take reconnaissance pictures or release a nuclear weapon, and then make a runway landing back at its launch site. Because it was to employ hypersonic dynamic soaring, it was dubbed 'Dynasoar'. Although development began in the late 1950s, with a view to its entering service in the mid-1960s, because there was no way to insulate its surface from the heat of re-entry its viability was severely undermined, and it was cancelled in 1963, even before sub-orbital trials with models could be initiated. The concept was sound, but it placed too great a demand on contemporary technology.

X-SAR (X-band SAR) A European-built imaging radar which flew alongside the SIR-C on the Shuttle Radar Laboratory.

ZCGF (Zeolite Crystal Growth Facility) An apparatus in which to grow crystals of zeolite – an amalgam of silica and alumina – whose selectively porous crystalline structure allows them to be used as filters, catalysts and absorbents. On its test (on STS-73) it was accompanied by its designer, Albert Sacco.

Bibliography

Any study of the development and operation of the Space Transportation System will ultimately rely upon NASA sources and, fortunately, it has been characteristically prolific in documenting its activities. There is also an excellent contemporary record in *Flight International, Aviation Week & Space Technology* and *Spaceflight*, the magazine of the British Interplanetary Society. Although 'current events' are now exceptionally well documented on the Internet, this network's historical coverage is rather patchy. In addition, therefore, the following magazines also proved invaluable in preparing this book: *Sky & Telescope, Astronomy, Astronomy Now, Astronomy & Space, New Scientist, Scientific American, Earth* and *National Geographic*. Frequent reference was also made to *Jane's (Interavia) Space Directory*, *Asimov's Guide to Science*, and the *McGraw-Hill Dictionary of Scientific and Technical Terms*.

The following books have provided useful 'snapshots of interpretation'. They have been listed in order of publication to provide a sense of advancement in the state-of-the-art.

Arthur C. Clarke *The Exploration Of Space*, Temple Press, 1951

Wernher von Braun *Across The Space Frontier*, a series of articles (with artwork by Chesley Bonestell) published in *Collier's Magazine*, New York, 1952; subsequently reprinted in *Across The Space Frontier* and *Conquest Of The Moon*, both edited by Cornelius Ryan and published by Viking Press in 1952 and 1953 respectively

Walter Dornberger *V-2*, Viking, 1954

Lynn Poole *Your Trip Into Space*, Lutterworth Press, 1954

D. R. Bates (Ed.) *Space Research And Exploration*, Eyre & Spottiswoode, 1957

Eric Burgess *Satellites And Spaceflight*, Chapman & Hall, 1957

R. B. Beard and A. C. Rotherham *Space Flight And Satellite Vehicles*, Newnes, 1957

Walter Sullivan *Assault On The Unknown: The International Geophysical Year*, McGraw-Hill, 1961

Wilfred Burchett and Anthony Purdy *Cosmonaut Yuri Gagarin: First Man In Space*, Panther, 1961

Eugene Emme *A History Of Space Flight*, Holt, Rinehart and Winston, 1965

Virgil 'Gus' Grissom *Gemini: A Personal Account Of Man's Venture Into Space*, Macmillan, 1968

Hugo Young, Bryan Silcock and Peter Dunn *Journey To Tranquillity: The History Of Man's Assault On The Moon*, Jonathan Cape, 1969
William Shelton *Soviet Space Exploration: The First Decade*, Arthur Barker, 1969
James Nobel Wilford *We Reach The Moon*, Corgi, 1969
Neil Armstrong, Michael Collins and Edwin 'Buzz' Aldrin *First On The Moon*, Michael Joseph, 1970
Peter Smolders *Soviets In Space: The Story Of The Salyut And The Soviet Approach To Present And Future Space Travel*, Lutterworth Press, 1973
Buzz Aldrin and Wayne Warga *Return To Earth*, Bantam, 1973
James Irwin and William Emerson *To Rule The Night: The Discovery Voyage Of Astronaut Jim Irwin*, Hodder And Stoughton, 1973
Gerard O'Neill *The Colonisation Of Space*, in Physics Today, September 1974
Sybil Parker (Ed.) *McGraw-Hill Dictionary Of Scientific And Technical Terms*, McGraw-Hill, 1974
Michael Collins *Carrying The Fire: An Astronaut's Autobiography*, Allen, 1975
Isaac Asimov *Eyes On The Universe: A History Of The Telescope*, Quartet, 1975
Peter Francis *Volcanoes*, Penguin, 1976
US Senate *Soviet Space Programs: 1971–75*, Senate Committee Aeronautical And Space Sciences, US Government Printing Office, 1976
Gerard O'Neill *The High Frontier: Human Colonies In Space*, Jonathan Cape, 1977
Louise Young *Earth's Aura: A Layman's Guide To The Atmosphere*, Penguin, 1977
T. A. Heppenheimer *Colonies In Space*, Warner Books, 1978
Henry Cooper *A House In Space: The First True Account Of The Skylab Experience*, Panther, 1978
Robert Powers *Planetary Encounters: The Future Of Unmanned Spaceflight*, Warner, 1979
Tom Wolfe *The Right Stuff*, Bantam, 1981
David Baker *The History Of Manned Space Flight*, Cavendish, 1981
Peter Francis *The Planets: A Decade Of Discovery*, Penguin, 1981
Isaac Asimov *Exploring The Earth And The Cosmos: The Growth And Future Of Human Knowledge*, Penguin, 1982
Tim Furniss *The Story Of The Space Shuttle*, Hodder & Stoughton, 1982
Nigel Macknight *Shuttle*, Macknight International, 1984
John and Nancy DeWaard *History Of NASA: America's Voyage To The Stars*, Bison, 1984
David Shapland and Michael Rycroft *Spacelab: Research In Earth Orbit*, Cambridge University Press, 1984
Iain Nicolson *Sputnik To Space Shuttle: The Complete Story Of Space Flight*, Dodd & Mead, 1985
Chuck Yeager and Leo Janos *Yeager: The Personal Story Of The Greatest Test Pilot Of All*, Arrow, 1985
Tim Furniss *Spaceflight: The Records*, Guinness, 1985
Melvyn Smith *Space Shuttle*, Foulis–Haynes, 1985
US Senate *Soviet Space Programs: 1976-80*, Senate Committee On Commerce, Science And Transportation, US Government Printing Office, 1985

Tim Furniss *Manned Spaceflight Log*, Jane's, 1986
John Gribbin *In Search Of The Big Bang: Quantum Physics And Cosmology*, Heinemann, 1986
Curtis Peebles *Guardians: Strategic Reconnaissance Satellites*, Ian Allan, 1987
John Gribbin *The Omega Point: The Search For The Missing Mass And The Ultimate Fate Of The Universe*, Heinemann, 1987
Malcolm McConnell *Challenger: A Major Malfunction*, Simon & Schuster, 1987
Peter Bond *Heroes In Space: From Gagarin To Challenger*, Blackwell, 1987
Yuri Shkolenko *The Space Age*, Progress Publishing (Moscow), 1987
James and Alcestis Oberg *Pioneering Space: Living On The Next Frontier*, McGraw-Hill, 1987
Joseph Trento *Prescription For Disaster: From The Glory Of Apollo To The Betrayal Of The Shuttle*, Harrap, 1987
Isaac Asimov *Asimov's New Guide To Science*, Penguin, 1987
John Gribbin *The Hole In The Sky: Man's Threat To The Ozone Layer*, Corgi, 1988
Joseph Allen *Entering Space: An Astronaut's Odyssey*, Macdonald, 1988
Gene Gurney & Jeff Forte *Space Shuttle Log: The First 25 Flights*, Aero, 1988
Valentin Lebedev *Diary Of A Cosmonaut: 211 Days In Space*, Phytoresource Research, 1988
Phillip Clark *The Soviet Manned Space Programme*, Salamander, 1988
Brian Harvey *Race Into Space: The Soviet Space Programme*, Ellis Horwood, 1988
Michael Collins *Liftoff: The Story Of America's Adventure In Space*, Grove, 1988
John Gribbin *In Search Of The Double Helix: Darwin, DNA And Beyond*, Corgi, 1988
Henry Hurt *For All Mankind*, Atlantic, 1988
Patrick Moore *The Planet Neptune*, Ellis Horwood, 1988
Bruce Bolt *Earthquakes*, Freeman, 1988
Barry Parker *Invisible Matter And The Fate Of The Universe*, Plenum, 1989
Antony Milne *Earth's Changing Climate: The Cosmic Connection*, Prism, 1989
Robert and Barbara Decker *Volcanoes*, Freeman, 1989
Bruce Murray *Journey Into Space: The First Thirty Years Of Space Exploration*, Norton, 1989
Fred Pearce *Turning Up The Heat: Our Perilous Future In The Global Greenhouse*, Paladin, 1989
Donat Wentzel *The Restless Sun*, Smithsonian, 1989
Clark Chapman and David Morrison *Cosmic Catastrophes*, Plenum, 1989
Kenneth Gatland *The Illustrated Encyclopedia Of Space Technology*, Salamander, 1989
Buzz Aldrin and Malcolm McConnell *Men From Earth*, Bantam, 1989
Ray Spangerburg and Diane Moser *Space Exploration: Opening The Space Frontier*, Facts-On-File, 1989
Ray Spangerburg and Diane Moser *Space Exploration: Living And Working In Space*, Facts-On-File, 1989
Isaac Asimov *Beginnings: The Story Of Origins – Of Mankind, Life, The Earth, The Universe*, Berkley, 1989
Patrick Moore *Mission To The Planets: The Illustrated Story Of Man's Exploration Of The Solar System*, Cassell, 1990

John Gribbin *Hothouse Earth: The Greenhouse Effect and Gaia*, Bantam, 1990
Ellis Miner *Uranus: The Planet, Rings And Satellites*, Ellis Horwood, 1990
Mark Littmann *Planets Beyond: Discovering The Outer Solar System*, Wiley, 1990
Louis Brewer Hall *Searching For Comets*, McGraw-Hill, 1990
Donald Goldsmith *Supernova: The Violent Death Of A Star*, Oxford, 1990
K. J. Gregory (Ed.) *The Restless Earth: The Landscapes Of Our Planet And The Forces That Shaped Them*, Guinness, 1990
Frank Winter *Rockets Into Space*, Harvard, 1990
Gordon Hooper *The Soviet Cosmonaut Team*, GRH Publications, 1990
James Baker *Planet Earth: The View From Space*, Harvard University Press, 1990
Dennis Newkirk *Almanac Of Soviet Manned Space Flight*, Gulf Publishing Company, 1990
Michael Rycroft (Ed.) *The Cambridge Encyclopedia Of Space*, Cambridge University Press, 1990
Nigel Calder *Spaceship Earth*, Channel 4 Books, 1991
John Gribbin *Blinded By The Light: The Secret Life Of The Sun*, Bantam, 1991
NASA *Space Station Freedom Media Handbook*, NASA, 1992
J K Davies *Space Exploration*, Chambers, 1992
Mike Gray *Angle Of Attack: Harrison Storms And The Race To The Moon*, Norton, 1992
Helen Sharman and Christopher Priest *Seize The Moment: An Autobiography Of Britain's First Astronaut*, Victor Gollancz, 1993
Peter Bond *Reaching For The Stars: The Illustrated History Of Manned Spaceflight*, Cassell, 1993
Michael Rowan-Robinson *Ripples In The Cosmos: A View Behind The Scenes Of The New Cosmology*, Freeman, 1993
Marcus Chown *Afterglow Of Creation: From The Fireball To The Discovery Of Cosmic Ripples*, Arrow, 1993
David Raup *Extinction: Bad Genes Or Bad Luck?*, Oxford, 1993
Wayne Matson *Cosmonautics: A Colorful History*, Cosmos Books, 1994
Henry Cooper *The Evening Star: Venus Observed*, Johns Hopkins, 1994
Vladimir Pivnyuk and Mark Bockman *Space Station Handbook: Mir User's Manual*, Cosmos Books, 1994
Nicholas Johnson *The Soviet Reach For The Moon*, Cosmos Books, 1994
Andrew Chaikin *A Man On The Moon: The Voyages Of The Apollo Astronauts*, Michael Joseph, 1994
Alan Shepard and Deke Slayton *Moonshot: The Inside Story Of America's Race To The Moon*, Virgin, 1994
Reta Beebe *Jupiter: The Giant Planet*, Smithsonian, 1994
David Levy *Impact Jupiter: The Crash Of Comet Shoemaker–Levy 9*, Plenum, 1995
Kenneth Phillips *Guide To The Sun*, Cambridge, 1995
Thomas Watters *Planets*, Macmillan-Smithsonian, 1995
US Senate *US–Russian Cooperation In Space*, US Congressional Office of Technology Assessment, US Government Printing Office, 1995
Duncan Steel *Rogue Asteroids And Doomsday Comets*, Wiley, 1995
Philip Charles and Frederick Seward *Exploring The X-Ray Universe*, Cambridge, 1995

Michael Neufeld *The Rockets And The Reich: Peenemunde And The Coming Of The Ballistic Missile Era*, Free Press, 1995
Dennis Jenkins *Space Shuttle: The History Of Developing The National Space Transportation System*, Jenkins, 1996
Claus Jensen *Contest For The Heavens: The Road To The Challenger Disaster*, Harvill, 1996
Stanley Schmidt and Robert Zubrin (Eds.) *Islands In The Sky: Bold New Ideas For Colonizing Space*, Wiley, 1996; articles from Analog Magazine
Piers Bizony *Island In The Sky: Building The International Space Station*, Aurum Press, 1996
John and Mary Gribbin *Fire On Earth: In Search Of The Doomsday Asteroid*, Simon & Schuster, 1996
K Ya Kondratyev, A A Buznikov & O M Pokrovsky *Global Change And Remote Sensing*, Wiley-Praxis, 1996
Brian Harvey *The New Russian Space Programme: From Competition To Collaboration*, Wiley-Praxis, 1996
Patrick Moore *The Planet Neptune: An Historical Survey Before Voyager,* 2nd edition, Wiley-Praxis, 1996
Patrick Moore and Peter Cattermole *Atlas Of Venus*, Cambridge, 1997
Ronald Greeley & Raymond Batson *The NASA Atlas Of The Solar System*, Cambridge, 1997
T. A. Heppenheimer *Countdown: A History Of Space Flight*, Wiley, 1997
Timothy Ferris *The Whole Shebang: A State-of-the-Universe(s) Report*, Weidenfeld & Nicholson, 1997
David Harland *The Mir Space Station: A Precursor To Space Colonisation*, Wiley-Praxis, 1997
Ellis Miner *Uranus: The Planet, Rings and Satellites*, 2nd edition, Wiley-Praxis, 1988.

Index

A bold reference is to a page with either an illustration or a substantial reference.

WILEY-PRAXIS SERIES IN SPACE SCIENCE AND TECHNOLOGY

Forthcoming Titles

THE SPACE DEBRIS ENVIRONMENT: Hazard and Risk Assessment
Nicholas L. Johnson, NASA Johnson Space Center, Houston, Texas, USA, and others

SOLAR SAILINIG: Technology, Dynamics and Mission Applications
Colin R. McInnes, Department of Aerospace Engineering, University of Glasgow, UK

SPACE GOVERNANCE: A Blueprint for Future Activities
George S. Robinson, Attorney-at-Law, President, Ocean-Space Services, Adjunct Professor, George Mason University, Institute of International Transactions, Virginia, USA; and Declan O'Donnell, Attorney-at-Law, President of the World Bar Association, President and Founder of the United Societies in Space, USA

THE MOON: RESOURCES, FUTURE DEVELOPMENT AND COLONIZATION
David G. Schrunk, formerly Radiation Safety Officer and Head of Nuclear Medicine, Polomar-Pomerado Hospital District, Escondido, California, Founder and Chairman, Science of Laws Institute, San Diego, California, USA; Burton L. Sharpe, System Sales Engineer, Communications Corporation, St Louis, MO, formerly Resident Site Engineer, NASA/Jet Propulsion Laboratory, supporiting US Transportation Command, Scott AFB ILUSA; and Bonnie L. Cooper, Scientist, Oceaneering Space Systems, Houston, Texas, USA

ROCKET AND SPACECRAFT PROPULSION: Principles, Practice and New Developments
M.J.L. Turner, Principal Research Fellow, Department of Physics and Astronomy, University of Leicester, UK